Introduction to Data Science

Science

Data Analysis and Prediction
Algorithms with R

CHAPMAN & HALL/CRC DATA SCIENCE SERIES

Reflecting the interdisciplinary nature of the field, this book series brings together researchers, practitioners, and instructors from statistics, computer science, machine learning, and analytics. The series will publish cutting-edge research, industry applications, and textbooks in data science.

The inclusion of concrete examples, applications, and methods is highly encouraged. The scope of the series includes titles in the areas of machine learning, pattern recognition, predictive analytics, business analytics, Big Data, visualization, programming, software, learning analytics, data wrangling, interactive graphics, and reproducible research.

Published Titles

Probability and Statistics for Data Science: Math + R + Data
Norman Matloff

Feature Engineering and Selection: A Practical Approach for Predictive Models
Max Kuhn and Kjell Johnson

Introduction to Data Science: Data Analysis and Prediction Algorithms with R
Rafael A. Irizarry

Introduction to Data Science

Science
Data Analysis and Prediction
Algorithms with R

Rafael A. Irizarry

Dana-Farber Cancer Institute and Harvard University

CRC Press
Taylor & Francis Group
Boca Raton London New York

CRC Press is an imprint of the
Taylor & Francis Group, an **informa** business

CRC Press
Taylor & Francis Group
6000 Broken Sound Parkway NW, Suite 300
Boca Raton, FL 33487-2742

International Standard Book Number-13: 978-0-367-35798-6 (Hardback)

Library of Congress Cataloging-in-Publication Data

Names: Irizarry, Rafael A., author.
Title: Introduction to data science : data analysis and prediction
algorithms with R / Rafael A. Irizarry.
Description: [Boca Raton] : [CRC Press], [2019] | Summary: "The book begins
by going over the basics of R and the tidyverse. You learn R throughout
the book, but in the first part we go over the building blocks needed to
keep learning during the rest of the book"-- Provided by publisher.
Identifiers: LCCN 2019025160 (print) | LCCN 2019025161 (ebook) | ISBN
9780367357986 (hardback) | ISBN 9780429341830 (ebook)
Subjects: LCSH: R (Computer program language) | Information visualization.
| Data mining. | Statistics--Data processing. | Probabilities--Data
processing. | Computer algorithms. | Quantitative research.
Classification: LCC QA276.45.R3 I75 2019 (print) | LCC QA276.45.R3
(ebook) | DDC 005.362--dc23
LC record available at https://lccn.loc.gov/2019025160
LC ebook record available at https://lccn.loc.gov/2019025161

Visit the Taylor & Francis Web site at
http://www.taylorandfrancis.com

and the CRC Press Web site at
http://www.crcpress.com

Contents

III Statistics with R   213

Preface

This book started out as the class notes used in the HarvardX Data Science Series[1].

The link for the online version of the book is https://rafalab.github.io/dsbook/

The R markdown code used to generate the book is available on GitHub[2]. Note that, the graphical theme used for plots throughout the book can be recreated using the `ds_theme_set()` function from **dslabs** package.

This work is licensed under the Creative Commons Attribution-NonCommercial-ShareAlike 4.0 International (CC BY-NC-SA 4.0)[3].

We make announcements related to the book on Twitter. For updates follow @rafalab[4]

---

[1] https://www.edx.org/professional-certificate/harvardx-data-science
[2] https://github.com/rafalab/dsbook
[3] https://creativecommons.org/licenses/by-nc-sa/4.0
[4] https://twitter.com/rafalab

Acknowledgments

This book is dedicated to all the people involved in building and maintaining R and the R packages we use in this book. A special thanks to the developers and maintainers of R base, the tidyverse, and the caret package.

A special thanks to my tidyverse guru David Robinson and Amy Gill for dozens of comments, edits, and suggestions. Also, many thanks to Stephanie Hicks who twice served as a co-instructor in my data science classes and Yihui Xie who patiently put up with my many questions about bookdown. Thanks also to Karl Broman, from whom I borrowed ideas for the Data Visualization and Productivity Tools parts, and to Hector Corrada-Bravo, for advice on how to best teach machine learning. Thanks to Peter Aldhous from whom I borrowed ideas for the principles of data visualization section and Jenny Bryan for writing *Happy Git and GitHub for the useR*, which influenced our Git chapters. Thanks to Alyssa Frazee for helping create the homework problem that became the Recommendation Systems chapter and to Amanda Cox for providing the New York Regents exams data. Also, many thanks to Jeff Leek, Roger Peng, and Brian Caffo, whose class inspired the way this book is divided and to Garrett Grolemund and Hadley Wickham for making the bookdown code for their R for Data Science book open. Finally, thanks to Alex Nones for proofreading the manuscript during its various stages.

This book was conceived during the teaching of several applied statistics courses, starting over fifteen years ago. The teaching assistants working with me throughout the years made important indirect contributions to this book. The latest iteration of this course is a HarvardX series coordinated by Heather Sternshein and Zzofia Gajdos. We thank them for their contributions. We are also grateful to all the students whose questions and comments helped us improve the book. The courses were partially funded by NIH grant R25GM114818. We are very grateful to the National Institutes of Health for its support.

A special thanks goes to all those who edited the book via GitHub pull requests or made suggestions by creating an *issue*: nickyfoto (Huang Qiang), desautm (Marc-André Désautels), michaschwab (Michail Schwab), alvarolarreategui (Alvaro Larreategui), jakevc (Jake VanCampen), omerta (Guillermo Lengemann), espinielli (Enrico Spinielli), asimumba (Aaron Simumba), braunschweig (Maldewar), gwierzchowski (Grzegorz Wierzchowski), technocrat (Richard Careaga), atzakas, defeit (David Emerson Feit), shiraamitchell (Shira Mitchell), Nathalie-S, andreashandel (Andreas Handel), berkowitze (Elias Berkowitz), Dean-Webb (Dean Webber), mohayusuf, jimrothstein, mPloenzke (Matthew Ploenzke), and David D. Kane.

Introduction

The demand for skilled data science practitioners in industry, academia, and government is rapidly growing. This book introduces concepts and skills that can help you tackle real-world data analysis challenges. It covers concepts from probability, statistical inference, linear regression, and machine learning. It also helps you develop skills such as R programming, data wrangling with **dplyr**, data visualization with **ggplot2**, algorithm building with **caret**, file organization with UNIX/Linux shell, version control with Git and GitHub, and reproducible document preparation with **knitr** and R markdown. The book is divided into six parts: **R**, **Data Visualization**, **Statistics with R**, **Data Wrangling**, **Machine Learning**, and **Productivity Tools**. Each part has several chapters meant to be presented as one lecture and includes dozens of exercises distributed across chapters.

Case studies

Throughout the book, we use motivating case studies. In each case study, we try to realistically mimic a data scientist's experience. For each of the concepts covered, we start by asking specific questions and answer these through data analysis. We learn the concepts as a means to answer the questions. Examples of the case studies included in the book are:

Case Study	Concept
US murder rates by state	R Basics
Student heights	Statistical Summaries
Trends in world health and economics	Data Visualization
The impact of vaccines on infectious disease rates	Data Visualization
The financial crisis of 2007-2008	Probability
Election forecasting	Statistical Inference
Reported student heights	Data Wrangling
Money Ball: Building a baseball team	Linear Regression
MNIST: Image processing hand-written digits	Machine Learning
Movie recommendation systems	Machine Learning

Who will find this book useful?

This book is meant to be a textbook for a first course in Data Science. No previous knowledge of R is necessary, although some experience with programming may be helpful. The statistical concepts used to answer the case study questions are only briefly introduced, so a Probability and Statistics textbook is highly recommended for in-depth understanding of these concepts. If you read and understand all the chapters and complete all the exercises, you will be well-positioned to perform basic data analysis tasks and you will be prepared to learn the more advanced concepts and skills needed to become an expert.

What does this book cover?

We start by going over the **basics of R** and the **tidyverse**. You learn R throughout the book, but in the first part we go over the building blocks needed to keep learning.

The growing availability of informative datasets and software tools has led to increased reliance on **data visualizations** in many fields. In the second part we demonstrate how to use **ggplot2** to generate graphs and describe important data visualization principles.

In the third part we demonstrate the importance of statistics in data analysis by answering case study questions using **probability, inference, and regression** with R.

The fourth part uses several examples to familiarize the reader with **data wrangling**. Among the specific skills we learn are web scrapping, using regular expressions, and joining and reshaping data tables. We do this using **tidyverse** tools.

In the fifth part we present several challenges that lead us to introduce **machine learning**. We learn to use the **caret** package to build prediction algorithms including K-nearest neighbors and random forests.

In the final part, we provide a brief introduction to the **productivity tools** we use on a day-to-day basis in data science projects. These are RStudio, UNIX/Linux shell, Git and GitHub, and **knitr** and R Markdown.

What is not covered by this book?

This book focuses on the data analysis aspects of data science. We therefore do not cover aspects related to data management or engineering. Although R programming is an essential part of the book, we do not teach more advanced computer science topics such as data structures, optimization, and algorithm theory. Similarly, we do not cover topics such as web services, interactive graphics, parallel computing, and data streaming processing. The statistical concepts are presented mainly as tools to solve problems and in-depth theoretical descriptions are not included in this book.

1

Getting started with R and RStudio

1.1 Why R?

R is not a programming language like C or Java. It was not created by software engineers for software development. Instead, it was developed by statisticians as an interactive environment for data analysis. You can read the full history in the paper A Brief History of S[1]. The interactivity is an indispensable feature in data science because, as you will soon learn, the ability to quickly explore data is a necessity for success in this field. However, like in other programming languages, you can save your work as scripts that can be easily executed at any moment. These scripts serve as a record of the analysis you performed, a key feature that facilitates reproducible work. If you are an expert programmer, you should not expect R to follow the conventions you are used to since you will be disappointed. If you are patient, you will come to appreciate the unequal power of R when it comes to data analysis and, specifically, data visualization.

Other attractive features of R are:

1. R is free and open source[2].
2. It runs on all major platforms: Windows, Mac Os, UNIX/Linux.
3. Scripts and data objects can be shared seamlessly across platforms.
4. There is a large, growing, and active community of R users and, as a result, there are numerous resources for learning and asking questions[3] [4] [5].
5. It is easy for others to contribute add-ons which enables developers to share software implementations of new data science methodologies. This gives R users early access to the latest methods and to tools which are developed for a wide variety of disciplines, including ecology, molecular biology, social sciences, and geography, just to name a few examples.

1.2 The R console

Interactive data analysis usually occurs on the *R console* that executes commands as you type them. There are several ways to gain access to an R console. One way is to simply start R on your computer. The console looks something like this:

---

[1] https://pdfs.semanticscholar.org/9b48/46f192aa37ca122cfabb1ed1b59866d8bfda.pdf
[2] https://opensource.org/history
[3] https://stats.stackexchange.com/questions/138/free-resources-for-learning-r
[4] https://www.r-project.org/help.html
[5] https://stackoverflow.com/documentation/r/topics

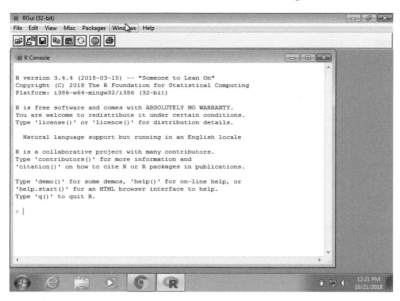

As a quick example, try using the console to calculate a 15% tip on a meal that cost $19.71:

```
0.15 * 19.71
#> [1] 2.96
```

Note that in this book, grey boxes are used to show R code typed into the R console. The symbol #> is used to denote what the R console outputs.

1.3   Scripts

One of the great advantages of R over point-and-click analysis software is that you can save your work as scripts. You can edit and save these scripts using a text editor. The material in this book was developed using the interactive *integrated development environment* (IDE) RStudio[6]. RStudio includes an editor with many R specific features, a console to execute your code, and other useful panes, including one to show figures.

---

[6]https://www.rstudio.com/

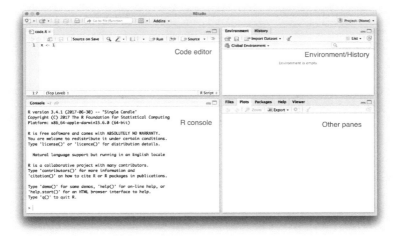

Most web-based R consoles also provide a pane to edit scripts, but not all permit you to save the scripts for later use.

All the R scripts used to generate this book can be found on GitHub[7].

1.4 RStudio

RStudio will be our launching pad for data science projects. It not only provides an editor for us to create and edit our scripts but also provides many other useful tools. In this section, we go over some of the basics.

1.4.1 The panes

When you start RStudio for the first time, you will see three panes. The left pane shows the R console. On the right, the top pane includes tabs such as *Environment* and *History*, while the bottom pane shows five tabs: *File*, *Plots*, *Packages*, *Help*, and *Viewer* (these tabs may change in new versions). You can click on each tab to move across the different features.

---

[7]https://github.com/rafalab/dsbook

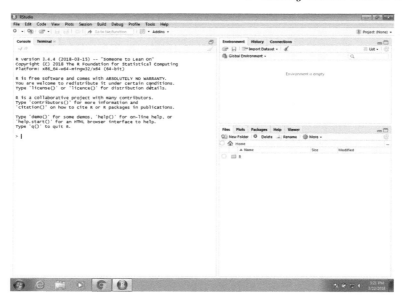

To start a new script, you can click on File, the New File, then R Script.

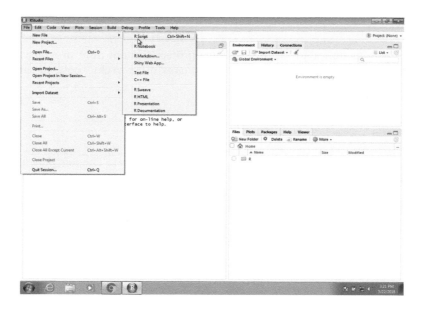

This starts a new pane on the left and it is here where you can start writing your script.

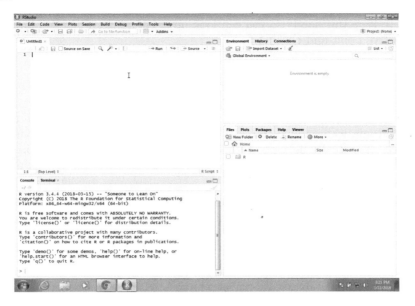

1.4.2   Key bindings

Many tasks we perform with the mouse can be achieved with a combination of key strokes instead. These keyboard versions for performing tasks are referred to as *key bindings*. For example, we just showed how to use the mouse to start a new script, but you can also use a key binding: Ctrl+Shift+N on Windows and command+shift+N on the Mac.

Although in this tutorial we often show how to use the mouse, **we highly recommend that you memorize key bindings for the operations you use most**. RStudio provides a useful cheat sheet with the most widely used commands. You can get it from RStudio directly:

You might want to keep this handy so you can look up key-bindings when you find yourself performing repetitive point-and-clicking.

1.4.3   Running commands while editing scripts

There are many editors specifically made for coding. These are useful because color and indentation are automatically added to make code more readable. RStudio is one of these editors, and it was specifically developed for R. One of the main advantages provided by RStudio over other editors is that we can test our code easily as we edit our scripts. Below we show an example.

Let's start by opening a new script as we did before. A next step is to give the script a name. We can do this through the editor by saving the current new unnamed script. To do this, click on the save icon or use the key binding Ctrl+S on Windows and command+S on the Mac.

When you ask for the document to be saved for the first time, RStudio will prompt you for a name. A good convention is to use a descriptive name, with lower case letters, no spaces, only hyphens to separate words, and then followed by the suffix *.R*. We will call this script *my-first-script.R*.

Now we are ready to start editing our first script. The first lines of code in an R script are dedicated to loading the libraries we will use. Another useful RStudio feature is that once we type `library()` it starts auto-completing with libraries that we have installed. Note what happens when we type `library(ti)`:

Another feature you may have noticed is that when you type `library(` the second parenthesis is automatically added. This will help you avoid one of the most common errors in coding: forgetting to close a parenthesis.

Now we can continue to write code. As an example, we will make a graph showing murder totals versus population totals by state. Once you are done writing the code needed to make this plot, you can try it out by *executing* the code. To do this, click on the *Run* button on the upper right side of the editing pane. You can also use the key binding: Ctrl+Shift+Enter on Windows or command+shift+return on the Mac.

Once you run the code, you will see it appear in the R console and, in this case, the generated plot appears in the plots console. Note that the plot console has a useful interface that permits you to click back and forward across different plots, zoom in to the plot, or save the plots as files.

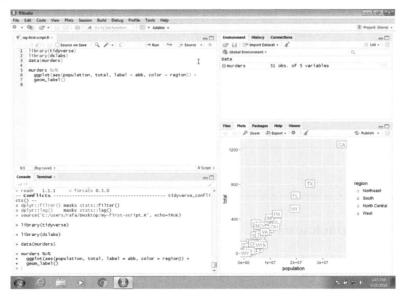

To run one line at a time instead of the entire script, you can use Control-Enter on Windows and command-return on the Mac.

1.4.4   Changing global options

You can change the look and functionality of RStudio quite a bit.

To change the global options you click on *Tools* then *Global Options....*

As an example we show how to make a change that we **highly recommend**. This is to change the *Save workspace to .RData on exit* to *Never* and uncheck the *Restore .RData into workspace at start*. By default, when you exit R saves all the objects you have created into a file called .RData. This is done so that when you restart the session in the same folder, it will load these objects. We find that this causes confusion especially when we share code with colleagues and assume they have this .RData file. To change these options, make your *General* settings look like this:

1.5   Installing R packages

The functionality provided by a fresh install of R is only a small fraction of what is possible. In fact, we refer to what you get after your first install as *base R*. The extra functionality comes from add-ons available from developers. There are currently hundreds of these available from

CRAN and many others shared via other repositories such as GitHub. However, because not everybody needs all available functionality, R instead makes different components available via *packages*. R makes it very easy to install packages from within R. For example, to install the **dslabs** package, which we use to share datasets and code related to this book, you would type:

```
install.packages("dslabs")
```

In RStudio, you can navigate to the *Tools* tab and select install packages. We can then load the package into our R sessions using the `library` function:

```
library(dslabs)
```

As you go through this book, you will see that we load packages without installing them. This is because once you install a package, it remains installed and only needs to be loaded with `library`. The package remains loaded until we quit the R session. If you try to load a package and get an error, it probably means you need to install it first.

We can install more than one package at once by feeding a character vector to this function:

```
install.packages(c("tidyverse", "dslabs"))
```

Note that installing **tidyverse** actually installs several packages. This commonly occurs when a package has *dependencies*, or uses functions from other packages. When you load a package using `library`, you also load its dependencies.

Once packages are installed, you can load them into R and you do not need to install them again, unless you install a fresh version of R. Remember packages are installed in R not RStudio.

It is helpful to keep a list of all the packages you need for your work in a script because if you need to perform a fresh install of R, you can re-install all your packages by simply running a script.

You can see all the packages you have installed using the following function:

```
installed.packages()
```

Part I

R

2

R basics

In this book, we will be using the R software environment for all our analysis. You will learn R and data analysis techniques simultaneously. To follow along you will therefore need access to R. We also recommend the use of an *integrated development environment* (IDE), such as RStudio, to save your work. Note that it is common for a course or workshop to offer access to an R environment and an IDE through your web browser, as done by RStudio cloud[1]. If you have access to such a resource, you don't need to install R and RStudio. However, if you intend on becoming an advanced data analyst, we highly recommend installing these tools on your computer[2]. Both R and RStudio are free and available online.

2.1 Case study: US Gun Murders

Imagine you live in Europe and are offered a job in a US company with many locations across all states. It is a great job, but news with headlines such as **US Gun Homicide Rate Higher Than Other Developed Countries**[3] have you worried. Charts like this may concern you even more:

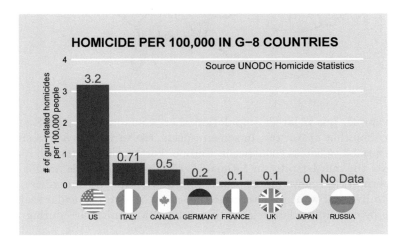

---

[1]https://rstudio.cloud

[2]https://rafalab.github.io/dsbook/installing-r-rstudio.html

[3]http://abcnews.go.com/blogs/headlines/2012/12/us-gun-ownership-homicide-rate-higher-than-other-developed-countries/

Or even worse, this version from everytown.org:

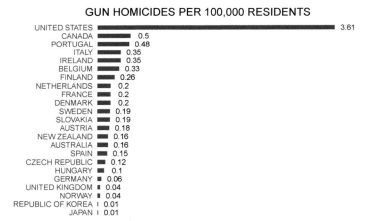

But then you remember that the US is a large and diverse country with 50 very different states as well as the District of Columbia (DC).

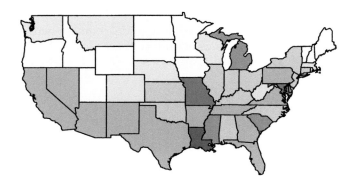

California, for example, has a larger population than Canada, and 20 US states have populations larger than that of Norway. In some respects, the variability across states in the US is akin to the variability across countries in Europe. Furthermore, although not included in the charts above, the murder rates in Lithuania, Ukraine, and Russia are higher than 4 per 100,000. So perhaps the news reports that worried you are too superficial. You have options of where to live and want to determine the safety of each particular state. We will gain some insights by examining data related to gun homicides in the US during 2010 using R.

Before we get started with our example, we need to cover logistics as well as some of the very basic building blocks that are required to gain more advanced R skills. Be aware that the usefulness of some of these building blocks may not be immediately obvious, but later in the book you will appreciate having mastered these skills.

2.2 The very basics

Before we get started with the motivating dataset, we need to cover the very basics of R.

2.2.1 Objects

Suppose a high school student asks us for help solving several quadratic equations of the form $ax^2 + bx + c = 0$. The quadratic formula gives us the solutions:

$$\frac{-b - \sqrt{b^2 - 4ac}}{2a} \text{ and } \frac{-b + \sqrt{b^2 - 4ac}}{2a}$$

which of course change depending on the values of a, b, and c. One advantage of programming languages is that we can define variables and write expressions with these variables, similar to how we do so in math, but obtain a numeric solution. We will write out general code for the quadratic equation below, but if we are asked to solve $x^2 + x - 1 = 0$, then we define:

```
a <- 1
b <- 1
c <- -1
```

which stores the values for later use. We use `<-` to assign values to the variables.

We can also assign values using `=` instead of `<-`, but we recommend against using `=` to avoid confusion.

Copy and paste the code above into your console to define the three variables. Note that R does not print anything when we make this assignment. This means the objects were defined successfully. Had you made a mistake, you would have received an error message.

To see the value stored in a variable, we simply ask R to evaluate `a` and it shows the stored value:

```
a
#> [1] 1
```

A more explicit way to ask R to show us the value stored in `a` is using `print` like this:

```
print(a)
#> [1] 1
```

We use the term *object* to describe stuff that is stored in R. Variables are examples, but objects can also be more complicated entities such as functions, which are described later.

2.2.2 The workspace

As we define objects in the console, we are actually changing the *workspace*. You can see all the variables saved in your workspace by typing:

```
ls()
#> [1] "a"            "b"            "c"
```

In RStudio, the *Environment* tab shows the values:

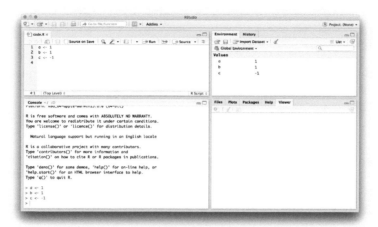

We should see a, b, and c. If you try to recover the value of a variable that is not in your workspace, you receive an error. For example, if you type x you will receive the following message: `Error: object 'x' not found`.

Now since these values are saved in variables, to obtain a solution to our equation, we use the quadratic formula:

```
(-b + sqrt(b^2 - 4*a*c) ) / ( 2*a )
#> [1] 0.618
(-b - sqrt(b^2 - 4*a*c) ) / ( 2*a )
#> [1] -1.62
```

2.2.3 Functions

Once you define variables, the data analysis process can usually be described as a series of *functions* applied to the data. R includes several predefined functions and most of the analysis pipelines we construct make extensive use of these.

We already used the `install.packages`, `library`, and `ls` functions. We also used the function `sqrt` to solve the quadratic equation above. There are many more prebuilt functions and even more can be added through packages. These functions do not appear in the workspace because you did not define them, but they are available for immediate use.

In general, we need to use parentheses to evaluate a function. If you type `ls`, the function is not evaluated and instead R shows you the code that defines the function. If you type `ls()` the function is evaluated and, as seen above, we see objects in the workspace.

Unlike `ls`, most functions require one or more *arguments*. Below is an example of how we assign an object to the argument of the function `log`. Remember that we earlier defined a to be 1:

```
log(8)
#> [1] 2.08
log(a)
#> [1] 0
```

You can find out what the function expects and what it does by reviewing the very useful manuals included in R. You can get help by using the `help` function like this:

```
help("log")
```

For most functions, we can also use this shorthand:

```
?log
```

The help page will show you what arguments the function is expecting. For example, `log` needs x and `base` to run. However, some arguments are required and others are optional. You can determine which arguments are optional by noting in the help document that a default value is assigned with =. Defining these is optional. For example, the base of the function `log` defaults to `base = exp(1)` making `log` the natural log by default.

If you want a quick look at the arguments without opening the help system, you can type:

```
args(log)
#> function (x, base = exp(1))
#> NULL
```

You can change the default values by simply assigning another object:

```
log(8, base = 2)
#> [1] 3
```

Note that we have not been specifying the argument x as such:

```
log(x = 8, base = 2)
#> [1] 3
```

The above code works, but we can save ourselves some typing: if no argument name is used, R assumes you are entering arguments in the order shown in the help file or by `args`. So by not using the names, it assumes the arguments are x followed by `base`:

```
log(8,2)
#> [1] 3
```

If using the arguments' names, then we can include them in whatever order we want:

```
log(base = 2, x = 8)
#> [1] 3
```

To specify arguments, we must use =, and cannot use <-.

There are some exceptions to the rule that functions need the parentheses to be evaluated. Among these, the most commonly used are the arithmetic and relational operators. For example:

```
2 ^ 3
#> [1] 8
```

You can see the arithmetic operators by typing:

```
help("+")
```

or

```
?"+"
```

and the relational operators by typing:

```
help(">")
```

or

```
?">"
```

2.2.4 Other prebuilt objects

There are several datasets that are included for users to practice and test out functions. You can see all the available datasets by typing:

```
data()
```

This shows you the object name for these datasets. These datasets are objects that can be used by simply typing the name. For example, if you type:

```
co2
```

R will show you Mauna Loa atmospheric CO_2 concentration data.

Other prebuilt objects are mathematical quantities, such as the constant π and ∞:

```
pi
#> [1] 3.14
Inf+1
#> [1] Inf
```

2.2.5 Variable names

We have used the letters a, b, and c as variable names, but variable names can be almost anything. Some basic rules in R are that variable names have to start with a letter, can't contain spaces, and should not be variables that are predefined in R. For example, don't name one of your variables `install.packages` by typing something like `install.packages <- 2`.

A nice convention to follow is to use meaningful words that describe what is stored, use only lower case, and use underscores as a substitute for spaces. For the quadratic equations, we could use something like this:

```
solution_1 <- (-b + sqrt(b^2 - 4*a*c)) / (2*a)
solution_2 <- (-b - sqrt(b^2 - 4*a*c)) / (2*a)
```

For more advice, we highly recommend studying Hadley Wickham's style guide[4].

2.2.6 Saving your workspace

Values remain in the workspace until you end your session or erase them with the function `rm`. But workspaces also can be saved for later use. In fact, when you quit R, the program asks you if you want to save your workspace. If you do save it, the next time you start R, the program will restore the workspace.

We actually recommend against saving the workspace this way because, as you start working on different projects, it will become harder to keep track of what is saved. Instead, we recommend you assign the workspace a specific name. You can do this by using the function `save` or `save.image`. To load, use the function `load`. When saving a workspace, we recommend the suffix `rda` or `RData`. In RStudio, you can also do this by navigating to the *Session* tab and choosing *Save Workspace as*. You can later load it using the *Load Workspace* options in the same tab. You can read the help pages on `save`, `save.image`, and `load` to learn more.

2.2.7 Motivating scripts

To solve another equation such as $3x^2 + 2x - 1$, we can copy and paste the code above and then redefine the variables and recompute the solution:

```
a <- 3
b <- 2
c <- -1
(-b + sqrt(b^2 - 4*a*c)) / (2*a)
(-b - sqrt(b^2 - 4*a*c)) / (2*a)
```

By creating and saving a script with the code above, we would not need to retype everything each time and, instead, simply change the variable names. Try writing the script above into an editor and notice how easy it is to change the variables and receive an answer.

---

[4]http://adv-r.had.co.nz/Style.html

2.2.8 Commenting your code

If a line of R code starts with the symbol #, it is not evaluated. We can use this to write reminders of why we wrote particular code. For example, in the script above we could add:

```
## Code to compute solution to quadratic equation of the form ax^2 + bx + c
## define the variables
a <- 3
b <- 2
c <- -1

## now compute the solution
(-b + sqrt(b^2 - 4*a*c)) / (2*a)
(-b - sqrt(b^2 - 4*a*c)) / (2*a)
```

2.3 Exercises

1. What is the sum of the first 100 positive integers? The formula for the sum of integers 1 through n is $n(n + 1)/2$. Define $n = 100$ and then use R to compute the sum of 1 through 100 using the formula. What is the sum?

2. Now use the same formula to compute the sum of the integers from 1 through 1,000.

3. Look at the result of typing the following code into R:

```
n <- 1000
x <- seq(1, n)
sum(x)
```

Based on the result, what do you think the functions seq and sum do? You can use help.

    a. sum creates a list of numbers and seq adds them up.
    b. seq creates a list of numbers and sum adds them up.
    c. seq creates a random list and sum computes the sum of 1 through 1,000.
    d. sum always returns the same number.

4. In math and programming, we say that we evaluate a function when we replace the argument with a given number. So if we type sqrt(4), we evaluate the sqrt function. In R, you can evaluate a function inside another function. The evaluations happen from the inside out. Use one line of code to compute the log, in base 10, of the square root of 100.

5. Which of the following will always return the numeric value stored in x? You can try out examples and use the help system if you want.

    a. `log(10^x)`
    b. `log10(x^10)`
    c. `log(exp(x))`
    d. `exp(log(x, base = 2))`

2.4 Data types

Variables in R can be of different types. For example, we need to distinguish numbers from character strings and tables from simple lists of numbers. The function `class` helps us determine what type of object we have:

```
a <- 2
class(a)
#> [1] "numeric"
```

To work efficiently in R, it is important to learn the different types of variables and what we can do with these.

2.4.1 Data frames

Up to now, the variables we have defined are just one number. This is not very useful for storing data. The most common way of storing a dataset in R is in a *data frame*. Conceptually, we can think of a data frame as a table with rows representing observations and the different variables reported for each observation defining the columns. Data frames are particularly useful for datasets because we can combine different data types into one object.

A large proportion of data analysis challenges start with data stored in a data frame. For example, we stored the data for our motivating example in a data frame. You can access this dataset by loading the **dslabs** library and loading the `murders` dataset using the `data` function:

```
library(dslabs)
data(murders)
```

To see that this is in fact a data frame, we type:

```
class(murders)
#> [1] "data.frame"
```

2.4.2 Examining an object

The function `str` is useful for finding out more about the structure of an object:

```
str(murders)
#> 'data.frame':    51 obs. of  5 variables:
#> $ state : chr "Alabama" "Alaska" "Arizona" "Arkansas" ...
#> $ abb : chr "AL" "AK" "AZ" "AR" ...
#> $ region : Factor w/ 4 levels "Northeast","South",..: 2 4 4 2 4 4 1 2 2
#>    2 ...
#> $ population: num 4779736 710231 6392017 2915918 37253956 ...
#> $ total : num 135 19 232 93 1257 ...
```

This tells us much more about the object. We see that the table has 51 rows (50 states plus DC) and five variables. We can show the first six lines using the function `head`:

```
head(murders)
#>          state abb region population total
#> 1      Alabama  AL  South     4779736   135
#> 2       Alaska  AK   West      710231    19
#> 3      Arizona  AZ   West     6392017   232
#> 4     Arkansas  AR  South     2915918    93
#> 5   California  CA   West    37253956  1257
#> 6     Colorado  CO   West     5029196    65
```

In this dataset, each state is considered an observation and five variables are reported for each state.

Before we go any further in answering our original question about different states, let's learn more about the components of this object.

2.4.3 The accessor: $

For our analysis, we will need to access the different variables represented by columns included in this data frame. To do this, we use the accessor operator $ in the following way:

```
murders$population
#>  [1]  4779736      710231   6392017   2915918 37253956   5029196   3574097
#>  [8]   897934      601723  19687653   9920000   1360301   1567582  12830632
#> [15]  6483802     3046355   2853118   4339367   4533372   1328361   5773552
#> [22]  6547629     9883640   5303925   2967297   5988927    989415   1826341
#> [29]  2700551     1316470   8791894   2059179 19378102   9535483    672591
#> [36] 11536504     3751351   3831074 12702379   1052567   4625364    814180
#> [43]  6346105    25145561   2763885    625741   8001024   6724540   1852994
#> [50]  5686986      563626
```

But how did we know to use `population`? Previously, by applying the function `str` to the object `murders`, we revealed the names for each of the five variables stored in this table. We can quickly access the variable names using:

```
names(murders)
#> [1] "state"      "abb"        "region"     "population" "total"
```

It is important to know that the order of the entries in `murders$population` preserves the order of the rows in our data table. This will later permit us to manipulate one variable based on the results of another. For example, we will be able to order the state names by the number of murders.

Tip: R comes with a very nice auto-complete functionality that saves us the trouble of typing out all the names. Try typing `murders$p` then hitting the *tab* key on your keyboard. This functionality and many other useful auto-complete features are available when working in RStudio.

2.4.4 Vectors: numerics, characters, and logical

The object `murders$population` is not one number but several. We call these types of objects *vectors*. A single number is technically a vector of length 1, but in general we use the term vectors to refer to objects with several entries. The function `length` tells you how many entries are in the vector:

```
pop <- murders$population
length(pop)
#> [1] 51
```

This particular vector is *numeric* since population sizes are numbers:

```
class(pop)
#> [1] "numeric"
```

In a numeric vector, every entry must be a number.

To store character strings, vectors can also be of class *character*. For example, the state names are characters:

```
class(murders$state)
#> [1] "character"
```

As with numeric vectors, all entries in a character vector need to be a character.

Another important type of vectors are *logical vectors*. These must be either TRUE or FALSE.

```
z <- 3 == 2
z
#> [1] FALSE
class(z)
#> [1] "logical"
```

Here the == is a relational operator asking if 3 is equal to 2. In R, if you just use one =, you actually assign a variable, but if you use two == you test for equality.

You can see the other *relational operators* by typing:

```
?Comparison
```

In future sections, you will see how useful relational operators can be.

We discuss more important features of vectors after the next set of exercises.

Advanced: Mathematically, the values in `pop` are integers and there is an integer class in R. However, by default, numbers are assigned class numeric even when they are round integers. For example, `class(1)` returns numeric. You can turn them into class integer with the `as.integer()` function or by adding an L like this: 1L. Note the class by typing: `class(1L)`

2.4.5   Factors

In the `murders` dataset, we might expect the region to also be a character vector. However, it is not:

```
class(murders$region)
#> [1] "factor"
```

It is a *factor*. Factors are useful for storing categorical data. We can see that there are only 4 regions by using the `levels` function:

```
levels(murders$region)
#> [1] "Northeast"      "South"          "North Central" "West"
```

In the background, R stores these *levels* as integers and keeps a map to keep track of the labels. This is more memory efficient than storing all the characters.

Note that the levels have an order that is different from the order of appearance in the factor object. The default is for the levels to follow alphabetical order. However, often we want the levels to follow a different order. We will see several examples of this in the Data Visualization part of the book. The function `reorder` lets us change the order of the levels of a factor variable based on a summary computed on a numeric vector. We will demonstrate this with a simple example.

Suppose we want the levels of the region by the total number of murders rather than alphabetical order. If there are values associated with each level, we can use the `reorder` and specify a data summary to determine the order. The following code takes the sum of the total murders in each region, and reorders the factor following these sums.

```
region <- murders$region
value <- murders$total
region <- reorder(region, value, FUN = sum)
levels(region)
#> [1] "Northeast"      "North Central" "West"          "South"
```

The new order is in agreement with the fact that the Northeast has the least murders and the South has the most.

Warning: Factors can be a source of confusion since sometimes they behave like characters and sometimes they do not. As a result, confusing factors and characters are a common source of bugs.

2.4.6   Lists

Data frames are a special case of *lists*. We will cover lists in more detail later, but know that they are useful because you can store any combination of different types. Below is an example of a list we created for you:

```
record
#> $name
#> [1] "John Doe"
#>
#> $student_id
#> [1] 1234
#>
#> $grades
#> [1] 95 82 91 97 93
#>
#> $final_grade
#> [1] "A"
class(record)
#> [1] "list"
```

As with data frames, you can extract the components of a list with the accessor $. In fact, data frames are a type of list.

```
record$student_id
#> [1] 1234
```

We can also use double square brackets ([[) like this:

```
record[["student_id"]]
#> [1] 1234
```

You should get used to the fact that in R, there are often several ways to do the same thing, such as accessing entries.

You might also encounter lists without variable names.

```
record2
#> [[1]]
#> [1] "John Doe"
#>
#> [[2]]
#> [1] 1234
```

If a list does not have names, you cannot extract the elements with $, but you can still use the brackets method and instead of providing the variable name, you provide the list index, like this:

```
record2[[1]]
#> [1] "John Doe"
```

We won't be using lists until later, but you might encounter one in your own exploration of R. For this reason, we show you some basics here.

2.4.7   Matrices

Matrices are another type of object that are common in R. Matrices are similar to data frames in that they are two-dimensional: they have rows and columns. However, like numeric, character and logical vectors, entries in matrices have to be all the same type. For this reason data frames are much more useful for storing data, since we can have characters, factors, and numbers in them.

Yet matrices have a major advantage over data frames: we can perform matrix algebra operations, a powerful type of mathematical technique. We do not describe these operations in this book, but much of what happens in the background when you perform a data analysis involves matrices. We cover matrices in more detail in Chapter 33.1 but describe them briefly here since some of the functions we will learn return matrices.

We can define a matrix using the `matrix` function. We need to specify the number of rows and columns.

```
mat <- matrix(1:12, 4, 3)
mat
#>      [,1] [,2] [,3]
#> [1,]    1    5    9
#> [2,]    2    6   10
#> [3,]    3    7   11
#> [4,]    4    8   12
```

You can access specific entries in a matrix using square brackets ([). If you want the second row, third column, you use:

```
mat[2, 3]
#> [1] 10
```

If you want the entire second row, you leave the column spot empty:

```
mat[2, ]
#> [1]  2  6 10
```

Notice that this returns a vector, not a matrix.

Similarly, if you want the entire third column, you leave the row spot empty:

```
mat[, 3]
#> [1]  9 10 11 12
```

This is also a vector, not a matrix.

You can access more than one column or more than one row if you like. This will give you a new matrix.

```
mat[, 2:3]
#>      [,1] [,2]
#> [1,]    5    9
#> [2,]    6   10
```

```
#> [3,]    7    11
#> [4,]    8    12
```

You can subset both rows and columns:

```
mat[1:2, 2:3]
#>        [,1] [,2]
#> [1,]    5    9
#> [2,]    6   10
```

We can convert matrices into data frames using the function `as.data.frame`:

```
as.data.frame(mat)
#>    V1 V2 V3
#> 1   1  5  9
#> 2   2  6 10
#> 3   3  7 11
#> 4   4  8 12
```

You can also use single square brackets ([) to access rows and columns of a data frame:

```
data("murders")
murders[25, 1]
#> [1] "Mississippi"
murders[2:3, ]
#>       state abb region population total
#> 2    Alaska  AK   West     710231    19
#> 3   Arizona  AZ   West    6392017   232
```

2.5   Exercises

1. Load the US murders dataset.

```
library(dslabs)
data(murders)
```

Use the function `str` to examine the structure of the `murders` object. Which of the following best describes the variables represented in this data frame?

   a.   The 51 states.
   b.   The murder rates for all 50 states and DC.
   c.   The state name, the abbreviation of the state name, the state's region, and the state's population and total number of murders for 2010.
   d.   `str` shows no relevant information.

2. What are the column names used by the data frame for these five variables?

3. Use the accessor $ to extract the state abbreviations and assign them to the object a. What is the class of this object?

4. Now use the square brackets to extract the state abbreviations and assign them to the object b. Use the identical function to determine if a and b are the same.

5. We saw that the region column stores a factor. You can corroborate this by typing:

```
class(murders$region)
```

With one line of code, use the function levels and length to determine the number of regions defined by this dataset.

6. The function table takes a vector and returns the frequency of each element. You can quickly see how many states are in each region by applying this function. Use this function in one line of code to create a table of states per region.

2.6 Vectors

In R, the most basic objects available to store data are *vectors*. As we have seen, complex datasets can usually be broken down into components that are vectors. For example, in a data frame, each column is a vector. Here we learn more about this important class.

2.6.1 Creating vectors

We can create vectors using the function c, which stands for *concatenate*. We use c to concatenate entries in the following way:

```
codes <- c(380, 124, 818)
codes
#> [1] 380 124 818
```

We can also create character vectors. We use the quotes to denote that the entries are characters rather than variable names.

```
country <- c("italy", "canada", "egypt")
```

In R you can also use single quotes:

```
country <- c('italy', 'canada', 'egypt')
```

But be careful not to confuse the single quote ' with the *back quote* `.

By now you should know that if you type:

```
country <- c(italy, canada, egypt)
```

you receive an error because the variables `italy`, `canada`, and `egypt` are not defined. If we do not use the quotes, R looks for variables with those names and returns an error.

2.6.2 Names

Sometimes it is useful to name the entries of a vector. For example, when defining a vector of country codes, we can use the names to connect the two:

```
codes <- c(italy = 380, canada = 124, egypt = 818)
codes
#>  italy canada  egypt
#>    380    124    818
```

The object `codes` continues to be a numeric vector:

```
class(codes)
#> [1] "numeric"
```

but with names:

```
names(codes)
#> [1] "italy"  "canada" "egypt"
```

If the use of strings without quotes looks confusing, know that you can use the quotes as well:

```
codes <- c("italy" = 380, "canada" = 124, "egypt" = 818)
codes
#>  italy canada  egypt
#>    380    124    818
```

There is no difference between this function call and the previous one. This is one of the many ways in which R is quirky compared to other languages.

We can also assign names using the **names** functions:

```
codes <- c(380, 124, 818)
country <- c("italy","canada","egypt")
names(codes) <- country
codes
#>  italy canada  egypt
#>    380    124    818
```

2.6.3 Sequences

Another useful function for creating vectors generates sequences:

```
seq(1, 10)
#>  [1]  1  2  3  4  5  6  7  8  9 10
```

The first argument defines the start, and the second defines the end which is included. The default is to go up in increments of 1, but a third argument lets us tell it how much to jump by:

```
seq(1, 10, 2)
#> [1] 1 3 5 7 9
```

If we want consecutive integers, we can use the following shorthand:

```
1:10
#>  [1]  1  2  3  4  5  6  7  8  9 10
```

When we use these functions, R produces integers, not numerics, because they are typically used to index something:

```
class(1:10)
#> [1] "integer"
```

However, if we create a sequence including non-integers, the class changes:

```
class(seq(1, 10, 0.5))
#> [1] "numeric"
```

2.6.4   Subsetting

We use square brackets to access specific elements of a vector. For the vector `codes` we defined above, we can access the second element using:

```
codes[2]
#> canada
#>    124
```

You can get more than one entry by using a multi-entry vector as an index:

```
codes[c(1,3)]
#> italy egypt
#>   380   818
```

The sequences defined above are particularly useful if we want to access, say, the first two elements:

```
codes[1:2]
#>  italy canada
#>    380    124
```

If the elements have names, we can also access the entries using these names. Below are two examples.

```
codes["canada"]
#> canada
```

```
#>    124
codes[c("egypt","italy")]
#> egypt italy
#>   818   380
```

2.7 Coercion

In general, *coercion* is an attempt by R to be flexible with data types. When an entry does not match the expected, some of the prebuilt R functions try to guess what was meant before throwing an error. This can also lead to confusion. Failing to understand *coercion* can drive programmers crazy when attempting to code in R since it behaves quite differently from most other languages in this regard. Let's learn about it with some examples.

We said that vectors must be all of the same type. So if we try to combine, say, numbers and characters, you might expect an error:

```
x <- c(1, "canada", 3)
```

But we don't get one, not even a warning! What happened? Look at x and its class:

```
x
#> [1] "1"      "canada" "3"
class(x)
#> [1] "character"
```

R *coerced* the data into characters. It guessed that because you put a character string in the vector, you meant the 1 and 3 to actually be character strings "1" and "3". The fact that not even a warning is issued is an example of how coercion can cause many unnoticed errors in R.

R also offers functions to change from one type to another. For example, you can turn numbers into characters with:

```
x <- 1:5
y <- as.character(x)
y
#> [1] "1" "2" "3" "4" "5"
```

You can turn it back with as.numeric:

```
as.numeric(y)
#> [1] 1 2 3 4 5
```

This function is actually quite useful since datasets that include numbers as character strings are common.

2.7.1   Not availables (NA)

When a function tries to coerce one type to another and encounters an impossible case, it usually gives us a warning and turns the entry into a special value called an NA for "not available". For example:

```
x <- c("1", "b", "3")
as.numeric(x)
#> Warning: NAs introduced by coercion
#> [1]  1 NA  3
```

R does not have any guesses for what number you want when you type b, so it does not try.

As a data scientist you will encounter the NAs often as they are generally used for missing data, a common problem in real-world datasets.

2.8   Exercises

1. Use the function c to create a vector with the average high temperatures in January for Beijing, Lagos, Paris, Rio de Janeiro, San Juan, and Toronto, which are 35, 88, 42, 84, 81, and 30 degrees Fahrenheit. Call the object temp.

2. Now create a vector with the city names and call the object city.

3. Use the names function and the objects defined in the previous exercises to associate the temperature data with its corresponding city.

4. Use the [and : operators to access the temperature of the first three cities on the list.

5. Use the [operator to access the temperature of Paris and San Juan.

6. Use the : operator to create a sequence of numbers $12, 13, 14, \ldots, 73$.

7. Create a vector containing all the positive odd numbers smaller than 100.

8. Create a vector of numbers that starts at 6, does not pass 55, and adds numbers in increments of $4/7$: 6, 6 + 4/7, 6 + 8/7, and so on. How many numbers does the list have? Hint: use seq and length.

9. What is the class of the following object a <- seq(1, 10, 0.5)?

10. What is the class of the following object a <- seq(1, 10)?

11. The class of class(a<-1) is numeric, not integer. R defaults to numeric and to force an integer, you need to add the letter L. Confirm that the class of 1L is integer.

12. Define the following vector:

```
x <- c("1", "3", "5")
```

and coerce it to get integers.

2.9 Sorting

Now that we have mastered some basic R knowledge, let's try to gain some insights into the safety of different states in the context of gun murders.

2.9.1 sort

Say we want to rank the states from least to most gun murders. The function `sort` sorts a vector in increasing order. We can therefore see the largest number of gun murders by typing:

```
library(dslabs)
data(murders)
sort(murders$total)
#>  [1]    2    4    5    5    7    8   11   12   12   16   19   21   22
#> [14]   27   32   36   38   53   63   65   67   84   93   93   97   97
#> [27]   99  111  116  118  120  135  142  207  219  232  246  250  286
#> [40]  293  310  321  351  364  376  413  457  517  669  805 1257
```

However, this does not give us information about which states have which murder totals. For example, we don't know which state had 1257.

2.9.2 order

The function `order` is closer to what we want. It takes a vector as input and returns the vector of indexes that sorts the input vector. This may sound confusing so let's look at a simple example. We can create a vector and sort it:

```
x <- c(31, 4, 15, 92, 65)
sort(x)
#> [1]  4 15 31 65 92
```

Rather than sort the input vector, the function `order` returns the index that sorts input vector:

```
index <- order(x)
x[index]
#> [1]  4 15 31 65 92
```

This is the same output as that returned by `sort(x)`. If we look at this index, we see why it works:

```
x
#> [1] 31  4 15 92 65
order(x)
#> [1] 2 3 1 5 4
```

The second entry of x is the smallest, so `order(x)` starts with 2. The next smallest is the third entry, so the second entry is 3 and so on.

How does this help us order the states by murders? First, remember that the entries of vectors you access with $ follow the same order as the rows in the table. For example, these two vectors containing state names and abbreviations, respectively, are matched by their order:

```
murders$state[1:6]
#> [1] "Alabama"     "Alaska"       "Arizona"      "Arkansas"     "California"
#> [6] "Colorado"
murders$abb[1:6]
#> [1] "AL" "AK" "AZ" "AR" "CA" "CO"
```

This means we can order the state names by their total murders. We first obtain the index that orders the vectors according to murder totals and then index the state names vector:

```
ind <- order(murders$total)
murders$abb[ind]
#>  [1] "VT" "ND" "NH" "WY" "HI" "SD" "ME" "ID" "MT" "RI" "AK" "IA" "UT"
#> [14] "WV" "NE" "OR" "DE" "MN" "KS" "CO" "NM" "NV" "AR" "WA" "CT" "WI"
#> [27] "DC" "OK" "KY" "MA" "MS" "AL" "IN" "SC" "TN" "AZ" "NJ" "VA" "NC"
#> [40] "MD" "OH" "MO" "LA" "IL" "GA" "MI" "PA" "NY" "FL" "TX" "CA"
```

According to the above, California had the most murders.

2.9.3 max and which.max

If we are only interested in the entry with the largest value, we can use `max` for the value:

```
max(murders$total)
#> [1] 1257
```

and `which.max` for the index of the largest value:

```
i_max <- which.max(murders$total)
murders$state[i_max]
#> [1] "California"
```

For the minimum, we can use `min` and `which.min` in the same way.

Does this mean California is the most dangerous state? In an upcoming section, we argue that we should be considering rates instead of totals. Before doing that, we introduce one last order-related function: `rank`.

2.9.4 rank

Although not as frequently used as `order` and `sort`, the function `rank` is also related to order and can be useful. For any given vector it returns a vector with the rank of the first entry, second entry, etc., of the input vector. Here is a simple example:

```
x <- c(31, 4, 15, 92, 65)
rank(x)
#> [1] 3 1 2 5 4
```

To summarize, let's look at the results of the three functions we have introduced:

original	sort	order	rank
31	4	2	3
4	15	3	1
15	31	1	2
92	65	5	5
65	92	4	4

2.9.5 Beware of recycling

Another common source of unnoticed errors in R is the use of *recycling*. We saw that vectors are added elementwise. So if the vectors don't match in length, it is natural to assume that we should get an error. But we don't. Notice what happens:

```
x <- c(1,2,3)
y <- c(10, 20, 30, 40, 50, 60, 70)
x+y
#> Warning in x + y: longer object length is not a multiple of shorter
#> object length
#> [1] 11 22 33 41 52 63 71
```

We do get a warning, but no error. For the output, R has recycled the numbers in x. Notice the last digit of numbers in the output.

2.10 Exercises

For these exercises we will use the US murders dataset. Make sure you load it prior to starting.

```
library(dslabs)
data("murders")
```

1. Use the $ operator to access the population size data and store it as the object pop. Then use the sort function to redefine pop so that it is sorted. Finally, use the [operator to report the smallest population size.

2. Now instead of the smallest population size, find the index of the entry with the smallest population size. Hint: use order instead of sort.

3. We can actually perform the same operation as in the previous exercise using the function which.min. Write one line of code that does this.

4. Now we know how small the smallest state is and we know which row represents it. Which

state is it? Define a variable `states` to be the state names from the `murders` data frame.
Report the name of the state with the smallest population.

5. You can create a data frame using the `data.frame` function. Here is a quick example:

```
temp <- c(35, 88, 42, 84, 81, 30)
city <- c("Beijing", "Lagos", "Paris", "Rio de Janeiro",
          "San Juan", "Toronto")
city_temps <- data.frame(name = city, temperature = temp)
```

Use the `rank` function to determine the population rank of each state from smallest population
size to biggest. Save these ranks in an object called `ranks`, then create a data frame with
the state name and its rank. Call the data frame `my_df`.

6. Repeat the previous exercise, but this time order `my_df` so that the states are ordered
from least populous to most populous. Hint: create an object `ind` that stores the indexes
needed to order the population values. Then use the bracket operator `[` to re-order each
column in the data frame.

7. The `na_example` vector represents a series of counts. You can quickly examine the object
using:

```
data("na_example")
str(na_example)
#>  int [1:1000] 2 1 3 2 1 3 1 4 3 2 ...
```

However, when we compute the average with the function `mean`, we obtain an `NA`:

```
mean(na_example)
#> [1] NA
```

The `is.na` function returns a logical vector that tells us which entries are `NA`. Assign this
logical vector to an object called `ind` and determine how many `NA`s does `na_example` have.

8. Now compute the average again, but only for the entries that are not `NA`. Hint: remember
the `!` operator.

2.11   Vector arithmetics

California had the most murders, but does this mean it is the most dangerous state? What if
it just has many more people than any other state? We can quickly confirm that California
indeed has the largest population:

```
library(dslabs)
data("murders")
murders$state[which.max(murders$population)]
#> [1] "California"
```

with over 37 million inhabitants. It is therefore unfair to compare the totals if we are
interested in learning how safe the state is. What we really should be computing is the

murders per capita. The reports we describe in the motivating section used murders per 100,000 as the unit. To compute this quantity, the powerful vector arithmetic capabilities of R come in handy.

2.11.1 Rescaling a vector

In R, arithmetic operations on vectors occur *element-wise*. For a quick example, suppose we have height in inches:

```
inches <- c(69, 62, 66, 70, 70, 73, 67, 73, 67, 70)
```

and want to convert to centimeters. Notice what happens when we multiply `inches` by 2.54:

```
inches * 2.54
#>  [1] 175 157 168 178 178 185 170 185 170 178
```

In the line above, we multiplied each element by 2.54. Similarly, if for each entry we want to compute how many inches taller or shorter than 69 inches, the average height for males, we can subtract it from every entry like this:

```
inches - 69
#>  [1]  0 -7 -3  1  1  4 -2  4 -2  1
```

2.11.2 Two vectors

If we have two vectors of the same length, and we sum them in R, they will be added entry by entry as follows:

$$\begin{pmatrix} a \\ b \\ c \\ d \end{pmatrix} + \begin{pmatrix} e \\ f \\ g \\ h \end{pmatrix} = \begin{pmatrix} a+e \\ b+f \\ c+g \\ d+h \end{pmatrix}$$

The same holds for other mathematical operations, such as -, * and /.

This implies that to compute the murder rates we can simply type:

```
murder_rate <- murders$total / murders$population * 100000
```

Once we do this, we notice that California is no longer near the top of the list. In fact, we can use what we have learned to order the states by murder rate:

```
murders$abb[order(murder_rate)]
#>  [1] "VT" "NH" "HI" "ND" "IA" "ID" "UT" "ME" "WY" "OR" "SD" "MN" "MT"
#> [14] "CO" "WA" "WV" "RI" "WI" "NE" "MA" "IN" "KS" "NY" "KY" "AK" "OH"
#> [27] "CT" "NJ" "AL" "IL" "OK" "NC" "NV" "VA" "AR" "TX" "NM" "CA" "FL"
#> [40] "TN" "PA" "AZ" "GA" "MS" "MI" "DE" "SC" "MD" "MO" "LA" "DC"
```

2.12   Exercises

1. Previously we created this data frame:

```
temp <- c(35, 88, 42, 84, 81, 30)
city <- c("Beijing", "Lagos", "Paris", "Rio de Janeiro",
          "San Juan", "Toronto")
city_temps <- data.frame(name = city, temperature = temp)
```

Remake the data frame using the code above, but add a line that converts the temperature from Fahrenheit to Celsius. The conversion is $C = \frac{5}{9} \times (F - 32)$.

2. What is the following sum $1 + 1/2^2 + 1/3^2 + \ldots 1/100^2$? Hint: thanks to Euler, we know it should be close to $\pi^2/6$.

3. Compute the per 100,000 murder rate for each state and store it in the object `murder_rate`. Then compute the average murder rate for the US using the function `mean`. What is the average?

2.13   Indexing

R provides a powerful and convenient way of indexing vectors. We can, for example, subset a vector based on properties of another vector. In this section, we continue working with our US murders example, which we can load like this:

```
library(dslabs)
data("murders")
```

2.13.1   Subsetting with logicals

We have now calculated the murder rate using:

```
murder_rate <- murders$total / murders$population * 100000
```

Imagine you are moving from Italy where, according to an ABC news report, the murder rate is only 0.71 per 100,000. You would prefer to move to a state with a similar murder rate. Another powerful feature of R is that we can use logicals to index vectors. If we compare a vector to a single number, it actually performs the test for each entry. The following is an example related to the question above:

```
ind <- murder_rate < 0.71
```

If we instead want to know if a value is less or equal, we can use:

```
ind <- murder_rate <= 0.71
```

Note that we get back a logical vector with `TRUE` for each entry smaller than or equal to 0.71. To see which states these are, we can leverage the fact that vectors can be indexed with logicals.

```
murders$state[ind]
#> [1] "Hawaii"        "Iowa"          "New Hampshire" "North Dakota"
#> [5] "Vermont"
```

In order to count how many are TRUE, the function `sum` returns the sum of the entries of a vector and logical vectors get *coerced* to numeric with `TRUE` coded as 1 and `FALSE` as 0. Thus we can count the states using:

```
sum(ind)
#> [1] 5
```

2.13.2   Logical operators

Suppose we like the mountains and we want to move to a safe state in the western region of the country. We want the murder rate to be at most 1. In this case, we want two different things to be true. Here we can use the logical operator *and*, which in R is represented with `&`. This operation results in `TRUE` only when both logicals are `TRUE`. To see this, consider this example:

```
TRUE & TRUE
#> [1] TRUE
TRUE & FALSE
#> [1] FALSE
FALSE & FALSE
#> [1] FALSE
```

For our example, we can form two logicals:

```
west <- murders$region == "West"
safe <- murder_rate <= 1
```

and we can use the `&` to get a vector of logicals that tells us which states satisfy both conditions:

```
ind <- safe & west
murders$state[ind]
#> [1] "Hawaii"  "Idaho"   "Oregon"  "Utah"    "Wyoming"
```

2.13.3   which

Suppose we want to look up California's murder rate. For this type of operation, it is convenient to convert vectors of logicals into indexes instead of keeping long vectors of logicals. The function `which` tells us which entries of a logical vector are TRUE. So we can type:

```
ind <- which(murders$state == "California")
murder_rate[ind]
#> [1] 3.37
```

2.13.4 `match`

If instead of just one state we want to find out the murder rates for several states, say New York, Florida, and Texas, we can use the function `match`. This function tells us which indexes of a second vector match each of the entries of a first vector:

```
ind <- match(c("New York", "Florida", "Texas"), murders$state)
ind
#> [1] 33 10 44
```

Now we can look at the murder rates:

```
murder_rate[ind]
#> [1] 2.67 3.40 3.20
```

2.13.5 `%in%`

If rather than an index we want a logical that tells us whether or not each element of a first vector is in a second, we can use the function `%in%`. Let's imagine you are not sure if Boston, Dakota, and Washington are states. You can find out like this:

```
c("Boston", "Dakota", "Washington") %in% murders$state
#> [1] FALSE FALSE  TRUE
```

Note that we will be using `%in%` often throughout the book.

Advanced: There is a connection between `match` and `%in%` through `which`. To see this, notice that the following two lines produce the same index (although in different order):

```
match(c("New York", "Florida", "Texas"), murders$state)
#> [1] 33 10 44
which(murders$state%in%c("New York", "Florida", "Texas"))
#> [1] 10 33 44
```

2.14 Exercises

Start by loading the library and data.

```
library(dslabs)
data(murders)
```

1. Compute the per 100,000 murder rate for each state and store it in an object called `murder_rate`. Then use logical operators to create a logical vector named `low` that tells us which entries of `murder_rate` are lower than 1.

2. Now use the results from the previous exercise and the function `which` to determine the indices of `murder_rate` associated with values lower than 1.

3. Use the results from the previous exercise to report the names of the states with murder rates lower than 1.

4. Now extend the code from exercises 2 and 3 to report the states in the Northeast with murder rates lower than 1. Hint: use the previously defined logical vector `low` and the logical operator `&`.

5. In a previous exercise we computed the murder rate for each state and the average of these numbers. How many states are below the average?

6. Use the match function to identify the states with abbreviations AK, MI, and IA. Hint: start by defining an index of the entries of `murders$abb` that match the three abbreviations, then use the `[` operator to extract the states.

7. Use the `%in%` operator to create a logical vector that answers the question: which of the following are actual abbreviations: MA, ME, MI, MO, MU?

8. Extend the code you used in exercise 7 to report the one entry that is **not** an actual abbreviation. Hint: use the `!` operator, which turns `FALSE` into `TRUE` and vice versa, then `which` to obtain an index.

2.15 Basic plots

In Chapter 7 we describe an add-on package that provides a powerful approach to producing plots in R. We then have an entire part on Data Visualization in which we provide many examples. Here we briefly describe some of the functions that are available in a basic R installation.

2.15.1 plot

The `plot` function can be used to make scatterplots. Here is a plot of total murders versus population.

```
x <- murders$population / 10^6
y <- murders$total
plot(x, y)
```

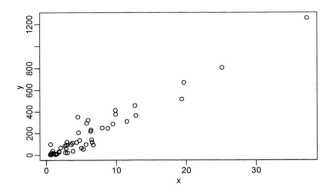

For a quick plot that avoids accessing variables twice, we can use the `with` function:

```
with(murders, plot(population, total))
```

The function `with` lets us use the `murders` column names in the `plot` function. It also works with any data frames and any function.

2.15.2 hist

We will describe histograms as they relate to distributions in the Data Visualization part of the book. Here we will simply note that histograms are a powerful graphical summary of a list of numbers that gives you a general overview of the types of values you have. We can make a histogram of our murder rates by simply typing:

```
x <- with(murders, total / population * 100000)
hist(x)
```

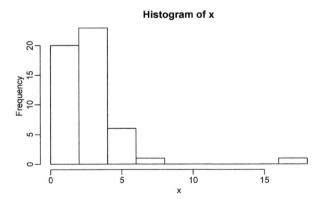

We can see that there is a wide range of values with most of them between 2 and 3 and one very extreme case with a murder rate of more than 15:

```
murders$state[which.max(x)]
#> [1] "District of Columbia"
```

2.15.3 boxplot

Boxplots will also be described in the Data Visualization part of the book. They provide a more terse summary than histograms, but they are easier to stack with other boxplots. For example, here we can use them to compare the different regions:

```
murders$rate <- with(murders, total / population * 100000)
boxplot(rate~region, data = murders)
```

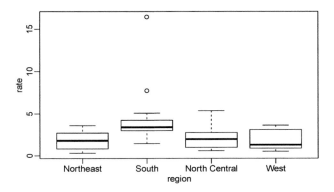

We can see that the South has higher murder rates than the other three regions.

2.15.4 image

The image function displays the values in a matrix using color. Here is a quick example:

```
x <- matrix(1:120, 12, 10)
image(x)
```

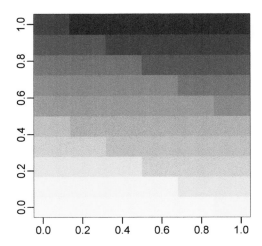

2.16   Exercises

1. We made a plot of total murders versus population and noted a strong relationship. Not surprisingly, states with larger populations had more murders.

```
library(dslabs)
data(murders)
population_in_millions <- murders$population/10^6
total_gun_murders <- murders$total
plot(population_in_millions, total_gun_murders)
```

Keep in mind that many states have populations below 5 million and are bunched up. We may gain further insights from making this plot in the log scale. Transform the variables using the `log10` transformation and then plot them.

2. Create a histogram of the state populations.

3. Generate boxplots of the state populations by region.

3

Programming basics

We teach R because it greatly facilitates data analysis, the main topic of this book. By coding in R, we can efficiently perform exploratory data analysis, build data analysis pipelines, and prepare data visualization to communicate results. However, R is not just a data analysis environment but a programming language. Advanced R programmers can develop complex packages and even improve R itself, but we do not cover advanced programming in this book. Nonetheless, in this section, we introduce three key programming concepts: conditional expressions, for-loops, and functions. These are not just key building blocks for advanced programming, but are sometimes useful during data analysis. We also note that there are several functions that are widely used to program in R but that we will not cover in this book. These include `split`, `cut`, `do.call`, and `Reduce`, as well as the **data.table** package. These are worth learning if you plan to become an expert R programmer.

3.1 Conditional expressions

Conditional expressions are one of the basic features of programming. They are used for what is called *flow control*. The most common conditional expression is the if-else statement. In R, we can actually perform quite a bit of data analysis without conditionals. However, they do come up occasionally, and you will need them once you start writing your own functions and packages.

Here is a very simple example showing the general structure of an if-else statement. The basic idea is to print the reciprocal of `a` unless `a` is 0:

```
a <- 0

if(a!=0){
  print(1/a)
} else{
  print("No reciprocal for 0.")
}
#> [1] "No reciprocal for 0."
```

Let's look at one more example using the US murders data frame:

```
library(dslabs)
data(murders)
murder_rate <- murders$total / murders$population*100000
```

Here is a very simple example that tells us which states, if any, have a murder rate lower than 0.5 per 100,000. The `if` statement protects us from the case in which no state satisfies the condition.

```
ind <- which.min(murder_rate)

if(murder_rate[ind] < 0.5){
  print(murders$state[ind])
} else{
  print("No state has murder rate that low")
}
#> [1] "Vermont"
```

If we try it again with a rate of 0.25, we get a different answer:

```
if(murder_rate[ind] < 0.25){
  print(murders$state[ind])
} else{
  print("No state has a murder rate that low.")
}
#> [1] "No state has a murder rate that low."
```

A related function that is very useful is `ifelse`. This function takes three arguments: a logical and two possible answers. If the logical is `TRUE`, the value in the second argument is returned and if `FALSE`, the value in the third argument is returned. Here is an example:

```
a <- 0
ifelse(a > 0, 1/a, NA)
#> [1] NA
```

The function is particularly useful because it works on vectors. It examines each entry of the logical vector and returns elements from the vector provided in the second argument, if the entry is `TRUE`, or elements from the vector provided in the third argument, if the entry is `FALSE`.

```
a <- c(0, 1, 2, -4, 5)
result <- ifelse(a > 0, 1/a, NA)
```

This table helps us see what happened:

a	is_a_positive	answer1	answer2	result
0	FALSE	Inf	NA	NA
1	TRUE	1.00	NA	1.0
2	TRUE	0.50	NA	0.5
-4	FALSE	-0.25	NA	NA
5	TRUE	0.20	NA	0.2

Here is an example of how this function can be readily used to replace all the missing values in a vector with zeros:

```
data(na_example)
no_nas <- ifelse(is.na(na_example), 0, na_example)
sum(is.na(no_nas))
#> [1] 0
```

Two other useful functions are `any` and `all`. The `any` function takes a vector of logicals and returns `TRUE` if any of the entries is `TRUE`. The `all` function takes a vector of logicals and returns `TRUE` if all of the entries are `TRUE`. Here is an example:

```
z <- c(TRUE, TRUE, FALSE)
any(z)
#> [1] TRUE
all(z)
#> [1] FALSE
```

3.2 Defining functions

As you become more experienced, you will find yourself needing to perform the same operations over and over. A simple example is computing averages. We can compute the average of a vector x using the `sum` and `length` functions: `sum(x)/length(x)`. Because we do this repeatedly, it is much more efficient to write a function that performs this operation. This particular operation is so common that someone already wrote the `mean` function and it is included in base R. However, you will encounter situations in which the function does not already exist, so R permits you to write your own. A simple version of a function that computes the average can be defined like this:

```
avg <- function(x){
  s <- sum(x)
  n <- length(x)
  s/n
}
```

Now `avg` is a function that computes the mean:

```
x <- 1:100
identical(mean(x), avg(x))
#> [1] TRUE
```

Notice that variables defined inside a function are not saved in the workspace. So while we use `s` and `n` when we call `avg`, the values are created and changed only during the call. Here is an illustrative example:

```
s <- 3
avg(1:10)
#> [1] 5.5
s
#> [1] 3
```

Note how s is still 3 after we call `avg`.

In general, functions are objects, so we assign them to variable names with `<-`. The function `function` tells R you are about to define a function. The general form of a function definition looks like this:

```
my_function <- function(VARIABLE_NAME){
  perform operations on VARIABLE_NAME and calculate VALUE
  VALUE
}
```

The functions you define can have multiple arguments as well as default values. For example, we can define a function that computes either the arithmetic or geometric average depending on a user defined variable like this:

```
avg <- function(x, arithmetic = TRUE){
  n <- length(x)
  ifelse(arithmetic, sum(x)/n, prod(x)^(1/n))
}
```

We will learn more about how to create functions through experience as we face more complex tasks.

3.3   Namespaces

Once you start becoming more of an R expert user, you will likely need to load several add-on packages for some of your analysis. Once you start doing this, it is likely that two packages use the same name for two different functions. And often these functions do completely different things. In fact, you have already encountered this because both **dplyr** and the R-base **stats** package define a `filter` function. There are five other examples in **dplyr**. We know this because when we first load **dplyr** we see the following message:

```
The following objects are masked from 'package:stats':

    filter, lag
```

```
The following objects are masked from 'package:base':

    intersect, setdiff, setequal, union
```

So what does R do when we type `filter`? Does it use the **dplyr** function or the **stats** function? From our previous work we know it uses the **dplyr** one. But what if we want to use the **stats** version?

These functions live in different *namespaces*. R will follow a certain order when searching for a function in these *namespaces*. You can see the order by typing:

```
search()
```

The first entry in this list is the global environment which includes all the objects you define.

So what if we want to use the **stats filter** instead of the **dplyr** filter but **dplyr** appears first in the search list? You can force the use of a specific name space by using double colons (::) like this:

```
stats::filter
```

If we want to be absolutely sure we use the **dplyr filter** we can use

```
dplyr::filter
```

Also note that if we want to use a function in a package without loading the entire package, we can use the double colon as well.

For more on this more advanced topic we recommend the R packages book[1].

3.4 For-loops

The formula for the sum of the series $1 + 2 + \cdots + n$ is $n(n+1)/2$. What if we weren't sure that was the right function? How could we check? Using what we learned about functions we can create one that computes the S_n:

```
compute_s_n <- function(n){
   x <- 1:n
   sum(x)
}
```

How can we compute S_n for various values of n, say $n = 1, \ldots, 25$? Do we write 25 lines of code calling compute_s_n? No, that is what for-loops are for in programming. In this case, we are performing exactly the same task over and over, and the only thing that is changing is the value of n. For-loops let us define the range that our variable takes (in our example $n = 1, \ldots, 10$), then change the value and evaluate expression as you *loop*.

Perhaps the simplest example of a for-loop is this useless piece of code:

```
for(i in 1:5){
   print(i)
}
#> [1] 1
#> [1] 2
#> [1] 3
#> [1] 4
#> [1] 5
```

Here is the for-loop we would write for our S_n example:

```
m <- 25
s_n <- vector(length = m) # create an empty vector
```

---

[1] http://r-pkgs.had.co.nz/namespace.html

```
for(n in 1:m){
  s_n[n] <- compute_s_n(n)
}
```

In each iteration $n = 1$, $n = 2$, etc..., we compute S_n and store it in the nth entry of `s_n`.

Now we can create a plot to search for a pattern:

```
n <- 1:m
plot(n, s_n)
```

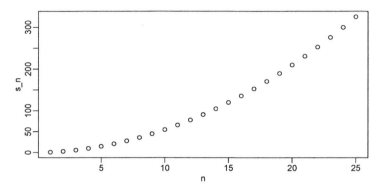

If you noticed that it appears to be a quadratic, you are on the right track because the formula is $n(n + 1)/2$.

3.5   Vectorization and functionals

Although for-loops are an important concept to understand, in R we rarely use them. As you learn more R, you will realize that *vectorization* is preferred over for-loops since it results in shorter and clearer code. We already saw examples in the Vector Arithmetic section. A *vectorized* function is a function that will apply the same operation on each of the vectors.

```
x <- 1:10
sqrt(x)
#>  [1] 1.00 1.41 1.73 2.00 2.24 2.45 2.65 2.83 3.00 3.16
y <- 1:10
x*y
#>  [1]    1    4    9   16   25   36   49   64   81  100
```

To make this calculation, there is no need for for-loops. However, not all functions work this way. For instance, the function we just wrote, `compute_s_n`, does not work element-wise since it is expecting a scalar. This piece of code does not run the function on each entry of `n`:

```
n <- 1:25
compute_s_n(n)
```

Functionals are functions that help us apply the same function to each entry in a vector, matrix, data frame, or list. Here we cover the functional that operates on numeric, logical, and character vectors: `sapply`.

The function `sapply` permits us to perform element-wise operations on any function. Here is how it works:

```
x <- 1:10
sapply(x, sqrt)
#>  [1] 1.00 1.41 1.73 2.00 2.24 2.45 2.65 2.83 3.00 3.16
```

Each element of x is passed on to the function `sqrt` and the result is returned. These results are concatenated. In this case, the result is a vector of the same length as the original x. This implies that the for-loop above can be written as follows:

```
n <- 1:25
s_n <- sapply(n, compute_s_n)
```

Other functionals are `apply`, `lapply`, `tapply`, `mapply`, `vapply`, and `replicate`. We mostly use `sapply`, `apply`, and `replicate` in this book, but we recommend familiarizing yourselves with the others as they can be very useful.

3.6 Exercises

1. What will this conditional expression return?

```
x <- c(1,2,-3,4)

if(all(x>0)){
  print("All Postives")
} else{
  print("Not all positives")
}
```

2. Which of the following expressions is always **FALSE** when at least one entry of a logical vector x is TRUE?

    a. `all(x)`
    b. `any(x)`
    c. `any(!x)`
    d. `all(!x)`

3. The function `nchar` tells you how many characters long a character vector is. Write a line of code that assigns to the object `new_names` the state abbreviation when the state name is longer than 8 characters.

4. Create a function `sum_n` that for any given value, say n, computes the sum of the integers from 1 to n (inclusive). Use the function to determine the sum of integers from 1 to 5,000.

5. Create a function `altman_plot` that takes two arguments, x and y, and plots the difference against the sum.

6. After running the code below, what is the value of x?

```
x <- 3
my_func <- function(y){
  x <- 5
  y+5
}
```

7. Write a function `compute_s_n` that for any given n computes the sum $S_n = 1^2 + 2^2 + 3^2 + \ldots n^2$. Report the value of the sum when $n = 10$.

8. Define an empty numerical vector `s_n` of size 25 using `s_n <- vector("numeric", 25)` and store in the results of $S_1, S_2, \ldots S_{25}$ using a for-loop.

9. Repeat exercise 8, but this time use `sapply`.

10. Repeat exercise 8, but this time use `map_dbl`.

11. Plot S_n versus n. Use points defined by $n = 1, \ldots, 25$.

12. Confirm that the formula for this sum is $S_n = n(n+1)(2n+1)/6$.

4

The tidyverse

Up to now we have been manipulating vectors by reordering and subsetting them through indexing. However, once we start more advanced analyses, the preferred unit for data storage is not the vector but the data frame. In this chapter we learn to work directly with data frames, which greatly facilitate the organization of information. We will be using data frames for the majority of this book. We will focus on a specific data format referred to as *tidy* and on specific collection of packages that are particularly helpful for working with *tidy* data referred to as the *tidyverse*.

We can load all the tidyverse packages at once by installing and loading the **tidyverse** package:

```
library(tidyverse)
```

We will learn how to implement the tidyverse approach throughout the book, but before delving into the details, in this chapter we introduce some of the most widely used tidyverse functionality, starting with the **dplyr** package for manipulating data frames and the **purrr** package for working with functions. Note that the tidyverse also includes a graphing package, **ggplot2**, which we introduce later in Chapter 7 in the Data Visualization part of the book; the **readr** package discussed in Chapter 5; and many others. In this chapter, we first introduce the concept of *tidy data* and then demonstrate how we use the tidyverse to work with data frames in this format.

4.1 Tidy data

We say that a data table is in *tidy* format if each row represents one observation and columns represent the different variables available for each of these observations. The `murders` dataset is an example of a tidy data frame.

```
#>          state abb region population total
#> 1      Alabama  AL  South    4779736   135
#> 2       Alaska  AK   West     710231    19
#> 3      Arizona  AZ   West    6392017   232
#> 4     Arkansas  AR  South    2915918    93
#> 5   California  CA   West   37253956  1257
#> 6     Colorado  CO   West    5029196    65
```

Each row represent a state with each of the five columns providing a different variable related to these states: name, abbreviation, region, population, and total murders.

To see how the same information can be provided in different formats, consider the following example:

```
#>          country year fertility
#> 1        Germany 1960       2.41
#> 2 South Korea 1960       6.16
#> 3        Germany 1961       2.44
#> 4 South Korea 1961       5.99
#> 5        Germany 1962       2.47
#> 6 South Korea 1962       5.79
```

This tidy dataset provides fertility rates for two countries across the years. This is a tidy dataset because each row presents one observation with the three variables being country, year, and fertility rate. However, this dataset originally came in another format and was reshaped for the **dslabs** package. Originally, the data was in the following format:

```
#>          country 1960 1961 1962
#> 1        Germany 2.41 2.44 2.47
#> 2 South Korea 6.16 5.99 5.79
```

The same information is provided, but there are two important differences in the format: 1) each row includes several observations and 2) one of the variables, year, is stored in the header. For the tidyverse packages to be optimally used, data need to be reshaped into `tidy` format, which you will learn to do in the Data Wrangling part of the book. Until then, we will use example datasets that are already in tidy format.

Although not immediately obvious, as you go through the book you will start to appreciate the advantages of working in a framework in which functions use tidy formats for both inputs and outputs. You will see how this permits the data analyst to focus on more important aspects of the analysis rather than the format of the data.

4.2   Exercises

1. Examine the built-in dataset `co2`. Which of the following is true:

    a. co2 is tidy data: it has one year for each row.
    b. co2 is not tidy: we need at least one column with a character vector.
    c. co2 is not tidy: it is a matrix instead of a data frame.
    d. co2 is not tidy: to be tidy we would have to wrangle it to have three columns (year, month and value), then each co2 observation would have a row.

2. Examine the built-in dataset `ChickWeight`. Which of the following is true:

    a. `ChickWeight` is not tidy: each chick has more than one row.
    b. `ChickWeight` is tidy: each observation (a weight) is represented by one row. The chick from which this measurement came is one of the variables.
    c. `ChickWeight` is not tidy: we are missing the year column.
    d. `ChickWeight` is tidy: it is stored in a data frame.

3. Examine the built-in dataset `BOD`. Which of the following is true:

a. BOD is not tidy: it only has six rows.
b. BOD is not tidy: the first column is just an index.
c. BOD is tidy: each row is an observation with two values (time and demand)
d. BOD is tidy: all small datasets are tidy by definition.

4. Which of the following built-in datasets is tidy (you can pick more than one):

a. BJsales
b. EuStockMarkets
c. DNase
d. Formaldehyde
e. Orange
f. UCBAdmissions

4.3 Manipulating data frames

The **dplyr** package from the **tidyverse** introduces functions that perform some of the most common operations when working with data frames and uses names for these functions that are relatively easy to remember. For instance, to change the data table by adding a new column, we use `mutate`. To filter the data table to a subset of rows, we use `filter`. Finally, to subset the data by selecting specific columns, we use `select`.

4.3.1 Adding a column with `mutate`

We want all the necessary information for our analysis to be included in the data table. So the first task is to add the murder rates to our murders data frame. The function `mutate` takes the data frame as a first argument and the name and values of the variable as a second argument using the convention `name = values`. So, to add murder rates, we use:

```
library(dslabs)
data("murders")
murders <- mutate(murders, rate = total / population * 100000)
```

Notice that here we used `total` and `population` inside the function, which are objects that are **not** defined in our workspace. But why don't we get an error?

This is one of **dplyr**'s main features. Functions in this package, such as `mutate`, know to look for variables in the data frame provided in the first argument. In the call to mutate above, `total` will have the values in `murders$total`. This approach makes the code much more readable.

We can see that the new column is added:

```
head(murders)
#>       state abb region population total rate
#> 1   Alabama  AL  South    4779736   135 2.82
#> 2    Alaska  AK   West     710231    19 2.68
```

```
#> 3     Arizona  AZ   West     6392017   232 3.63
#> 4    Arkansas  AR  South     2915918    93 3.19
#> 5  California  CA   West    37253956  1257 3.37
#> 6    Colorado  CO   West     5029196    65 1.29
```

Although we have overwritten the original `murders` object, this does not change the object that loaded with `data(murders)`. If we load the `murders` data again, the original will overwrite our mutated version.

4.3.2   Subsetting with `filter`

Now suppose that we want to filter the data table to only show the entries for which the murder rate is lower than 0.71. To do this we use the `filter` function, which takes the data table as the first argument and then the conditional statement as the second. Like `mutate`, we can use the unquoted variable names from `murders` inside the function and it will know we mean the columns and not objects in the workspace.

```
filter(murders, rate <= 0.71)
#>                state abb        region population total  rate
#> 1            Hawaii  HI          West    1360301     7 0.515
#> 2              Iowa  IA North Central    3046355    21 0.689
#> 3 New Hampshire  NH      Northeast    1316470     5 0.380
#> 4   North Dakota  ND North Central     672591     4 0.595
#> 5           Vermont  VT      Northeast     625741     2 0.320
```

4.3.3   Selecting columns with `select`

Although our data table only has six columns, some data tables include hundreds. If we want to view just a few, we can use the **dplyr** `select` function. In the code below we select three columns, assign this to a new object and then filter the new object:

```
new_table <- select(murders, state, region, rate)
filter(new_table, rate <= 0.71)
#>                state        region  rate
#> 1            Hawaii          West 0.515
#> 2              Iowa North Central 0.689
#> 3 New Hampshire      Northeast 0.380
#> 4   North Dakota North Central 0.595
#> 5           Vermont      Northeast 0.320
```

In the call to `select`, the first argument `murders` is an object, but `state`, `region`, and `rate` are variable names.

4.4 Exercises

1. Load the **dplyr** package and the murders dataset.

```
library(dplyr)
library(dslabs)
data(murders)
```

You can add columns using the **dplyr** function `mutate`. This function is aware of the column names and inside the function you can call them unquoted:

```
murders <- mutate(murders, population_in_millions = population / 10^6)
```

We can write `population` rather than `murders$population`. The function `mutate` knows we are grabbing columns from `murders`.

Use the function `mutate` to add a murders column named `rate` with the per 100,000 murder rate as in the example code above. Make sure you redefine `murders` as done in the example code above (murders <- [your code]) so we can keep using this variable.

2. If `rank(x)` gives you the ranks of x from lowest to highest, `rank(-x)` gives you the ranks from highest to lowest. Use the function `mutate` to add a column `rank` containing the rank, from highest to lowest murder rate. Make sure you redefine `murders` so we can keep using this variable.

3. With **dplyr**, we can use `select` to show only certain columns. For example, with this code we would only show the states and population sizes:

```
select(murders, state, population) %>% head()
```

Use `select` to show the state names and abbreviations in `murders`. Do not redefine `murders`, just show the results.

4. The **dplyr** function `filter` is used to choose specific rows of the data frame to keep. Unlike `select` which is for columns, `filter` is for rows. For example, you can show just the New York row like this:

```
filter(murders, state == "New York")
```

You can use other logical vectors to filter rows.

Use `filter` to show the top 5 states with the highest murder rates. After we add murder rate and rank, do not change the murders dataset, just show the result. Remember that you can filter based on the `rank` column.

5. We can remove rows using the `!=` operator. For example, to remove Florida, we would do this:

```
no_florida <- filter(murders, state != "Florida")
```

Create a new data frame called `no_south` that removes states from the South region. How many states are in this category? You can use the function `nrow` for this.

6. We can also use `%in%` to filter with **dplyr**. You can therefore see the data from New York and Texas like this:

```
filter(murders, state %in% c("New York", "Texas"))
```

Create a new data frame called `murders_nw` with only the states from the Northeast and the West. How many states are in this category?

7. Suppose you want to live in the Northeast or West **and** want the murder rate to be less than 1. We want to see the data for the states satisfying these options. Note that you can use logical operators with `filter`. Here is an example in which we filter to keep only small states in the Northeast region.

```
filter(murders, population < 5000000 & region == "Northeast")
```

Make sure `murders` has been defined with `rate` and `rank` and still has all states. Create a table called `my_states` that contains rows for states satisfying both the conditions: it is in the Northeast or West and the murder rate is less than 1. Use `select` to show only the state name, the rate, and the rank.

4.5   The pipe: %>%

With **dplyr** we can perform a series of operations, for example `select` and then `filter`, by sending the results of one function to another using what is called the *pipe operator*: `%>%`. Some details are included below.

We wrote code above to show three variables (state, region, rate) for states that have murder rates below 0.71. To do this, we defined the intermediate object `new_table`. In **dplyr** we can write code that looks more like a description of what we want to do without intermediate objects:

$$\text{original data} \rightarrow \text{select} \rightarrow \text{filter}$$

For such an operation, we can use the pipe `%>%`. The code looks like this:

```
murders %>% select(state, region, rate) %>% filter(rate <= 0.71)
#>                state        region  rate
#> 1            Hawaii          West 0.515
#> 2              Iowa North Central 0.689
#> 3 New Hampshire     Northeast 0.380
#> 4 North Dakota North Central 0.595
#> 5           Vermont     Northeast 0.320
```

This line of code is equivalent to the two lines of code above. What is going on here?

In general, the pipe *sends* the result of the left side of the pipe to be the first argument of the function on the right side of the pipe. Here is a very simple example:

```
16 %>% sqrt()
#> [1] 4
```

We can continue to pipe values along:

```
16 %>% sqrt() %>% log2()
#> [1] 2
```

The above statement is equivalent to `log2(sqrt(16))`.

Remember that the pipe sends values to the first argument, so we can define other arguments as if the first argument is already defined:

```
16 %>% sqrt() %>% log(base = 2)
#> [1] 2
```

Therefore, when using the pipe with data frames and **dplyr**, we no longer need to specify the required first argument since the **dplyr** functions we have described all take the data as the first argument. In the code we wrote:

```
murders %>% select(state, region, rate) %>% filter(rate <= 0.71)
```

`murders` is the first argument of the `select` function, and the new data frame (formerly `new_table`) is the first argument of the `filter` function.

Note that the pipe works well with functions where the first argument is the input data. Functions in **tidyverse** packages like **dplyr** have this format and can be used easily with the pipe.

4.6 Exercises

1. The pipe `%>%` can be used to perform operations sequentially without having to define intermediate objects. Start by redefining murder to include rate and rank.

```
murders <- mutate(murders, rate =  total / population * 100000,
                  rank = rank(-rate))
```

In the solution to the previous exercise, we did the following:

```
my_states <- filter(murders, region %in% c("Northeast", "West") &
                    rate < 1)
```

```
select(my_states, state, rate, rank)
```

The pipe `%>%` permits us to perform both operations sequentially without having to define an intermediate variable `my_states`. We therefore could have mutated and selected in the same line like this:

```
mutate(murders, rate =  total / population * 100000,
       rank = rank(-rate)) %>%
  select(state, rate, rank)
```

Notice that `select` no longer has a data frame as the first argument. The first argument is assumed to be the result of the operation conducted right before the `%>%`.

Repeat the previous exercise, but now instead of creating a new object, show the result and only include the state, rate, and rank columns. Use a pipe `%>%` to do this in just one line.

2. Reset `murders` to the original table by using `data(murders)`. Use a pipe to create a new data frame called `my_states` that considers only states in the Northeast or West which have a murder rate lower than 1, and contains only the state, rate and rank columns. The pipe should also have four components separated by three `%>%`. The code should look something like this:

```
my_states <- murders %>%
  mutate SOMETHING %>%
  filter SOMETHING %>%
  select SOMETHING
```

4.7 Summarizing data

An important part of exploratory data analysis is summarizing data. The average and standard deviation are two examples of widely used summary statistics. More informative summaries can often be achieved by first splitting data into groups. In this section, we cover two new **dplyr** verbs that make these computations easier: `summarize` and `group_by`. We learn to access resulting values using the `pull` function.

4.7.1 summarize

The `summarize` function in **dplyr** provides a way to compute summary statistics with intuitive and readable code. We start with a simple example based on heights. The `heights` dataset includes heights and sex reported by students in an in-class survey.

```
library(dplyr)
library(dslabs)
data(heights)
```

The following code computes the average and standard deviation for females:

```
s <- heights %>%
  filter(sex == "Female") %>%
  summarize(average = mean(height), standard_deviation = sd(height))
s
#>    average standard_deviation
#> 1   64.9                 3.76
```

This takes our original data table as input, filters it to keep only females, and then produces a new summarized table with just the average and the standard deviation of heights. We get to choose the names of the columns of the resulting table. For example, above we decided to use `average` and `standard_deviation`, but we could have used other names just the same.

Because the resulting table stored in `s` is a data frame, we can access the components with the accessor `$`:

```
s$average
#> [1] 64.9
s$standard_deviation
#> [1] 3.76
```

As with most other **dplyr** functions, `summarize` is aware of the variable names and we can use them directly. So when inside the call to the `summarize` function we write `mean(height)`, the function is accessing the column with the name "height" and then computing the average of the resulting numeric vector. We can compute any other summary that operates on vectors and returns a single value. For example, we can add the median, minimum, and maximum heights like this:

```
heights %>%
    filter(sex == "Female") %>%
    summarize(median = median(height), minimum = min(height),
              maximum = max(height))
#>    median minimum maximum
#> 1      65      51      79
```

We can obtain these three values with just one line using the `quantile` function: for example, `quantile(x, c(0,0.5,1))` returns the min (0th percentile), median (50th percentile), and max (100th percentile) of the vector `x`. However, if we attempt to use a function like this that returns two or more values inside `summarize`:

```
heights %>%
    filter(sex == "Female") %>%
    summarize(range = quantile(height, c(0, 0.5, 1)))
```

we will receive an error: `Error: expecting result of length one, got : 2`. With the function `summarize`, we can only call functions that return a single value. In Section 4.12, we will learn how to deal with functions that return more than one value.

For another example of how we can use the `summarize` function, let's compute the average murder rate for the United States. Remember our data table includes total murders and population size for each state and we have already used **dplyr** to add a murder rate column:

```
murders <- murders %>% mutate(rate = total/population*100000)
```

Remember that the US murder rate is **not** the average of the state murder rates:

```
summarize(murders, mean(rate))
#>    mean(rate)
#> 1        2.78
```

This is because in the computation above the small states are given the same weight as the large ones. The US murder rate is the total number of murders in the US divided by the total US population. So the correct computation is:

```
us_murder_rate <- murders %>%
  summarize(rate = sum(total) / sum(population) * 100000)
us_murder_rate
#>    rate
#> 1 3.03
```

This computation counts larger states proportionally to their size which results in a larger value.

4.7.2   pull

The `us_murder_rate` object defined above represents just one number. Yet we are storing it in a data frame:

```
class(us_murder_rate)
#> [1] "data.frame"
```

since, as most **dplyr** functions, `summarize` always returns a data frame.

This might be problematic if we want to use this result with functions that require a numeric value. Here we show a useful trick for accessing values stored in data when using pipes: when a data object is piped that object and its columns can be accessed using the `pull` function. To understand what we mean take a look at this line of code:

```
us_murder_rate %>% pull(rate)
#> [1] 3.03
```

This returns the value in the `rate` column of `us_murder_rate` making it equivalent to `us_murder_rate$rate`.

To get a number from the original data table with one line of code we can type:

```
us_murder_rate <- murders %>%
  summarize(rate = sum(total) / sum(population) * 100000) %>%
  pull(rate)

us_murder_rate
#> [1] 3.03
```

which is now a numeric:

```
class(us_murder_rate)
#> [1] "numeric"
```

4.7.3   Group then summarize with `group_by`

A common operation in data exploration is to first split data into groups and then compute summaries for each group. For example, we may want to compute the average and standard deviation for men's and women's heights separately. The `group_by` function helps us do this.

If we type this:

```
heights %>% group_by(sex)
#> # A tibble: 1,050 x 2
#> # Groups:    sex [2]
#>    sex    height
#>    <fct>  <dbl>
#> 1 Male     75
#> 2 Male     70
#> 3 Male     68
#> 4 Male     74
#> 5 Male     61
#> # ... with 1,045 more rows
```

The result does not look very different from `heights`, except we see `Groups: sex [2]` when we print the object. Although not immediately obvious from its appearance, this is now a special data frame called a *grouped data frame* and **dplyr** functions, in particular `summarize`, will behave differently when acting on this object. Conceptually, you can think of this table as many tables, with the same columns but not necessarily the same number of rows, stacked together in one object. When we summarize the data after grouping, this is what happens:

```
heights %>%
  group_by(sex) %>%
  summarize(average = mean(height), standard_deviation = sd(height))
#> # A tibble: 2 x 3
#>   sex     average standard_deviation
#>   <fct>   <dbl>          <dbl>
#> 1 Female   64.9          3.76
#> 2 Male     69.3          3.61
```

The `summarize` function applies the summarization to each group separately.

For another example, let's compute the median murder rate in the four regions of the country:

```
murders %>%
  group_by(region) %>%
  summarize(median_rate = median(rate))
#> # A tibble: 4 x 2
#>   region         median_rate
#>   <fct>              <dbl>
#> 1 Northeast           1.80
#> 2 South               3.40
#> 3 North Central       1.97
#> 4 West                1.29
```

4.8   Sorting data frames

When examining a dataset, it is often convenient to sort the table by the different columns. We know about the `order` and `sort` function, but for ordering entire tables, the **dplyr** function `arrange` is useful. For example, here we order the states by population size:

```
murders %>%
  arrange(population) %>%
  head()
#>                     state abb       region population total   rate
#> 1                 Wyoming  WY         West     563626     5  0.887
#> 2 District of Columbia  DC        South     601723    99 16.453
#> 3                 Vermont  VT    Northeast     625741     2  0.320
#> 4            North Dakota  ND North Central   672591     4  0.595
#> 5                  Alaska  AK         West     710231    19  2.675
#> 6            South Dakota  SD North Central   814180     8  0.983
```

With `arrange` we get to decide which column to sort by. To see the states by population, from smallest to largest, we arrange by `rate` instead:

```
murders %>%
  arrange(rate) %>%
  head()
#>               state abb       region population total  rate
#> 1           Vermont  VT    Northeast     625741     2 0.320
#> 2 New Hampshire  NH    Northeast    1316470     5 0.380
#> 3            Hawaii  HI         West    1360301     7 0.515
#> 4      North Dakota  ND North Central   672591     4 0.595
#> 5              Iowa  IA North Central  3046355    21 0.689
#> 6             Idaho  ID         West    1567582    12 0.766
```

Note that the default behavior is to order in ascending order. In **dplyr**, the function `desc` transforms a vector so that it is in descending order. To sort the table in descending order, we can type:

```
murders %>%
  arrange(desc(rate))
```

4.8.1   Nested sorting

If we are ordering by a column with ties, we can use a second column to break the tie. Similarly, a third column can be used to break ties between first and second and so on. Here we order by `region`, then within region we order by murder rate:

```
murders %>%
  arrange(region, rate) %>%
  head()
```

```
#>          state abb    region population total  rate
#> 1       Vermont  VT Northeast      625741     2 0.320
#> 2 New Hampshire  NH Northeast     1316470     5 0.380
#> 3         Maine  ME Northeast     1328361    11 0.828
#> 4   Rhode Island RI Northeast     1052567    16 1.520
#> 5 Massachusetts  MA Northeast     6547629   118 1.802
#> 6      New York  NY Northeast    19378102   517 2.668
```

4.8.2 The top n

In the code above, we have used the function **head** to avoid having the page fill up with the entire dataset. If we want to see a larger proportion, we can use the **top_n** function. This function takes a data frame as its first argument, the number of rows to show in the second, and the variable to filter by in the third. Here is an example of how to see the top 5 rows:

```
murders %>% top_n(5, rate)
#>                  state abb        region population total   rate
#> 1 District of Columbia  DC         South     601723    99  16.45
#> 2            Louisiana  LA         South    4533372   351   7.74
#> 3             Maryland  MD         South    5773552   293   5.07
#> 4             Missouri  MO North Central    5988927   321   5.36
#> 5       South Carolina  SC         South    4625364   207   4.48
```

Note that rows are not sorted by **rate**, only filtered. If we want to sort, we need to use **arrange**. Note that if the third argument is left blank, **top_n**, filters by the last column.

4.9 Exercises

For these exercises, we will be using the data from the survey collected by the United States National Center for Health Statistics (NCHS). This center has conducted a series of health and nutrition surveys since the 1960's. Starting in 1999, about 5,000 individuals of all ages have been interviewed every year and they complete the health examination component of the survey. Part of the data is made available via the **NHANES** package. Once you install the **NHANES** package, you can load the data like this:

```
library(NHANES)
data(NHANES)
```

The **NHANES** data has many missing values. Remember that the main summarization function in R will return `NA` if any of the entries of the input vector is an `NA`. Here is an example:

```
library(dslabs)
data(na_example)
mean(na_example)
#> [1] NA
```

```
sd(na_example)
#> [1] NA
```

To ignore the NAs we can use the `na.rm` argument:

```
mean(na_example, na.rm = TRUE)
#> [1] 2.3
sd(na_example, na.rm = TRUE)
#> [1] 1.22
```

Let's now explore the NHANES data.

1. We will provide some basic facts about blood pressure. First let's select a group to set the standard. We will use 20-to-29-year-old females. `AgeDecade` is a categorical variable with these ages. Note that the category is coded like " 20-29", with a space in front! What is the average and standard deviation of systolic blood pressure as saved in the `BPSysAve` variable? Save it to a variable called `ref`.

Hint: Use `filter` and `summarize` and use the `na.rm = TRUE` argument when computing the average and standard deviation. You can also filter the NA values using `filter`.

2. Using a pipe, assign the average to a numeric variable `ref_avg`. Hint: Use the code similar to above and then `pull`.

3. Now report the min and max values for the same group.

4. Compute the average and standard deviation for females, but for each age group separately rather than a selected decade as in question 1. Note that the age groups are defined by `AgeDecade`. Hint: rather than filtering by age and gender, filter by `Gender` and then use `group_by`.

5. Repeat exercise 4 for males.

6. We can actually combine both summaries for exercises 4 and 5 into one line of code. This is because `group_by` permits us to group by more than one variable. Obtain one big summary table using `group_by(AgeDecade, Gender)`.

7. For males between the ages of 40-49, compare systolic blood pressure across race as reported in the `Race1` variable. Order the resulting table from lowest to highest average systolic blood pressure.

4.10   Tibbles

Tidy data must be stored in data frames. We introduced the data frame in Section 2.4.1 and have been using the `murders` data frame throughout the book. In Section 4.7.3 we introduced the `group_by` function, which permits stratifying data before computing summary statistics. But where is the group information stored in the data frame?

```
murders %>% group_by(region)
#> # A tibble: 51 x 6
#> # Groups:   region [4]
```

```
#>     state      abb    region  population  total   rate
#>     <chr>      <chr>  <fct>         <dbl>  <dbl>  <dbl>
#> 1 Alabama      AL     South       4779736    135   2.82
#> 2 Alaska       AK     West         710231     19   2.68
#> 3 Arizona      AZ     West        6392017    232   3.63
#> 4 Arkansas     AR     South       2915918     93   3.19
#> 5 California   CA     West       37253956   1257   3.37
#> # ... with 46 more rows
```

Notice that there are no columns with this information. But, if you look closely at the output above, you see the line `A tibble` followd by dimensions. We can learn the class of the returned object using:

```
murders %>% group_by(region) %>% class()
#> [1] "grouped_df" "tbl_df"      "tbl"          "data.frame"
```

The `tbl`, pronounced tibble, is a special kind of data frame. The functions `group_by` and `summarize` always return this type of data frame. The `group_by` function returns a special kind of `tbl`, the `grouped_df`. We will say more about these later. For consistency, the **dplyr** manipulation verbs (`select`, `filter`, `mutate`, and `arrange`) preserve the class of the input: if they receive a regular data frame they return a regular data frame, while if they receive a tibble they return a tibble. But tibbles are the preferred format in the tidyverse and as a result tidyverse functions that produce a data frame from scratch return a tibble. For example, in Chapter 5 we will see that tidyverse functions used to import data create tibbles.

Tibbles are very similar to data frames. In fact, you can think of them as a modern version of data frames. Nonetheless there are three important differences which we describe in the next.

4.10.1 Tibbles display better

The print method for tibbles is more readable than that of a data frame. To see this, compare the outputs of typing `murders` and the output of murders if we convert it to a tibble. We can do this using `as_tibble(murders)`. If using RStudio, output for a tibble adjusts to your window size. To see this, change the width of your R console and notice how more/less columns are shown.

4.10.2 Subsets of tibbles are tibbles

If you subset the columns of a data frame, you may get back an object that is not a data frame, such as a vector or scalar. For example:

```
class(murders[,4])
#> [1] "numeric"
```

is not a data frame. With tibbles this does not happen:

```
class(as_tibble(murders)[,4])
#> [1] "tbl_df"      "tbl"          "data.frame"
```

This is useful in the tidyverse since functions require data frames as input.

With tibbles, if you want to access the vector that defines a column, and not get back a data frame, you need to use the accessor `$`:

```
class(as_tibble(murders)$population)
#> [1] "numeric"
```

A related feature is that tibbles will give you a warning if you try to access a column that does not exist. If we accidentally write `Population` instead of `population` this:

```
murders$Population
#> NULL
```

returns a `NULL` with no warning, which can make it harder to debug. In contrast, if we try this with a tibble we get an informative warning:

```
as_tibble(murders)$Population
#> Warning: Unknown or uninitialised column: 'Population'.
#> NULL
```

4.10.3   Tibbles can have complex entries

While data frame columns need to be vectors of numbers, strings, or logical values, tibbles can have more complex objects, such as lists or functions. Also, we can create tibbles with functions:

```
tibble(id = c(1, 2, 3), func = c(mean, median, sd))
#> # A tibble: 3 x 2
#>      id func
#>   <dbl> <list>
#> 1     1 <fn>
#> 2     2 <fn>
#> 3     3 <fn>
```

4.10.4   Tibbles can be grouped

The function `group_by` returns a special kind of tibble: a grouped tibble. This class stores information that lets you know which rows are in which groups. The tidyverse functions, in particular the `summarize` function, are aware of the group information.

4.10.5   Create a tibble using `tibble` instead of `data.frame`

It is sometimes useful for us to create our own data frames. To create a data frame in the tibble format, you can do this by using the `tibble` function.

```
grades <- tibble(names = c("John", "Juan", "Jean", "Yao"),
                 exam_1 = c(95, 80, 90, 85),
                 exam_2 = c(90, 85, 85, 90))
```

Note that base R (without packages loaded) has a function with a very similar name, `data.frame`, that can be used to create a regular data frame rather than a tibble. One other important difference is that by default `data.frame` coerces characters into factors without providing a warning or message:

```
grades <- data.frame(names = c("John", "Juan", "Jean", "Yao"),
                     exam_1 = c(95, 80, 90, 85),
                     exam_2 = c(90, 85, 85, 90))
class(grades$names)
#> [1] "factor"
```

To avoid this, we use the rather cumbersome argument `stringsAsFactors`:

```
grades <- data.frame(names = c("John", "Juan", "Jean", "Yao"),
                     exam_1 = c(95, 80, 90, 85),
                     exam_2 = c(90, 85, 85, 90),
                     stringsAsFactors = FALSE)
class(grades$names)
#> [1] "character"
```

To convert a regular data frame to a tibble, you can use the `as_tibble` function.

```
as_tibble(grades) %>% class()
#> [1] "tbl_df"     "tbl"          "data.frame"
```

4.11 The dot operator

One of the advantages of using the pipe %>% is that we do not have to keep naming new objects as we manipulate the data frame. As a quick reminder, if we want to compute the median murder rate for states in the southern states, instead of typing:

```
tab_1 <- filter(murders, region == "South")
tab_2 <- mutate(tab_1, rate = total / population * 10^5)
rates <- tab_2$rate
median(rates)
#> [1] 3.4
```

We can avoid defining any new intermediate objects by instead typing:

```
filter(murders, region == "South") %>%
  mutate(rate = total / population * 10^5) %>%
  summarize(median = median(rate)) %>%
```

```
  pull(median)
#> [1] 3.4
```

We can do this because each of these functions takes a data frame as the first argument. But what if we want to access a component of the data frame. For example, what if the pull function was not available and we wanted to access `tab_2$rate`? What data frame name would we use? The answer is the dot operator.

For example to access the rate vector without the `pull` function we could use

```
rates <-filter(murders, region == "South") %>%
  mutate(rate = total / population * 10^5) %>%
  .$rate
median(rates)
#> [1] 3.4
```

In the next section, we will see other instances in which using the . is useful.

4.12   do

The tidyverse functions know how to interpret grouped tibbles. Furthermore, to facilitate stringing commands through the pipe %>%, tidyverse functions consistently return data frames, since this assures that the output of a function is accepted as the input of another. But most R functions do not recognize grouped tibbles nor do they return data frames. The `quantile` function is an example we described in Section 4.7.1. The do function serves as a bridge between R functions such as `quantile` and the tidyverse. The do function understands grouped tibbles and always returns a data frame.

In Section 4.7.1, we noted that if we attempt to use `quantile` to obtain the min, median and max in one call, we will receive an error: `Error: expecting result of length one, got : 2`.

```
data(heights)
heights %>%
  filter(sex == "Female") %>%
  summarize(range = quantile(height, c(0, 0.5, 1)))
```

We can use the do function to fix this.

First we have to write a function that fits into the tidyverse approach: that is, it receives a data frame and returns a data frame.

```
my_summary <- function(dat){
  x <- quantile(dat$height, c(0, 0.5, 1))
  tibble(min = x[1], median = x[2], max = x[3])
}
```

We can now apply the function to the heights dataset to obtain the summaries:

```
heights %>%
  group_by(sex) %>%
  my_summary
#> # A tibble: 1 x 3
#>    min median   max
#>   <dbl>  <dbl> <dbl>
#> 1    50   68.5  82.7
```

But this is not what we want. We want a summary for each sex and the code returned just one summary. This is because `my_summary` is not part of the tidyverse and does not know how to handled grouped tibbles. `do` makes this connection:

```
heights %>%
  group_by(sex) %>%
  do(my_summary(.))
#> # A tibble: 2 x 4
#> # Groups:   sex [2]
#>   sex       min median   max
#>   <fct>   <dbl>  <dbl> <dbl>
#> 1 Female     51   65.0    79
#> 2 Male       50   69    82.7
```

Note that here we need to use the dot operator. The tibble created by `group_by` is piped to `do`. Within the call to `do`, the name of this tibble is `.` and we want to send it to `my_summary`. If you do not use the dot, then `my_summary` has ___no argument and returns an error telling us that `argument "dat"` is missing. You can see the error by typing:

```
heights %>%
  group_by(sex) %>%
  do(my_summary())
```

If you do not use the parenthesis, then the function is not executed and instead `do` tries to return the function. This gives an error because `do` must always return a data frame. You can see the error by typing:

```
heights %>%
  group_by(sex) %>%
  do(my_summary)
```

4.13 The purrr package

In Section 3.5 we learned about the `sapply` function, which permitted us to apply the same function to each element of a vector. We constructed a function and used `sapply` to compute the sum of the first n integers for several values of n like this:

```
compute_s_n <- function(n){
  x <- 1:n
```

```
   sum(x)
}
n <- 1:25
s_n <- sapply(n, compute_s_n)
```

This type of operation, applying the same function or procedure to elements of an object, is quite common in data analysis. The **purrr** package includes functions similar to `sapply` but that better interact with other tidyverse functions. The main advantage is that we can better control the output type of functions. In contrast, `sapply` can return several different object types; for example, we might expect a numeric result from a line of code, but `sapply` might convert our result to character under some circumstances. **purrr** functions will never do this: they will return objects of a specified type or return an error if this is not possible.

The first **purrr** function we will learn is `map`, which works very similar to `sapply` but always, without exception, returns a list:

```
library(purrr)
s_n <- map(n, compute_s_n)
class(s_n)
#> [1] "list"
```

If we want a numeric vector, we can instead use `map_dbl` which always returns a vector of numeric values.

```
s_n <- map_dbl(n, compute_s_n)
class(s_n)
#> [1] "numeric"
```

This produces the same results as the `sapply` call shown above.

A particularly useful **purrr** function for interacting with the rest of the tidyverse is `map_df`, which always returns a tibble data frame. However, the function being called needs to return a vector or a list with names. For this reason, the following code would result in a `Argument 1 must have names` error:

```
s_n <- map_df(n, compute_s_n)
```

We need to change the function to make this work:

```
compute_s_n <- function(n){
   x <- 1:n
   tibble(sum = sum(x))
}
s_n <- map_df(n, compute_s_n)
```

The **purrr** package provides much more functionality not covered here. For more details you can consult this online resource.

4.14 Tidyverse conditionals

A typical data analysis will often involve one or more conditional operations. In Section 3.1 we described the `ifelse` function, which we will use extensively in this book. In this section we present two **dplyr** functions that provide further functionality for performing conditional operations.

4.14.1 `case_when`

The `case_when` function is useful for vectorizing conditional statements. It is similar to `ifelse` but can output any number of values, as opposed to just `TRUE` or `FALSE`. Here is an example splitting numbers into negative, positive, and 0:

```
x <- c(-2, -1, 0, 1, 2)
case_when(x < 0 ~ "Negative", x > 0 ~ "Positive", TRUE ~ "Zero")
#> [1] "Negative" "Negative" "Zero"     "Positive" "Positive"
```

A common use for this function is to define categorical variables based on existing variables. For example, suppose we want to compare the murder rates in three groups of states: *New England*, *West Coast*, *South*, and *other*. For each state, we need to ask if it is in New England, if it is not we ask if it is in the West Coast, if not we ask if it is in the South, and if not we assign *other*. Here is how we use `case_when` to do this:

```
murders %>%
  mutate(group = case_when(
    abb %in% c("ME", "NH", "VT", "MA", "RI", "CT") ~ "New England",
    abb %in% c("WA", "OR", "CA") ~ "West Coast",
    region == "South" ~ "South",
    TRUE ~ "Other")) %>%
  group_by(group) %>%
  summarize(rate = sum(total) / sum(population) * 10^5)
#> # A tibble: 4 x 2
#>   group       rate
#>   <chr>      <dbl>
#> 1 New England 1.72
#> 2 Other       2.71
#> 3 South       3.63
#> 4 West Coast  2.90
```

4.14.2 `between`

A common operation in data analysis is to determine if a value falls inside an interval. We can check this using conditionals. For example to check if the elements of a vector `x` are between `a` and `b` we can type

```
x >= a & x <= b
```

However, this can become cumbersome, especially within the tidyverse approach. The `between` function performs the same operation.

```
between(x, a, b)
```

4.15   Exercises

1. Load the `murders` dataset. Which of the following is true?

    a. `murders` is in tidy format and is stored in a tibble.
    b. `murders` is in tidy format and is stored in a data frame.
    c. `murders` is not in tidy format and is stored in a tibble.
    d. `murders` is not in tidy format and is stored in a data frame.

2. Use `as_tibble` to convert the `murders` data table into a tibble and save it in an object called `murders_tibble`.

3. Use the `group_by` function to convert murders into a tibble that is grouped by region.

4. Write tidyverse code that is equivalent to this code:

```
exp(mean(log(murders$population)))
```

Write it using the pipe so that each function is called without arguments. Use the dot operator to access the population. Hint: The code should start with `murders %>%`.

5. Use the `map_df` to create a data frame with three columns named n, s_n, and s_n_2. The first column should contain the numbers 1 through 100. The second and third columns should each contain the sum of 1 through n with n the row number.

5

Importing data

We have been using data sets already stored as R objects. A data scientist will rarely have such luck and will have to import data into R from either a file, a database, or other sources. Currently, one of the most common ways of storing and sharing data for analysis is through electronic spreadsheets. A spreadsheet stores data in rows and columns. It is basically a file version of a data frame. When saving such a table to a computer file, one needs a way to define when a new row or column ends and the other begins. This in turn defines the cells in which single values are stored.

When creating spreadsheets with text files, like the ones created with a simple text editor, a new row is defined with return and columns are separated with some predefined special character. The most common characters are comma (,), semicolon (;), space ( ), and tab (a preset number of spaces or \t). Here is an example of what a comma separated file looks like if we open it with a basic text editor:

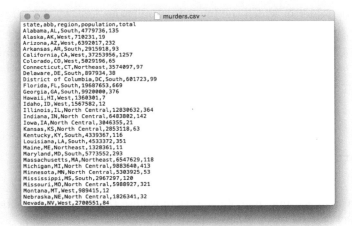

The first row contains column names rather than data. We call this a *header*, and when we read-in data from a spreadsheet it is important to know if the file has a header or not. Most reading functions assume there is a header. To know if the file has a header, it helps to look at the file before trying to read it. This can be done with a text editor or with RStudio. In RStudio, we can do this by either opening the file in the editor or navigating to the file location, double clicking on the file, and hitting *View File*.

However, not all spreadsheet files are in a text format. Google Sheets, which are rendered on a browser, are an example. Another example is the proprietary format used by Microsoft Excel. These can't be viewed with a text editor. Despite this, due to the widespread use of Microsoft Excel software, this format is widely used.

We start this chapter by describing the difference between text (ASCII), Unicode, and binary files and how this affects how we import them. We then explain the concepts of file paths and working directories, which are essential to understand how to import data effectively. We then introduce the **readr** and **readxl** package and the functions that are available to import spreadsheets into R. Finally, we provide some recommendations on how to store and organize data in files. More complex challenges such as extracting data from web pages or PDF documents are left for the Data Wrangling part of the book.

5.1   Paths and the working directory

The first step when importing data from a spreadsheet is to locate the file containing the data. Although we do not recommend it, you can use an approach similar to what you do to open files in Microsoft Excel by clicking on the RStudio "File" menu, clicking "Import Dataset", then clicking through folders until you find the file. We want to be able to write code rather than use the point-and-click approach. The keys and concepts we need to learn to do this are described in detail in the Productivity Tools part of this book. Here we provide an overview of the very basics.

The main challenge in this first step is that we need to let the R functions doing the importing know where to look for the file containing the data. The simplest way to do this is to have a copy of the file in the folder in which the importing functions look by default. Once we do this, all we have to supply to the importing function is the filename.

A spreadsheet containing the US murders data is included as part of the **dslabs** package. Finding this file is not straightforward, but the following lines of code copy the file to the folder in which R looks in by default. We explain how these lines work below.

```
filename <- "murders.csv"
dir <- system.file("extdata", package = "dslabs")
fullpath <- file.path(dir, filename)
file.copy(fullpath, "murders.csv")
```

This code does not read the data into R, it just copies a file. But once the file is copied, we can import the data with a simple line of code. Here we use the `read_csv` function from the **readr** package, which is part of the tidyverse.

```
library(tidyverse)
dat <- read_csv(filename)
```

The data is imported and stored in `dat`. The rest of this section defines some important concepts and provides an overview of how we write code that tells R how to find the files we want to import. Chapter 36 provides more details on this topic.

5.1.1   The filesystem

You can think of your computer's filesystem as a series of nested folders, each containing other folders and files. Data scientists refer to folders as *directories*. We refer to the folder that contains all other folders as the *root directory*. We refer to the directory in which we are currently located as the *working directory*. The working directory therefore changes as you move through folders: think of it as your current location.

5.1.2 Relative and full paths

The *path* of a file is a list of directory names that can be thought of as instructions on what folders to click on, and in what order, to find the file. If these instructions are for finding the file from the root directory we refer to it as the *full path*. If the instructions are for finding the file starting in the working directory we refer to it as a *relative path*. Section 36.3 provides more details on this topic.

To see an example of a full path on your system type the following:

```
system.file(package = "dslabs")
```

The strings separated by slashes are the directory names. The first slash represents the root directory and we know this is a full path because it starts with a slash. If the first directory name appears without a slash in front, then the path is assumed to be relative. We can use the function `list.files` to see examples of relative paths.

```
dir <- system.file(package = "dslabs")
list.files(path = dir)
#>  [1] "data"        "DESCRIPTION" "extdata"     "help"
#>  [5] "html"        "INDEX"       "Meta"        "NAMESPACE"
#>  [9] "R"           "script"
```

These relative paths give us the location of the files or directories if we start in the directory with the full path. For example, the full path to the `help` directory in the example above is `/Library/Frameworks/R.framework/Versions/3.5/Resources/library/dslabs/help`.

Note: You will probably not make much use of the `system.file` function in your day-to-day data analysis work. We introduce it in this section because it facilitates the sharing of spreadsheets by including them in the **dslabs** package. You will rarely have the luxury of data being included in packages you already have installed. However, you will frequently need to navigate full and relative paths and import spreadsheet formatted data.

5.1.3 The working directory

We highly recommend only writing relative paths in your code. The reason is that full paths are unique to your computer and you want your code to be portable. You can get the full path of your working directory without writing out explicitly by using the `getwd` function.

```
wd <- getwd()
```

If you need to change your working directory, you can use the function `setwd` or you can change it through RStudio by clicking on "Session".

5.1.4   Generating path names

Another example of obtaining a full path without writing out explicitly was given above
when we created the object `fullpath` like this:

```
filename <- "murders.csv"
dir <- system.file("extdata", package = "dslabs")
fullpath <- file.path(dir, filename)
```

The function `system.file` provides the full path of the folder containing all the files and
directories relevant to the package specified by the `package` argument. By exploring the
directories in `dir` we find that the `extdata` contains the file we want:

```
dir <- system.file(package = "dslabs")
filename %in% list.files(file.path(dir, "extdata"))
#> [1] TRUE
```

The `system.file` function permits us to provide a subdirectory as a first argument, so we
can obtain the fullpath of the `extdata` directory like this:

```
dir <- system.file("extdata", package = "dslabs")
```

The function `file.path` is used to combine directory names to produce the full path of the
file we want to import.

```
fullpath <- file.path(dir, filename)
```

5.1.5   Copying files using paths

The final line of code we used to copy the file into our home directory used
the function `file.copy`. This function takes two arguments: the file to copy and the name
to give it in the new directory.

```
file.copy(fullpath, "murders.csv")
#> [1] TRUE
```

If a file is copied successfully, the `file.copy` function returns `TRUE`. Note that we are giving
the file the same name, `murders.csv`, but we could have named it anything. Also note that
by not starting the string with a slash, R assumes this is a relative path and copies the file
to the working directory.

You should be able to see the file in your working directory and can check by using:

```
list.files()
```

5.2 The readr and readxl packages

In this section we introduce the main tidyverse data importing functions. We will use the murders.csv file provided by the **dslabs** package as an example. To simplify the illustration we will copy the file to our working directory using the following code:

```
filename <- "murders.csv"
dir <- system.file("extdata", package = "dslabs")
fullpath <- file.path(dir, filename)
file.copy(fullpath, "murders.csv")
```

5.2.1 readr

The **readr** library includes functions for reading data stored in text file spreadsheets into R. **readr** is part of the **tidyverse** package, or you can load it directly:

```
library(readr)
```

The following functions are available to read-in spreadsheets:

Function	Format	Typical suffix
read_table	white space separated values	txt
read_csv	comma separated values	csv
read_csv2	semicolon separated values	csv
read_tsv	tab delimited separated values	tsv
read_delim	general text file format, must define delimiter	txt

Although the suffix usually tells us what type of file it is, there is no guarantee that these always match. We can open the file to take a look or use the function read_lines to look at a few lines:

```
read_lines("murders.csv", n_max = 3)
#> [1] "state,abb,region,population,total"
#> [2] "Alabama,AL,South,4779736,135"
#> [3] "Alaska,AK,West,710231,19"
```

This also shows that there is a header. Now we are ready to read-in the data into R. From the .csv suffix and the peek at the file, we know to use **read_csv**:

```
dat <- read_csv(filename)
#> Parsed with column specification:
#> cols(
#>   state = col_character(),
#>   abb = col_character(),
#>   region = col_character(),
```

```
#>    population = col_double(),
#>    total = col_double()
#> )
```

Note that we receive a message letting us know what data types were used for each column. Also note that `dat` is a `tibble`, not just a data frame. This is because `read_csv` is a **tidyverse** parser. We can confirm that the data has in fact been read-in with:

```
View(dat)
```

Finally, note that we can also use the full path for the file:

```
dat <- read_csv(fullpath)
```

5.2.2   readxl

You can load the **readxl** package using

```
library(readxl)
```

The package provides functions to read-in Microsoft Excel formats:

Function	Format	Typical suffix
read_excel	auto detect the format	xls, xlsx
read_xls	original format	xls
read_xlsx	new format	xlsx

The Microsoft Excel formats permit you to have more than one spreadsheet in one file. These are referred to as *sheets*. The functions listed above read the first sheet by default, but we can also read the others. The `excel_sheets` function gives us the names of all the sheets in an Excel file. These names can then be passed to the `sheet` argument in the three functions above to read sheets other than the first.

5.3   Exercises

1. Use the `read_csv` function to read each of the files that the following code saves in the `files` object:

```
path <- system.file("extdata", package = "dslabs")
files <- list.files(path)
files
```

2. Note that the last one, the `olive` file, gives us a warning. This is because the first line of the file is missing the header for the first column.

Read the help file for `read_csv` to figure out how to read in the file without reading this header. If you skip the header, you should not get this warning. Save the result to an object called `dat`.

3. A problem with the previous approach is that we don't know what the columns represent. Type:

```
names(dat)
```

to see that the names are not informative.

Use the `readLines` function to read in just the first line (we later learn how to extract values from the output).

5.4 Downloading files

Another common place for data to reside is on the internet. When these data are in files, we can download them and then import them or even read them directly from the web. For example, we note that because our **dslabs** package is on GitHub, the file we downloaded with the package has a url:

```
url <- "https://raw.githubusercontent.com/rafalab/dslabs/master/inst/
extdata/murders.csv"
```

The `read_csv` file can read these files directly:

```
dat <- read_csv(url)
```

If you want to have a local copy of the file, you can use the `download.file` function:

```
download.file(url, "murders.csv")
```

This will download the file and save it on your system with the name `murders.csv`. You can use any name here, not necessarily `murders.csv`. Note that when using `download.file` you should be careful as it will overwrite existing files without warning.

Two functions that are sometimes useful when downloading data from the internet are `tempdir` and `tempfile`. The first creates a directory with a random name that is very likely to be unique. Similarly, `tempfile` creates a character string, not a file, that is likely to be a unique filename. So you can run a command like this which erases the temporary file once it imports the data:

```
tmp_filename <- tempfile()
download.file(url, tmp_filename)
dat <- read_csv(tmp_filename)
file.remove(tmp_filename)
```

5.5   R-base importing functions

R-base also provides import functions. These have similar names to those in the **tidyverse**, for example `read.table`, `read.csv` and `read.delim`. However, there are a couple of important differences. To show this we read-in the data with an R-base function:

```
dat2 <- read.csv(filename)
```

An important difference is that the characters are converted to factors:

```
class(dat2$abb)
#> [1] "factor"
class(dat2$region)
#> [1] "factor"
```

This can be avoided by setting the argument `stringsAsFactors` to `FALSE`.

```
dat <- read.csv("murders.csv", stringsAsFactors = FALSE)
class(dat$state)
#> [1] "character"
```

In our experience this can be a cause for confusion since a variable that was saved as characters in file is converted to factors regardless of what the variable represents. In fact, we **highly** recommend setting `stringsAsFactors=FALSE` to be your default approach when using the R-base parsers. You can easily convert the desired columns to factors after importing data.

5.5.1   scan

When reading in spreadsheets many things can go wrong. The file might have a multiline header, be missing cells, or it might use an unexpected encoding[1]. We recommend you read this post about common issues found here: https://www.joelonsoftware.com/2003/10/08/the-absolute-minimum-every-software-developer-absolutely-positively-must-know-about-unicode-and-character-sets-no-excuses/.

With experience you will learn how to deal with different challenges. Carefully reading the help files for the functions discussed here will be useful. Two other functions that are helpful are `scan`. With scan you can read-in each cell of a file. Here is an example:

```
path <- system.file("extdata", package = "dslabs")
filename <- "murders.csv"
x <- scan(file.path(path, filename), sep=",", what = "c")
x[1:10]
#>  [1] "state"      "abb"       "region"    "population" "total"
#>  [6] "Alabama"    "AL"        "South"     "4779736"    "135"
```

Note that the tidyverse provides `read_lines`, a similarly useful function.

---

[1]https://en.wikipedia.org/wiki/Character_encoding

5.6 Text versus binary files

For data science purposes, files can generally be classified into two categories: text files (also known as ASCII files) and binary files. You have already worked with text files. All your R scripts are text files and so are the R markdown files used to create this book. The csv tables you have read are also text files. One big advantage of these files is that we can easily "look" at them without having to purchase any kind of special software or follow complicated instructions. Any text editor can be used to examine a text file, including freely available editors such as RStudio, Notepad, textEdit, vi, emacs, nano, and pico. To see this, try opening a csv file using the "Open file" RStudio tool. You should be able to see the content right on your editor. However, if you try to open, say, an Excel xls file, jpg or png file, you will not be able to see anything immediately useful. These are binary files. Excel files are actually compressed folders with several text files inside. But the main distinction here is that text files can be easily examined.

Although R includes tools for reading widely used binary files, such as xls files, in general you will want to find data sets stored in text files. Similarly, when sharing data you want to make it available as text files as long as storage is not an issue (binary files are much more efficient at saving space on your disk). In general, plain-text formats make it easier to share data since commercial software is not required for working with the data.

Extracting data from a spreadsheet stored as a text file is perhaps the easiest way to bring data from a file to an R session. Unfortunately, spreadsheets are not always available and the fact that you can look at text files does not necessarily imply that extracting data from them will be straightforward. In the Data Wrangling part of the book we learn to extract data from more complex text files such as html files.

5.7 Unicode versus ASCII

A pitfall in data science is assuming a file is an ASCII text file when, in fact, it is something else that can look a lot like an ASCII text file: a Unicode text file.

To understand the difference between these, remember that everything on a computer needs to eventually be converted to 0s and 1s. ASCII is an *encoding* that maps characters to numbers. ASCII uses 7 bits (0s and 1s) which results in $2^7 = 128$ unique items, enough to encode all the characters on an English language keyboard. However, other languages use characters not included in this encoding. For example, the é in México is not encoded by ASCII. For this reason, a new encoding, using more than 7 bits, was defined: Unicode. When using Unicode, one can chose between 8, 16, and 32 bits abbreviated UTF-8, UTF-16, and UTF-32 respectively. RStudio actually defaults to UTF-8 encoding.

Although we do not go into the details of how to deal with the different encodings here, it is important that you know these different encodings exist so that you can better diagnose a problem if you encounter it. One way problems manifest themselves is when you see "weird looking" characters you were not expecting. This StackOverflow discussion is an example: https://stackoverflow.com/questions/18789330/r-on-windows-character-encoding-hell.

5.8   Organizing data with spreadsheets

Although there are R packages designed to read this format, if you are choosing a file format to save your own data, you generally want to avoid Microsoft Excel. We recommend Google Sheets as a free software tool for organizing data. We provide more recommendations in the section Data Organization with Spreadsheets. This book focuses on data analysis. Yet often a data scientist needs to collect data or work with others collecting data. Filling out a spreadsheet by hand is a practice we highly discourage and instead recommend that the process be automatized as much as possible. But sometimes you just have to do it. In this section, we provide recommendations on how to store data in a spreadsheet. We summarize a paper by Karl Broman and Kara Woo[2]. Below are their general recommendations. Please read the paper for important details.

- **Be Consistent** - Before you commence entering data, have a plan. Once you have a plan, be consistent and stick to it.
- **Choose Good Names for Things** - You want the names you pick for objects, files, and directories to be memorable, easy to spell, and descriptive. This is actually a hard balance to achieve and it does require time and thought. One important rule to follow is **do not use spaces**, use underscores _ or dashes instead -. Also, avoid symbols; stick to letters and numbers.
- **Write Dates as YYYY-MM-DD** - To avoid confusion, we strongly recommend using this global ISO 8601 standard.
- **No Empty Cells** - Fill in all cells and use some common code for missing data.
- **Put Just One Thing in a Cell** - It is better to add columns to store the extra information rather than having more than one piece of information in one cell.
- **Make It a Rectangle** - The spreadsheet should be a rectangle.
- **Create a Data Dictionary** - If you need to explain things, such as what the columns are or what the labels used for categorical variables are, do this in a separate file.
- **No Calculations in the Raw Data Files** - Excel permits you to perform calculations. Do not make this part of your spreadsheet. Code for calculations should be in a script.
- **Do Not Use Font Color or Highlighting as Data** - Most import functions are not able to import this information. Encode this information as a variable instead.
- **Make Backups** - Make regular backups of your data.
- **Use Data Validation to Avoid Errors** - Leverage the tools in your spreadsheet software so that the process is as error-free and repetitive-stress-injury-free as possible.
- **Save the Data as Text Files** - Save files for sharing in comma or tab delimited format.

5.9   Exercises

1. Pick a measurement you can take on a regular basis. For example, your daily weight or how long it takes you to run 5 miles. Keep a spreadsheet that includes the date, the hour, the measurement, and any other informative variable you think is worth keeping. Do this for 2 weeks. Then make a plot.

---

[2]https://www.tandfonline.com/doi/abs/10.1080/00031305.2017.1375989

Part II

Data Visualization

6

Introduction to data visualization

Looking at the numbers and character strings that define a dataset is rarely useful. To convince yourself, print and stare at the US murders data table:

```
library(dslabs)
data(murders)
head(murders)
#>        state abb region population total
#> 1    Alabama  AL  South    4779736   135
#> 2     Alaska  AK   West     710231    19
#> 3    Arizona  AZ   West    6392017   232
#> 4   Arkansas  AR  South    2915918    93
#> 5 California  CA   West   37253956  1257
#> 6   Colorado  CO   West    5029196    65
```

What do you learn from staring at this table? How quickly can you determine which states have the largest populations? Which states have the smallest? How large is a typical state? Is there a relationship between population size and total murders? How do murder rates vary across regions of the country? For most human brains, it is quite difficult to extract this information just by looking at the numbers. In contrast, the answer to all the questions above are readily available from examining this plot:

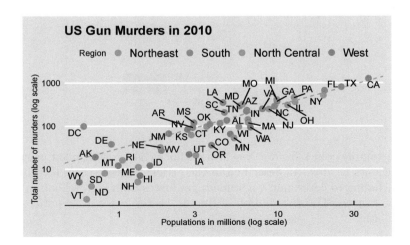

We are reminded of the saying "a picture is worth a thousand words". Data visualization provides a powerful way to communicate a data-driven finding. In some cases, the visualization is so convincing that no follow-up analysis is required.

The growing availability of informative datasets and software tools has led to increased reliance on data visualizations across many industries, academia, and government. A salient example is news organizations, which are increasingly embracing *data journalism* and including effective *infographics* as part of their reporting.

A particularly effective example is a Wall Street Journal article[1] showing data related to the impact of vaccines on battling infectious diseases. One of the graphs shows measles cases by US state through the years with a vertical line demonstrating when the vaccine was introduced.

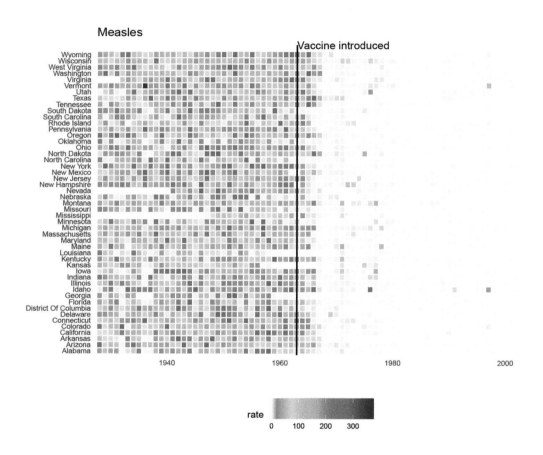

Another striking example comes from a New York Times chart[2], which summarizes scores from the NYC Regents Exams. As described in the article[3], these scores are collected for several reasons, including to determine if a student graduates from high school. In New York City you need a 65 to pass. The distribution of the test scores forces us to notice something somewhat problematic:

---

[1]http://graphics.wsj.com/infectious-diseases-and-vaccines/?mc_cid=711ddeb86e

[2]http://graphics8.nytimes.com/images/2011/02/19/nyregion/19schoolsch/19schoolsch-popup.gif

[3]https://www.nytimes.com/2011/02/19/nyregion/19schools.html

Scraping by

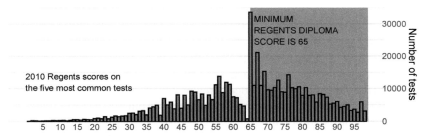

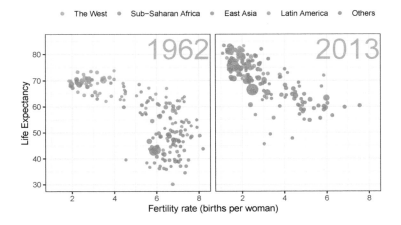

The most common test score is the minimum passing grade, with very few scores just below the threshold. This unexpected result is consistent with students close to passing having their scores bumped up.

This is an example of how data visualization can lead to discoveries which would otherwise be missed if we simply subjected the data to a battery of data analysis tools or procedures. Data visualization is the strongest tool of what we call *exploratory data analysis* (EDA). John W. Tukey[4], considered the father of EDA, once said,

> "The greatest value of a picture is when it forces us to notice what we never expected to see."

Many widely used data analysis tools were initiated by discoveries made via EDA. EDA is perhaps the most important part of data analysis, yet it is one that is often overlooked.

Data visualization is also now pervasive in philanthropic and educational organizations. In the talks New Insights on Poverty[5] and The Best Stats You've Ever Seen[6], Hans Rosling forces us to notice the unexpected with a series of plots related to world health and economics. In his videos, he uses animated graphs to show us how the world is changing and how old narratives are no longer true.

---

[4]https://en.wikipedia.org/wiki/John_Tukey
[5]https://www.ted.com/talks/hans_rosling_reveals_new_insights_on_poverty?language=en
[6]https://www.ted.com/talks/hans_rosling_shows_the_best_stats_you_ve_ever_seen

It is also important to note that mistakes, biases, systematic errors and other unexpected problems often lead to data that should be handled with care. Failure to discover these problems can give rise to flawed analyses and false discoveries. As an example, consider that measurement devices sometimes fail and that most data analysis procedures are not designed to detect these. Yet these data analysis procedures will still give you an answer. The fact that it can be difficult or impossible to notice an error just from the reported results makes data visualization particularly important.

In this part of the book, we will learn the basics of data visualization and exploratory data analysis by using three motivating examples. We will use the **ggplot2** package to code. To learn the very basics, we will start with a somewhat artificial example: heights reported by students. Then we will cover the two examples mentioned above: 1) world health and economics and 2) infectious disease trends in the United States.

Of course, there is much more to data visualization than what we cover here. The following are references for those who wish to learn more:

- ER Tufte (1983) The visual display of quantitative information. Graphics Press.
- ER Tufte (1990) Envisioning information. Graphics Press.
- ER Tufte (1997) Visual explanations. Graphics Press.
- WS Cleveland (1993) Visualizing data. Hobart Press.
- WS Cleveland (1994) The elements of graphing data. CRC Press.
- A Gelman, C Pasarica, R Dodhia (2002) Let's practice what we preach: Turning tables into graphs. The American Statistician 56:121-130.
- NB Robbins (2004) Creating more effective graphs. Wiley.
- A Cairo (2013) The functional art: An introduction to information graphics and visualization. New Riders.
- N Yau (2013) Data points: Visualization that means something. Wiley.

We also do not cover interactive graphics, a topic that is too advanced for this book. Some useful resources for those interested in learning more can be found below:

- https://shiny.rstudio.com/
- https://d3js.org/

7

ggplot2

Exploratory data visualization is perhaps the greatest strength of R. One can quickly go from idea to data to plot with a unique balance of flexibility and ease. For example, Excel may be easier than R for some plots, but it is nowhere near as flexible. D3.js may be more flexible and powerful than R, but it takes much longer to generate a plot.

Throughout the book, we will be creating plots using the **ggplot2**[1] package.

```
library(dplyr)
library(ggplot2)
```

Many other approaches are available for creating plots in R. In fact, the plotting capabilities that come with a basic installation of R are already quite powerful. There are also other packages for creating graphics such as **grid** and **lattice**. We chose to use **ggplot2** in this book because it breaks plots into components in a way that permits beginners to create relatively complex and aesthetically pleasing plots using syntax that is intuitive and comparatively easy to remember.

One reason **ggplot2** is generally more intuitive for beginners is that it uses a grammar of graphics[2], the _gg_ in **ggplot2**. This is analogous to the way learning grammar can help a beginner construct hundreds of different sentences by learning just a handful of verbs, nouns and adjectives without having to memorize each specific sentence. Similarly, by learning a handful of **ggplot2** building blocks and its grammar, you will be able to create hundreds of different plots.

Another reason **ggplot2** is easy for beginners is that its default behavior is carefully chosen to satisfy the great majority of cases and is visually pleasing. As a result, it is possible to create informative and elegant graphs with relatively simple and readable code.

One limitation is that **ggplot2** is designed to work exclusively with data tables in tidy format (where rows are observations and columns are variables). However, a substantial percentage of datasets that beginners work with are in, or can be converted into, this format. An advantage of this approach is that, assuming that our data is tidy, **ggplot2** simplifies plotting code and the learning of grammar for a variety of plots.

To use **ggplot2** you will have to learn several functions and arguments. These are hard to memorize, so we highly recommend you have the ggplot2 cheat sheet handy. You can get a copy here: https://www.rstudio.com/wp-content/uploads/2015/03/ggplot2-cheatsheet.pdf or simply perform an internet search for "ggplot2 cheat sheet".

---

[1] https://ggplot2.tidyverse.org/
[2] http://www.springer.com/us/book/9780387245447

7.1   The components of a graph

We will construct a graph that summarizes the US murders dataset that looks like this:

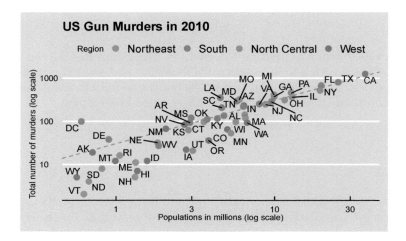

We can clearly see how much states vary across population size and the total number of murders. Not surprisingly, we also see a clear relationship between murder totals and population size. A state falling on the dashed grey line has the same murder rate as the US average. The four geographic regions are denoted with color, which depicts how most southern states have murder rates above the average.

This data visualization shows us pretty much all the information in the data table. The code needed to make this plot is relatively simple. We will learn to create the plot part by part.

The first step in learning **ggplot2** is to be able to break a graph apart into components. Let's break down the plot above and introduce some of the **ggplot2** terminology. The main three components to note are:

- **Data**: The US murders data table is being summarized. We refer to this as the **data** component.
- **Geometry**: The plot above is a scatterplot. This is referred to as the **geometry** component. Other possible geometries are barplot, histogram, smooth densities, qqplot, and boxplot. We will learn more about these in the Data Visualization part of the book.
- **Aesthetic mapping**: The plot uses several visual cues to represent the information provided by the dataset. The two most important cues in this plot are the point positions on the x-axis and y-axis, which represent population size and the total number of murders, respectively. Each point represents a different observation, and we *map* data about these observations to visual cues like x- and y-scale. Color is another visual cue that we map to region. We refer to this as the **aesthetic mapping** component. How we define the mapping depends on what **geometry** we are using.

We also note that:

- The points are labeled with the state abbreviations.

- The range of the x-axis and y-axis appears to be defined by the range of the data. They are both on log-scales.
- There are labels, a title, a legend, and we use the style of The Economist magazine.

We will now construct the plot piece by piece.

We start by loading the dataset:

```
library(dslabs)
data(murders)
```

7.2   ggplot objects

The first step in creating a **ggplot2** graph is to define a **ggplot** object. We do this with the function `ggplot`, which initializes the graph. If we read the help file for this function, we see that the first argument is used to specify what data is associated with this object:

```
ggplot(data = murders)
```

We can also pipe the data in as the first argument. So this line of code is equivalent to the one above:

```
murders %>% ggplot()
```

It renders a plot, in this case a blank slate since no geometry has been defined. The only style choice we see is a grey background.

What has happened above is that the object was created and, because it was not assigned, it was automatically evaluated. But we can assign our plot to an object, for example like this:

```
p <- ggplot(data = murders)
```

```
class(p)
#> [1] "gg"      "ggplot"
```

To render the plot associated with this object, we simply print the object **p**. The following two lines of code each produce the same plot we see above:

```
print(p)
p
```

7.3 Geometries

In `ggplot2` we create graphs by adding *layers*. Layers can define geometries, compute summary statistics, define what scales to use, or even change styles. To add layers, we use the symbol `+`. In general, a line of code will look like this:

DATA %>% `ggplot()` + LAYER 1 + LAYER 2 + ... + LAYER N

Usually, the first added layer defines the geometry. We want to make a scatterplot. What geometry do we use?

Taking a quick look at the cheat sheet, we see that the function used to create plots with this geometry is `geom_point`.

(Image courtesy of RStudio[3]. CC-BY-4.0 license[4].)

Geometry function names follow the pattern: `geom_X` where X is the name of the geometry. Some examples include `geom_point`, `geom_bar`, and `geom_histogram`.

For `geom_point` to run properly we need to provide data and a mapping. We have already connected the object **p** with the `murders` data table, and if we add the layer `geom_point` it defaults to using this data. To find out what mappings are expected, we read the **Aesthetics** section of the help file `geom_point` help file:

---

[3]https://github.com/rstudio/cheatsheets
[4]https://github.com/rstudio/cheatsheets/blob/master/LICENSE

```
> Aesthetics
>
> geom_point understands the following aesthetics (required aesthetics are
  in bold):
>
> x
>
> y
>
> alpha
>
> colour
```

and, as expected, we see that at least two arguments are required x and y.

7.4 Aesthetic mappings

Aesthetic mappings describe how properties of the data connect with features of the graph, such as distance along an axis, size, or color. The **aes** function connects data with what we see on the graph by defining aesthetic mappings and will be one of the functions you use most often when plotting. The outcome of the **aes** function is often used as the argument of a geometry function. This example produces a scatterplot of total murders versus population in millions:

```
murders %>% ggplot() +
  geom_point(aes(x = population/10^6, y = total))
```

We can drop the x = and y = if we wanted to since these are the first and second expected arguments, as seen in the help page.

Instead of defining our plot from scratch, we can also add a layer to the p object that was defined above as p <- ggplot(data = murders):

```
p + geom_point(aes(population/10^6, total))
```

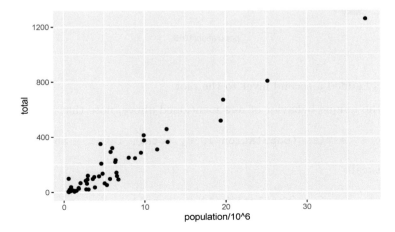

The scale and labels are defined by default when adding this layer. Like **dplyr** functions, `aes` also uses the variable names from the object component: we can use `population` and `total` without having to call them as `murders$population` and `murders$total`. The behavior of recognizing the variables from the data component is quite specific to `aes`. With most functions, if you try to access the values of `population` or `total` outside of `aes` you receive an error.

7.5   Layers

A second layer in the plot we wish to make involves adding a label to each point to identify the state. The `geom_label` and `geom_text` functions permit us to add text to the plot with and without a rectangle behind the text, respectively.

Because each point (each state in this case) has a label, we need an aesthetic mapping to make the connection between points and labels. By reading the help file, we learn that we supply the mapping between point and label through the `label` argument of `aes`. So the code looks like this:

```
p + geom_point(aes(population/10^6, total)) +
  geom_text(aes(population/10^6, total, label = abb))
```

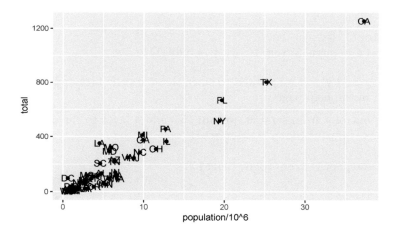

We have successfully added a second layer to the plot.

As an example of the unique behavior of `aes` mentioned above, note that this call:

```
p_test <- p + geom_text(aes(population/10^6, total, label = abb))
```

is fine, whereas this call:

```
p_test <- p + geom_text(aes(population/10^6, total), label = abb)
```

will give you an error since `abb` is not found because it is outside of the `aes` function. The layer `geom_text` does not know where to find `abb` since it is a column name and not a global variable.

7.5.1   Tinkering with arguments

Each geometry function has many arguments other than `aes` and `data`. They tend to be specific to the function. For example, in the plot we wish to make, the points are larger than the default size. In the help file we see that `size` is an aesthetic and we can change it like this:

```
p + geom_point(aes(population/10^6, total), size = 3) +
  geom_text(aes(population/10^6, total, label = abb))
```

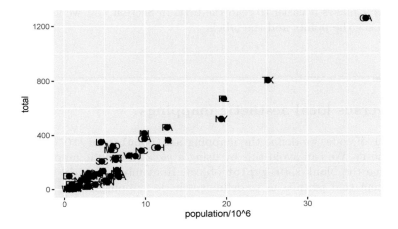

`size` is **not** a mapping: whereas mappings use data from specific observations and need to be inside `aes()`, operations we want to affect all the points the same way do not need to be included inside `aes`.

Now because the points are larger it is hard to see the labels. If we read the help file for `geom_text`, we see the `nudge_x` argument, which moves the text slightly to the right or to the left:

```
p + geom_point(aes(population/10^6, total), size = 3) +
  geom_text(aes(population/10^6, total, label = abb), nudge_x = 1.5)
```

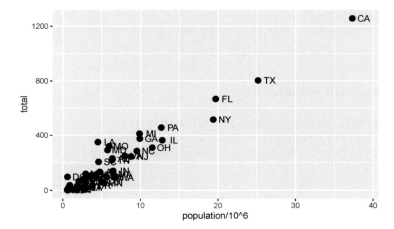

This is preferred as it makes it easier to read the text. In Section 7.11 we learn a better way of assuring we can see the points and the labels.

7.6   Global versus local aesthetic mappings

In the previous line of code, we define the mapping `aes(population/10^6, total)` twice, once in each geometry. We can avoid this by using a *global* aesthetic mapping. We can do this when we define the blank slate `ggplot` object. Remember that the function `ggplot` contains an argument that permits us to define aesthetic mappings:

```
args(ggplot)
#> function (data = NULL, mapping = aes(), ..., environment = parent.frame())
#> NULL
```

If we define a mapping in `ggplot`, all the geometries that are added as layers will default to this mapping. We redefine `p`:

```
p <- murders %>% ggplot(aes(population/10^6, total, label = abb))
```

and then we can simply write the following code to produce the previous plot:

```
p + geom_point(size = 3) +
  geom_text(nudge_x = 1.5)
```

We keep the `size` and `nudge_x` arguments in `geom_point` and `geom_text`, respectively, because we want to only increase the size of points and only nudge the labels. If we put those arguments in `aes` then they would apply to both plots. Also note that the `geom_point` function does not need a `label` argument and therefore ignores that aesthetic.

If necessary, we can override the global mapping by defining a new mapping within each layer. These *local* definitions override the *global*. Here is an example:

```
p + geom_point(size = 3) +
  geom_text(aes(x = 10, y = 800, label = "Hello there!"))
```

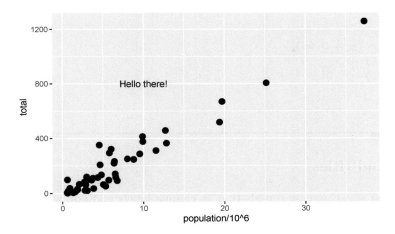

Clearly, the second call to geom_text does not use population and total.

7.7   Scales

First, our desired scales are in log-scale. This is not the default, so this change needs to be added through a *scales* layer. A quick look at the cheat sheet reveals the scale_x_continuous function lets us control the behavior of scales. We use them like this:

```
p + geom_point(size = 3) +
  geom_text(nudge_x = 0.05) +
  scale_x_continuous(trans = "log10") +
  scale_y_continuous(trans = "log10")
```

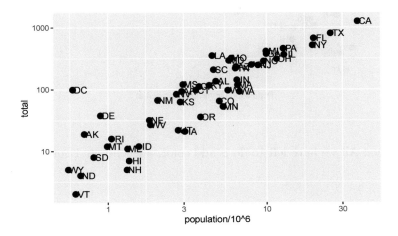

Because we are in the log-scale now, the *nudge* must be made smaller.

This particular transformation is so common that **ggplot2** provides the specialized functions `scale_x_log10` and `scale_y_log10`, which we can use to rewrite the code like this:

```
p + geom_point(size = 3) +
  geom_text(nudge_x = 0.05) +
  scale_x_log10() +
  scale_y_log10()
```

7.8   Labels and titles

Similarly, the cheat sheet quickly reveals that to change labels and add a title, we use the following functions:

```
p + geom_point(size = 3) +
  geom_text(nudge_x = 0.05) +
  scale_x_log10() +
  scale_y_log10() +
  xlab("Populations in millions (log scale)") +
  ylab("Total number of murders (log scale)") +
  ggtitle("US Gun Murders in 2010")
```

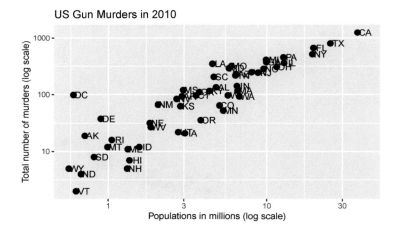

We are almost there! All we have left to do is add color, a legend, and optional changes to the style.

7.9   Categories as colors

We can change the color of the points using the `col` argument in the `geom_point` function. To facilitate demonstration of new features, we will redefine `p` to be everything except the points layer:

```
p <-  murders %>% ggplot(aes(population/10^6, total, label = abb)) +
  geom_text(nudge_x = 0.05) +
  scale_x_log10() +
  scale_y_log10() +
  xlab("Populations in millions (log scale)") +
  ylab("Total number of murders (log scale)") +
  ggtitle("US Gun Murders in 2010")
```

and then test out what happens by adding different calls to `geom_point`. We can make all the points blue by adding the `color` argument:

```
p + geom_point(size = 3, color ="blue")
```

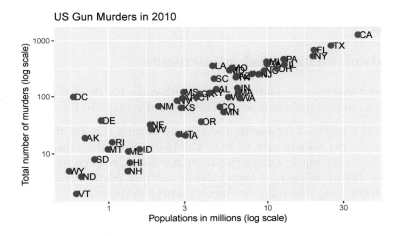

This, of course, is not what we want. We want to assign color depending on the geographical region. A nice default behavior of **ggplot2** is that if we assign a categorical variable to color, it automatically assigns a different color to each category and also adds a legend.

Since the choice of color is determined by a feature of each observation, this is an aesthetic mapping. To map each point to a color, we need to use **aes**. We use the following code:

```
p + geom_point(aes(col=region), size = 3)
```

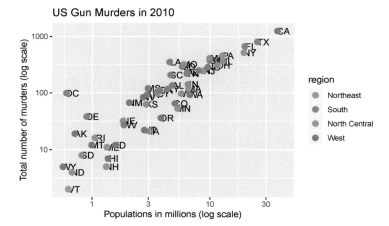

The x and y mappings are inherited from those already defined in p, so we do not redefine them. We also move aes to the first argument since that is where mappings are expected in this function call.

Here we see yet another useful default behavior: **ggplot2** automatically adds a legend that maps color to region. To avoid adding this legend we set the geom_point argument show.legend = FALSE.

7.10   Annotation, shapes, and adjustments

We often want to add shapes or annotation to figures that are not derived directly from the aesthetic mapping; examples include labels, boxes, shaded areas, and lines.

Here we want to add a line that represents the average murder rate for the entire country. Once we determine the per million rate to be r, this line is defined by the formula: $y = rx$, with y and x our axes: total murders and population in millions, respectively. In the log-scale this line turns into: $\log(y) = \log(r) + \log(x)$. So in our plot it's a line with slope 1 and intercept $\log(r)$. To compute this value, we use our **dplyr** skills:

```
r <- murders %>%
  summarize(rate = sum(total) /  sum(population) * 10^6) %>%
  pull(rate)
```

To add a line we use the geom_abline function. **ggplot2** uses ab in the name to remind us we are supplying the intercept (a) and slope (b). The default line has slope 1 and intercept 0 so we only have to define the intercept:

```
p + geom_point(aes(col=region), size = 3) +
  geom_abline(intercept = log10(r))
```

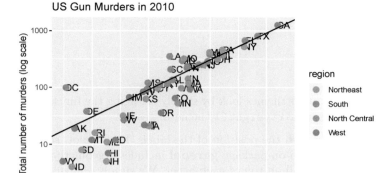

Here `geom_abline` does not use any information from the data object.

We can change the line type and color of the lines using arguments. Also, we draw it first so it doesn't go over our points.

```
p <- p + geom_abline(intercept = log10(r), lty = 2, color = "darkgrey") +
  geom_point(aes(col=region), size = 3)
```

Note that we have redefined p and used this new p below and in the next section.

The default plots created by **ggplot2** are already very useful. However, we frequently need to make minor tweaks to the default behavior. Although it is not always obvious how to make these even with the cheat sheet, **ggplot2** is very flexible.

For example, we can make changes to the legend via the `scale_color_discrete` function. In our plot the word *region* is capitalized and we can change it like this:

```
p <- p + scale_color_discrete(name = "Region")
```

7.11   Add-on packages

The power of **ggplot2** is augmented further due to the availability of add-on packages. The remaining changes needed to put the finishing touches on our plot require the **ggthemes** and **ggrepel** packages.

The style of a **ggplot2** graph can be changed using the **theme** functions. Several themes are included as part of the **ggplot2** package. In fact, for most the plots in this book, we use a function in the **dslabs** package that automatically sets a default theme:

```
ds_theme_set()
```

Many other themes are added by the package **ggthemes**. Among those are the `theme_economist` theme that we used. After installing the package, you can change the style by adding a layer like this:

```
library(ggthemes)
p + theme_economist()
```

You can see how some of the other themes look by simply changing the function. For instance, you might try the `theme_fivethirtyeight()` theme instead.

The final difference has to do with the position of the labels. In our plot, some of the labels fall on top of each other. The add-on package **ggrepel** includes a geometry that adds labels while ensuring that they don't fall on top of each other. We simply change `geom_text` with `geom_text_repel`.

7.12   Putting it all together

Now that we are done testing, we can write one piece of code that produces our desired plot from scratch.

```
library(ggthemes)
library(ggrepel)

r <- murders %>%
  summarize(rate = sum(total) /  sum(population) * 10^6) %>%
  pull(rate)

murders %>% ggplot(aes(population/10^6, total, label = abb)) +
  geom_abline(intercept = log10(r), lty = 2, color = "darkgrey") +
  geom_point(aes(col=region), size = 3) +
  geom_text_repel() +
  scale_x_log10() +
  scale_y_log10() +
  xlab("Populations in millions (log scale)") +
  ylab("Total number of murders (log scale)") +
  ggtitle("US Gun Murders in 2010") +
  scale_color_discrete(name = "Region") +
  theme_economist()
```

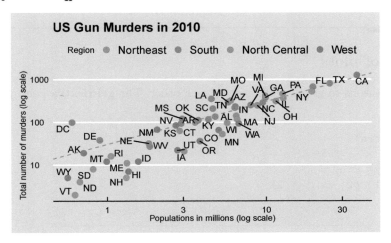

7.13 Quick plots with qplot

We have learned the powerful approach to generating visualization with ggplot. However, there are instances in which all we want is to make a quick plot of, for example, a histogram of the values in a vector, a scatterplot of the values in two vectors, or a boxplot using categorical and numeric vectors. We demonstrated how to generate these plots with `hist`, `plot`, and `boxplot`. However, if we want to keep consistent with the ggplot style, we can use the function `qplot`.

If we have values in two vectors, say:

```
data(murders)
x <- log10(murders$population)
y <- murders$total
```

and we want to make a scatterplot with ggplot, we would have to type something like:

```
data.frame(x = x, y = y) %>%
  ggplot(aes(x, y)) +
  geom_point()
```

This seems like too much code for such a simple plot. The `qplot` function sacrifices the flexibility provided by the `ggplot` approach, but allows us to generate a plot quickly.

```
qplot(x, y)
```

We will learn more about `qplot` in Section 8.16

7.14 Grids of plots

There are often reasons to graph plots next to each other. The **gridExtra** package permits us to do that:

```
library(gridExtra)
p1 <- qplot(x)
p2 <- qplot(x,y)
grid.arrange(p1, p2, ncol = 2)
```

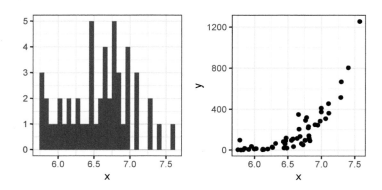

7.15 Exercises

Start by loading the **dplyr** and **ggplot2** library as well as the murders and heights data.

```
library(dplyr)
library(ggplot2)
library(dslabs)
data(heights)
data(murders)
```

1. With **ggplot2** plots can be saved as objects. For example we can associate a dataset with a plot object like this

```
p <- ggplot(data = murders)
```

Because data is the first argument we don't need to spell it out

```
p <- ggplot(murders)
```

and we can also use the pipe:

```
p <- murders %>% ggplot()
```

What is class of the object p?

2. Remember that to print an object you can use the command **print** or simply type the object. Print the object p defined in exercise one and describe what you see.

    a. Nothing happens.
    b. A blank slate plot.
    c. A scatterplot.
    d. A histogram.

3. Using the pipe %>%, create an object p but this time associated with the **heights** dataset instead of the **murders** dataset.

4. What is the class of the object p you have just created?

5. Now we are going to add a layer and the corresponding aesthetic mappings. For the murders data we plotted total murders versus population sizes. Explore the **murders** data frame to remind yourself what are the names for these two variables and select the correct answer. **Hint**: Look at ?murders.

    a. **state** and **abb**.
    b. **total_murders** and **population_size**.
    c. **total** and **population**.
    d. **murders** and **size**.

6. To create the scatterplot we add a layer with **geom_point**. The aesthetic mappings require us to define the x-axis and y-axis variables, respectively. So the code looks like this:

```
murders %>% ggplot(aes(x = , y = )) +
  geom_point()
```

except we have to define the two variables x and y. Fill this out with the correct variable names.

7. Note that if we don't use argument names, we can obtain the same plot by making sure we enter the variable names in the right order like this:

```
murders %>% ggplot(aes(population, total)) +
  geom_point()
```

Remake the plot but now with total in the x-axis and population in the y-axis.

8. If instead of points we want to add text, we can use the **geom_text()** or **geom_label()** geometries. The following code

```
murders %>% ggplot(aes(population, total)) + geom_label()
```

will give us the error message: **Error: geom_label requires the following missing aesthetics: label**

Why is this?

a. We need to map a character to each point through the label argument in aes.
b. We need to let `geom_label` know what character to use in the plot.
c. The `geom_label` geometry does not require x-axis and y-axis values.
d. `geom_label` is not a ggplot2 command.

9. Rewrite the code above to abbreviation as the label through `aes`

10. Change the color of the labels through blue. How will we do this?

a. Adding a column called `blue` to `murders`.
b. Because each label needs a different color we map the colors through `aes`.
c. Use the `color` argument in `ggplot`.
d. Because we want all colors to be blue, we do not need to map colors, just use the color argument in `geom_label`.

11. Rewrite the code above to make the labels blue.

12. Now suppose we want to use color to represent the different regions. In this case which of the following is most appropriate:

a. Adding a column called `color` to `murders` with the color we want to use.
b. Because each label needs a different color we map the colors through the color argument of `aes` .
c. Use the `color` argument in `ggplot`.
d. Because we want all colors to be blue, we do not need to map colors, just use the color argument in `geom_label`.

13. Rewrite the code above to make the labels' color be determined by the state's region.

14. Now we are going to change the x-axis to a log scale to account for the fact the distribution of population is skewed. Let's start by defining an object `p` holding the plot we have made up to now

```
p <- murders %>%
  ggplot(aes(population, total, label = abb, color = region)) +
  geom_label()
```

To change the y-axis to a log scale we learned about the `scale_x_log10()` function. Add this layer to the object `p` to change the scale and render the plot.

15. Repeat the previous exercise but now change both axes to be in the log scale.

16. Now edit the code above to add the title "Gun murder data" to the plot. Hint: use the `ggtitle` function.

8

Visualizing data distributions

You may have noticed that numerical data is often summarized with the *average* value. For example, the quality of a high school is sometimes summarized with one number: the average score on a standardized test. Occasionally, a second number is reported: the *standard deviation*. For example, you might read a report stating that scores were 680 plus or minus 50 (the standard deviation). The report has summarized an entire vector of scores with just two numbers. Is this appropriate? Is there any important piece of information that we are missing by only looking at this summary rather than the entire list?

Our first data visualization building block is learning to summarize lists of factors or numeric vectors. More often than not, the best way to share or explore this summary is through data visualization. The most basic statistical summary of a list of objects or numbers is its distribution. Once a vector has been summarized as a distribution, there are several data visualization techniques to effectively relay this information.

In this chapter, we first discuss properties of a variety of distributions and how to visualize distributions using a motivating example of student heights. We then discuss the **ggplot2** geometries for these visualizations in Section 8.16.

8.1   Variable types

We will be working with two types of variables: categorical and numeric. Each can be divided into two other groups: categorical can be ordinal or not, whereas numerical variables can be discrete or continuous.

When each entry in a vector comes from one of a small number of groups, we refer to the data as *categorical data*. Two simple examples are sex (male or female) and regions (Northeast, South, North Central, West). Some categorical data can be ordered even if they are not numbers per se, such as spiciness (mild, medium, hot). In statistics textbooks, ordered categorical data are referred to as *ordinal* data.

Examples of numerical data are population sizes, murder rates, and heights. Some numerical data can be treated as ordered categorical. We can further divide numerical data into continuous and discrete. Continuous variables are those that can take any value, such as heights, if measured with enough precision. For example, a pair of twins may be 68.12 and 68.11 inches, respectively. Counts, such as population sizes, are discrete because they have to be round numbers.

Keep in mind that discrete numeric data can be considered ordinal. Although this is technically true, we usually reserve the term ordinal data for variables belonging to a small number of different groups, with each group having many members. In contrast, when we have many groups with few cases in each group, we typically refer to them as discrete numerical variables. So, for example, the number of packs of cigarettes a person smokes a day, rounded to the closest pack, would be considered ordinal, while the actual number of cigarettes would be considered a numerical variable. But, indeed, there are examples that can be considered both numerical and ordinal when it comes to visualizing data.

8.2 Case study: describing student heights

Here we introduce a new motivating problem. It is an artificial one, but it will help us illustrate the concepts needed to understand distributions.

Pretend that we have to describe the heights of our classmates to ET, an extraterrestrial that has never seen humans. As a first step, we need to collect data. To do this, we ask students to report their heights in inches. We ask them to provide sex information because we know there are two different distributions by sex. We collect the data and save it in the `heights` data frame:

```
library(tidyverse)
library(dslabs)
data(heights)
```

One way to convey the heights to ET is to simply send him this list of 1050 heights. But there are much more effective ways to convey this information, and understanding the concept of a distribution will help. To simplify the explanation, we first focus on male heights. We examine the female height data in Section 8.14.

8.3 Distribution function

It turns out that, in some cases, the average and the standard deviation are pretty much all we need to understand the data. We will learn data visualization techniques that will help us determine when this two number summary is appropriate. These same techniques will serve as an alternative for when two numbers are not enough.

The most basic statistical summary of a list of objects or numbers is its distribution. The simplest way to think of a distribution is as a compact description of a list with many entries. This concept should not be new for readers of this book. For example, with categorical data, the distribution simply describes the proportion of each unique category. The sex represented in the heights dataset is:

```
#>
#> Female    Male
#>  0.227   0.773
```

This two-category *frequency table* is the simplest form of a distribution. We don't really need to visualize it since one number describes everything we need to know: 23% are females and the rest are males. When there are more categories, then a simple barplot describes the distribution. Here is an example with US state regions:

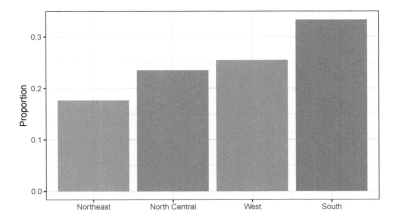

This particular plot simply shows us four numbers, one for each category. We usually use barplots to display a few numbers. Although this particular plot does not provide much more insight than a frequency table itself, it is a first example of how we convert a vector into a plot that succinctly summarizes all the information in the vector. When the data is numerical, the task of displaying distributions is more challenging.

8.4 Cumulative distribution functions

Numerical data that are not categorical also have distributions. In general, when data is not categorical, reporting the frequency of each entry is not an effective summary since most entries are unique. In our case study, while several students reported a height of 68 inches, only one student reported a height of 68.503937007874 inches and only one student reported a height 68.8976377952756 inches. We assume that they converted from 174 and 175 centimeters, respectively.

Statistics textbooks teach us that a more useful way to define a distribution for numeric data is to define a function that reports the proportion of the data below a for all possible values of a. This function is called the cumulative distribution function (CDF). In statistics, the following notation is used:

$$F(a) = \Pr(x \leq a)$$

Here is a plot of F for the male height data:

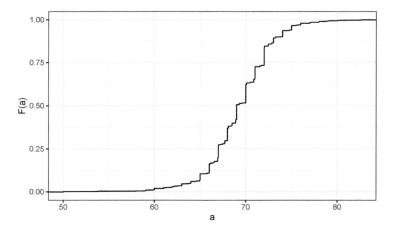

Similar to what the frequency table does for categorical data, the CDF defines the distribution for numerical data. From the plot, we can see that 16% of the values are below 65, since $F(66) = 0.164$, or that 84% of the values are below 72, since $F(72) = 0.841$, and so on. In fact, we can report the proportion of values between any two heights, say a and b, by computing $F(b) - F(a)$. This means that if we send this plot above to ET, he will have all the information needed to reconstruct the entire list. Paraphrasing the expression "a picture is worth a thousand words", in this case, a picture is as informative as 812 numbers.

A final note: because CDFs can be defined mathematically the word *empirical* is added to make the distinction when data is used. We therefore use the term empirical CDF (eCDF).

8.5   Histograms

Although the CDF concept is widely discussed in statistics textbooks, the plot is actually not very popular in practice. The main reason is that it does not easily convey characteristics of interest such as: at what value is the distribution centered? Is the distribution symmetric? What ranges contain 95% of the values? Histograms are much preferred because they greatly facilitate answering such questions. Histograms sacrifice just a bit of information to produce plots that are much easier to interpret.

The simplest way to make a histogram is to divide the span of our data into non-overlapping bins of the same size. Then, for each bin, we count the number of values that fall in that interval. The histogram plots these counts as bars with the base of the bar defined by the intervals. Here is the histogram for the height data splitting the range of values into one inch intervals: $[49.5, 50.5], [51.5, 52.5], (53.5, 54.5], ..., (82.5, 83.5]$

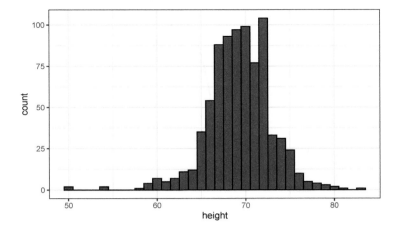

As you can see in the figure above, a histogram is similar to a barplot, but it differs in that the x-axis is numerical, not categorical.

If we send this plot to ET, he will immediately learn some important properties about our data. First, the range of the data is from 50 to 84 with the majority (more than 95%) between 63 and 75 inches. Second, the heights are close to symmetric around 69 inches. Also, by adding up counts, ET could obtain a very good approximation of the proportion of the data in any interval. Therefore, the histogram above is not only easy to interpret, but also provides almost all the information contained in the raw list of 812 heights with about 30 bin counts.

What information do we lose? Note that all values in each interval are treated the same when computing bin heights. So, for example, the histogram does not distinguish between 64, 64.1, and 64.2 inches. Given that these differences are almost unnoticeable to the eye, the practical implications are negligible and we were able to summarize the data to just 23 numbers.

We discuss how to code histograms in Section 8.16.

8.6   Smoothed density

Smooth density plots are aesthetically more appealing than histograms. Here is what a smooth density plot looks like for our heights data:

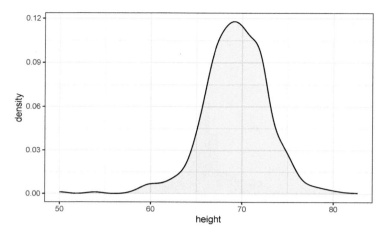

In this plot, we no longer have sharp edges at the interval boundaries and many of the local peaks have been removed. Also, the scale of the y-axis changed from counts to *density*.

To understand the smooth densities, we have to understand *estimates*, a topic we don't cover until later. However, we provide a heuristic explanation to help you understand the basics so you can use this useful data visualization tool.

The main new concept you must understand is that we assume that our list of observed values is a subset of a much larger list of unobserved values. In the case of heights, you can imagine that our list of 812 male students comes from a hypothetical list containing all the heights of all the male students in all the world measured very precisely. Let's say there are 1,000,000 of these measurements. This list of values has a distribution, like any list of values, and this larger distribution is really what we want to report to ET since it is much more general. Unfortunately, we don't get to see it.

However, we make an assumption that helps us perhaps approximate it. If we had 1,000,000 values, measured very precisely, we could make a histogram with very, very small bins. The assumption is that if we show this, the height of consecutive bins will be similar. This is what we mean by smooth: we don't have big jumps in the heights of consecutive bins. Below we have a hypothetical histogram with bins of size 1:

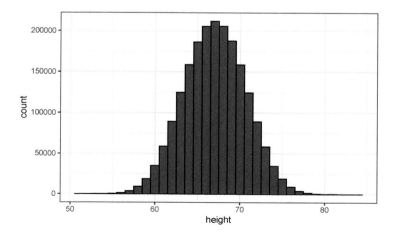

The smaller we make the bins, the smoother the histogram gets. Here are the histograms with bin width of 1, 0.5, and 0.1:

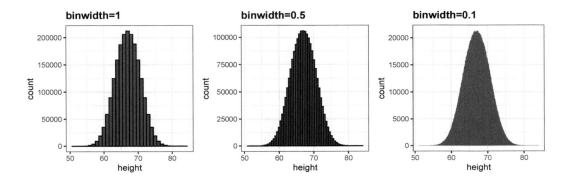

The smooth density is basically the curve that goes through the top of the histogram bars when the bins are very, very small. To make the curve not depend on the hypothetical size of the hypothetical list, we compute the curve on frequencies rather than counts:

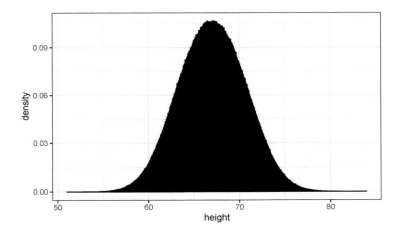

Now, back to reality. We don't have millions of measurements. Instead, we have 812 and we can't make a histogram with very small bins.

We therefore make a histogram, using bin sizes appropriate for our data and computing frequencies rather than counts, and we draw a smooth curve that goes through the tops of the histogram bars. The following plots demonstrate the steps that lead to a smooth density:

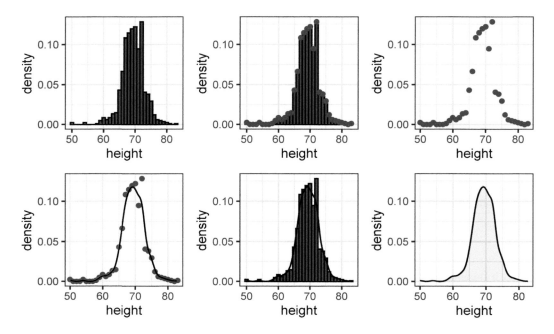

However, remember that *smooth* is a relative term. We can actually control the *smoothness* of the curve that defines the smooth density through an option in the function that computes the smooth density curve. Here are two examples using different degrees of smoothness on the same histogram:

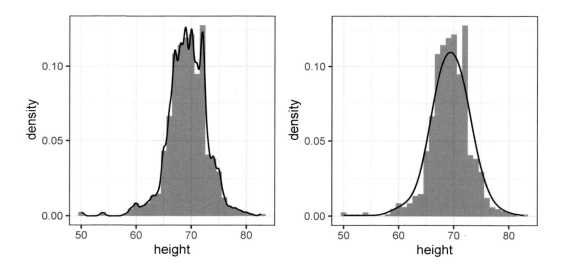

We need to make this choice with care as the resulting visualizations can change our interpretation of the data. We should select a degree of smoothness that we can defend as being representative of the underlying data. In the case of height, we really do have reason to believe that the proportion of people with similar heights should be the same. For example, the proportion that is 72 inches should be more similar to the proportion that is 71 than to the proportion that is 78 or 65. This implies that the curve should be pretty smooth; that is, the curve should look more like the example on the right than on the left.

While the histogram is an assumption-free summary, the smoothed density is based on some assumptions.

8.6.1 Interpreting the y-axis

Note that interpreting the y-axis of a smooth density plot is not straightforward. It is scaled so that the area under the density curve adds up to 1. If you imagine we form a bin with a base 1 unit in length, the y-axis value tells us the proportion of values in that bin. However, this is only true for bins of size 1. For other size intervals, the best way to determine the proportion of data in that interval is by computing the proportion of the total area contained in that interval. For example, here are the proportion of values between 65 and 68:

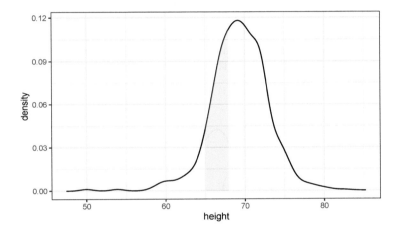

The proportion of this area is about 0.3, meaning that about that proportion is between 65 and 68 inches.

By understanding this, we are ready to use the smooth density as a summary. For this dataset, we would feel quite comfortable with the smoothness assumption, and therefore with sharing this aesthetically pleasing figure with ET, which he could use to understand our male heights data:

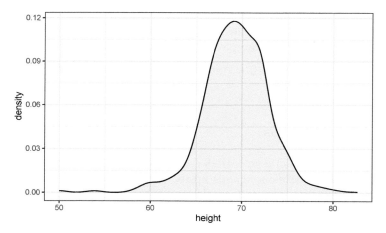

8.6.2   Densities permit stratification

As a final note, we point out that an advantage of smooth densities over histograms for visualization purposes is that densities make it easier to compare two distributions. This is in large part because the jagged edges of the histogram add clutter. Here is an example comparing male and female heights:

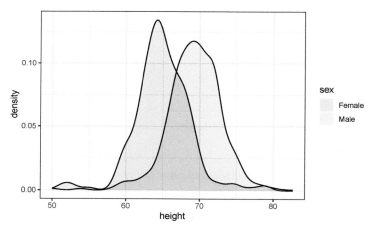

With the right argument, `ggplot` automatically shades the intersecting region with a different color. We will show examples of **ggplot2** code for densities in Section 9 as well as Section 8.16.

8.7   Exercises

1. In the `murders` dataset, the region is a categorical variable and the following is its distribution:

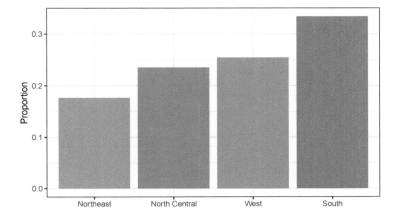

To the closet 5%, what proportion of the states are in the North Central region?

2. Which of the following is true:

    a. The graph above is a histogram.
    b. The graph above shows only four numbers with a bar plot.
    c. Categories are not numbers, so it does not make sense to graph the distribution.
    d. The colors, not the height of the bars, describe the distribution.

3. The plot below shows the eCDF for male heights:

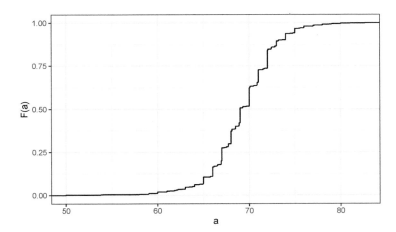

Based on the plot, what percentage of males are shorter than 75 inches?

    a. 100%
    b. 95%
    c. 80%
    d. 72 inches

4. To the closest inch, what height m has the property that 1/2 of the male students are taller than m and 1/2 are shorter?

a. 61 inches
b. 64 inches
c. 69 inches
d. 74 inches

5. Here is an eCDF of the murder rates across states:

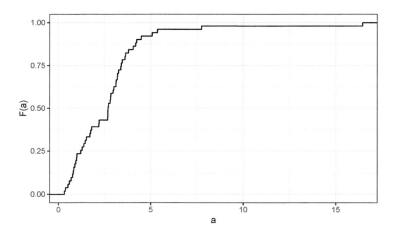

Knowing that there are 51 states (counting DC) and based on this plot, how many states have murder rates larger than 10 per 100,000 people?

a. 1
b. 5
c. 10
d. 50

6. Based on the eCDF above, which of the following statements are true:

a. About half the states have murder rates above 7 per 100,000 and the other half below.
b. Most states have murder rates below 2 per 100,000.
c. All the states have murder rates above 2 per 100,000.
d. With the exception of 4 states, the murder rates are below 5 per 100,000.

7. Below is a histogram of male heights in our `heights` dataset:

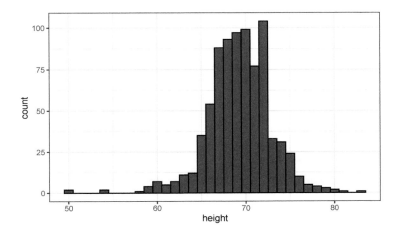

Based on this plot, how many males are between 63.5 and 65.5?

    a. 10
    b. 24
    c. 34
    d. 100

8. About what **percentage** are shorter than 60 inches?

    a. 1%
    b. 10%
    c. 25%
    d. 50%

9. Based on the density plot below, about what proportion of US states have populations larger than 10 million?

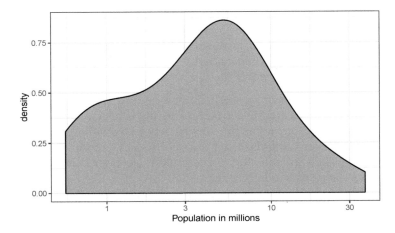

    a. 0.02
    b. 0.15

   c. 0.50

   d. 0.55

10. Below are three density plots. Is it possible that they are from the same dataset?

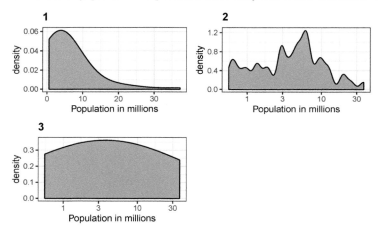

Which of the following statements is true:

   a. It is impossible that they are from the same dataset.

   b. They are from the same dataset, but the plots are different due to code errors.

   c. They are the same dataset, but the first and second plot undersmooth and the third oversmooths.

   d. They are the same dataset, but the first is not in the log scale, the second undersmooths, and the third oversmooths.

8.8   The normal distribution

Histograms and density plots provide excellent summaries of a distribution. But can we summarize even further? We often see the average and standard deviation used as summary statistics: a two-number summary! To understand what these summaries are and why they are so widely used, we need to understand the normal distribution.

The normal distribution, also known as the bell curve and as the Gaussian distribution, is one of the most famous mathematical concepts in history. A reason for this is that approximately normal distributions occur in many situations, including gambling winnings, heights, weights, blood pressure, standardized test scores, and experimental measurement errors. There are explanations for this, but we describe these later. Here we focus on how the normal distribution helps us summarize data.

Rather than using data, the normal distribution is defined with a mathematical formula. For any interval (a, b), the proportion of values in that interval can be computed using this formula:

$$\Pr(a < x < b) = \int_a^b \frac{1}{\sqrt{2\pi}s} e^{-\frac{1}{2}\left(\frac{x-m}{s}\right)^2} dx$$

You don't need to memorize or understand the details of the formula. But note that it is completely defined by just two parameters: m and s. The rest of the symbols in the formula represent the interval ends that we determine, a and b, and known mathematical constants π and e. These two parameters, m and s, are referred to as the *average* (also called the *mean*) and the *standard deviation* (SD) of the distribution, respectively.

The distribution is symmetric, centered at the average, and most values (about 95%) are within 2 SDs from the average. Here is what the normal distribution looks like when the average is 0 and the SD is 1:

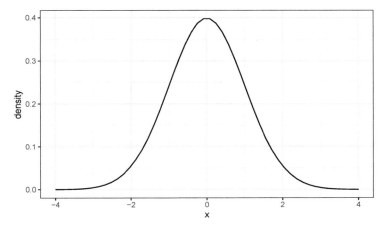

The fact that the distribution is defined by just two parameters implies that if a dataset is approximated by a normal distribution, all the information needed to describe the distribution can be encoded in just two numbers: the average and the standard deviation. We now define these values for an arbitrary list of numbers.

For a list of numbers contained in a vector x, the average is defined as:

```
m <- sum(x) / length(x)
```

and the SD is defined as:

```
s <- sqrt(sum((x-mu)^2) / length(x))
```

which can be interpreted as the average distance between values and their average.

Let's compute the values for the height for males which we will store in the object x:

```
index <- heights$sex=="Male"
x <- heights$height[index]
```

The pre-built functions `mean` and `sd` (note that for reasons explained in Section 16.2, `sd` divides by `length(x)-1` rather than `length(x)`) can be used here:

```
m <- mean(x)
s <- sd(x)
c(average = m, sd = s)
#> average       sd
#>  69.31     3.61
```

Here is a plot of the smooth density and the normal distribution with mean = 69.3 and SD = 3.6 plotted as a black line with our student height smooth density in blue:

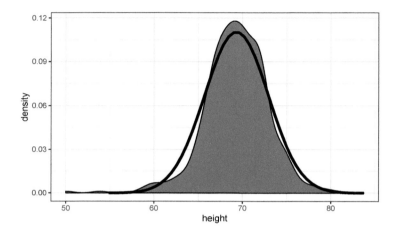

The normal distribution does appear to be quite a good approximation here. We now will see how well this approximation works at predicting the proportion of values within intervals.

8.9   Standard units

For data that is approximately normally distributed, it is convenient to think in terms of *standard units*. The standard unit of a value tells us how many standard deviations away from the average it is. Specifically, for a value x from a vector X, we define the value of x in standard units as z = (x - m)/s with m and s the average and standard deviation of X, respectively. Why is this convenient?

First look back at the formula for the normal distribution and note that what is being exponentiated is $-z^2/2$ with z equivalent to x in standard units. Because the maximum of $e^{-z^2/2}$ is when $z = 0$, this explains why the maximum of the distribution occurs at the average. It also explains the symmetry since $-z^2/2$ is symmetric around 0. Second, note that if we convert the normally distributed data to standard units, we can quickly know if, for example, a person is about average ($z = 0$), one of the largest ($z \approx 2$), one of the smallest ($z \approx -2$), or an extremely rare occurrence ($z > 3$ or $z < -3$). Remember that it does not matter what the original units are, these rules apply to any data that is approximately normal.

In R, we can obtain standard units using the function `scale`:

```
z <- scale(x)
```

Now to see how many men are within 2 SDs from the average, we simply type:

```
mean(abs(z) < 2)
#> [1] 0.95
```

The proportion is about 95%, which is what the normal distribution predicts! To further confirm that, in fact, the approximation is a good one, we can use quantile-quantile plots.

8.10   Quantile-quantile plots

A systematic way to assess how well the normal distribution fits the data is to check if the observed and predicted proportions match. In general, this is the approach of the quantile-quantile plot (QQ-plot).

First let's define the theoretical quantiles for the normal distribution. In statistics books we use the symbol $\Phi(x)$ to define the function that gives us the probability of a standard normal distribution being smaller than x. So, for example, $\Phi(-1.96) = 0.025$ and $\Phi(1.96) = 0.975$. In R, we can evaluate Φ using the pnorm function:

```
pnorm(-1.96)
#> [1] 0.025
```

The inverse function $\Phi^{-1}(x)$ gives us the *theoretical quantiles* for the normal distribution. So, for example, $\Phi^{-1}(0.975) = 1.96$. In R, we can evaluate the inverse of Φ using the qnorm function.

```
qnorm(0.975)
#> [1] 1.96
```

Note that these calculations are for the standard normal distribution by default (mean = 0, standard deviation = 1), but we can also define these for any normal distribution. We can do this using the mean and sd arguments in the pnorm and qnorm function. For example, we can use qnorm to determine quantiles of a distribution with a specific average and standard deviation

```
qnorm(0.975, mean = 5, sd = 2)
#> [1] 8.92
```

For the normal distribution, all the calculations related to quantiles are done without data, thus the name *theoretical quantiles*. But quantiles can be defined for any distribution, including an empirical one. So if we have data in a vector x, we can define the quantile associated with any proportion p as the q for which the proportion of values below q is p. Using R code, we can define q as the value for which mean(x <= q) = p. Notice that not all p have a q for which the proportion is exactly p. There are several ways of defining the best q as discussed in the help for the quantile function.

To give a quick example, for the male heights data, we have that:

```
mean(x <= 69.5)
#> [1] 0.515
```

So about 50% are shorter or equal to 69 inches. This implies that if $p = 0.50$ then $q = 69.5$.

The idea of a QQ-plot is that if your data is well approximated by normal distribution then the quantiles of your data should be similar to the quantiles of a normal distribution. To construct a QQ-plot, we do the following:

1. Define a vector of m proportions $p_1, p_2, \ldots, p_m$.
2. Define a vector of quantiles $q_1, \ldots, q_m$ for your data for the proportions $p_1, \ldots, p_m$. We refer to these as the *sample quantiles*.
3. Define a vector of theoretical quantiles for the proportions $p_1, \ldots, p_m$ for a normal distribution with the same average and standard deviation as the data.
4. Plot the sample quantiles versus the theoretical quantiles.

Let's construct a QQ-plot using R code. Start by defining the vector of proportions.

```
p <- seq(0.05, 0.95, 0.05)
```

To obtain the quantiles from the data, we can use the `quantile` function like this:

```
sample_quantiles <- quantile(x, p)
```

To obtain the theoretical normal distribution quantiles with the corresponding average and SD, we use the `qnorm` function:

```
theoretical_quantiles <- qnorm(p, mean = mean(x), sd = sd(x))
```

To see if they match or not, we plot them against each other and draw the identity line:

```
qplot(theoretical_quantiles, sample_quantiles) + geom_abline()
```

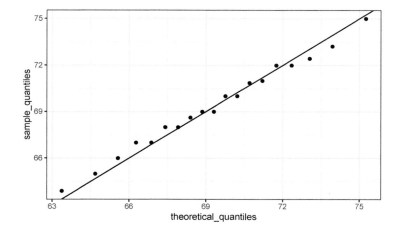

Notice that this code becomes much cleaner if we use standard units:

```
sample_quantiles <- quantile(z, p)
theoretical_quantiles <- qnorm(p)
qplot(theoretical_quantiles, sample_quantiles) + geom_abline()
```

The above code is included to help describe QQ-plots. However, in practice it is easier to use the **ggplot2** code described in Section 8.16:

```
heights %>% filter(sex=="Male") %>%
  ggplot(aes(sample = scale(height))) +
  geom_qq() +
  geom_abline()
```

While for the illustration above we used 20 quantiles, the default from the geom_qq function is to use as many quantiles as data points.

8.11   Percentiles

Before we move on, let's define some terms that are commonly used in exploratory data analysis.

Percentiles are special cases of *quantiles* that are commonly used. The percentiles are the quantiles you obtain when setting the p at $0.01, 0.02, ..., 0.99$. We call, for example, the case of $p = 0.25$ the 25th percentile, which gives us a number for which 25% of the data is below. The most famous percentile is the 50th, also known as the *median.*

For the normal distribution the *median* and average are the same, but this is generally not the case.

Another special case that receives a name are the *quartiles*, which are obtained when setting $p = 0.25, 0.50$, and 0.75.

8.12   Boxplots

To introduce boxplots we will go back to the US murder data. Suppose we want to summarize the murder rate distribution. Using the data visualization technique we have learned, we can quickly see that the normal approximation does not apply here:

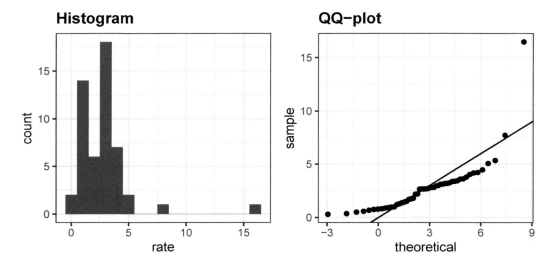

In this case, the histogram above or a smooth density plot would serve as a relatively succinct summary.

Now suppose those used to receiving just two numbers as summaries ask us for a more compact numerical summary.

Here Tukey offered some advice. Provide a five-number summary composed of the range along with the quartiles (the 25th, 50th, and 75th percentiles). Tukey further suggested that we ignore *outliers* when computing the range and instead plot these as independent points. We provide a detailed explanation of outliers later. Finally, he suggested we plot these numbers as a "box" with "whiskers" like this:

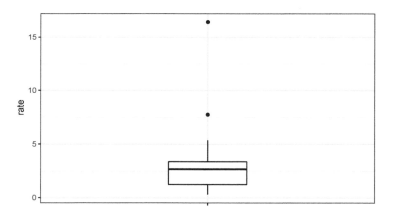

with the box defined by the 25% and 75% percentile and the whiskers showing the range. The distance between these two is called the *interquartile* range. The two points are outliers according to Tukey's definition. The median is shown with a horizontal line. Today, we call these *boxplots*.

From just this simple plot, we know that the median is about 2.5, that the distribution is not symmetric, and that the range is 0 to 5 for the great majority of states with two exceptions.

We discuss how to make boxplots in Section 8.16.

8.13 Stratification

In data analysis we often divide observations into groups based on the values of one or more variables associated with those observations. For example in the next section we divide the height values into groups based on a sex variable: females and males. We call this procedure *stratification* and refer to the resulting groups as *strata*.

Stratification is common in data visualization because we are often interested in how the distribution of variables differs across different subgroups. We will see several examples throughout this part of the book. We will revisit the concept of stratification when we learn regression in Chapter 17 and in the Machine Learning part of the book.

8.14 Case study: describing student heights (continued)

Using the histogram, density plots, and QQ-plots, we have become convinced that the male height data is well approximated with a normal distribution. In this case, we report back to ET a very succinct summary: male heights follow a normal distribution with an average of 69.3 inches and a SD of 3.6 inches. With this information, ET will have a good idea of what to expect when he meets our male students. However, to provide a complete picture we need to also provide a summary of the female heights.

We learned that boxplots are useful when we want to quickly compare two or more distributions. Here are the heights for men and women:

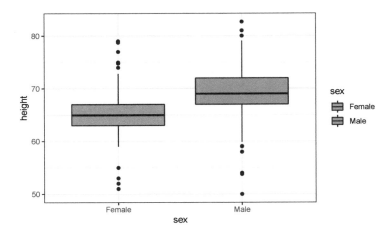

The plot immediately reveals that males are, on average, taller than females. The standard deviations appear to be similar. But does the normal approximation also work for the female height data collected by the survey? We expect that they will follow a normal distribution, just like males. However, exploratory plots reveal that the approximation is not as useful:

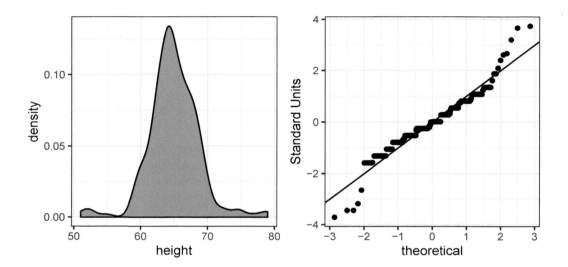

We see something we did not see for the males: the density plot has a second "bump". Also, the QQ-plot shows that the highest points tend to be taller than expected by the normal distribution. Finally, we also see five points in the QQ-plot that suggest shorter than expected heights for a normal distribution. When reporting back to ET, we might need to provide a histogram rather than just the average and standard deviation for the female heights.

However, go back and read Tukey's quote. We have noticed what we didn't expect to see. If we look at other female height distributions, we do find that they are well approximated with a normal distribution. So why are our female students different? Is our class a requirement for the female basketball team? Are small proportions of females claiming to be taller than they are? Another, perhaps more likely, explanation is that in the form students used to enter their heights, FEMALE was the default sex and some males entered their heights, but forgot to change the sex variable. In any case, data visualization has helped discover a potential flaw in our data.

Regarding the five smallest values, note that these values are:

```
heights %>% filter(sex=="Female") %>%
  top_n(5, desc(height)) %>%
  pull(height)
#> [1] 51 53 55 52 52
```

Because these are reported heights, a possibility is that the student meant to enter 5'1", 5'2", 5'3" or 5'5".

8.15   Exercises

1. Define variables containing the heights of males and females like this:

```
library(dslabs)
data(heights)
male <- heights$height[heights$sex=="Male"]
female <- heights$height[heights$sex=="Female"]
```

How many measurements do we have for each?

2. Suppose we can't make a plot and want to compare the distributions side by side. We can't just list all the numbers. Instead, we will look at the percentiles. Create a five row table showing `female_percentiles` and `male_percentiles` with the 10th, 30th, 50th, ..., 90th percentiles for each sex. Then create a data frame with these two as columns.

3. Study the following boxplots showing population sizes by country:

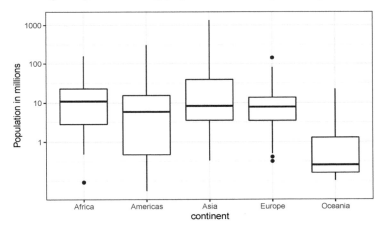

Which continent has the country with the biggest population size?

4. What continent has the largest median population size?

5. What is median population size for Africa to the nearest million?

6. What proportion of countries in Europe have populations below 14 million?

    a. 0.99
    b. 0.75
    c. 0.50
    d. 0.25

7. If we use a log transformation, which continent shown above has the largest interquartile range?

8. Load the height data set and create a vector x with just the male heights:

```
library(dslabs)
data(heights)
x <- heights$height[heights$sex=="Male"]
```

What proportion of the data is between 69 and 72 inches (taller than 69, but shorter or equal to 72)? Hint: use a logical operator and **mean**.

9. Suppose all you know about the data is the average and the standard deviation. Use the normal approximation to estimate the proportion you just calculated. Hint: start by computing the average and standard deviation. Then use the **pnorm** function to predict the proportions.

10. Notice that the approximation calculated in question two is very close to the exact calculation in the first question. Now perform the same task for more extreme values. Compare the exact calculation and the normal approximation for the interval (79,81]. How many times bigger is the actual proportion than the approximation?

11. Approximate the distribution of adult men in the world as normally distributed with an average of 69 inches and a standard deviation of 3 inches. Using this approximation, estimate the proportion of adult men that are 7 feet tall or taller, referred to as *seven footers*. Hint: use the **pnorm** function.

12. There are about 1 billion men between the ages of 18 and 40 in the world. Use your answer to the previous question to estimate how many of these men (18-40 year olds) are seven feet tall or taller in the world?

13. There are about 10 National Basketball Association (NBA) players that are 7 feet tall or higher. Using the answer to the previous two questions, what proportion of the world's 18-to-40-year-old *seven footers* are in the NBA?

14. Repeat the calculations performed in the previous question for Lebron James' height: 6 feet 8 inches. There are about 150 players that are at least that tall.

15. In answering the previous questions, we found that it is not at all rare for a seven footer to become an NBA player. What would be a fair critique of our calculations:

    a.   Practice and talent are what make a great basketball player, not height.
    b.   The normal approximation is not appropriate for heights.
    c.   As seen in question 3, the normal approximation tends to underestimate the extreme values. It's possible that there are more seven footers than we predicted.
    d.   As seen in question 3, the normal approximation tends to overestimate the extreme values. It's possible that there are fewer seven footers than we predicted.

8.16   ggplot2 geometries

In Chapter 7, we introduced the **ggplot2** package for data visualization. Here we demonstrate how to generate plots related to distributions, specifically the plots shown earlier in this chapter.

8.16.1 Barplots

To generate a barplot we can use the `geom_bar` geometry. The default is to count the number of each category and draw a bar. Here is the plot for the regions of the US.

```
murders %>% ggplot(aes(region)) + geom_bar()
```

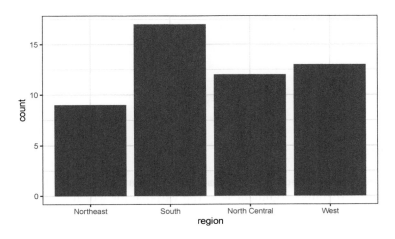

We often already have a table with a distribution that we want to present as a barplot. Here is an example of such a table:

```
data(murders)
tab <- murders %>%
  count(region) %>%
  mutate(proportion = n/sum(n))
tab
#> # A tibble: 4 x 3
#>   region          n proportion
#>   <fct>       <int>      <dbl>
#> 1 Northeast       9      0.176
#> 2 South          17      0.333
#> 3 North Central  12      0.235
#> 4 West           13      0.255
```

We no longer want `geom_bar` to count, but rather just plot a bar to the height provided by the `proportion` variable. For this we need to provide x (the categories) and y (the values) and use the `stat="identity"` option.

```
tab %>% ggplot(aes(region, proportion)) + geom_bar(stat = "identity")
```

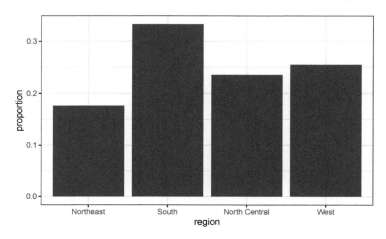

8.16.2   Histograms

To generate histograms we use `geom_histogram`. By looking at the help file for this function, we learn that the only required argument is `x`, the variable for which we will construct a histogram. We dropped the `x` because we know it is the first argument. The code looks like this:

```
heights %>%
  filter(sex == "Female") %>%
  ggplot(aes(height)) +
  geom_histogram()
```

If we run the code above, it gives us a message:

    `stat_bin()` using `bins` = 30. Pick better value with `binwidth`.

We previously used a bin size of 1 inch, so the code looks like this:

```
heights %>%
  filter(sex == "Female") %>%
  ggplot(aes(height)) +
  geom_histogram(binwidth = 1)
```

Finally, if for aesthetic reasons we want to add color, we use the arguments described in the help file. We also add labels and a title:

```
heights %>%
  filter(sex == "Female") %>%
  ggplot(aes(height)) +
  geom_histogram(binwidth = 1, fill = "blue", col = "black") +
  xlab("Male heights in inches") +
  ggtitle("Histogram")
```

Histogram

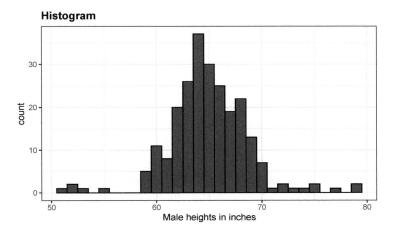

8.16.3   Density plots

To create a smooth density, we use the `geom_density`. To make a smooth density plot with the data previously shown as a histogram we can use this code:

```
heights %>%
  filter(sex == "Female") %>%
  ggplot(aes(height)) +
  geom_density()
```

To fill in with color, we can use the `fill` argument.

```
heights %>%
  filter(sex == "Female") %>%
  ggplot(aes(height)) +
  geom_density(fill="blue")
```

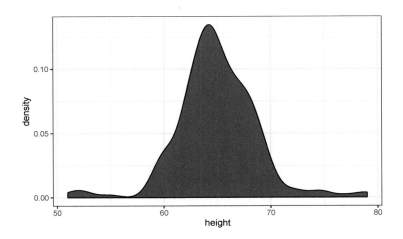

To change the smoothness of the density, we use the `adjust` argument to multiply the default value by that `adjust`. For example, if we want the bandwidth to be twice as big we use:

```
heights %>%
  filter(sex == "Female") +
  geom_density(fill="blue", adjust = 2)
```

8.16.4 Boxplots

The geometry for boxplot is `geom_boxplot`. As discussed, boxplots are useful for comparing distributions. For example, below are the previously shown heights for women, but compared to men. For this geometry, we need arguments `x` as the categories, and `y` as the values.

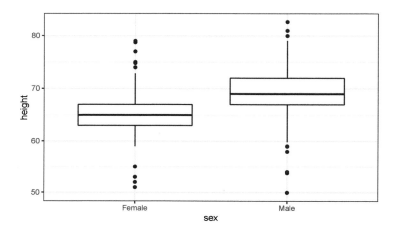

8.16.5 QQ-plots

For qq-plots we use the `geom_qq` geometry. From the help file, we learn that we need to specify the `sample` (we will learn about samples in a later chapter). Here is the qqplot for men heights.

```
heights %>% filter(sex=="Male") %>%
  ggplot(aes(sample = height)) +
  geom_qq()
```

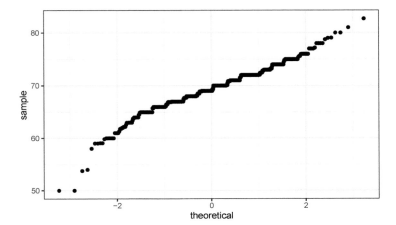

By default, the sample variable is compared to a normal distribution with average 0 and standard deviation 1. To change this, we use the **dparams** arguments based on the help file. Adding an identity line is as simple as assigning another layer. For straight lines, we use the **geom_abline** function. The default line is the identity line (slope $= 1$, intercept $= 0$).

```
params <- heights %>% filter(sex=="Male") %>%
  summarize(mean = mean(height), sd = sd(height))

heights %>% filter(sex=="Male") %>%
  ggplot(aes(sample = height)) +
  geom_qq(dparams = params) +
  geom_abline()
```

Another option here is to scale the data first and then make a qqplot against the standard normal.

```
heights %>%
  filter(sex=="Male") %>%
  ggplot(aes(sample = scale(height))) +
  geom_qq() +
  geom_abline()
```

8.16.6   Images

Images were not needed for the concepts described in this chapter, but we will use images in Section 10.14, so we introduce the two geometries used to create images: **geom_tile** and **geom_raster**. They behave similarly; to see how they differ, please consult the help file. To create an image in **ggplot2** we need a data frame with the x and y coordinates as well as the values associated with each of these. Here is a data frame.

```
x <- expand.grid(x = 1:12, y = 1:10) %>%
  mutate(z = 1:120)
```

Note that this is the tidy version of a matrix, `matrix(1:120, 12, 10)`. To plot the image we use the following code:

```
x %>% ggplot(aes(x, y, fill = z)) +
  geom_raster()
```

With these images you will often want to change the color scale. This can be done through the `scale_fill_gradientn` layer.

```
x %>% ggplot(aes(x, y, fill = z)) +
  geom_raster() +
  scale_fill_gradientn(colors =  terrain.colors(10))
```

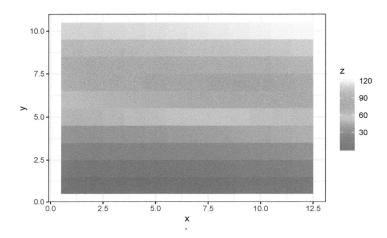

8.16.7   Quick plots

In Section 7.13 we introduced `qplot` as a useful function when we need to make a quick scatterplot. We can also use `qplot` to make histograms, density plots, boxplot, qqplots and more. Although it does not provide the level of control of `ggplot`, `qplot` is definitely useful as it permits us to make a plot with a short snippet of code.

Suppose we have the female heights in an object `x`:

```
x <- heights %>%
  filter(sex=="Male") %>%
  pull(height)
```

To make a quick histogram we can use:

```
qplot(x)
```

The function guesses that we want to make a histogram because we only supplied one variable. In Section 7.13 we saw that if we supply `qplot` two variables, it automatically makes a scatterplot.

To make a quick qqplot you have to use the `sample` argument. Note that we can add layers just as we do with `ggplot`.

```
qplot(sample = scale(x)) + geom_abline()
```

If we supply a factor and a numeric vector, we obtain a plot like the one below. Note that in the code below we are using the `data` argument. Because the data frame is not the first argument in `qplot`, we have to use the dot operator.

```
heights %>% qplot(sex, height, data = .)
```

We can also select a specific geometry by using the `geom` argument. So to convert the plot above to a boxplot, we use the following code:

```
heights %>% qplot(sex, height, data = ., geom = "boxplot")
```

We can also use the `geom` argument to generate a density plot instead of a histogram:

```
qplot(x, geom = "density")
```

Although not as much as with `ggplot`, we do have some flexibility to improve the results of `qplot`. Looking at the help file we see several ways in which we can improve the look of the histogram above. Here is an example:

```
qplot(x, bins=15, color = I("black"), xlab = "Population")
```

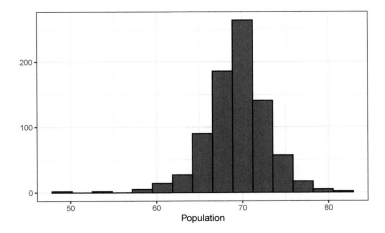

Technical note: The reason we use `I("black")` is because we want `qplot` to treat `"black"` as a character rather than convert it to a factor, which is the default behavior within `aes`, which is internally called here. In general, the function `I` is used in R to say "keep it as it is".

8.17   Exercises

1. Now we are going to use the `geom_histogram` function to make a histogram of the heights in the `height` data frame. When reading the documentation for this function we see that it requires just one mapping, the values to be used for the histogram. Make a histogram of all the plots.

What is the variable containing the heights?

    a. `sex`
    b. `heights`
    c. `height`
    d. `heights$height`

2. Now create a ggplot object using the pipe to assign the heights data to a ggplot object. Assign `height` to the x values through the `aes` function.

3. Now we are ready to add a layer to actually make the histogram. Use the object created in the previous exercise and the `geom_histogram` function to make the histogram.

4. Note that when we run the code in the previous exercise we get the warning: `stat_bin()` using `bins = 30`. Pick better value with `binwidth`.'

Use the `binwidth` argument to change the histogram made in the previous exercise to use bins of size 1 inch.

5. Instead of a histogram, we are going to make a smooth density plot. In this case we will not make an object, but instead render the plot with one line of code. Change the geometry in the code previously used to make a smooth density instead of a histogram.

6. Now we are going to make a density plot for males and females separately. We can do this using the `group` argument. We assign groups via the aesthetic mapping as each point needs to a group before making the calculations needed to estimate a density.

7. We can also assign groups through the `color` argument. This has the added benefit that it uses color to distinguish the groups. Change the code above to use color.

8. We can also assign groups through the `fill` argument. This has the added benefit that it uses colors to distinguish the groups, like this:

```
heights %>%
  ggplot(aes(height, fill = sex)) +
  geom_density()
```

However, here the second density is drawn over the other. We can make the curves more visible by using alpha blending to add transparency. Set the alpha parameter to 0.2 in the `geom_density` function to make this change.

9

Data visualization in practice

In this chapter, we will demonstrate how relatively simple **ggplot2** code can create insightful and aesthetically pleasing plots. As motivation we will create plots that help us better understand trends in world health and economics. We will implement what we learned in Chapters 7 and 8.16 and learn how to augment the code to perfect the plots. As we go through our case study, we will describe relevant general data visualization principles and learn concepts such as *faceting*, *time series plots*, *transformations*, and *ridge plots*.

9.1 Case study: new insights on poverty

Hans Rosling[1] was the co-founder of the Gapminder Foundation[2], an organization dedicated to educating the public by using data to dispel common myths about the so-called developing world. The organization uses data to show how actual trends in health and economics contradict the narratives that emanate from sensationalist media coverage of catastrophes, tragedies, and other unfortunate events. As stated in the Gapminder Foundation's website:

> Journalists and lobbyists tell dramatic stories. That's their job. They tell stories about extraordinary events and unusual people. The piles of dramatic stories pile up in peoples' minds into an over-dramatic worldview and strong negative stress feelings: "The world is getting worse!", "It's we vs. them!", "Other people are strange!", "The population just keeps growing!" and "Nobody cares!"

Hans Rosling conveyed actual data-based trends in a dramatic way of his own, using effective data visualization. This section is based on two talks that exemplify this approach to education: [New Insights on Poverty][3] and The Best Stats You've Ever Seen[4]. Specifically, in this section, we use data to attempt to answer the following two questions:

1. Is it a fair characterization of today's world to say it is divided into western rich nations and the developing world in Africa, Asia, and Latin America?
2. Has income inequality across countries worsened during the last 40 years?

To answer these questions, we will be using the `gapminder` dataset provided in **dslabs**. This dataset was created using a number of spreadsheets available from the Gapminder Foundation. You can access the table like this:

---

[1] https://en.wikipedia.org/wiki/Hans_Rosling
[2] http://www.gapminder.org/
[3] https://www.ted.com/talks/hans_rosling_reveals_new_insights_on_poverty?language=en
[4] https://www.ted.com/talks/hans_rosling_shows_the_best_stats_you_ve_ever_seen

```
library(tidyverse)
library(dslabs)
data(gapminder)
gapminder %>% as_tibble()
#> # A tibble: 10,545 x 9
#>    country  year infant_mortality life_expectancy fertility population
#>    <fct>   <int>            <dbl>           <dbl>     <dbl>      <dbl>
#> 1 Albania  1960             115.            62.9      6.19    1636054
#> 2 Algeria  1960             148.            47.5      7.65   11124892
#> 3 Angola   1960             208             36.0      7.32    5270844
#> 4 Antigu~  1960              NA             63.0      4.43      54681
#> 5 Argent~  1960              59.9           65.4      3.11   20619075
#> # ... with 1.054e+04 more rows, and 3 more variables: gdp <dbl>,
#> #   continent <fct>, region <fct>
```

9.1.1   Hans Rosling's quiz

As done in the *New Insights on Poverty* video, we start by testing our knowledge regarding differences in child mortality across different countries. For each of the six pairs of countries below, which country do you think had the highest child mortality rates in 2015? Which pairs do you think are most similar?

1. Sri Lanka or Turkey
2. Poland or South Korea
3. Malaysia or Russia
4. Pakistan or Vietnam
5. Thailand or South Africa

When answering these questions without data, the non-European countries are typically picked as having higher child mortality rates: Sri Lanka over Turkey, South Korea over Poland, and Malaysia over Russia. It is also common to assume that countries considered to be part of the developing world: Pakistan, Vietnam, Thailand, and South Africa, have similarly high mortality rates.

To answer these questions **with data**, we can use **dplyr**. For example, for the first comparison we see that:

```
gapminder %>%
  filter(year == 2015 & country %in% c("Sri Lanka","Turkey")) %>%
  select(country, infant_mortality)
#>     country infant_mortality
#> 1 Sri Lanka              8.4
#> 2    Turkey             11.6
```

Turkey has the higher infant mortality rate.

We can use this code on all comparisons and find the following:

country	infant mortality	country	infant mortality
Sri Lanka	8.4	Turkey	11.6
Poland	4.5	South Korea	2.9
Malaysia	6.0	Russia	8.2
Pakistan	65.8	Vietnam	17.3
Thailand	10.5	South Africa	33.6

We see that the European countries on this list have higher child mortality rates: Poland has a higher rate than South Korea, and Russia has a higher rate than Malaysia. We also see that Pakistan has a much higher rate than Vietnam, and South Africa has a much higher rate than Thailand. It turns out that when Hans Rosling gave this quiz to educated groups of people, the average score was less than 2.5 out of 5, worse than what they would have obtained had they guessed randomly. This implies that more than ignorant, we are misinformed. In this chapter we see how data visualization helps inform us.

9.2   Scatterplots

The reason for this stems from the preconceived notion that the world is divided into two groups: the western world (Western Europe and North America), characterized by long life spans and small families, versus the developing world (Africa, Asia, and Latin America) characterized by short life spans and large families. But do the data support this dichotomous view?

The necessary data to answer this question is also available in our `gapminder` table. Using our newly learned data visualization skills, we will be able to tackle this challenge.

In order to analyze this world view, our first plot is a scatterplot of life expectancy versus fertility rates (average number of children per woman). We start by looking at data from about 50 years ago, when perhaps this view was first cemented in our minds.

```
filter(gapminder, year == 1962) %>%
  ggplot(aes(fertility, life_expectancy)) +
  geom_point()
```

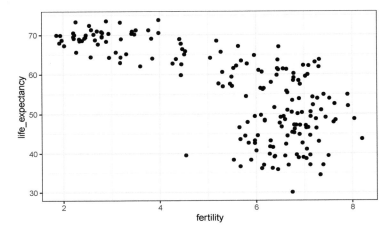

Most points fall into two distinct categories:

1. Life expectancy around 70 years and 3 or fewer children per family.
2. Life expectancy lower than 65 years and more than 5 children per family.

To confirm that indeed these countries are from the regions we expect, we can use color to represent continent.

```
filter(gapminder, year == 1962) %>%
  ggplot( aes(fertility, life_expectancy, color = continent)) +
  geom_point()
```

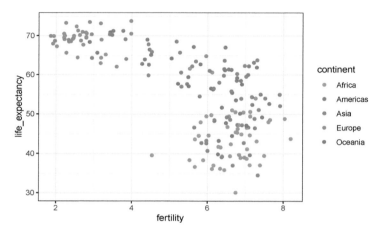

In 1962, "the West versus developing world" view was grounded in some reality. Is this still the case 50 years later?

9.3   Faceting

We could easily plot the 2012 data in the same way we did for 1962. To make comparisons, however, side by side plots are preferable. In **ggplot2**, we can achieve this by *faceting* variables: we stratify the data by some variable and make the same plot for each strata.

To achieve faceting, we add a layer with the function `facet_grid`, which automatically separates the plots. This function lets you facet by up to two variables using columns to represent one variable and rows to represent the other. The function expects the row and column variables to be separated by a ~. Here is an example of a scatterplot with `facet_grid` added as the last layer:

```
filter(gapminder, year%in%c(1962, 2012)) %>%
  ggplot(aes(fertility, life_expectancy, col = continent)) +
  geom_point() +
  facet_grid(continent~year)
```

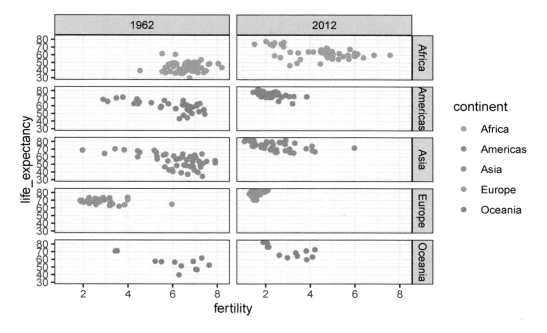

We see a plot for each continent/year pair. However, this is just an example and more than what we want, which is simply to compare 1962 and 2012. In this case, there is just one variable and we use . to let facet know that we are not using one of the variables:

```
filter(gapminder, year%in%c(1962, 2012)) %>%
  ggplot(aes(fertility, life_expectancy, col = continent)) +
  geom_point() +
  facet_grid(. ~ year)
```

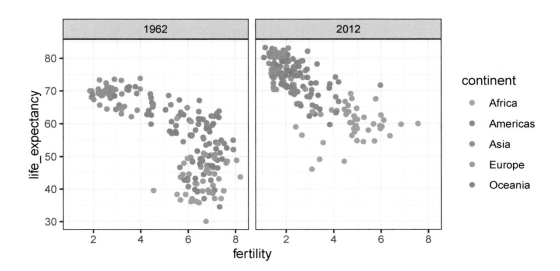

This plot clearly shows that the majority of countries have moved from the *developing world* cluster to the *western world* one. In 2012, the western versus developing world view no longer makes sense. This is particularly clear when comparing Europe to Asia, the latter of which includes several countries that have made great improvements.

9.3.1 `facet_wrap`

To explore how this transformation happened through the years, we can make the plot for several years. For example, we can add 1970, 1980, 1990, and 2000. If we do this, we will not want all the plots on the same row, the default behavior of `facet_grid`, since they will become too thin to show the data. Instead, we will want to use multiple rows and columns. The function `facet_wrap` permits us to do this by automatically wrapping the series of plots so that each display has viewable dimensions:

```
years <- c(1962, 1980, 1990, 2000, 2012)
continents <- c("Europe", "Asia")
gapminder %>%
    filter(year %in% years & continent %in% continents) %>%
    ggplot( aes(fertility, life_expectancy, col = continent)) +
    geom_point() +
    facet_wrap(~year)
```

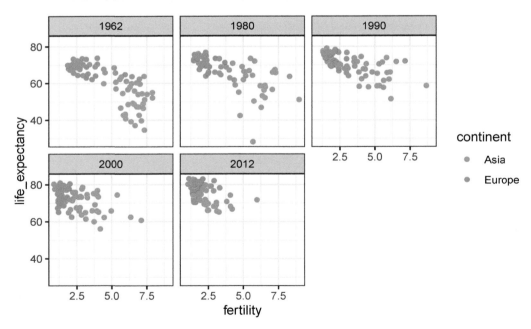

This plot clearly shows how most Asian countries have improved at a much faster rate than European ones.

9.3.2 Fixed scales for better comparisons

The default choice of the range of the axes is important. When not using `facet`, this range is determined by the data shown in the plot. When using `facet`, this range is determined by the data shown in all plots and therefore kept fixed across plots. This makes comparisons across plots much easier. For example, in the above plot, we can see that life expectancy has increased and the fertility has decreased across most countries. We see this because the cloud of points moves. This is not the case if we adjust the scales:

```
filter(gapminder, year%in%c(1962, 2012)) %>%
  ggplot(aes(fertility, life_expectancy, col = continent)) +
  geom_point() +
  facet_wrap(. ~ year, scales = "free")
```

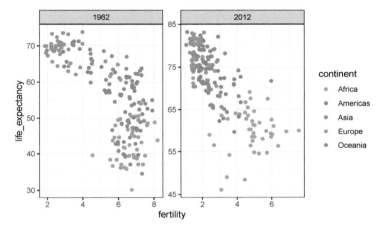

In the plot above, we have to pay special attention to the range to notice that the plot on the right has a larger life expectancy.

9.4 Time series plots

The visualizations above effectively illustrate that data no longer supports the western versus developing world view. Once we see these plots, new questions emerge. For example, which countries are improving more and which ones less? Was the improvement constant during the last 50 years or was it more accelerated during certain periods? For a closer look that may help answer these questions, we introduce *time series plots*.

Time series plots have time in the x-axis and an outcome or measurement of interest on the y-axis. For example, here is a trend plot of United States fertility rates:

```
gapminder %>%
  filter(country == "United States") %>%
  ggplot(aes(year, fertility)) +
  geom_point()
```

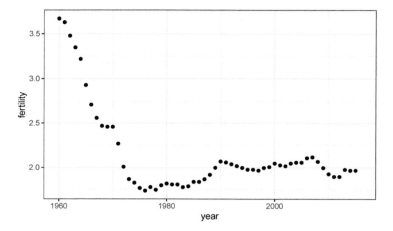

We see that the trend is not linear at all. Instead there is sharp drop during the 1960s and 1970s to below 2. Then the trend comes back to 2 and stabilizes during the 1990s.

When the points are regularly and densely spaced, as they are here, we create curves by joining the points with lines, to convey that these data are from a single series, here a country. To do this, we use the `geom_line` function instead of `geom_point`.

```
gapminder %>%
    filter(country == "United States") %>%
    ggplot(aes(year, fertility)) +
    geom_line()
```

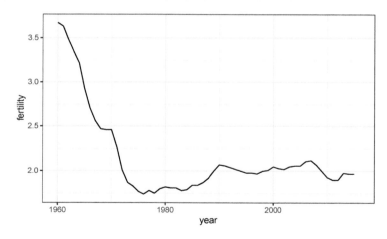

This is particularly helpful when we look at two countries. If we subset the data to include two countries, one from Europe and one from Asia, then adapt the code above:

```
countries <- c("South Korea","Germany")

gapminder %>% filter(country %in% countries) %>%
    ggplot(aes(year,fertility)) +
    geom_line()
```

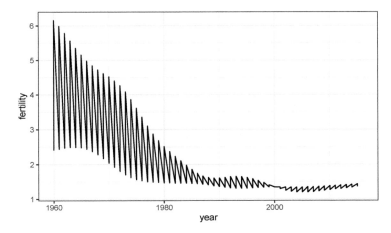

Unfortunately, this is **not** the plot that we want. Rather than a line for each country, the points for both countries are joined. This is actually expected since we have not told `ggplot` anything about wanting two separate lines. To let `ggplot` know that there are two curves that need to be made separately, we assign each point to a `group`, one for each country:

```
countries <- c("South Korea","Germany")
```

```
gapminder %>% filter(country %in% countries & !is.na(fertility)) %>%
  ggplot(aes(year, fertility, group = country)) +
  geom_line()
```

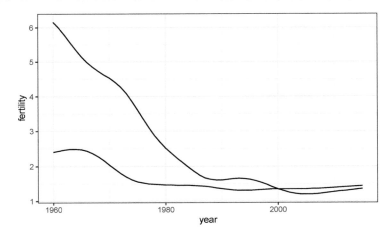

But which line goes with which country? We can assign colors to make this distinction. A useful side-effect of using the `color` argument to assign different colors to the different countries is that the data is automatically grouped:

```
countries <- c("South Korea","Germany")
```

```
gapminder %>% filter(country %in% countries & !is.na(fertility)) %>%
  ggplot(aes(year,fertility, col = country)) +
  geom_line()
```

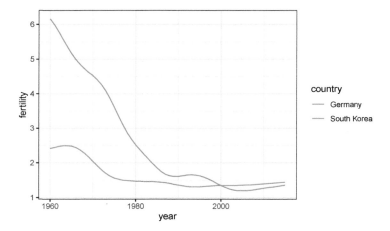

The plot clearly shows how South Korea's fertility rate dropped drastically during the 1960s and 1970s, and by 1990 had a similar rate to that of Germany.

9.4.1 Labels instead of legends

For trend plots we recommend labeling the lines rather than using legends since the viewer can quickly see which line is which country. This suggestion actually applies to most plots: labeling is usually preferred over legends.

We demonstrate how we can do this using the life expectancy data. We define a data table with the label locations and then use a second mapping just for these labels:

```
labels <- data.frame(country = countries, x = c(1975,1965), y = c(60,72))

gapminder %>%
  filter(country %in% countries) %>%
  ggplot(aes(year, life_expectancy, col = country)) +
  geom_line() +
  geom_text(data = labels, aes(x, y, label = country), size = 5) +
  theme(legend.position = "none")
```

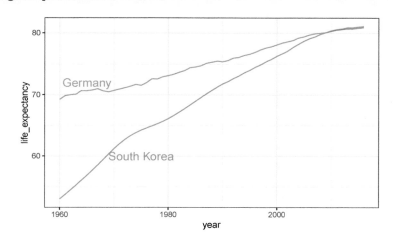

The plot clearly shows how an improvement in life expectancy followed the drops in fertility rates. In 1960, Germans lived 15 years longer than South Koreans, although by 2010 the gap is completely closed. It exemplifies the improvement that many non-western countries have achieved in the last 40 years.

9.5 Data transformations

We now shift our attention to the second question related to the commonly held notion that wealth distribution across the world has become worse during the last decades. When general audiences are asked if poor countries have become poorer and rich countries become richer, the majority answers yes. By using stratification, histograms, smooth densities, and boxplots, we will be able to understand if this is in fact the case. First we learn how transformations can sometimes help provide more informative summaries and plots.

The `gapminder` data table includes a column with the countries' gross domestic product (GDP). GDP measures the market value of goods and services produced by a country in a year. The GDP per person is often used as a rough summary of a country's wealth. Here we divide this quantity by 365 to obtain the more interpretable measure *dollars per day*. Using current US dollars as a unit, a person surviving on an income of less than $2 a day is defined to be living in *absolute poverty*. We add this variable to the data table:

```
gapminder <- gapminder %>%  mutate(dollars_per_day = gdp/population/365)
```

The GDP values are adjusted for inflation and represent current US dollars, so these values are meant to be comparable across the years. Of course, these are country averages and within each country there is much variability. All the graphs and insights described below relate to country averages and not to individuals.

9.5.1 Log transformation

Here is a histogram of per day incomes from 1970:

```
past_year <- 1970
gapminder %>%
  filter(year == past_year & !is.na(gdp)) %>%
  ggplot(aes(dollars_per_day)) +
  geom_histogram(binwidth = 1, color = "black")
```

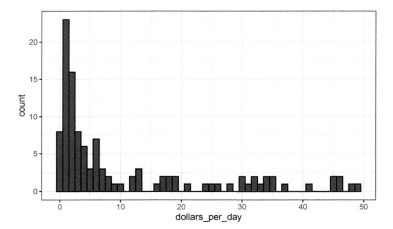

We use the `color = "black"` argument to draw a boundary and clearly distinguish the bins.

In this plot, we see that for the majority of countries, averages are below $10 a day. However, the majority of the x-axis is dedicated to the 35 countries with averages above $10. So the plot is not very informative about countries with values below $10 a day.

It might be more informative to quickly be able to see how many countries have average daily incomes of about $1 (extremely poor), $2 (very poor), $4 (poor), $8 (middle), $16 (well off), $32 (rich), $64 (very rich) per day. These changes are multiplicative and log transformations convert multiplicative changes into additive ones: when using base 2, a doubling of a value turns into an increase by 1.

Here is the distribution if we apply a log base 2 transform:

```
gapminder %>%
  filter(year == past_year & !is.na(gdp)) %>%
  ggplot(aes(log2(dollars_per_day))) +
  geom_histogram(binwidth = 1, color = "black")
```

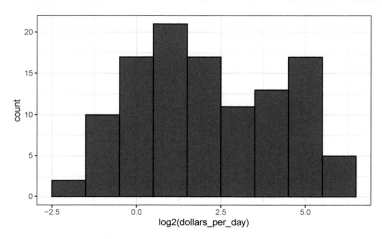

In a way this provides a *close-up* of the mid to lower income countries.

9.5.2 Which base?

In the case above, we used base 2 in the log transformations. Other common choices are base e (the natural log) and base 10.

In general, we do not recommend using the natural log for data exploration and visualization. This is because while $2^2, 2^3, 2^4, \ldots$ or $10^2, 10^3, \ldots$ are easy to compute in our heads, the same is not true for $e^2, e^3, \ldots$, so the scale is not intuitive or easy to interpret.

In the dollars per day example, we used base 2 instead of base 10 because the resulting range is easier to interpret. The range of the values being plotted is 0.327, 48.885.

In base 10, this turns into a range that includes very few integers: just 0 and 1. With base two, our range includes -2, -1, 0, 1, 2, 3, 4, and 5. It is easier to compute 2^x and 10^x when x is an integer and between -10 and 10, so we prefer to have smaller integers in the scale. Another consequence of a limited range is that choosing the binwidth is more challenging. With log base 2, we know that a binwidth of 1 will translate to a bin with range x to $2x$.

For an example in which base 10 makes more sense, consider population sizes. A log base 10 is preferable since the range for these is:

```
filter(gapminder, year == past_year) %>%
  summarize(min = min(population), max = max(population))
#>     min       max
#> 1 46075 8.09e+08
```

Here is the histogram of the transformed values:

```
gapminder %>%
  filter(year == past_year) %>%
  ggplot(aes(log10(population))) +
  geom_histogram(binwidth = 0.5, color = "black")
```

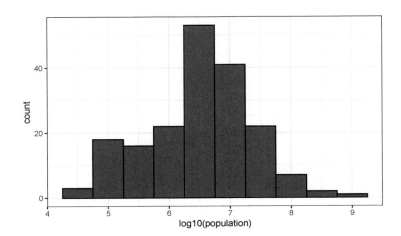

In the above, we quickly see that country populations range between ten thousand and ten billion.

9.5.3   Transform the values or the scale?

There are two ways we can use log transformations in plots. We can log the values before
plotting them or use log scales in the axes. Both approaches are useful and have different
strengths. If we log the data, we can more easily interpret intermediate values in the scale.
For example, if we see:

```
----1----x----2--------3----
```

for log transformed data, we know that the value of x is about 1.5. If the scales are logged:

```
----1----x----10------100---
```

then, to determine x, we need to compute $10^{1.5}$, which is not easy to do in our heads. The
advantage of using logged scales is that we see the original values on the axes. However,
the advantage of showing logged scales is that the original values are displayed in the plot,
which are easier to interpret. For example, we would see "32 dollars a day" instead of "5 log
base 2 dollars a day".

As we learned earlier, if we want to scale the axis with logs, we can use the
`scale_x_continuous` function. Instead of logging the values first, we apply this layer:

```
gapminder %>%
  filter(year == past_year & !is.na(gdp)) %>%
  ggplot(aes(dollars_per_day)) +
  geom_histogram(binwidth = 1, color = "black") +
  scale_x_continuous(trans = "log2")
```

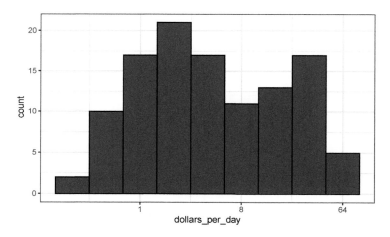

Note that the log base 10 transformation has its own function: `scale_x_log10()`, but
currently base 2 does not, although we could easily define our own.

There are other transformations available through the `trans` argument. As we learn later
on, the square root (`sqrt`) transformation is useful when considering counts. The logistic
transformation (`logit`) is useful when plotting proportions between 0 and 1. The `reverse`
transformation is useful when we want smaller values to be on the right or on top.

9.6 Visualizing multimodal distributions

In the histogram above we see two *bumps*: one at about 4 and another at about 32. In statistics these bumps are sometimes referred to as *modes*. The mode of a distribution is the value with the highest frequency. The mode of the normal distribution is the average. When a distribution, like the one above, doesn't monotonically decrease from the mode, we call the locations where it goes up and down again *local modes* and say that the distribution has *multiple modes*.

The histogram above suggests that the 1970 country income distribution has two modes: one at about 2 dollars per day (1 in the log 2 scale) and another at about 32 dollars per day (5 in the log 2 scale). This *bimodality* is consistent with a dichotomous world made up of countries with average incomes less than \$8 (3 in the log 2 scale) a day and countries above that.

9.7 Comparing multiple distributions with boxplots and ridge plots

A histogram showed us that the 1970 income distribution values show a dichotomy. However, the histogram does not show us if the two groups of countries are *west* versus the *developing* world.

Let's start by quickly examining the data by region. We reorder the regions by the median value and use a log scale.

```
gapminder %>%
  filter(year == past_year & !is.na(gdp)) %>%
  mutate(region = reorder(region, dollars_per_day, FUN = median)) %>%
  ggplot(aes(dollars_per_day, region)) +
  geom_point() +
  scale_x_continuous(trans = "log2")
```

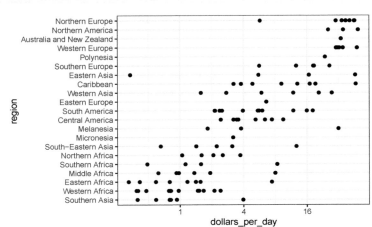

We can already see that there is indeed a "west versus the rest" dichotomy: we see two clear groups, with the rich group composed of North America, Northern and Western Europe, New Zealand and Australia. We define groups based on this observation:

```
gapminder <- gapminder %>%
  mutate(group = case_when(
    region %in% c("Western Europe", "Northern Europe","Southern Europe",
                    "Northern America",
                    "Australia and New Zealand") ~ "West",
    region %in% c("Eastern Asia", "South-Eastern Asia") ~ "East Asia",
    region %in% c("Caribbean", "Central America",
                    "South America") ~ "Latin America",
    continent == "Africa" &
      region != "Northern Africa" ~ "Sub-Saharan",
    TRUE ~ "Others"))
```

We turn this `group` variable into a factor to control the order of the levels:

```
gapminder <- gapminder %>%
  mutate(group = factor(group, levels = c("Others", "Latin America",
                                          "East Asia", "Sub-Saharan",
                                          "West")))
```

In the next section we demonstrate how to visualize and compare distributions across groups.

9.7.1   Boxplots

The exploratory data analysis above has revealed two characteristics about average income distribution in 1970. Using a histogram, we found a bimodal distribution with the modes relating to poor and rich countries. We now want to compare the distribution across these five groups to confirm the "west versus the rest" dichotomy. The number of points in each category is large enough that a summary plot may be useful. We could generate five histograms or five density plots, but it may be more practical to have all the visual summaries in one plot. We therefore start by stacking boxplots next to each other. Note that we add the layer `theme(axis.text.x = element_text(angle = 90, hjust = 1))` to turn the group labels vertical, since they do not fit if we show them horizontally, and remove the axis label to make space.

```
p <- gapminder %>%
  filter(year == past_year & !is.na(gdp)) %>%
  ggplot(aes(group, dollars_per_day)) +
  geom_boxplot() +
  scale_y_continuous(trans = "log2") +
  xlab("") +
  theme(axis.text.x = element_text(angle = 90, hjust = 1))
p
```

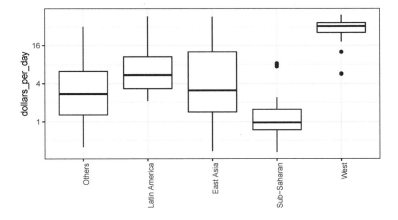

Boxplots have the limitation that by summarizing the data into five numbers, we might miss important characteristics of the data. One way to avoid this is by showing the data.

```
p + geom_point(alpha = 0.5)
```

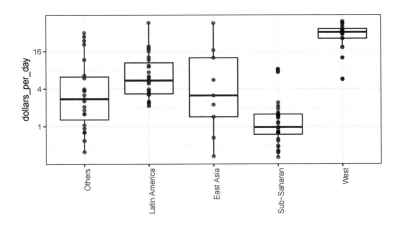

9.7.2 Ridge plots

Showing each individual point does not always reveal important characteristics of the distribution. Although not the case here, when the number of data points is so large that there is over-plotting, showing the data can be counterproductive. Boxplots help with this by providing a five-number summary, but this has limitations too. For example, boxplots will not permit us to discover bimodal distributions. To see this, note that the two plots below are summarizing the same dataset:

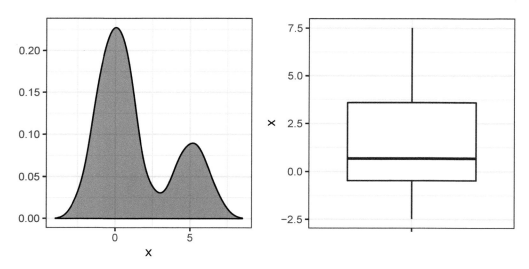

In cases in which we are concerned that the boxplot summary is too simplistic, we can show stacked smooth densities or histograms. We refer to these as *ridge plots*. Because we are used to visualizing densities with values in the x-axis, we stack them vertically. Also, because more space is needed in this approach, it is convenient to overlay them. The package **ggridges** provides a convenient function for doing this. Here is the income data shown above with boxplots but with a *ridge plot*.

```
library(ggridges)
p <- gapminder %>%
  filter(year == past_year & !is.na(dollars_per_day)) %>%
  ggplot(aes(dollars_per_day, group)) +
  scale_x_continuous(trans = "log2")
p  + geom_density_ridges()
```

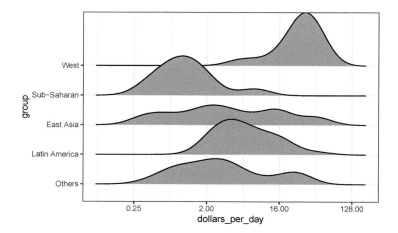

Note that we have to invert the x and y used for the boxplot. A useful `geom_density_ridges` parameter is `scale`, which lets you determine the amount of overlap, with `scale = 1` meaning no overlap and larger values resulting in more overlap.

If the number of data points is small enough, we can add them to the ridge plot using the following code:

```
p + geom_density_ridges(jittered_points = TRUE)
```

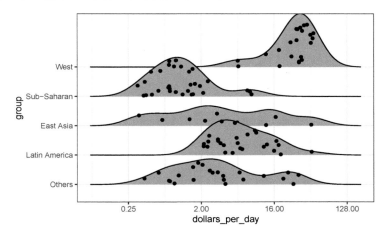

By default, the height of the points is jittered and should not be interpreted in any way. To show data points, but without using jitter we can use the following code to add what is referred to as a *rug representation* of the data.

```
p + geom_density_ridges(jittered_points = TRUE,
                        position = position_points_jitter(height = 0),
                        point_shape = '|', point_size = 3,
                        point_alpha = 1, alpha = 0.7)
```

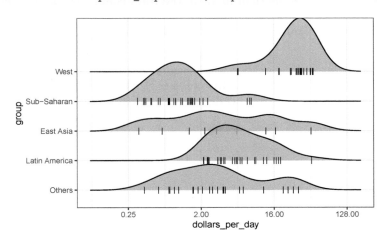

9.7.3 Example: 1970 versus 2010 income distributions

Data exploration clearly shows that in 1970 there was a "west versus the rest" dichotomy. But does this dichotomy persist? Let's use `facet_grid` see how the distributions have changed. To start, we will focus on two groups: the west and the rest. We make four histograms.

```
past_year <- 1970
present_year <- 2010
years <- c(past_year, present_year)
gapminder %>%
  filter(year %in% years & !is.na(gdp)) %>%
  mutate(west = ifelse(group == "West", "West", "Developing")) %>%
  ggplot(aes(dollars_per_day)) +
  geom_histogram(binwidth = 1, color = "black") +
  scale_x_continuous(trans = "log2") +
  facet_grid(year ~ west)
```

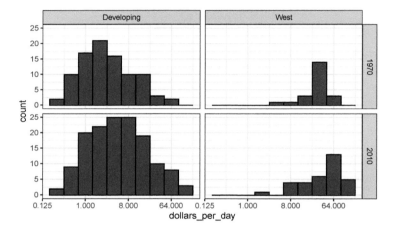

Before we interpret the findings of this plot, we notice that there are more countries represented in the 2010 histograms than in 1970: the total counts are larger. One reason for this is that several countries were founded after 1970. For example, the Soviet Union divided into several countries during the 1990s. Another reason is that data was available for more countries in 2010.

We remake the plots using only countries with data available for both years. In the data wrangling part of this book, we will learn **tidyverse** tools that permit us to write efficient code for this, but here we can use simple code using the `intersect` function:

```
country_list_1 <- gapminder %>%
  filter(year == past_year & !is.na(dollars_per_day)) %>%
  pull(country)

country_list_2 <- gapminder %>%
  filter(year == present_year & !is.na(dollars_per_day)) %>%
  pull(country)

country_list <- intersect(country_list_1, country_list_2)
```

These 108 account for 86% of the world population, so this subset should be representative.

Let's remake the plot, but only for this subset by simply adding `country %in% country_list` to the `filter` function:

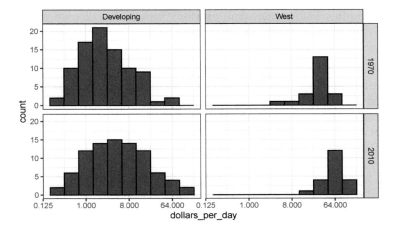

We now see that the rich countries have become a bit richer, but percentage-wise, the poor countries appear to have improved more. In particular, we see that the proportion of *developing* countries earning more than $16 a day increased substantially.

To see which specific regions improved the most, we can remake the boxplots we made above, but now adding the year 2010 and then using facet to compare the two years.

```
gapminder %>%
    filter(year %in% years & country %in% country_list) %>%
    ggplot(aes(group, dollars_per_day)) +
    geom_boxplot() +
    theme(axis.text.x = element_text(angle = 90, hjust = 1)) +
    scale_y_continuous(trans = "log2") +
    xlab("") +
    facet_grid(. ~ year)
```

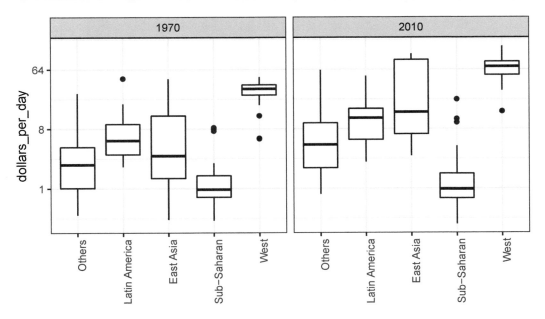

Here, we pause to introduce another powerful **ggplot2** feature. Because we want to compare each region before and after, it would be convenient to have the 1970 boxplot next to the 2010 boxplot for each region. In general, comparisons are easier when data are plotted next to each other.

So instead of faceting, we keep the data from each year together and ask to color (or fill) them depending on the year. Note that groups are automatically separated by year and each pair of boxplots drawn next to each other. Because year is a number, we turn it into a factor since **ggplot2** automatically assigns a color to each category of a factor. Note that we have to convert the year columns from numeric to factor.

```
gapminder %>%
    filter(year %in% years & country %in% country_list) %>%
    mutate(year = factor(year)) %>%
    ggplot(aes(group, dollars_per_day, fill = year)) +
    geom_boxplot() +
    theme(axis.text.x = element_text(angle = 90, hjust = 1)) +
    scale_y_continuous(trans = "log2") +
    xlab("")
```

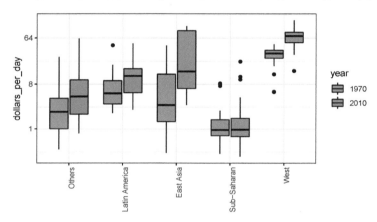

Finally, we point out that if what we are most interested in is comparing before and after values, it might make more sense to plot the percentage increases. We are still not ready to learn to code this, but here is what the plot would look like:

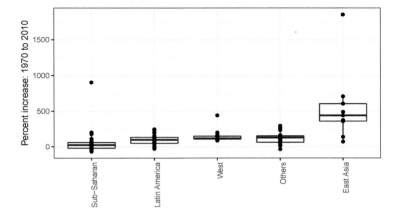

The previous data exploration suggested that the income gap between rich and poor countries has narrowed considerably during the last 40 years. We used a series of histograms and boxplots to see this. We suggest a succinct way to convey this message with just one plot.

Let's start by noting that density plots for income distribution in 1970 and 2010 deliver the message that the gap is closing:

```
gapminder %>%
    filter(year %in% years & country %in% country_list) %>%
    ggplot(aes(dollars_per_day)) +
    geom_density(fill = "grey") +
    scale_x_continuous(trans = "log2") +
    facet_grid(. ~ year)
```

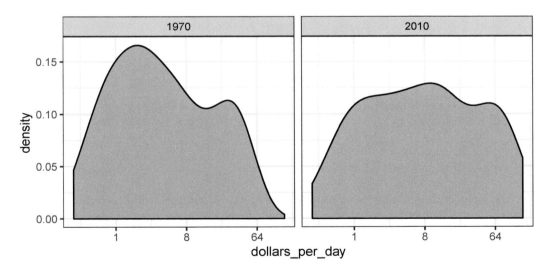

In the 1970 plot, we see two clear modes: poor and rich countries. In 2010, it appears that some of the poor countries have shifted towards the right, closing the gap.

The next message we need to convey is that the reason for this change in distribution is that several poor countries became richer, rather than some rich countries becoming poorer. To do this, we can assign a color to the groups we identified during data exploration.

However, we first need to learn how to make these smooth densities in a way that preserves information on the number of countries in each group. To understand why we need this, note the discrepancy in the size of each group:

Developing	West
87	21

But when we overlay two densities, the default is to have the area represented by each distribution add up to 1, regardless of the size of each group:

```
gapminder %>%
  filter(year %in% years & country %in% country_list) %>%
  mutate(group = ifelse(group == "West", "West", "Developing")) %>%
  ggplot(aes(dollars_per_day, fill = group)) +
  scale_x_continuous(trans = "log2") +
  geom_density(alpha = 0.2) +
  facet_grid(year ~ .)
```

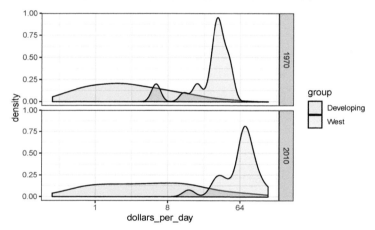

This makes it appear as if there are the same number of countries in each group. To change this, we will need to learn to access computed variables with `geom_density` function.

9.7.4 Accessing computed variables

To have the areas of these densities be proportional to the size of the groups, we can simply multiply the y-axis values by the size of the group. From the `geom_density` help file, we see that the functions compute a variable called `count` that does exactly this. We want this variable to be on the y-axis rather than the density.

In **ggplot2**, we access these variables by surrounding the name with two dots. We will therefore use the following mapping:

```
aes(x = dollars_per_day, y = ..count..)
```

We can now create the desired plot by simply changing the mapping in the previous code chunk. We will also expand the limits of the x-axis.

```
p <- gapminder %>%
  filter(year %in% years & country %in% country_list) %>%
  mutate(group = ifelse(group == "West", "West", "Developing")) %>%
  ggplot(aes(dollars_per_day, y = ..count.., fill = group)) +
  scale_x_continuous(trans = "log2", limit = c(0.125, 300))

p + geom_density(alpha = 0.2) +
  facet_grid(year ~ .)
```

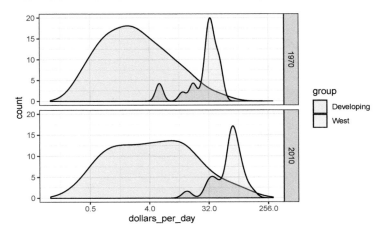

If we want the densities to be smoother, we use the `bw` argument so that the same bandwidth is used in each density. We selected 0.75 after trying out several values.

```
p + geom_density(alpha = 0.2, bw = 0.75) + facet_grid(year ~ .)
```

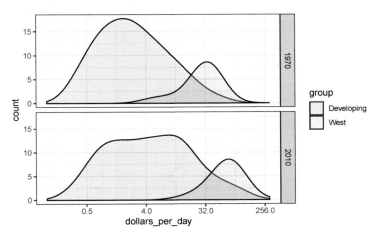

This plot now shows what is happening very clearly. The developing world distribution is changing. A third mode appears consisting of the countries that most narrowed the gap.

To visualize if any of the groups defined above are driving this we can quickly make a ridge plot:

```
gapminder %>%
    filter(year %in% years & !is.na(dollars_per_day)) %>%
    ggplot(aes(dollars_per_day, group)) +
    scale_x_continuous(trans = "log2") +
    geom_density_ridges(adjust = 1.5) +
    facet_grid(. ~ year)
```

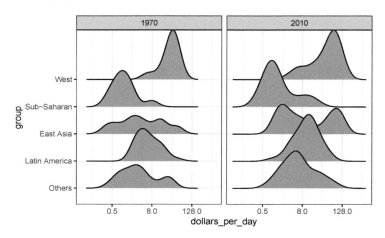

Another way to achieve this is by stacking the densities on top of each other:

```
gapminder %>%
    filter(year %in% years & country %in% country_list) %>%
    group_by(year) %>%
    mutate(weight = population/sum(population)*2) %>%
    ungroup() %>%
    ggplot(aes(dollars_per_day, fill = group)) +
    scale_x_continuous(trans = "log2", limit = c(0.125, 300)) +
    geom_density(alpha = 0.2, bw = 0.75, position = "stack") +
    facet_grid(year ~ .)
```

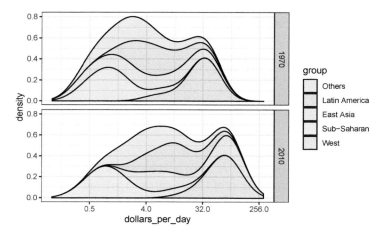

Here we can clearly see how the distributions for East Asia, Latin America, and others shift markedly to the right. While Sub-Saharan Africa remains stagnant.

Notice that we order the levels of the group so that the West's density is plotted first, then Sub-Saharan Africa. Having the two extremes plotted first allows us to see the remaining bimodality better.

9.7.5 Weighted densities

As a final point, we note that these distributions weigh every country the same. So if most of the population is improving, but living in a very large country, such as China, we might not appreciate this. We can actually weight the smooth densities using the `weight` mapping argument. The plot then looks like this:

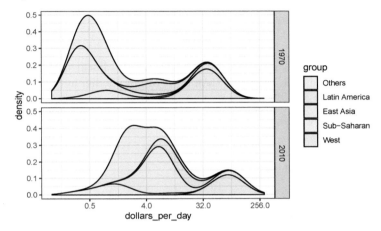

This particular figure shows very clearly how the income distribution gap is closing with most of the poor remaining in Sub-Saharan Africa.

9.8 The ecological fallacy and importance of showing the data

Throughout this section, we have been comparing regions of the world. We have seen that, on average, some regions do better than others. In this section, we focus on describing the importance of variability within the groups when examining the relationship between a country's infant mortality rates and average income.

We define a few more regions and compare the averages across regions:

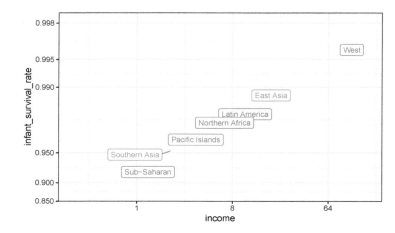

The relationship between these two variables is almost perfectly linear and the graph shows a dramatic difference. While in the West less than 0.5% of infants die, in Sub-Saharan Africa the rate is higher than 6%!

Note that the plot uses a new transformation, the logistic transformation.

9.8.1   Logistic transformation

The logistic or logit transformation for a proportion or rate p is defined as:

$$f(p) = \log\left(\frac{p}{1-p}\right)$$

When p is a proportion or probability, the quantity that is being logged, $p/(1-p)$, is called the *odds*. In this case p is the proportion of infants that survived. The odds tell us how many more infants are expected to survive than to die. The log transformation makes this symmetric. If the rates are the same, then the log odds is 0. Fold increases or decreases turn into positive and negative increments, respectively.

This scale is useful when we want to highlight differences near 0 or 1. For survival rates this is important because a survival rate of 90% is unacceptable, while a survival of 99% is relatively good. We would much prefer a survival rate closer to 99.9%. We want our scale to highlight these difference and the logit does this. Note that 99.9/0.1 is about 10 times bigger than 99/1 which is about 10 times larger than 90/10. By using the log, these fold changes turn into constant increases.

9.8.2   Show the data

Now, back to our plot. Based on the plot above, do we conclude that a country with a low income is destined to have low survival rate? Do we conclude that survival rates in Sub-Saharan Africa are all lower than in Southern Asia, which in turn are lower than in the Pacific Islands, and so on?

Jumping to this conclusion based on a plot showing averages is referred to as the *ecological fallacy*. The almost perfect relationship between survival rates and income is only observed for the averages at the region level. Once we show all the data, we see a somewhat more complicated story:

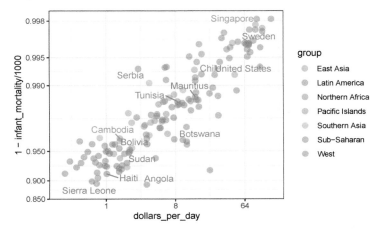

Specifically, we see that there is a large amount of variability. We see that countries from the same regions can be quite different and that countries with the same income can have different survival rates. For example, while on average Sub-Saharan Africa had the worse health and economic outcomes, there is wide variability within that group. Mauritius and Botswana are doing better than Angola and Sierra Leone, with Mauritius comparable to Western countries.

10

Data visualization principles

We have already provided some rules to follow as we created plots for our examples. Here, we aim to provide some general principles we can use as a guide for effective data visualization. Much of this section is based on a talk by Karl Broman[1] titled "Creating Effective Figures and Tables"[2] and includes some of the figures which were made with code that Karl makes available on his GitHub repository[3], as well as class notes from Peter Aldhous' Introduction to Data Visualization course[4]. Following Karl's approach, we show some examples of plot styles we should avoid, explain how to improve them, and use these as motivation for a list of principles. We compare and contrast plots that follow these principles to those that don't.

The principles are mostly based on research related to how humans detect patterns and make visual comparisons. The preferred approaches are those that best fit the way our brains process visual information. When deciding on a visualization approach, it is also important to keep our goal in mind. We may be comparing a viewable number of quantities, describing distributions for categories or numeric values, comparing the data from two groups, or describing the relationship between two variables. As a final note, we want to emphasize that for a data scientist it is important to adapt and optimize graphs to the audience. For example, an exploratory plot made for ourselves will be different than a chart intended to communicate a finding to a general audience.

We will be using these libraries:

```
library(tidyverse)
library(dslabs)
library(gridExtra)
```

10.1 Encoding data using visual cues

We start by describing some principles for encoding data. There are several approaches at our disposal including position, aligned lengths, angles, area, brightness, and color hue.

To illustrate how some of these strategies compare, let's suppose we want to report the results from two hypothetical polls regarding browser preference taken in 2000 and then 2015. For each year, we are simply comparing four quantities – the four percentages. A widely used graphical representation of percentages, popularized by Microsoft Excel, is the pie chart:

---

[1] http://kbroman.org/
[2] https://www.biostat.wisc.edu/~kbroman/presentations/graphs2017.pdf
[3] https://github.com/kbroman/Talk_Graphs
[4] http://paldhous.github.io/ucb/2016/dataviz/index.html

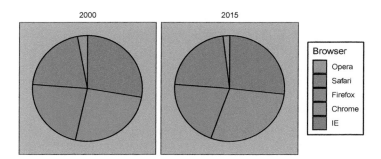

Here we are representing quantities with both areas and angles, since both the angle and area of each pie slice are proportional to the quantity the slice represents. This turns out to be a sub-optimal choice since, as demonstrated by perception studies, humans are not good at precisely quantifying angles and are even worse when area is the only available visual cue. The donut chart is an example of a plot that uses only area:

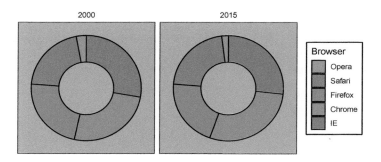

To see how hard it is to quantify angles and area, note that the rankings and all the percentages in the plots above changed from 2000 to 2015. Can you determine the actual percentages and rank the browsers' popularity? Can you see how the percentages changed from 2000 to 2015? It is not easy to tell from the plot. In fact, the `pie` R function help file states that:

> Pie charts are a very bad way of displaying information. The eye is good at judging linear measures and bad at judging relative areas. A bar chart or dot chart is a preferable way of displaying this type of data.

In this case, simply showing the numbers is not only clearer, but would also save on printing costs if printing a paper copy:

Browser	2000	2015
Opera	3	2
Safari	21	22
Firefox	23	21
Chrome	26	29
IE	28	27

The preferred way to plot these quantities is to use length and position as visual cues, since humans are much better at judging linear measures. The barplot uses this approach by using bars of length proportional to the quantities of interest. By adding horizontal lines at strategically chosen values, in this case at every multiple of 10, we ease the visual burden of quantifying through the position of the top of the bars. Compare and contrast the information we can extract from the two figures.

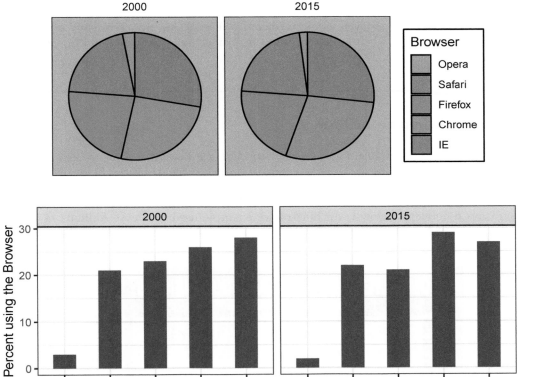

Notice how much easier it is to see the differences in the barplot. In fact, we can now determine the actual percentages by following a horizontal line to the x-axis.

If for some reason you need to make a pie chart, label each pie slice with its respective percentage so viewers do not have to infer them from the angles or area:

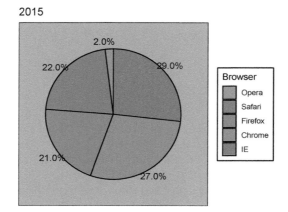

In general, when displaying quantities, position and length are preferred over angles and/or area. Brightness and color are even harder to quantify than angles. But, as we will see later, they are sometimes useful when more than two dimensions must be displayed at once.

10.2 Know when to include 0

When using barplots, it is misinformative not to start the bars at 0. This is because, by using a barplot, we are implying the length is proportional to the quantities being displayed. By avoiding 0, relatively small differences can be made to look much bigger than they actually are. This approach is often used by politicians or media organizations trying to exaggerate a difference. Below is an illustrative example used by Peter Aldhous in this lecture: http://paldhous.github.io/ucb/2016/dataviz/week2.html.

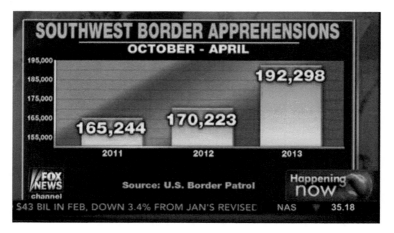

(Source: Fox News, via Media Matters[5].)

---

[5]http://mediamatters.org/blog/2013/04/05/fox-news-newest-dishonest-chart-immigration-enf/193507

From the plot above, it appears that apprehensions have almost tripled when, in fact, they have only increased by about 16%. Starting the graph at 0 illustrates this clearly:

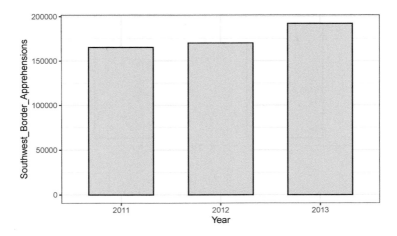

Here is another example, described in detail in a Flowing Data blog post:

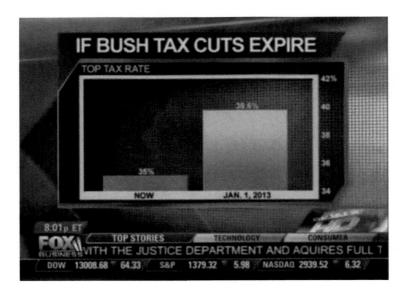

(Source: Fox News, via Flowing Data[6].)

This plot makes a 13% increase look like a five fold change. Here is the appropriate plot:

[6]http://flowingdata.com/2012/08/06/fox-news-continues-charting-excellence/

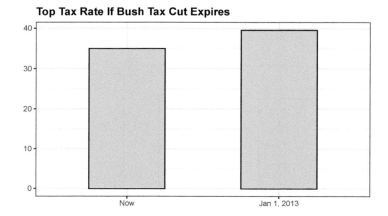

Finally, here is an extreme example that makes a very small difference of under 2% look like a 10-100 fold change:

(Source: Venezolana de Televisión via Pakistan Today[7] and Diego Mariano.)

Here is the appropriate plot:

[7]https://www.pakistantoday.com.pk/2018/05/18/whats-at-stake-in-venezuelan-presidential-vote

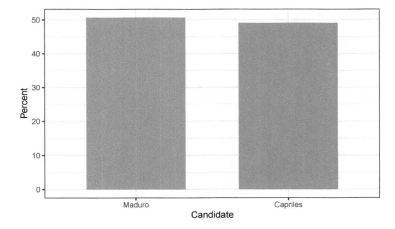

When using position rather than length, it is then not necessary to include 0. This is particularly the case when we want to compare differences between groups relative to the within-group variability. Here is an illustrative example showing country average life expectancy stratified across continents in 2012:

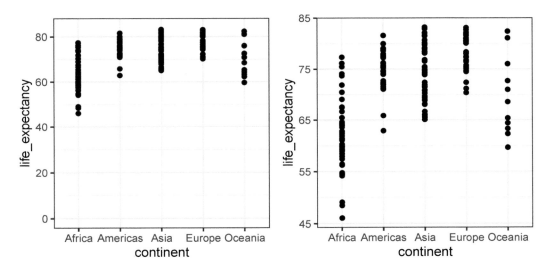

Note that in the plot on the left, which includes 0, the space between 0 and 43 adds no information and makes it harder to compare the between and within group variability.

10.3 Do not distort quantities

During President Barack Obama's 2011 State of the Union Address, the following chart was used to compare the US GDP to the GDP of four competing nations:

(Source: The 2011 State of the Union Address[8])

Judging by the area of the circles, the US appears to have an economy over five times larger than China's and over 30 times larger than France's. However, if we look at the actual numbers, we see that this is not the case. The actual ratios are 2.6 and 5.8 times bigger than China and France, respectively. The reason for this distortion is that the radius, rather than the area, was made to be proportional to the quantity, which implies that the proportion between the areas is squared: 2.6 turns into 6.5 and 5.8 turns into 34.1. Here is a comparison of the circles we get if we make the value proportional to the radius and to the area:

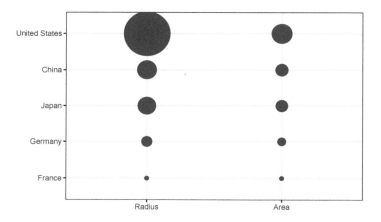

Not surprisingly, **ggplot2** defaults to using area rather than radius. Of course, in this case, we really should not be using area at all since we can use position and length:

---

[8]https://www.youtube.com/watch?v=kl2g40GoRxg

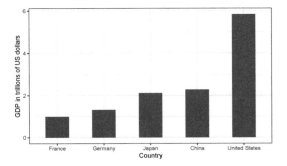

10.4 Order categories by a meaningful value

When one of the axes is used to show categories, as is done in barplots, the default **ggplot2** behavior is to order the categories alphabetically when they are defined by character strings. If they are defined by factors, they are ordered by the factor levels. We rarely want to use alphabetical order. Instead, we should order by a meaningful quantity. In all the cases above, the barplots were ordered by the values being displayed. The exception was the graph showing barplots comparing browsers. In this case, we kept the order the same across the barplots to ease the comparison. Specifically, instead of ordering the browsers separately in the two years, we ordered both years by the average value of 2000 and 2015.

We previously learned how to use the `reorder` function, which helps us achieve this goal. To appreciate how the right order can help convey a message, suppose we want to create a plot to compare the murder rate across states. We are particularly interested in the most dangerous and safest states. Note the difference when we order alphabetically (the default) versus when we order by the actual rate:

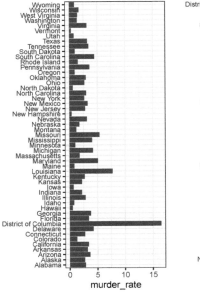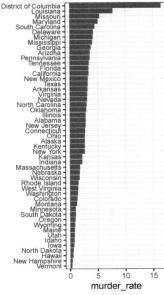

We can make the second plot like this:

```
data(murders)
murders %>% mutate(murder_rate = total / population * 100000) %>%
  mutate(state = reorder(state, murder_rate)) %>%
  ggplot(aes(state, murder_rate)) +
  geom_bar(stat="identity") +
  coord_flip() +
  theme(axis.text.y = element_text(size = 6)) +
  xlab("")
```

The `reorder` function lets us reorder groups as well. Earlier we saw an example related to income distributions across regions. Here are the two versions plotted against each other:

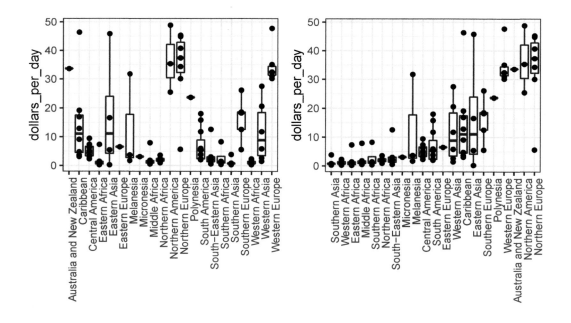

The first orders the regions alphabetically, while the second orders them by the group's median.

10.5 Show the data

We have focused on displaying single quantities across categories. We now shift our attention to displaying data, with a focus on comparing groups.

To motivate our first principle, "show the data", we go back to our artificial example of describing heights to ET, an extraterrestrial. This time let's assume ET is interested in the difference in heights between males and females. A commonly seen plot used for comparisons between groups, popularized by software such as Microsoft Excel, is the dynamite plot, which shows the average and standard errors (standard errors are defined in a later chapter, but do not confuse them with the standard deviation of the data). The plot looks like this:

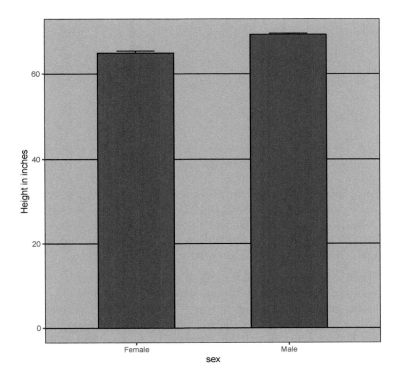

The average of each group is represented by the top of each bar and the antennae extend out from the average to the average plus two standard errors. If all ET receives is this plot, he will have little information on what to expect if he meets a group of human males and females. The bars go to 0: does this mean there are tiny humans measuring less than one foot? Are all males taller than the tallest females? Is there a range of heights? ET can't answer these questions since we have provided almost no information on the height distribution.

This brings us to our first principle: show the data. This simple **ggplot2** code already generates a more informative plot than the barplot by simply showing all the data points:

```
heights %>%
  ggplot(aes(sex, height)) +
  geom_point()
```

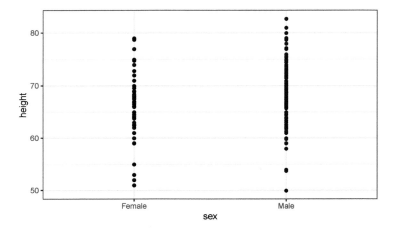

For example, this plot gives us an idea of the range of the data. However, this plot has limitations as well, since we can't really see all the 238 and 812 points plotted for females and males, respectively, and many points are plotted on top of each other. As we have previously described, visualizing the distribution is much more informative. But before doing this, we point out two ways we can improve a plot showing all the points.

The first is to add *jitter*, which adds a small random shift to each point. In this case, adding horizontal jitter does not alter the interpretation, since the point heights do not change, but we minimize the number of points that fall on top of each other and, therefore, get a better visual sense of how the data is distributed. A second improvement comes from using *alpha blending*: making the points somewhat transparent. The more points fall on top of each other, the darker the plot, which also helps us get a sense of how the points are distributed. Here is the same plot with jitter and alpha blending:

```
heights %>%
  ggplot(aes(sex, height)) +
  geom_jitter(width = 0.1, alpha = 0.2)
```

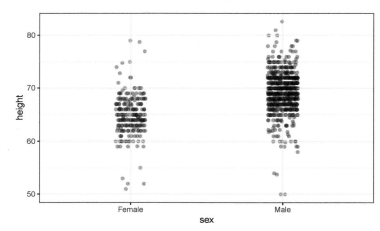

Now we start getting a sense that, on average, males are taller than females. We also note dark horizontal bands of points, demonstrating that many report values that are rounded to the nearest integer.

10.6   Ease comparisons

10.6.1   Use common axes

Since there are so many points, it is more effective to show distributions rather than individual points. We therefore show histograms for each group:

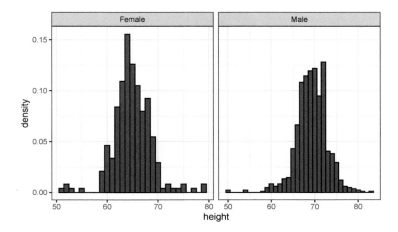

However, from this plot it is not immediately obvious that males are, on average, taller than females. We have to look carefully to notice that the x-axis has a higher range of values in the male histogram. An important principle here is to **keep the axes the same** when comparing data across two plots. Below we see how the comparison becomes easier:

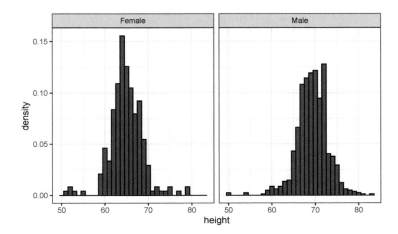

10.6.2 Align plots vertically to see horizontal changes and horizontally to see vertical changes

In these histograms, the visual cue related to decreases or increases in height are shifts to the left or right, respectively: horizontal changes. Aligning the plots vertically helps us see this change when the axes are fixed:

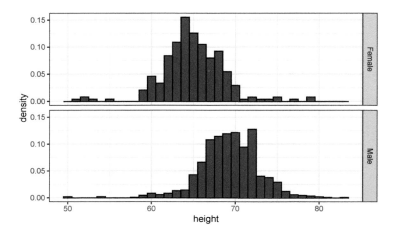

```
heights %>%
  ggplot(aes(height, ..density..)) +
  geom_histogram(binwidth = 1, color="black") +
  facet_grid(sex~.)
```

This plot makes it much easier to notice that men are, on average, taller.

If , we want the more compact summary provided by boxplots, we then align them horizontally since, by default, boxplots move up and down with changes in height. Following our *show the data* principle, we then overlay all the data points:

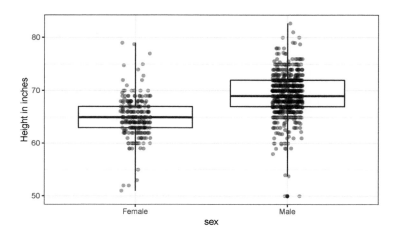

```
heights %>%
  ggplot(aes(sex, height)) +
  geom_boxplot(coef=3) +
  geom_jitter(width = 0.1, alpha = 0.2) +
  ylab("Height in inches")
```

Now contrast and compare these three plots, based on exactly the same data:

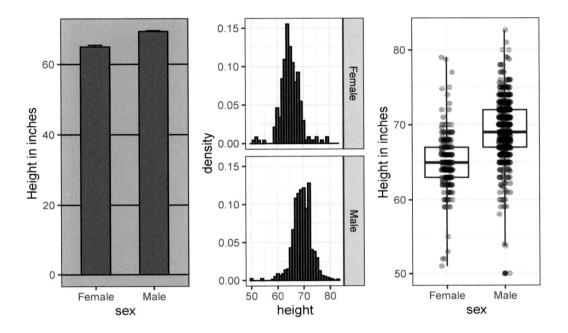

Notice how much more we learn from the two plots on the right. Barplots are useful for showing one number, but not very useful when we want to describe distributions.

10.6.3 Consider transformations

We have motivated the use of the log transformation in cases where the changes are multiplicative. Population size was an example in which we found a log transformation to yield a more informative transformation.

The combination of an incorrectly chosen barplot and a failure to use a log transformation when one is merited can be particularly distorting. As an example, consider this barplot showing the average population sizes for each continent in 2015:

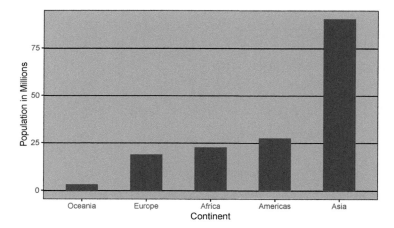

From this plot, one would conclude that countries in Asia are much more populous than in other continents. Following the *show the data* principle, we quickly notice that this is due to two very large countries, which we assume are India and China:

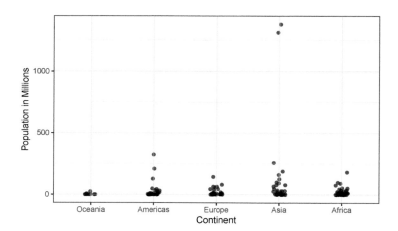

Using a log transformation here provides a much more informative plot. We compare the original barplot to a boxplot using the log scale transformation for the y-axis:

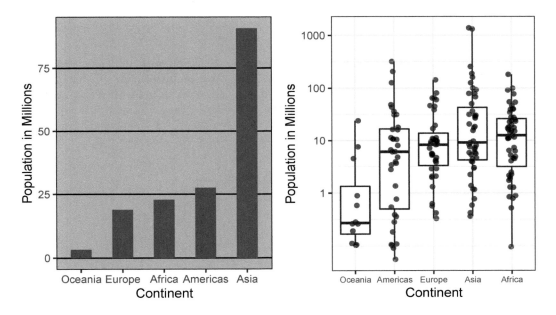

With the new plot, we realize that countries in Africa actually have a larger median population size than those in Asia.

Other transformations you should consider are the logistic transformation (`logit`), useful to better see fold changes in odds, and the square root transformation (`sqrt`), useful for count data.

10.6.4   Visual cues to be compared should be adjacent

For each continent, let's compare income in 1970 versus 2010. When comparing income data across regions between 1970 and 2010, we made a figure similar to the one below, but this time we investigate continents rather than regions.

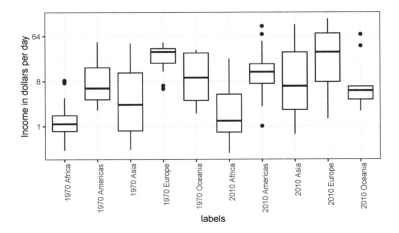

The default in **ggplot2** is to order labels alphabetically so the labels with 1970 come before the labels with 2010, making the comparisons challenging because a continent's distribution in 1970 is visually far from its distribution in 2010. It is much easier to make the comparison between 1970 and 2010 for each continent when the boxplots for that continent are next to each other:

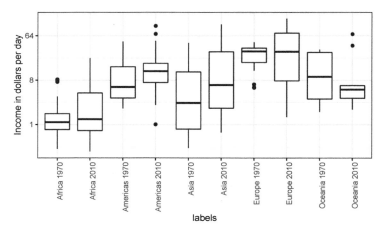

10.6.5 Use color

The comparison becomes even easier to make if we use color to denote the two things we want to compare:

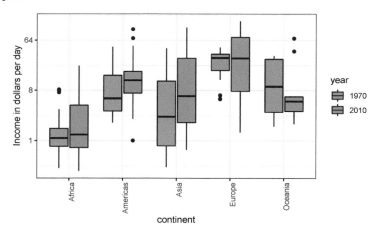

10.7 Think of the color blind

About 10% of the population is color blind. Unfortunately, the default colors used in **ggplot2** are not optimal for this group. However, **ggplot2** does make it easy to change the color palette used in the plots. An example of how we can use a color blind friendly palette is described here: http://www.cookbook-r.com/Graphs/Colors_(ggplot2)/#a-colorblind-friendly-palette:

```
color_blind_friendly_cols <-
  c("#999999", "#E69F00", "#56B4E9", "#009E73",
    "#F0E442", "#0072B2", "#D55E00", "#CC79A7")
```

Here are the colors

There are several resources that can help you select colors, for example this one: http://bconnelly.net/2013/10/creating-colorblind-friendly-figures/.

10.8 Plots for two variables

In general, you should use scatterplots to visualize the relationship between two variables. In every single instance in which we have examined the relationship between two variables, including total murders versus population size, life expectancy versus fertility rates, and infant mortality versus income, we have used scatterplots. This is the plot we generally recommend. However, there are some exceptions and we describe two alternative plots here: the *slope chart* and the *Bland-Altman plot*.

10.8.1 Slope charts

One exception where another type of plot may be more informative is when you are comparing variables of the same type, but at different time points and for a relatively small number of comparisons. For example, comparing life expectancy between 2010 and 2015. In this case, we might recommend a *slope chart*.

There is no geometry for slope charts in **ggplot2**, but we can construct one using `geom_line`. We need to do some tinkering to add labels. Below is an example comparing 2010 to 2015 for large western countries:

```
west <- c("Western Europe","Northern Europe","Southern Europe",
          "Northern America","Australia and New Zealand")

dat <- gapminder %>%
  filter(year%in% c(2010, 2015) & region %in% west &
           !is.na(life_expectancy) & population > 10^7)

dat %>%
  mutate(location = ifelse(year == 2010, 1, 2),
         location = ifelse(year == 2015 &
```

```
                                country %in% c("United Kingdom", "Portugal"),
                          location+0.22, location),
            hjust = ifelse(year == 2010, 1, 0)) %>%
  mutate(year = as.factor(year)) %>%
  ggplot(aes(year, life_expectancy, group = country)) +
  geom_line(aes(color = country), show.legend = FALSE) +
  geom_text(aes(x = location, label = country, hjust = hjust),
            show.legend = FALSE) +
  xlab("") + ylab("Life Expectancy")
```

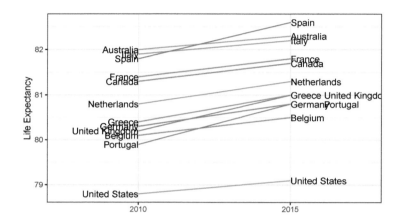

An advantage of the slope chart is that it permits us to quickly get an idea of changes based on the slope of the lines. Although we are using angle as the visual cue, we also have position to determine the exact values. Comparing the improvements is a bit harder with a scatterplot:

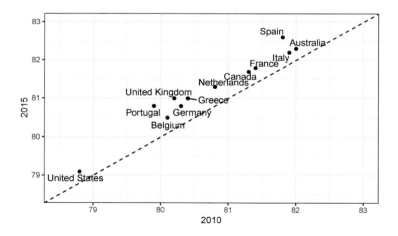

In the scatterplot, we have followed the principle *use common axes* since we are comparing these before and after. However, if we have many points, slope charts stop being useful as it becomes hard to see all the lines.

10.8.2   Bland-Altman plot

Since we are primarily interested in the difference, it makes sense to dedicate one of our axes to it. The Bland-Altman plot, also known as the Tukey mean-difference plot and the MA-plot, shows the difference versus the average:

```
library(ggrepel)
dat %>%
  mutate(year = paste0("life_expectancy_", year)) %>%
  select(country, year, life_expectancy) %>%
  spread(year, life_expectancy) %>%
  mutate(average = (life_expectancy_2015 + life_expectancy_2010)/2,
         difference = life_expectancy_2015 - life_expectancy_2010) %>%
  ggplot(aes(average, difference, label = country)) +
  geom_point() +
  geom_text_repel() +
  geom_abline(lty = 2) +
  xlab("Average of 2010 and 2015") +
  ylab("Difference between 2015 and 2010")
```

Here, by simply looking at the y-axis, we quickly see which countries have shown the most improvement. We also get an idea of the overall value from the x-axis.

10.9   Encoding a third variable

An earlier scatterplot showed the relationship between infant survival and average income. Below is a version of this plot that encodes three variables: OPEC membership, region, and population.

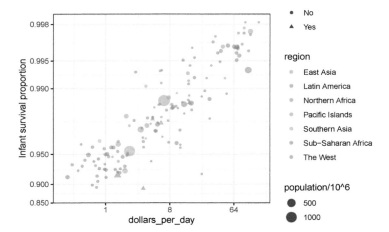

We encode categorical variables with color and shape. These shapes can be controlled with **shape** argument. Below are the shapes available for use in R. For the last five, the color goes inside.

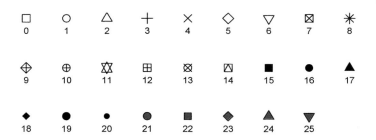

For continuous variables, we can use color, intensity, or size. We now show an example of how we do this with a case study.

When selecting colors to quantify a numeric variable, we choose between two options: sequential and diverging. Sequential colors are suited for data that goes from high to low. High values are clearly distinguished from low values. Here are some examples offered by the package RColorBrewer:

```
library(RColorBrewer)
display.brewer.all(type="seq")
```

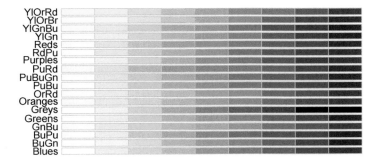

Diverging colors are used to represent values that diverge from a center. We put equal emphasis on both ends of the data range: higher than the center and lower than the center. An example of when we would use a divergent pattern would be if we were to show height in standard deviations away from the average. Here are some examples of divergent patterns:

```
library(RColorBrewer)
display.brewer.all(type="div")
```

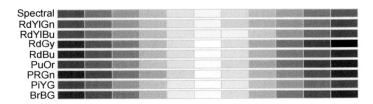

10.10 Avoid pseudo-three-dimensional plots

The figure below, taken from the scientific literature[9], shows three variables: dose, drug type and survival. Although your screen/book page is flat and two-dimensional, the plot tries to imitate three dimensions and assigned a dimension to each variable.

---

[9]https://projecteuclid.org/download/pdf_1/euclid.ss/1177010488

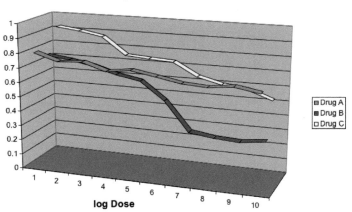

(Image courtesy of Karl Broman)

Humans are not good at seeing in three dimensions (which explains why it is hard to parallel park) and our limitation is even worse with regard to pseudo-three-dimensions. To see this, try to determine the values of the survival variable in the plot above. Can you tell when the purple ribbon intersects the red one? This is an example in which we can easily use color to represent the categorical variable instead of using a pseudo-3D:

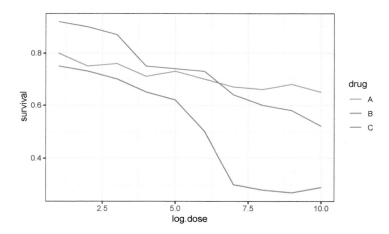

Notice how much easier it is to determine the survival values.

Pseudo-3D is sometimes used completely gratuitously: plots are made to look 3D even when the 3rd dimension does not represent a quantity. This only adds confusion and makes it harder to relay your message. Here are two examples:

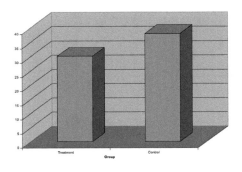

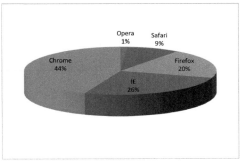

(Images courtesy of Karl Broman)

10.11 Avoid too many significant digits

By default, statistical software like R returns many significant digits. The default behavior in R is to show 7 significant digits. That many digits often adds no information and the added visual clutter can make it hard for the viewer to understand the message. As an example, here are the per 10,000 disease rates, computed from totals and population in R, for California across the five decades:

state	year	Measles	Pertussis	Polio
California	1940	37.8826320	18.3397861	18.3397861
California	1950	13.9124205	4.7467350	4.7467350
California	1960	14.1386471	0.0000000	0.0000000
California	1970	0.9767889	0.0000000	0.0000000
California	1980	0.3743467	0.0515466	0.0515466

We are reporting precision up to 0.00001 cases per 10,000, a very small value in the context of the changes that are occurring across the dates. In this case, two significant figures is more than enough and clearly makes the point that rates are decreasing:

state	year	Measles	Pertussis	Polio
California	1940	37.9	18.3	18.3
California	1950	13.9	4.7	4.7
California	1960	14.1	0.0	0.0
California	1970	1.0	0.0	0.0
California	1980	0.4	0.1	0.1

Useful ways to change the number of significant digits or to round numbers are `signif` and `round`. You can define the number of significant digits globally by setting options like this: `options(digits = 3)`.

Another principle related to displaying tables is to place values being compared on columns rather than rows. Note that our table above is easier to read than this one:

state	disease	1940	1950	1960	1970	1980
California	Measles	37.9	13.9	14.1	1	0.4
California	Pertussis	18.3	4.7	0.0	0	0.1
California	Polio	18.3	4.7	0.0	0	0.1

10.12   Know your audience

Graphs can be used for 1) our own exploratory data analysis, 2) to convey a message to experts, or 3) to help tell a story to a general audience. Make sure that the intended audience understands each element of the plot.

As a simple example, consider that for your own exploration it may be more useful to log-transform data and then plot it. However, for a general audience that is unfamiliar with converting logged values back to the original measurements, using a log-scale for the axis instead of log-transformed values will be much easier to digest.

10.13   Exercises

For these exercises, we will be using the vaccines data in the **dslabs** package:

```
library(dslabs)
data(us_contagious_diseases)
```

1. Pie charts are appropriate:

    a. When we want to display percentages.
    b. When **ggplot2** is not available.
    c. When I am in a bakery.
    d. Never. Barplots and tables are always better.

2. What is the problem with the plot below:

Results of Presidential Election 2016

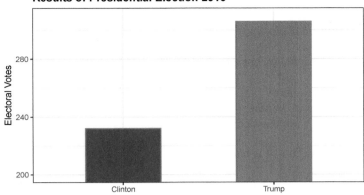

a. The values are wrong. The final vote was 306 to 232.
b. The axis does not start at 0. Judging by the length, it appears Trump received 3 times as many votes when, in fact, it was about 30% more.
c. The colors should be the same.
d. Percentages should be shown as a pie chart.

3. Take a look at the following two plots. They show the same information: 1928 rates of measles across the 50 states.

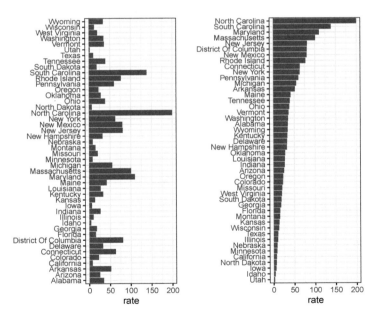

Which plot is easier to read if you are interested in determining which are the best and worst states in terms of rates, and why?

a. They provide the same information, so they are both equally as good.
b. The plot on the right is better because it orders the states alphabetically.
c. The plot on the right is better because alphabetical order has nothing to do with

the disease and by ordering according to actual rate, we quickly see the states
with most and least rates.

  d. Both plots should be a pie chart.

4. To make the plot on the left, we have to reorder the levels of the states' variables.

```
dat <- us_contagious_diseases %>%
  filter(year == 1967 & disease=="Measles" & !is.na(population)) %>%
  mutate(rate = count / population * 10000 * 52 / weeks_reporting)
```

Note what happens when we make a barplot:

```
dat %>% ggplot(aes(state, rate)) +
  geom_bar(stat="identity") +
  coord_flip()
```

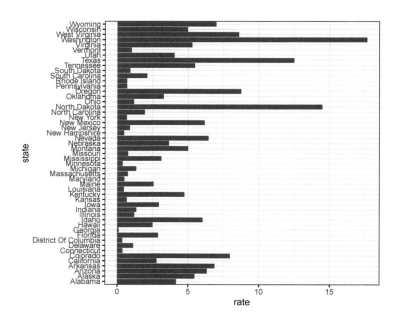

Define these objects:

```
state <- dat$state
rate <- dat$count/dat$population*10000*52/dat$weeks_reporting
```

Redefine the `state` object so that the levels are re-ordered. Print the new object `state` and
its levels so you can see that the vector is not re-ordered by the levels.

5. Now with one line of code, define the `dat` table as done above, but change the use mutate
to create a rate variable and re-order the state variable so that the levels are re-ordered by
this variable. Then make a barplot using the code above, but for this new `dat`.

6. Say we are interested in comparing gun homicide rates across regions of the US. We see
this plot:

```
library(dslabs)
data("murders")
murders %>% mutate(rate = total/population*100000) %>%
group_by(region) %>%
summarize(avg = mean(rate)) %>%
mutate(region = factor(region)) %>%
ggplot(aes(region, avg)) +
geom_bar(stat="identity") +
ylab("Murder Rate Average")
```

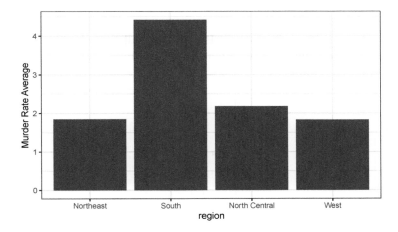

and decide to move to a state in the western region. What is the main problem with this interpretation?

   a. The categories are ordered alphabetically.
   b. The graph does not show standarad errors.
   c. It does not show all the data. We do not see the variability within a region and it's possible that the safest states are not in the West.
   d. The Northeast has the lowest average.

7. Make a boxplot of the murder rates defined as

```
data("murders")
murders %>% mutate(rate = total/population*100000)
```

by region, showing all the points and ordering the regions by their median rate.

8. The plots below show three continuous variables.

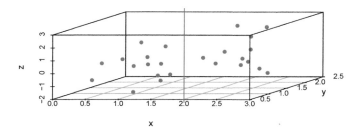

The line $x = 2$ appears to separate the points. But it is actually not the case, which we can see by plotting the data in a couple of two-dimensional points.

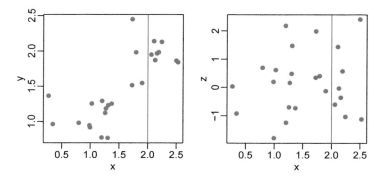

Why is this happening?

    a.   Humans are not good at reading pseudo-3D plots.
    b.   There must be an error in the code.
    c.   The colors confuse us.
    d.   Scatterplots should not be used to compare two variables when we have access to 3.

9. Reproduce the image plot we previously made but for smallpox. For this plot, do not include years in which cases were not reported in 10 or more weeks.

10. Now reproduce the time series plot we previously made, but this time following the instructions of the previous question.

11. For the state of California, make time series plots showing rates for all diseases. Include only years with 10 or more weeks reporting. Use a different color for each disease.

12. Now do the same for the rates for the US. Hint: compute the US rate by using summarize, the total divided by total population.

10.14   Case study: vaccines and infectious diseases

Vaccines have helped save millions of lives. In the 19th century, before herd immunization was achieved through vaccination programs, deaths from infectious diseases, such as smallpox and polio, were common. However, today vaccination programs have become somewhat controversial despite all the scientific evidence for their importance.

The controversy started with a paper[10] published in 1988 and led by Andrew Wakefield[11] claiming there was a link between the administration of the measles, mumps, and rubella (MMR) vaccine and the appearance of autism and bowel disease. Despite much scientific evidence contradicting this finding, sensationalist media reports and fear-mongering from conspiracy theorists led parts of the public into believing that vaccines were harmful. As a result, many parents ceased to vaccinate their children. This dangerous practice can be potentially disastrous given that the Centers for Disease Control (CDC) estimates that vaccinations will prevent more than 21 million hospitalizations and 732,000 deaths among children born in the last 20 years (see Benefits from Immunization during the Vaccines for Children Program Era — United States, 1994-2013, MMWR[12]). The 1988 paper has since been retracted[10] and Andrew Wakefield was eventually "struck off the UK medical register, with a statement identifying deliberate falsification in the research published in The Lancet, and was thereby barred from practicing medicine in the UK." (source: Wikipedia[11]). Yet misconceptions persist, in part due to self-proclaimed activists who continue to disseminate misinformation about vaccines.

Effective communication of data is a strong antidote to misinformation and fear-mongering. Earlier we used an example provided by a Wall Street Journal article[13] showing data related to the impact of vaccines on battling infectious diseases. Here we reconstruct that example.

The data used for these plots were collected, organized, and distributed by the Tycho Project[14]. They include weekly reported counts for seven diseases from 1928 to 2011, from all fifty states. We include the yearly totals in the **dslabs** package:

```
library(tidyverse)
library(RColorBrewer)
library(dslabs)
data(us_contagious_diseases)
names(us_contagious_diseases)
#> [1] "disease"         "state"          "year"
#> [4] "weeks_reporting" "count"          "population"
```

We create a temporary object `dat` that stores only the measles data, includes a per 100,000 rate, orders states by average value of disease and removes Alaska and Hawaii since they only became states in the late 1950s. Note that there is a `weeks_reporting` column that

---

[10] http://www.thelancet.com/journals/lancet/article/PIIS0140-6736(97)11096-0/abstract

[11] https://en.wikipedia.org/wiki/Andrew_Wakefield

[12] https://www.cdc.gov/mmwr/preview/mmwrhtml/mm6316a4.htm

[13] http://graphics.wsj.com/infectious-diseases-and-vaccines/

[14] http://www.tycho.pitt.edu/

tells us for how many weeks of the year data was reported. We have to adjust for that value when computing the rate.

```
the_disease <- "Measles"
dat <- us_contagious_diseases %>%
  filter(!state%in%c("Hawaii","Alaska") & disease == the_disease) %>%
  mutate(rate = count / population * 10000 * 52 / weeks_reporting) %>%
  mutate(state = reorder(state, rate))
```

We can now easily plot disease rates per year. Here are the measles data from California:

```
dat %>% filter(state == "California" & !is.na(rate)) %>%
  ggplot(aes(year, rate)) +
  geom_line() +
  ylab("Cases per 10,000")  +
  geom_vline(xintercept=1963, col = "blue")
```

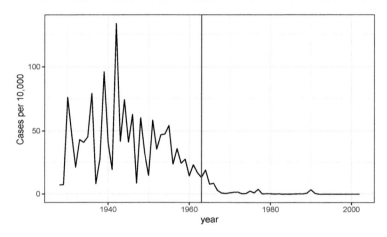

We add a vertical line at 1963 since this is when the vaccine was introduced [Control, Centers for Disease; Prevention (2014). CDC health information for international travel 2014 (the yellow book). p. 250. ISBN 9780199948505].

Now can we show data for all states in one plot? We have three variables to show: year, state, and rate. In the WSJ figure, they use the x-axis for year, the y-axis for state, and color hue to represent rates. However, the color scale they use, which goes from yellow to blue to green to orange to red, can be improved.

In our example, we want to use a sequential palette since there is no meaningful center, just low and high rates.

We use the geometry `geom_tile` to tile the region with colors representing disease rates. We use a square root transformation to avoid having the really high counts dominate the plot. Notice that missing values are shown in grey. Note that once a disease was pretty much eradicated, some states stopped reporting cases all together. This is why we see so much grey after 1980.

```
dat %>% ggplot(aes(year, state, fill = rate)) +
  geom_tile(color = "grey50") +
  scale_x_continuous(expand=c(0,0)) +
```

```
scale_fill_gradientn(colors = brewer.pal(9, "Reds"), trans = "sqrt") +
geom_vline(xintercept=1963, col = "blue") +
theme_minimal() +
theme(panel.grid = element_blank(),
      legend.position="bottom",
      text = element_text(size = 8)) +
ggtitle(the_disease) +
ylab("") + xlab("")
```

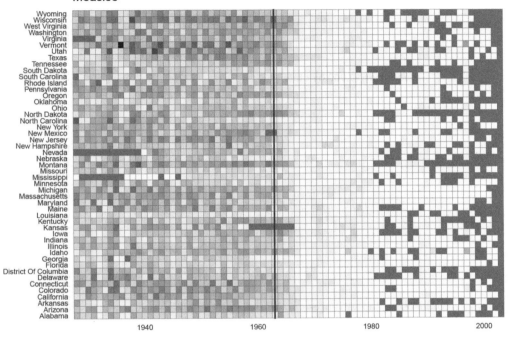

This plot makes a very striking argument for the contribution of vaccines. However, one limitation of this plot is that it uses color to represent quantity, which we earlier explained makes it harder to know exactly how high values are going. Position and lengths are better cues. If we are willing to lose state information, we can make a version of the plot that shows the values with position. We can also show the average for the US, which we compute like this:

```
avg <- us_contagious_diseases %>%
  filter(disease==the_disease) %>% group_by(year) %>%
  summarize(us_rate = sum(count, na.rm = TRUE) /
            sum(population, na.rm = TRUE) * 10000)
```

Now to make the plot we simply use the `geom_line` geometry:

```
dat %>%
  filter(!is.na(rate)) %>%
    ggplot() +
  geom_line(aes(year, rate, group = state),  color = "grey50",
            show.legend = FALSE, alpha = 0.2, size = 1) +
  geom_line(mapping = aes(year, us_rate),  data = avg, size = 1) +
  scale_y_continuous(trans = "sqrt", breaks = c(5, 25, 125, 300)) +
  ggtitle("Cases per 10,000 by state") +
  xlab("") + ylab("") +
  geom_text(data = data.frame(x = 1955, y = 50),
            mapping = aes(x, y, label="US average"),
            color="black") +
  geom_vline(xintercept=1963, col = "blue")
```

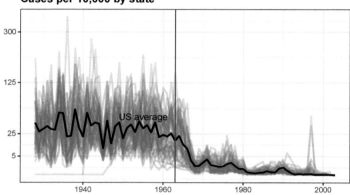

In theory, we could use color to represent the categorical value state, but it is hard to pick 50 distinct colors.

10.15   Exercises

1. Reproduce the image plot we previously made but for smallpox. For this plot, do not include years in which cases were not reported in 10 or more weeks.

2. Now reproduce the time series plot we previously made, but this time following the instructions of the previous question for smallpox.

3. For the state of California, make a time series plot showing rates for all diseases. Include only years with 10 or more weeks reporting. Use a different color for each disease.

4. Now do the same for the rates for the US. Hint: compute the US rate by using summarize: the total divided by total population.

11

Robust summaries

11.1 Outliers

We previously described how boxplots show *outliers*, but we did not provide a precise definition. Here we discuss outliers, approaches that can help detect them, and summaries that take into account their presence.

Outliers are very common in data science. Data recording can be complex and it is common to observe data points generated in error. For example, an old monitoring device may read out nonsensical measurements before completely failing. Human error is also a source of outliers, in particular when data entry is done manually. An individual, for instance, may mistakenly enter their height in centimeters instead of inches or put the decimal in the wrong place.

How do we distinguish an outlier from measurements that were too big or too small simply due to expected variability? This is not always an easy question to answer, but we try to provide some guidance. Let's begin with a simple case.

Suppose a colleague is charged with collecting demography data for a group of males. The data report height in feet and are stored in the object:

```
library(tidyverse)
library(dslabs)
data(outlier_example)
str(outlier_example)
#>  num [1:500] 5.59 5.8 5.54 6.15 5.83 5.54 5.87 5.93 5.89 5.67 ...
```

Our colleague uses the fact that heights are usually well approximated by a normal distribution and summarizes the data with average and standard deviation:

```
mean(outlier_example)
#> [1] 6.1
sd(outlier_example)
#> [1] 7.8
```

and writes a report on the interesting fact that this group of males is much taller than usual. The average height is over six feet tall! Using your data science skills, however, you notice something else that is unexpected: the standard deviation is over 7 feet. Adding and subtracting two standard deviations, you note that 95% of this population will have heights between -9.489, 21.697 feet, which does not make sense. A quick plot reveals the problem:

```
boxplot(outlier_example)
```

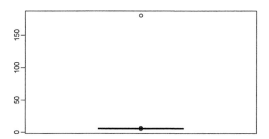

There appears to be at least one value that is nonsensical, since we know that a height of 180 feet is impossible. The boxplot detects this point as an outlier.

11.2   Median

When we have an outlier like this, the average can become very large. Mathematically, we can make the average as large as we want by simply changing one number: with 500 data points, we can increase the average by any amount Δ by adding $\Delta \times 500$ to a single number. The median, defined as the value for which half the values are smaller and the other half are bigger, is robust to such outliers. No matter how large we make the largest point, the median remains the same.

With this data the median is:

```
median(outlier_example)
#> [1] 5.74
```

which is about 5 feet and 9 inches.

The median is what boxplots display as a horizontal line.

11.3   The inter quartile range (IQR)

The box in boxplots is defined by the first and third quartile. These are meant to provide an idea of the variability in the data: 50% of the data is within this range. The difference between the 3rd and 1st quartile (or 75th and 25th percentiles) is referred to as the inter quartile range (IQR). As is the case with the median, this quantity will be robust to outliers as large values do not affect it. We can do some math to see that for normally distributed data, the IQR / 1.349 approximates the standard deviation of the data had an outlier not been present. We can see that this works well in our example since we get a standard deviation estimate of:

```
IQR(outlier_example) / 1.349
#> [1] 0.245
```

which is about 3 inches.

11.4   Tukey's definition of an outlier

In R, points falling outside the whiskers of the boxplot are referred to as *outliers*. This definition of outlier was introduced by Tukey. The top whisker ends at the 75th percentile plus $1.5 \times$ IQR. Similarly the bottom whisker ends at the 25th percentile minus $1.5\times$ IQR. If we define the first and third quartiles as Q_1 and Q_3, respectively, then an outlier is anything outside the range:

$$[Q_1 - 1.5 \times (Q_3 - Q1), Q_3 + 1.5 \times (Q_3 - Q1)].$$

When the data is normally distributed, the standard units of these values are:

```
q3 <- qnorm(0.75)
q1 <- qnorm(0.25)
iqr <- q3 - q1
r <- c(q1 - 1.5*iqr, q3 + 1.5*iqr)
r
#> [1] -2.7  2.7
```

Using the **pnorm** function, we see that 99.3% of the data falls in this interval.

Keep in mind that this is not such an extreme event: if we have 1000 data points that are normally distributed, we expect to see about 7 outside of this range. But these would not be outliers since we expect to see them under the typical variation.

If we want an outlier to be rarer, we can increase the 1.5 to a larger number. Tukey also used 3 and called these *far out* outliers. With a normal distribution, 100% of the data falls in this interval. This translates into about 2 in a million chance of being outside the range. In the **geom_boxplot** function, this can be controlled by the **outlier.size** argument, which defaults to 1.5.

The 180 inches measurement is well beyond the range of the height data:

```
max_height <- quantile(outlier_example, 0.75) + 3*IQR(outlier_example)
max_height
#>   75%
#> 6.91
```

If we take this value out, we can see that the data is in fact normally distributed as expected:

```
x <- outlier_example[outlier_example < max_height]
qqnorm(x)
qqline(x)
```

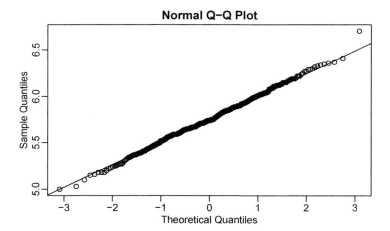

11.5 Median absolute deviation

Another way to robustly estimate the standard deviation in the presence of outliers is to use the median absolute deviation (MAD). To compute the MAD, we first compute the median, and then for each value we compute the distance between that value and the median. The MAD is defined as the median of these distances. For technical reasons not discussed here, this quantity needs to be multiplied by 1.4826 to assure it approximates the actual standard deviation. The `mad` function already incorporates this correction. For the height data, we get a MAD of:

```
mad(outlier_example)
#> [1] 0.237
```

which is about 3 inches.

11.6 Exercises

We are going to use the **HistData** package. If it is not installed you can install it like this:

```
install.packages("HistData")
```

Load the height data set and create a vector x with just the male heights used in Galton's data on the heights of parents and their children from his historic research on heredity.

```
library(HistData)
data(Galton)
x <- Galton$child
```

1. Compute the average and median of these data.

2. Compute the median and median absolute deviation of these data.

3. Now suppose Galton made a mistake when entering the first value and forgot to use the decimal point. You can imitate this error by typing:

```
x_with_error <- x
x_with_error[1] <- x_with_error[1]*10
```

How many inches does the average grow after this mistake?

4. How many inches does the SD grow after this mistake?

5. How many inches does the median grow after this mistake?

6. How many inches does the MAD grow after this mistake?

7. How could you use exploratory data analysis to detect that an error was made?

    a. Since it is only one value out of many, we will not be able to detect this.
    b. We would see an obvious shift in the distribution.
    c. A boxplot, histogram, or qq-plot would reveal a clear outlier.
    d. A scatterplot would show high levels of measurement error.

8. How much can the average accidentally grow with mistakes like this? Write a function called `error_avg` that takes a value k and returns the average of the vector x after the first entry changed to k. Show the results for k=10000 and k=-10000.

11.7   Case study: self-reported student heights

The heights we have been looking at are not the original heights reported by students. The original reported heights are also included in the **dslabs** package and can be loaded like this:

```
library(dslabs)
data("reported_heights")
```

Height is a character vector so we create a new column with the numeric version:

```
reported_heights <- reported_heights %>%
  mutate(original_heights = height, height = as.numeric(height))
#> Warning: NAs introduced by coercion
```

Note that we get a warning about NAs. This is because some of the self reported heights were not numbers. We can see why we get these:

```
reported_heights %>% filter(is.na(height)) %>%  head()
#>               time_stamp    sex height original_heights
#> 1 2014-09-02 15:16:28   Male     NA             5' 4"
#> 2 2014-09-02 15:16:37 Female     NA             165cm
#> 3 2014-09-02 15:16:52   Male     NA               5'7
#> 4 2014-09-02 15:16:56   Male     NA             >9000
#> 5 2014-09-02 15:16:56   Male     NA             5'7"
#> 6 2014-09-02 15:17:09 Female     NA             5'3"
```

Some students self-reported their heights using feet and inches rather than just inches.
Others used centimeters and others were just trolling. For now we will remove these entries:

```
reported_heights <- filter(reported_heights, !is.na(height))
```

If we compute the average and standard deviation, we notice that we obtain strange results.
The average and standard deviation are different from the median and MAD:

```
reported_heights %>%
  group_by(sex) %>%
  summarize(average = mean(height), sd = sd(height),
            median = median(height), MAD = mad(height))
#> # A tibble: 2 x 5
#>   sex     average    sd median    MAD
#>   <chr>     <dbl> <dbl>  <dbl>  <dbl>
#> 1 Female     63.4  27.9   64.2   4.05
#> 2 Male      103.  530.    70     4.45
```

This suggests that we have outliers, which is confirmed by creating a boxplot:

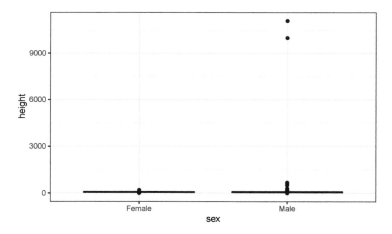

We can see some rather extreme values. To see what these values are, we can quickly look at
the largest values using the **arrange** function:

```
reported_heights %>% arrange(desc(height)) %>% top_n(10, height)
#>               time_stamp    sex height original_heights
#> 1  2014-09-03 23:55:37   Male  11111            11111
```

```
#> 2  2016-04-10 22:45:49   Male  10000              10000
#> 3  2015-08-10 03:10:01   Male    684                684
#> 4  2015-02-27 18:05:06   Male    612                612
#> 5  2014-09-02 15:16:41   Male    511                511
#> 6  2014-09-07 20:53:43   Male    300                300
#> 7  2014-11-28 12:18:40   Male    214                214
#> 8  2017-04-03 16:16:57   Male    210                210
#> 9  2015-11-24 10:39:45   Male    192                192
#> 10 2014-12-26 10:00:12   Male    190                190
#> 11 2016-11-06 10:21:02 Female    190                190
```

The first seven entries look like strange errors. However, the next few look like they were entered as centimeters instead of inches. Since 184 cm is equivalent to six feet tall, we suspect that 184 was actually meant to be 72 inches.

We can review all the nonsensical answers by looking at the data considered to be *far out* by Tukey:

```
whisker <- 3*IQR(reported_heights$height)
max_height <- quantile(reported_heights$height, .75) + whisker
min_height <- quantile(reported_heights$height, .25) - whisker
reported_heights %>%
  filter(!between(height, min_height, max_height)) %>%
  select(original_heights) %>%
  head(n=10) %>% pull(original_heights)
#>  [1] "6"      "5.3"    "511"    "6"      "2"      "5.25"   "5.5"    "11111"
#>  [9] "6"      "6.5"
```

Examining these heights carefully, we see two common mistakes: entries in centimeters, which turn out to be too large, and entries of the form x.y with x and y representing feet and inches, respectively, which turn out to be too small. Some of the even smaller values, such as 1.6, could be entries in meters.

In the Data Wrangling part of this book we will learn techniques for correcting these values and converting them into inches. Here we were able to detect this problem using careful data exploration to uncover issues with the data: the first step in the great majority of data science projects.

Part III

Statistics with R

12

Introduction to statistics with R

Data analysis is one of the main focuses of this book. While the computing tools we have introduced are relatively recent developments, data analysis has been around for over a century. Throughout the years, data analysts working on specific projects have come up with ideas and concepts that generalize across many applications. They have also identified common ways to get fooled by apparent patterns in the data and important mathematical realities that are not immediately obvious. The accumulation of these ideas and insights has given rise to the discipline of statistics, which provides a mathematical framework that greatly facilitates the description and formal evaluation of these ideas.

To avoid repeating common mistakes and wasting time reinventing the wheel, it is important for a data analyst to have an in-depth understanding of statistics. However, due to the maturity of the discipline, there are dozens of excellent books already published on this topic and we therefore do not focus on describing the mathematical framework here. Instead, we introduce concepts briefly and then provide detailed case studies demonstrating how statistics is used in data analysis along with R code implementing these ideas. We also use R code to help elucidate some of the main statistical concepts that are usually described using mathematics. We highly recommend complementing this part of the book with a basic statistics textbook. Two examples are *Statistics* by Freedman, Pisani, and Purves and *Statistical Inference* by Casella and Berger. The specific concepts covered in this part are Probability, Statistical Inference, Statistical Models, Regression, and Linear Models, which are major topics covered in a statistics course. The case studies we present relate to the financial crisis, forecasting election results, understanding heredity, and building a baseball team.

13

Probability

In games of chance, probability has a very intuitive definition. For instance, we know what it means that the chance of a pair of dice coming up seven is 1 in 6. However, this is not the case in other contexts. Today probability theory is being used much more broadly with the word *probability* commonly used in everyday language. Google's auto-complete of "What are the chances of" give us: "having twins", "rain today", "getting struck by lightning", and "getting cancer". One of the goals of this part of the book is to help us understand how probability is useful to understand and describe real-world events when performing data analysis.

Because knowing how to compute probabilities gives you an edge in games of chance, throughout history many smart individuals, including famous mathematicians such as Cardano, Fermat, and Pascal, spent time and energy thinking through the math of these games. As a result, Probability Theory was born. Probability continues to be highly useful in modern games of chance. For example, in poker, we can compute the probability of winning a hand based on the cards on the table. Also, casinos rely on probability theory to develop games that almost certainly guarantee a profit.

Probability theory is useful in many other contexts and, in particular, in areas that depend on data affected by chance in some way. All of the other chapters in this part build upon probability theory. Knowledge of probability is therefore indispensable for data science.

13.1 Discrete probability

We start by covering some basic principles related to categorical data. The subset of probability is referred to as *discrete probability*. It will help us understand the probability theory we will later introduce for numeric and continuous data, which is much more common in data science applications. Discrete probability is more useful in card games and therefore we use these as examples.

13.1.1 Relative frequency

The word probability is used in everyday language. Answering questions about probability is often hard, if not impossible. Here we discuss a mathematical definition of *probability* that does permit us to give precise answers to certain questions.

For example, if I have 2 red beads and 3 blue beads inside an urn[1] (most probability books use this archaic term, so we do too) and I pick one at random, what is the probability of

---

[1] https://en.wikipedia.org/wiki/Urn_problem

picking a red one? Our intuition tells us that the answer is 2/5 or 40%. A precise definition can be given by noting that there are five possible outcomes of which two satisfy the condition necessary for the event "pick a red bead". Since each of the five outcomes has the same chance of occurring, we conclude that the probability is .4 for red and .6 for blue.

A more tangible way to think about the probability of an event is as the proportion of times the event occurs when we repeat the experiment an infinite number of times, independently, and under the same conditions.

13.1.2   Notation

We use the notation $\Pr(A)$ to denote the probability of event A happening. We use the very general term *event* to refer to things that can happen when something occurs by chance. In our previous example, the event was "picking a red bead". In a political poll in which we call 100 likely voters at random, an example of an event is "calling 48 Democrats and 52 Republicans".

In data science applications, we will often deal with continuous variables. These events will often be things like "is this person taller than 6 feet". In this case, we write events in a more mathematical form: $X \geq 6$. We will see more of these examples later. Here we focus on categorical data.

13.1.3   Probability distributions

If we know the relative frequency of the different categories, defining a distribution for categorical outcomes is relatively straightforward. We simply assign a probability to each category. In cases that can be thought of as beads in an urn, for each bead type, their proportion defines the distribution.

If we are randomly calling likely voters from a population that is 44% Democrat, 44% Republican, 10% undecided, and 2% Green Party, these proportions define the probability for each group. The probability distribution is:

$\Pr(\text{picking a Republican})$	$=$	0.44
$\Pr(\text{picking a Democrat})$	$=$	0.44
$\Pr(\text{picking an undecided})$	$=$	0.10
$\Pr(\text{picking a Green})$	$=$	0.02

13.2   Monte Carlo simulations for categorical data

Computers provide a way to actually perform the simple random experiment described above: pick a bead at random from a bag that contains three blue beads and two red ones. Random number generators permit us to mimic the process of picking at random.

An example is the `sample` function in R. We demonstrate its use in the code below. First, we use the function `rep` to generate the urn:

```
beads <- rep(c("red", "blue"), times = c(2,3))
beads
#> [1] "red"  "red"  "blue" "blue" "blue"
```

and then use `sample` to pick a bead at random:

```
sample(beads, 1)
#> [1] "blue"
```

This line of code produces one random outcome. We want to repeat this experiment an infinite number of times, but it is impossible to repeat forever. Instead, we repeat the experiment a large enough number of times to make the results practically equivalent to repeating forever. **This is an example of a *Monte Carlo* simulation.**

Much of what mathematical and theoretical statisticians study, which we do not cover in this book, relates to providing rigorous definitions of "practically equivalent" as well as studying how close a large number of experiments gets us to what happens in the limit. Later in this section, we provide a practical approach to deciding what is "large enough".

To perform our first Monte Carlo simulation, we use the `replicate` function, which permits us to repeat the same task any number of times. Here, we repeat the random event $B = 10,000$ times:

```
B <- 10000
events <- replicate(B, sample(beads, 1))
```

We can now see if our definition actually is in agreement with this Monte Carlo simulation approximation. We can use `table` to see the distribution:

```
tab <- table(events)
tab
#> events
#> blue   red
#> 5920 4080
```

and `prop.table` gives us the proportions:

```
prop.table(tab)
#> events
#> blue   red
#> 0.592 0.408
```

The numbers above are the estimated probabilities provided by this Monte Carlo simulation. Statistical theory, not covered here, tells us that as B gets larger, the estimates get closer to $3/5=.6$ and $2/5=.4$.

Although this is a simple and not very useful example, we will use Monte Carlo simulations to estimate probabilities in cases in which it is harder to compute the exact ones. Before delving into more complex examples, we use simple ones to demonstrate the computing tools available in R.

13.2.1   Setting the random seed

Before we continue, we will briefly explain the following important line of code:

```
set.seed(1986)
```

Throughout this book, we use random number generators. This implies that many of the results presented can actually change by chance, which then suggests that a frozen version of the book may show a different result than what you obtain when you try to code as shown in the book. This is actually fine since the results are random and change from time to time. However, if you want to ensure that results are exactly the same every time you run them, you can set R's random number generation seed to a specific number. Above we set it to 1986. We want to avoid using the same seed everytime. A popular way to pick the seed is the year - month - day. For example, we picked 1986 on December 20, 2018: $2018 - 12 - 20 = 1986$.

You can learn more about setting the seed by looking at the documentation:

```
?set.seed
```

In the exercises, we may ask you to set the seed to assure that the results you obtain are exactly what we expect them to be.

13.2.2   With and without replacement

The function `sample` has an argument that permits us to pick more than one element from the urn. However, by default, this selection occurs *without replacement*: after a bead is selected, it is not put back in the bag. Notice what happens when we ask to randomly select five beads:

```
sample(beads, 5)
#> [1] "red"  "blue" "blue" "blue" "red"
sample(beads, 5)
#> [1] "red"  "red"  "blue" "blue" "blue"
sample(beads, 5)
#> [1] "blue" "red"  "blue" "red"  "blue"
```

This results in rearrangements that always have three blue and two red beads. If we ask that six beads be selected, we get an error:

```
sample(beads, 6)
```

```
Error in sample.int(length(x), size, replace, prob) :     cannot take a sample
larger than the population when 'replace = FALSE'
```

However, the `sample` function can be used directly, without the use of `replicate`, to repeat the same experiment of picking 1 out of the 5 beads, continually, under the same conditions. To do this, we sample *with replacement*: return the bead back to the urn after selecting it. We can tell `sample` to do this by changing the `replace` argument, which defaults to `FALSE`, to `replace = TRUE`:

```
events <- sample(beads, B, replace = TRUE)
prop.table(table(events))
#> events
#> blue    red
#> 0.602 0.398
```

Not surprisingly, we get results very similar to those previously obtained with `replicate`.

13.3 Independence

We say two events are independent if the outcome of one does not affect the other. The classic example is coin tosses. Every time we toss a fair coin, the probability of seeing heads is 1/2 regardless of what previous tosses have revealed. The same is true when we pick beads from an urn with replacement. In the example above, the probability of red is 0.40 regardless of previous draws.

Many examples of events that are not independent come from card games. When we deal the first card, the probability of getting a King is 1/13 since there are thirteen possibilities: Ace, Deuce, Three, ..., Ten, Jack, Queen, King, and Ace. Now if we deal a King for the first card, and don't replace it into the deck, the probabilities of a second card being a King is less because there are only three Kings left: the probability is 3 out of 51. These events are therefore **not independent**: the first outcome affected the next one.

To see an extreme case of non-independent events, consider our example of drawing five beads at random **without** replacement:

```
x <- sample(beads, 5)
```

If you have to guess the color of the first bead, you will predict blue since blue has a 60% chance. But if I show you the result of the last four outcomes:

```
x[2:5]
#> [1] "blue" "blue" "blue" "red"
```

would you still guess blue? Of course not. Now you know that the probability of red is 1 since the only bead left is red. The events are not independent, so the probabilities change.

13.4 Conditional probabilities

When events are not independent, *conditional probabilities* are useful. We already saw an example of a conditional probability: we computed the probability that a second dealt card is a King given that the first was a King. In probability, we use the following notation:

$$\Pr(\text{Card 2 is a king} \mid \text{Card 1 is a king}) = 3/51$$

We use the | as shorthand for "given that" or "conditional on".

When two events, say A and B, are independent, we have:

$$\Pr(A \mid B) = \Pr(A)$$

This is the mathematical way of saying: the fact that B happened does not affect the probability of A happening. In fact, this can be considered the mathematical definition of independence.

13.5 Addition and multiplication rules

13.5.1 Multiplication rule

If we want to know the probability of two events, say A and B, occurring, we can use the multiplication rule:

$$\Pr(A \text{ and } B) = \Pr(A)\Pr(B \mid A)$$

Let's use Blackjack as an example. In Blackjack, you are assigned two random cards. After you see what you have, you can ask for more. The goal is to get closer to 21 than the dealer, without going over. Face cards are worth 10 points and Aces are worth 11 or 1 (you choose).

So, in a Blackjack game, to calculate the chances of getting a 21 by drawing an Ace and then a face card, we compute the probability of the first being an Ace and multiply by the probability of drawing a face card or a 10 given that the first was an Ace: $1/13 \times 16/51 \approx 0.025$

The multiplication rule also applies to more than two events. We can use induction to expand for more events:

$$\Pr(A \text{ and } B \text{ and } C) = \Pr(A)\Pr(B \mid A)\Pr(C \mid A \text{ and } B)$$

13.5.2 Multiplication rule under independence

When we have independent events, then the multiplication rule becomes simpler:

$$\Pr(A \text{ and } B \text{ and } C) = \Pr(A)\Pr(B)\Pr(C)$$

But we have to be very careful before using this since assuming independence can result in very different and incorrect probability calculations when we don't actually have independence.

As an example, imagine a court case in which the suspect was described as having a mustache and a beard. The defendant has a mustache and a beard and the prosecution brings in an "expert" to testify that $1/10$ men have beards and $1/5$ have mustaches, so using the multiplication rule we conclude that only $1/10 \times 1/5$ or 0.02 have both.

But to multiply like this we need to assume independence! Say the conditional probability of a man having a mustache conditional on him having a beard is .95. So the correct calculation probability is much higher: $1/10 \times 95/100 = 0.095$.

The multiplication rule also gives us a general formula for computing conditional probabilities:

$$\Pr(B \mid A) = \frac{\Pr(A \text{ and } B)}{\Pr(A)}$$

To illustrate how we use these formulas and concepts in practice, we will use several examples related to card games.

13.5.3   Addition rule

The addition rule tells us that:

$$\Pr(A \text{ or } B) = \Pr(A) + \Pr(B) - \Pr(A \text{ and } B)$$

This rule is intuitive: think of a Venn diagram. If we simply add the probabilities, we count the intersection twice so we need to substract one instance.

13.6   Combinations and permutations

In our very first example, we imagined an urn with five beads. As a reminder, to compute the probability distribution of one draw, we simply listed out all the possibilities. There were 5 and so then, for each event, we counted how many of these possibilities were associated with the event. The resulting probability of choosing a blue bead is 3/5 because out of the five possible outcomes, three were blue.

For more complicated cases, the computations are not as straightforward. For instance, what is the probability that if I draw five cards without replacement, I get all cards of the same suit, what is known as a "flush" in poker? In a discrete probability course you learn theory on how to make these computations. Here we focus on how to use R code to compute the answers.

First, let's construct a deck of cards. For this, we will use the `expand.grid` and `paste` functions. We use `paste` to create strings by joining smaller strings. To do this, we take the number and suit of a card and create the card name like this:

```
number <- "Three"
suit <- "Hearts"
paste(number, suit)
#> [1] "Three Hearts"
```

`paste` also works on pairs of vectors performing the operation element-wise:

```
paste(letters[1:5], as.character(1:5))
#> [1] "a 1" "b 2" "c 3" "d 4" "e 5"
```

The function `expand.grid` gives us all the combinations of entries of two vectors. For example, if you have blue and black pants and white, grey, and plaid shirts, all your combinations are:

```
expand.grid(pants = c("blue", "black"), shirt = c("white", "grey", "plaid"))
#>    pants shirt
#> 1  blue white
#> 2 black white
#> 3  blue  grey
#> 4 black  grey
#> 5  blue plaid
#> 6 black plaid
```

Here is how we generate a deck of cards:

```
suits <- c("Diamonds", "Clubs", "Hearts", "Spades")
numbers <- c("Ace", "Deuce", "Three", "Four", "Five", "Six", "Seven",
             "Eight", "Nine", "Ten", "Jack", "Queen", "King")
deck <- expand.grid(number=numbers, suit=suits)
deck <- paste(deck$number, deck$suit)
```

With the deck constructed, we can double check that the probability of a King in the first card is $1/13$ by computing the proportion of possible outcomes that satisfy our condition:

```
kings <- paste("King", suits)
mean(deck %in% kings)
#> [1] 0.0769
```

Now, how about the conditional probability of the second card being a King given that the first was a King? Earlier, we deduced that if one King is already out of the deck and there are 51 left, then this probability is $3/51$. Let's confirm by listing out all possible outcomes.

To do this, we can use the `permutations` function from the **gtools** package. For any list of size `n`, this function computes all the different combinations we can get when we select `r` items. Here are all the ways we can choose two numbers from a list consisting of `1,2,3`:

```
library(gtools)
permutations(3, 2)
#>       [,1] [,2]
#> [1,]    1    2
#> [2,]    1    3
#> [3,]    2    1
#> [4,]    2    3
#> [5,]    3    1
#> [6,]    3    2
```

Notice that the order matters here: 3,1 is different than 1,3. Also, note that (1,1), (2,2), and (3,3) do not appear because once we pick a number, it can't appear again.

Optionally, we can add a vector. If you want to see five random seven digit phone numbers out of all possible phone numbers (without repeats), you can type:

```
all_phone_numbers <- permutations(10, 7, v = 0:9)
n <- nrow(all_phone_numbers)
index <- sample(n, 5)
all_phone_numbers[index,]
#>       [,1] [,2] [,3] [,4] [,5] [,6] [,7]
#> [1,]    1    3    8    0    6    7    5
#> [2,]    2    9    1    6    4    8    0
#> [3,]    5    1    6    0    9    8    2
#> [4,]    7    4    6    0    2    8    1
#> [5,]    4    6    5    9    2    8    0
```

Instead of using the numbers 1 through 10, the default, it uses what we provided through v: the digits 0 through 9.

To compute all possible ways we can choose two cards when the order matters, we type:

```
hands <- permutations(52, 2, v = deck)
```

This is a matrix with two columns and 2652 rows. With a matrix we can get the first and second cards like this:

```
first_card <- hands[,1]
second_card <- hands[,2]
```

Now the cases for which the first hand was a King can be computed like this:

```
kings <- paste("King", suits)
sum(first_card %in% kings)
#> [1] 204
```

To get the conditional probability, we compute what fraction of these have a King in the second card:

```
sum(first_card%in%kings & second_card%in%kings)/sum(first_card %in% kings)
#> [1] 0.0588
```

which is exactly 3/51, as we had already deduced. Notice that the code above is equivalent to:

```
mean(first_card%in%kings & second_card%in%kings)/mean(first_card %in% kings)
#> [1] 0.0588
```

which uses `mean` instead of `sum` and is an R version of:

$$\frac{\Pr(A \text{ and } B)}{\Pr(A)}$$

How about if the order doesn't matter? For example, in Blackjack if you get an Ace and a face card in the first draw, it is called a *Natural 21* and you win automatically. If we wanted to compute the probability of this happening, we would enumerate the *combinations*, not the permutations, since the order does not matter.

```
combinations(3,2)
#>      [,1] [,2]
#> [1,]    1    2
#> [2,]    1    3
#> [3,]    2    3
```

In the second line, the outcome does not include (2,1) because (1,2) already was enumerated. The same applies to (3,1) and (3,2).

So to compute the probability of a *Natural 21* in Blackjack, we can do this:

```
aces <- paste("Ace", suits)

facecard <- c("King", "Queen", "Jack", "Ten")
facecard <- expand.grid(number = facecard, suit = suits)
facecard <- paste(facecard$number, facecard$suit)

hands <- combinations(52, 2, v = deck)
mean(hands[,1] %in% aces & hands[,2] %in% facecard)
#> [1] 0.0483
```

In the last line, we assume the Ace comes first. This is only because we know the way `combination` enumerates possibilities and it will list this case first. But to be safe, we could have written this and produced the same answer:

```
mean((hands[,1] %in% aces & hands[,2] %in% facecard) |
        (hands[,2] %in% aces & hands[,1] %in% facecard))
#> [1] 0.0483
```

13.6.1 Monte Carlo example

Instead of using `combinations` to deduce the exact probability of a Natural 21, we can use a Monte Carlo to estimate this probability. In this case, we draw two cards over and over and keep track of how many 21s we get. We can use the function sample to draw two cards without replacements:

```
hand <- sample(deck, 2)
hand
#> [1] "Queen Clubs"  "Seven Spades"
```

And then check if one card is an Ace and the other a face card or a 10. Going forward, we include 10 when we say *face card*. Now we need to check both possibilities:

```
(hands[1] %in% aces & hands[2] %in% facecard) |
  (hands[2] %in% aces & hands[1] %in% facecard)
#> [1] FALSE
```

If we repeat this 10,000 times, we get a very good approximation of the probability of a Natural 21.

Let's start by writing a function that draws a hand and returns TRUE if we get a 21. The function does not need any arguments because it uses objects defined in the global environment.

```
blackjack <- function(){
   hand <- sample(deck, 2)
   (hand[1] %in% aces & hand[2] %in% facecard) |
     (hand[2] %in% aces & hand[1] %in% facecard)
}
```

Here we do have to check both possibilities: Ace first or Ace second because we are not using the `combinations` function. The function returns TRUE if we get a 21 and FALSE otherwise:

```
blackjack()
#> [1] FALSE
```

Now we can play this game, say, 10,000 times:

```
B <- 10000
results <- replicate(B, blackjack())
mean(results)
#> [1] 0.0475
```

13.7 Examples

In this section, we describe two discrete probability popular examples: the Monty Hall problem and the birthday problem. We use R to help illustrate the mathematical concepts.

13.7.1   Monty Hall problem

In the 1970s, there was a game show called "Let's Make a Deal" and Monty Hall was the host. At some point in the game, contestants were asked to pick one of three doors. Behind one door there was a prize. The other doors had a goat behind them to show the contestant they had lost. After the contestant picked a door, before revealing whether the chosen door contained a prize, Monty Hall would open one of the two remaining doors and show the contestant there was no prize behind that door. Then he would ask "Do you want to switch doors?" What would you do?

We can use probability to show that if you stick with the original door choice, your chances of winning a prize remain 1 in 3. However, if you switch to the other door, your chances of winning double to 2 in 3! This seems counterintuitive. Many people incorrectly think both chances are 1 in 2 since you are choosing between 2 options. You can watch a detailed mathematical explanation on Khan Academy[2] or read one on Wikipedia[3]. Below we use a Monte Carlo simulation to see which strategy is better. Note that this code is written longer than it should be for pedagogical purposes.

Let's start with the stick strategy:

```
B <- 10000
monty_hall <- function(strategy){
  doors <- as.character(1:3)
  prize <- sample(c("car", "goat", "goat"))
  prize_door <- doors[prize == "car"]
  my_pick  <- sample(doors, 1)
  show <- sample(doors[!doors %in% c(my_pick, prize_door)],1)
  stick <- my_pick
  stick == prize_door
  switch <- doors[!doors%in%c(my_pick, show)]
  choice <- ifelse(strategy == "stick", stick, switch)
  choice == prize_door
}
stick <- replicate(B, monty_hall("stick"))
mean(stick)
#> [1] 0.342
switch <- replicate(B, monty_hall("switch"))
mean(switch)
#> [1] 0.668
```

As we write the code, we note that the lines starting with `my_pick` and `show` have no influence on the last logical operation when we stick to our original choice anyway. From this we should realize that the chance is 1 in 3, what we began with. When we switch, the Monte Carlo estimate confirms the 2/3 calculation. This helps us gain some insight by showing that we are removing a door, `show`, that is definitely not a winner from our choices. We also see that unless we get it right when we first pick, you win: 1 - 1/3 = 2/3.

---

[2]https://www.khanacademy.org/math/precalculus/prob-comb/dependent-events-precalc/v/monty-hall-problem

[3]https://en.wikipedia.org/wiki/Monty_Hall_problem

13.7.2 Birthday problem

Suppose you are in a classroom with 50 people. If we assume this is a randomly selected group of 50 people, what is the chance that at least two people have the same birthday? Although it is somewhat advanced, we can deduce this mathematically. We will do this later. Here we use a Monte Carlo simulation. For simplicity, we assume nobody was born on February 29. This actually doesn't change the answer much.

First, note that birthdays can be represented as numbers between 1 and 365, so a sample of 50 birthdays can be obtained like this:

```
n <- 50
bdays <- sample(1:365, n, replace = TRUE)
```

To check if in this particular set of 50 people we have at least two with the same birthday, we can use the function `duplicated`, which returns `TRUE` whenever an element of a vector is a duplicate. Here is an example:

```
duplicated(c(1,2,3,1,4,3,5))
#> [1] FALSE FALSE FALSE  TRUE FALSE  TRUE FALSE
```

The second time 1 and 3 appear, we get a `TRUE`. So to check if two birthdays were the same, we simply use the `any` and `duplicated` functions like this:

```
any(duplicated(bdays))
#> [1] TRUE
```

In this case, we see that it did happen. At least two people had the same birthday.

To estimate the probability of a shared birthday in the group, we repeat this experiment by sampling sets of 50 birthdays over and over:

```
B <- 10000
same_birthday <- function(n){
  bdays <- sample(1:365, n, replace=TRUE)
  any(duplicated(bdays))
}
results <- replicate(B, same_birthday(50))
mean(results)
#> [1] 0.969
```

Were you expecting the probability to be this high?

People tend to underestimate these probabilities. To get an intuition as to why it is so high, think about what happens when the group size is close to 365. At this stage, we run out of days and the probability is one.

Say we want to use this knowledge to bet with friends about two people having the same birthday in a group of people. When are the chances larger than 50%? Larger than 75%?

Let's create a look-up table. We can quickly create a function to compute this for any group size:

```
compute_prob <- function(n, B=10000){
  results <- replicate(B, same_birthday(n))
  mean(results)
}
```

Using the function `sapply`, we can perform element-wise operations on any function:

```
n <- seq(1,60)
prob <- sapply(n, compute_prob)
```

We can now make a plot of the estimated probabilities of two people having the same birthday in a group of size n:

```
library(tidyverse)
prob <- sapply(n, compute_prob)
qplot(n, prob)
```

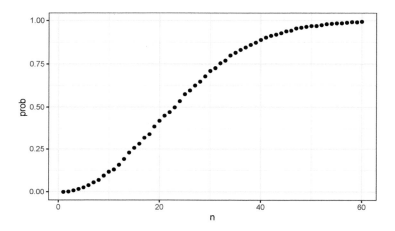

Now let's compute the exact probabilities rather than use Monte Carlo approximations. Not only do we get the exact answer using math, but the computations are much faster since we don't have to generate experiments.

To make the math simpler, instead of computing the probability of it happening, we will compute the probability of it not happening. For this, we use the multiplication rule.

Let's start with the first person. The probability that person 1 has a unique birthday is 1. The probability that person 2 has a unique birthday, given that person 1 already took one, is 364/365. Then, given that the first two people have unique birthdays, person 3 is left with 363 days to choose from. We continue this way and find the chances of all 50 people having a unique birthday is:

$$1 \times \frac{364}{365} \times \frac{363}{365} \cdots \frac{365 - n + 1}{365}$$

We can write a function that does this for any number:

```
exact_prob <- function(n){
  prob_unique <- seq(365,365-n+1)/365
  1 - prod( prob_unique)
}
eprob <- sapply(n, exact_prob)
qplot(n, prob) + geom_line(aes(n, eprob), col = "red")
```

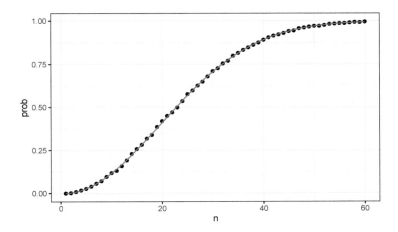

This plot shows that the Monte Carlo simulation provided a very good estimate of the exact probability. Had it not been possible to compute the exact probabilities, we would have still been able to accurately estimate the probabilities.

13.8 Infinity in practice

The theory described here requires repeating experiments over and over forever. In practice we can't do this. In the examples above, we used $B = 10,000$ Monte Carlo experiments and it turned out that this provided accurate estimates. The larger this number, the more accurate the estimate becomes until the approximaton is so good that your computer can't tell the difference. But in more complex calculations, 10,000 may not be nearly enough. Also, for some calculations, 10,000 experiments might not be computationally feasible. In practice, we won't know what the answer is, so we won't know if our Monte Carlo estimate is accurate. We know that the larger B, the better the approximation. But how big do we need it to be? This is actually a challenging question and answering it often requires advanced theoretical statistics training.

One practical approach we will describe here is to check for the stability of the estimate. The following is an example with the birthday problem for a group of 25 people.

```
B <- 10^seq(1, 5, len = 100)
compute_prob <- function(B, n=25){
  same_day <- replicate(B, same_birthday(n))
  mean(same_day)
```

```
}
prob <- sapply(B, compute_prob)
qplot(log10(B), prob, geom = "line")
```

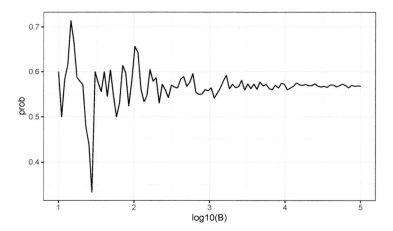

In this plot, we can see that the values start to stabilize (that is, they vary less than .01) around 1000. Note that the exact probability, which we know in this case, is 0.569.

13.9   Exercises

1. One ball will be drawn at random from a box containing: 3 cyan balls, 5 magenta balls, and 7 yellow balls. What is the probability that the ball will be cyan?

2. What is the probability that the ball will not be cyan?

3. Instead of taking just one draw, consider taking two draws. You take the second draw without returning the first draw to the box. We call this sampling **without** replacement. What is the probability that the first draw is cyan and that the second draw is not cyan?

4. Now repeat the experiment, but this time, after taking the first draw and recording the color, return it to the box and shake the box. We call this sampling **with** replacement. What is the probability that the first draw is cyan and that the second draw is not cyan?

5. Two events A and B are independent if $\Pr(A \text{ and } B) = \Pr(A)P(B)$. Under which situation are the draws independent?

    a. You don't replace the draw.
    b. You replace the draw.
    c. Neither
    d. Both

6. Say you've drawn 5 balls from the box, with replacement, and all have been yellow. What is the probability that the next one is yellow?

7. If you roll a 6-sided die six times, what is the probability of not seeing a 6?

8. Two teams, say the Celtics and the Cavs, are playing a seven game series. The Cavs are a better team and have a 60% chance of winning each game. What is the probability that the Celtics win **at least** one game?

9. Create a Monte Carlo simulation to confirm your answer to the previous problem. Use B <- 10000 simulations. Hint: use the following code to generate the results of the first four games:

```
celtic_wins <- sample(c(0,1), 4, replace = TRUE, prob = c(0.6, 0.4))
```

The Celtics must win one of these 4 games.

10. Two teams, say the Cavs and the Warriors, are playing a seven game championship series. The first to win four games, therefore, wins the series. The teams are equally good so they each have a 50-50 chance of winning each game. If the Cavs lose the first game, what is the probability that they win the series?

11. Confirm the results of the previous question with a Monte Carlo simulation.

12. Two teams, A and B, are playing a seven game series. Team A is better than team B and has a $p > 0.5$ chance of winning each game. Given a value p, the probability of winning the series for the underdog team B can be computed with the following function based on a Monte Carlo simulation:

```
prob_win <- function(p){
  B <- 10000
  result <- replicate(B, {
    b_win <- sample(c(1,0), 7, replace = TRUE, prob = c(1-p, p))
    sum(b_win)>=4
  })
  mean(result)
}
```

Use the function `sapply` to compute the probability, call it `Pr`, of winning for p <- seq(0.5, 0.95, 0.025). Then plot the result.

13. Repeat the exercise above, but now keep the probability fixed at p <- 0.75 and compute the probability for different series lengths: best of 1 game, 3 games, 5 games,... Specifically, N <- seq(1, 25, 2). Hint: use this function:

```
prob_win <- function(N, p=0.75){
  B <- 10000
  result <- replicate(B, {
    b_win <- sample(c(1,0), N, replace = TRUE, prob = c(1-p, p))
    sum(b_win)>=(N+1)/2
  })
  mean(result)
}
```

13.10   Continuous probability

In Section 8.4, we explained why when summarizing a list of numeric values, such as heights, it is not useful to construct a distribution that defines a proportion to each possible outcome. For example, if we measure every single person in a very large population of size n with extremely high precision, since no two people are exactly the same height, we need to assign the proportion $1/n$ to each observed value and attain no useful summary at all. Similarly, when defining probability distributions, it is not useful to assign a very small probability to every single height.

Just as when using distributions to summarize numeric data, it is much more practical to define a function that operates on intervals rather than single values. The standard way of doing this is using the *cumulative distribution function* (CDF).

We described empirical cumulative distribution function (eCDF) in Section 8.4 as a basic summary of a list of numeric values. As an example, we earlier defined the height distribution for adult male students. Here, we define the vector x to contain these heights:

```
library(tidyverse)
library(dslabs)
data(heights)
x <- heights %>% filter(sex=="Male") %>% pull(height)
```

We defined the empirical distribution function as:

```
F <- function(a) mean(x<=a)
```

which, for any value a, gives the proportion of values in the list x that are smaller or equal than a.

Keep in mind that we have not yet introduced probability in the context of CDFs. Let's do this by asking the following: if I pick one of the male students at random, what is the chance that he is taller than 70.5 inches? Because every student has the same chance of being picked, the answer to this is equivalent to the proportion of students that are taller than 70.5 inches. Using the CDF we obtain an answer by typing:

```
1 - F(70)
#> [1] 0.377
```

Once a CDF is defined, we can use this to compute the probability of any subset. For instance, the probability of a student being between height a and height b is:

```
F(b)-F(a)
```

Because we can compute the probability for any possible event this way, the cumulative probability function defines the probability distribution for picking a height at random from our vector of heights x.

13.11 Theoretical continuous distributions

In Section 8.8 we introduced the normal distribution as a useful approximation to many naturally occurring distributions, including that of height. The cumulative distribution for the normal distribution is defined by a mathematical formula which in R can be obtained with the function `pnorm`. We say that a random quantity is normally distributed with average `m` and standard deviation `s` if its probability distribution is defined by:

```
F(a) = pnorm(a, m, s)
```

This is useful because if we are willing to use the normal approximation for, say, height, we don't need the entire dataset to answer questions such as: what is the probability that a randomly selected student is taller then 70 inches? We just need the average height and standard deviation:

```
m <- mean(x)
s <- sd(x)
1 - pnorm(70.5, m, s)
#> [1] 0.371
```

13.11.1 Theoretical distributions as approximations

The normal distribution is derived mathematically: we do not need data to define it. For practicing data scientists, almost everything we do involves data. Data is always, technically speaking, discrete. For example, we could consider our height data categorical with each specific height a unique category. The probability distribution is defined by the proportion of students reporting each height. Here is a plot of that probability distribution:

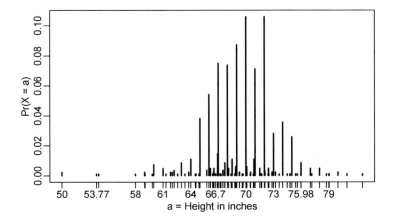

While most students rounded up their heights to the nearest inch, others reported values with more precision. One student reported his height to be 69.6850393700787, which is 177 centimeters. The probability assigned to this height is 0.001 or 1 in 812. The probability for

70 inches is much higher at 0.106, but does it really make sense to think of the probability of being exactly 70 inches as being different than 69.6850393700787? Clearly it is much more useful for data analytic purposes to treat this outcome as a continuous numeric variable, keeping in mind that very few people, or perhaps none, are exactly 70 inches, and that the reason we get more values at 70 is because people round to the nearest inch.

With continuous distributions, the probability of a singular value is not even defined. For example, it does not make sense to ask what is the probability that a normally distributed value is 70. Instead, we define probabilities for intervals. We thus could ask what is the probability that someone is between 69.5 and 70.5.

In cases like height, in which the data is rounded, the normal approximation is particularly useful if we deal with intervals that include exactly one round number. For example, the normal distribution is useful for approximating the proportion of students reporting values in intervals like the following three:

```
mean(x <= 68.5) - mean(x <= 67.5)
#> [1] 0.115
mean(x <= 69.5) - mean(x <= 68.5)
#> [1] 0.119
mean(x <= 70.5) - mean(x <= 69.5)
#> [1] 0.122
```

Note how close we get with the normal approximation:

```
pnorm(68.5, m, s) - pnorm(67.5, m, s)
#> [1] 0.103
pnorm(69.5, m, s) - pnorm(68.5, m, s)
#> [1] 0.11
pnorm(70.5, m, s) - pnorm(69.5, m, s)
#> [1] 0.108
```

However, the approximation is not as useful for other intervals. For instance, notice how the approximation breaks down when we try to estimate:

```
mean(x <= 70.9) - mean(x<=70.1)
#> [1] 0.0222
```

with

```
pnorm(70.9, m, s) - pnorm(70.1, m, s)
#> [1] 0.0836
```

In general, we call this situation *discretization*. Although the true height distribution is continuous, the reported heights tend to be more common at discrete values, in this case, due to rounding. As long as we are aware of how to deal with this reality, the normal approximation can still be a very useful tool.

13.11.2 The probability density

For categorical distributions, we can define the probability of a category. For example, a roll of a die, let's call it X, can be 1,2,3,4,5 or 6. The probability of 4 is defined as:

$$\Pr(X = 4) = 1/6$$

The CDF can then easily be defined:

$$F(4) = \Pr(X \le 4) = \Pr(X = 4) + \Pr(X = 3) + \Pr(X = 2) + \Pr(X = 1)$$

Although for continuous distributions the probability of a single value $\Pr(X = x)$ is not defined, there is a theoretical definition that has a similar interpretation. The probability density at x is defined as the function $f(a)$ such that:

$$F(a) = \Pr(X \le a) = \int_{-\infty}^{a} f(x)\, dx$$

For those that know calculus, remember that the integral is related to a sum: it is the sum of bars with widths approximating 0. If you don't know calculus, you can think of $f(x)$ as a curve for which the area under that curve up to the value a, gives you the probability $\Pr(X \le a)$.

For example, to use the normal approximation to estimate the probability of someone being taller than 76 inches, we use:

```
1 - pnorm(76, m, s)
#> [1] 0.0321
```

which mathematically is the grey area below:

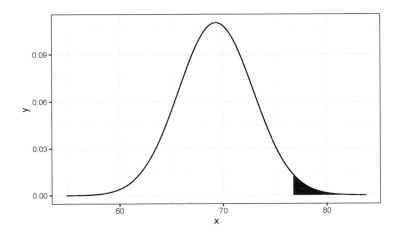

The curve you see is the probability density for the normal distribution. In R, we get this using the function `dnorm`.

Although it may not be immediately obvious why knowing about probability densities is useful, understanding this concept will be essential to those wanting to fit models to data for which predefined functions are not available.

13.12   Monte Carlo simulations for continuous variables

R provides functions to generate normally distributed outcomes. Specifically, the `rnorm` function takes three arguments: size, average (defaults to 0), and standard deviation (defaults to 1) and produces random numbers. Here is an example of how we could generate data that looks like our reported heights:

```
n <- length(x)
m <- mean(x)
s <- sd(x)
simulated_heights <- rnorm(n, m, s)
```

Not surprisingly, the distribution looks normal:

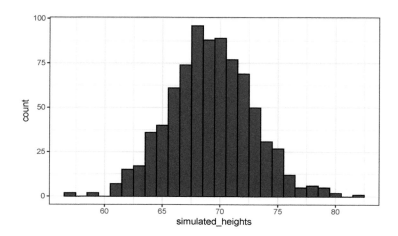

This is one of the most useful functions in R as it will permit us to generate data that mimics natural events and answers questions related to what could happen by chance by running Monte Carlo simulations.

If, for example, we pick 800 males at random, what is the distribution of the tallest person? How rare is a seven footer in a group of 800 males? The following Monte Carlo simulation helps us answer that question:

```
B <- 10000
tallest <- replicate(B, {
  simulated_data <- rnorm(800, m, s)
  max(simulated_data)
})
```

Having a seven footer is quite rare:

```
mean(tallest >= 7*12)
#> [1] 0.0199
```

Here is the resulting distribution:

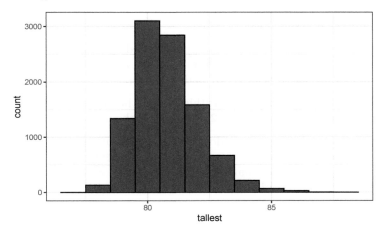

Note that it does not look normal.

13.13 Continuous distributions

We introduced the normal distribution in Section 8.8 and used it as an example above. The normal distribution is not the only useful theoretical distribution. Other continuous distributions that we may encounter are the student-t, Chi-square, exponential, gamma, beta, and beta-binomial. R provides functions to compute the density, the quantiles, the cumulative distribution functions and to generate Monte Carlo simulations. R uses a convention that lets us remember the names, namely using the letters d, q, p, and r in front of a shorthand for the distribution. We have already seen the functions dnorm, pnorm, and rnorm for the normal distribution. The functions qnorm gives us the quantiles. We can therefore draw a distribution like this:

```
x <- seq(-4, 4, length.out = 100)
qplot(x, f, geom = "line", data = data.frame(x, f = dnorm(x)))
```

For the student-t, described later in Section 16.10, the shorthand t is used so the functions are dt for the density, qt for the quantiles, pt for the cumulative distribution function, and rt for Monte Carlo simulation.

13.14 Exercises

1. Assume the distribution of female heights is approximated by a normal distribution with a mean of 64 inches and a standard deviation of 3 inches. If we pick a female at random, what is the probability that she is 5 feet or shorter?

2. Assume the distribution of female heights is approximated by a normal distribution with a mean of 64 inches and a standard deviation of 3 inches. If we pick a female at random, what is the probability that she is 6 feet or taller?

3. Assume the distribution of female heights is approximated by a normal distribution with a mean of 64 inches and a standard deviation of 3 inches. If we pick a female at random, what is the probability that she is between 61 and 67 inches?

4. Repeat the exercise above, but convert everything to centimeters. That is, multiply every height, including the standard deviation, by 2.54. What is the answer now?

5. Notice that the answer to the question does not change when you change units. This makes sense since the answer to the question should not be affected by what units we use. In fact, if you look closely, you notice that 61 and 64 are both 1 SD away from the average. Compute the probability that a randomly picked, normally distributed random variable is within 1 SD from the average.

6. To see the math that explains why the answers to questions 3, 4, and 5 are the same, suppose we have a random variable with average m and standard error s. Suppose we ask the probability of X being smaller or equal to a. Remember that, by definition, a is $(a - m)/s$ standard deviations s away from the average m. The probability is:

$$\Pr(X \leq a)$$

Now we subtract μ to both sides and then divide both sides by σ:

$$\Pr\left(\frac{X - m}{s} \leq \frac{a - m}{s}\right)$$

The quantity on the left is a standard normal random variable. It has an average of 0 and a standard error of 1. We will call it Z:

$$\Pr\left(Z \leq \frac{a - m}{s}\right)$$

So, no matter the units, the probability of $X \leq a$ is the same as the probability of a standard normal variable being less than $(a - m)/s$. If mu is the average and sigma the standard error, which of the following R code would give us the right answer in every situation:

```
a.   mean(X<=a)
b.   pnorm((a - m)/s)
c.   pnorm((a - m)/s, m, s)
d.   pnorm(a)
```

7. Imagine the distribution of male adults is approximately normal with an expected value of 69 and a standard deviation of 3. How tall is the male in the 99th percentile? Hint: use qnorm.

8. The distribution of IQ scores is approximately normally distributed. The average is 100 and the standard deviation is 15. Suppose you want to know the distribution of the highest IQ across all graduating classes if 10,000 people are born each in your school district. Run a Monte Carlo simulation with B=1000 generating 10,000 IQ scores and keeping the highest. Make a histogram.

14

Random variables

In data science, we often deal with data that is affected by chance in some way: the data comes from a random sample, the data is affected by measurement error, or the data measures some outcome that is random in nature. Being able to quantify the uncertainty introduced by randomness is one of the most important jobs of a data analyst. Statistical inference offers a framework, as well as several practical tools, for doing this. The first step is to learn how to mathematically describe random variables.

In this chapter, we introduce random variables and their properties starting with their application to games of chance. We then describe some of the events surrounding the financial crisis of 2007-2008[1] using probability theory. This financial crisis was in part caused by underestimating the risk of certain securities[2] sold by financial institutions. Specifically, the risks of mortgage-backed securities (MBS) and collateralized debt obligations (CDO) were grossly underestimated. These assets were sold at prices that assumed most homeowners would make their monthly payments, and the probability of this not occurring was calculated as being low. A combination of factors resulted in many more defaults than were expected, which led to a price crash of these securities. As a consequence, banks lost so much money that they needed government bailouts to avoid closing down completely.

14.1 Random variables

Random variables are numeric outcomes resulting from random processes. We can easily generate random variables using some of the simple examples we have shown. For example, define X to be 1 if a bead is blue and red otherwise:

```
beads <- rep( c("red", "blue"), times = c(2,3))
X <- ifelse(sample(beads, 1) == "blue", 1, 0)
```

Here X is a random variable: every time we select a new bead the outcome changes randomly. See below:

```
ifelse(sample(beads, 1) == "blue", 1, 0)
#> [1] 1
ifelse(sample(beads, 1) == "blue", 1, 0)
#> [1] 0
ifelse(sample(beads, 1) == "blue", 1, 0)
#> [1] 0
```

---

[1] https://en.wikipedia.org/w/index.php?title=Financial_crisis_of_2007%E2%80%932008
[2] https://en.wikipedia.org/w/index.php?title=Security_(finance)

Sometimes it's 1 and sometimes it's 0.

14.2   Sampling models

Many data generation procedures, those that produce the data we study, can be modeled quite well as draws from an urn. For instance, we can model the process of polling likely voters as drawing 0s (Republicans) and 1s (Democrats) from an urn containing the 0 and 1 code for all likely voters. In epidemiological studies, we often assume that the subjects in our study are a random sample from the population of interest. The data related to a specific outcome can be modeled as a random sample from an urn containing the outcome for the entire population of interest. Similarly, in experimental research, we often assume that the individual organisms we are studying, for example worms, flies, or mice, are a random sample from a larger population. Randomized experiments can also be modeled by draws from an urn given the way individuals are assigned into groups: when getting assigned, you draw your group at random. Sampling models are therefore ubiquitous in data science. Casino games offer a plethora of examples of real-world situations in which sampling models are used to answer specific questions. We will therefore start with such examples.

Suppose a very small casino hires you to consult on whether they should set up roulette wheels. To keep the example simple, we will assume that 1,000 people will play and that the only game you can play on the roulette wheel is to bet on red or black. The casino wants you to predict how much money they will make or lose. They want a range of values and, in particular, they want to know what's the chance of losing money. If this probability is too high, they will pass on installing roulette wheels.

We are going to define a random variable S that will represent the casino's total winnings. Let's start by constructing the urn. A roulette wheel has 18 red pockets, 18 black pockets and 2 green ones. So playing a color in one game of roulette is equivalent to drawing from this urn:

```
color <- rep(c("Black", "Red", "Green"), c(18, 18, 2))
```

The 1,000 outcomes from 1,000 people playing are independent draws from this urn. If red comes up, the gambler wins and the casino loses a dollar, so we draw a -\$1. Otherwise, the casino wins a dollar and we draw a \$1. To construct our random variable S, we can use this code:

```
n <- 1000
X <- sample(ifelse(color == "Red", -1, 1),  n, replace = TRUE)
X[1:10]
#>  [1] -1  1  1 -1 -1 -1  1  1  1  1
```

Because we know the proportions of 1s and -1s, we can generate the draws with one line of code, without defining `color`:

```
X <- sample(c(-1,1), n, replace = TRUE, prob=c(9/19, 10/19))
```

We call this a **sampling model** since we are modeling the random behavior of roulette with the sampling of draws from an urn. The total winnings S is simply the sum of these 1,000 independent draws:

```
X <- sample(c(-1,1), n, replace = TRUE, prob=c(9/19, 10/19))
S <- sum(X)
S
#> [1] 22
```

14.3 The probability distribution of a random variable

If you run the code above, you see that S changes every time. This is, of course, because S is a **random variable**. The probability distribution of a random variable tells us the probability of the observed value falling at any given interval. So, for example, if we want to know the probability that we lose money, we are asking the probability that S is in the interval $S < 0$.

Note that if we can define a cumulative distribution function $F(a) = \Pr(S \le a)$, then we will be able to answer any question related to the probability of events defined by our random variable S, including the event $S < 0$. We call this F the random variable's *distribution function*.

We can estimate the distribution function for the random variable S by using a Monte Carlo simulation to generate many realizations of the random variable. With this code, we run the experiment of having 1,000 people play roulette, over and over, specifically $B = 10,000$ times:

```
n <- 1000
B <- 10000
roulette_winnings <- function(n){
  X <- sample(c(-1,1), n, replace = TRUE, prob=c(9/19, 10/19))
  sum(X)
}
S <- replicate(B, roulette_winnings(n))
```

Now we can ask the following: in our simulations, how often did we get sums less than or equal to a?

```
mean(S <= a)
```

This will be a very good approximation of $F(a)$ and we can easily answer the casino's question: how likely is it that we will lose money? We can see it is quite low:

```
mean(S<0)
#> [1] 0.0456
```

We can visualize the distribution of S by creating a histogram showing the probability $F(b) - F(a)$ for several intervals $(a, b]$:

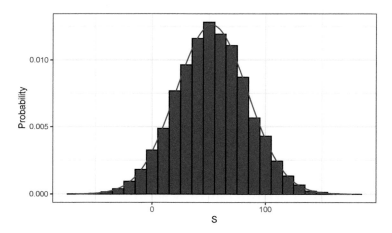

We see that the distribution appears to be approximately normal. A qq-plot will confirm that the normal approximation is close to a perfect approximation for this distribution. If, in fact, the distribution is normal, then all we need to define the distribution is the average and the standard deviation. Because we have the original values from which the distribution is created, we can easily compute these with `mean(S)` and `sd(S)`. The blue curve you see added to the histogram above is a normal density with this average and standard deviation.

This average and this standard deviation have special names. They are referred to as the *expected value* and *standard error* of the random variable S. We will say more about these in the next section.

Statistical theory provides a way to derive the distribution of random variables defined as independent random draws from an urn. Specifically, in our example above, we can show that $(S + n)/2$ follows a binomial distribution. We therefore do not need to run for Monte Carlo simulations to know the probability distribution of S. We did this for illustrative purposes.

We can use the function `dbinom` and `pbinom` to compute the probabilities exactly. For example, to compute $\Pr(S < 0)$ we note that:

$$\Pr(S < 0) = \Pr((S + n)/2 < (0 + n)/2)$$

and we can use the `pbinom` to compute

$$\Pr(S \leq 0)$$

```
n <- 1000
pbinom(n/2, size = n, prob = 10/19)
#> [1] 0.0511
```

Because this is a discrete probability function, to get $\Pr(S < 0)$ rather than $\Pr(S \leq 0)$, we write:

```
pbinom(n/2-1, size = n, prob = 10/19)
#> [1] 0.0448
```

For the details of the binomial distribution, you can consult any basic probability book or even Wikipedia[3].

Here we do not cover these details. Instead, we will discuss an incredibly useful approximation provided by mathematical theory that applies generally to sums and averages of draws from any urn: the Central Limit Theorem (CLT).

14.4 Distributions versus probability distributions

Before we continue, let's make an important distinction and connection between the distribution of a list of numbers and a probability distribution. In the visualization chapter, we described how any list of numbers $x_1, \ldots, x_n$ has a distribution. The definition is quite straightforward. We define $F(a)$ as the function that tells us what proportion of the list is less than or equal to a. Because they are useful summaries when the distribution is approximately normal, we define the average and standard deviation. These are defined with a straightforward operation of the vector containing the list of numbers x:

```
m <- sum(x)/length(x)
s <- sqrt(sum((x - m)^2) / length(x))
```

A random variable X has a distribution function. To define this, we do not need a list of numbers. It is a theoretical concept. In this case, we define the distribution as the $F(a)$ that answers the question: what is the probability that X is less than or equal to a? There is no list of numbers.

However, if X is defined by drawing from an urn with numbers in it, then there is a list: the list of numbers inside the urn. In this case, the distribution of that list is the probability distribution of X and the average and standard deviation of that list are the expected value and standard error of the random variable.

Another way to think about it that does not involve an urn is to run a Monte Carlo simulation and generate a very large list of outcomes of X. These outcomes are a list of numbers. The distribution of this list will be a very good approximation of the probability distribution of X. The longer the list, the better the approximation. The average and standard deviation of this list will approximate the expected value and standard error of the random variable.

14.5 Notation for random variables

In statistical textbooks, upper case letters are used to denote random variables and we follow this convention here. Lower case letters are used for observed values. You will see some notation that includes both. For example, you will see events defined as $X \leq x$. Here X is a random variable, making it a random event, and x is an arbitrary value and not random. So, for example, X might represent the number on a die roll and x will represent an actual value we see 1, 2, 3, 4, 5, or 6. So in this case, the probability of $X = x$ is 1/6 regardless of

---

[3]https://en.wikipedia.org/w/index.php?title=Binomial_distribution

the observed value x. This notation is a bit strange because, when we ask questions about probability, X is not an observed quantity. Instead, it's a random quantity that we will see in the future. We can talk about what we expect it to be, what values are probable, but not what it is. But once we have data, we do see a realization of X. So data scientists talk of what could have been after we see what actually happened.

14.6   The expected value and standard error

We have described sampling models for draws. We will now go over the mathematical theory that lets us approximate the probability distributions for the sum of draws. Once we do this, we will be able to help the casino predict how much money they will make. The same approach we use for the sum of draws will be useful for describing the distribution of averages and proportion which we will need to understand how polls work.

The first important concept to learn is the *expected value*. In statistics books, it is common to use letter E like this:

$$E[X]$$

to denote the expected value of the random variable X.

A random variable will vary around its expected value in a way that if you take the average of many, many draws, the average of the draws will approximate the expected value, getting closer and closer the more draws you take.

Theoretical statistics provides techniques that facilitate the calculation of expected values in different circumstances. For example, a useful formula tells us that the *expected value of a random variable defined by one draw is the average of the numbers in the urn*. In the urn used to model betting on red in roulette, we have 20 one dollars and 18 negative one dollars. The expected value is thus:

$$E[X] = (20 + -18)/38$$

which is about 5 cents. It is a bit counterintuitive to say that X varies around 0.05, when the only values it takes is 1 and -1. One way to make sense of the expected value in this context is by realizing that if we play the game over and over, the casino wins, on average, 5 cents per game. A Monte Carlo simulation confirms this:

```
B <- 10^6
x <- sample(c(-1,1), B, replace = TRUE, prob=c(9/19, 10/19))
mean(x)
#> [1] 0.0517
```

In general, if the urn has two possible outcomes, say a and b, with proportions p and $1 - p$ respectively, the average is:

$$E[X] = ap + b(1 - p)$$

To see this, notice that if there are n beads in the urn, then we have np as and $n(1-p)$ bs and because the average is the sum, $n \times a \times p + n \times b \times (1-p)$, divided by the total n, we get that the average is $ap + b(1-p)$.

Now the reason we define the expected value is because this mathematical definition turns out to be useful for approximating the probability distributions of sum, which then is useful for describing the distribution of averages and proportions. The first useful fact is that the *expected value of the sum of the draws* is:

$$\text{number of draws} \times \text{average of the numbers in the urn}$$

So if 1,000 people play roulette, the casino expects to win, on average, about $1,000 \times \$0.05$ = \$50. But this is an expected value. How different can one observation be from the expected value? The casino really needs to know this. What is the range of possibilities? If negative numbers are too likely, they will not install roulette wheels. Statistical theory once again answers this question. The *standard error* (SE) gives us an idea of the size of the variation around the expected value. In statistics books, it's common to use:

$$\text{SE}[X]$$

to denote the standard error of a random variable.

If our draws are independent, then the *standard error of the sum* is given by the equation:

$$\sqrt{\text{number of draws}} \times \text{standard deviation of the numbers in the urn}$$

Using the definition of standard deviation, we can derive, with a bit of math, that if an urn contains two values a and b with proportions p and $(1-p)$, respectively, the standard deviation is:

$$|\, b - a \,| \sqrt{p(1-p)}.$$

So in our roulette example, the standard deviation of the values inside the urn is: $|\, 1 - (-1) \,| \sqrt{10/19 \times 9/19}$ or:

```
2 * sqrt(90)/19
#> [1] 0.999
```

The standard error tells us the typical difference between a random variable and its expectation. Since one draw is obviously the sum of just one draw, we can use the formula above to calculate that the random variable defined by one draw has an expected value of 0.05 and a standard error of about 1. This makes sense since we either get 1 or -1, with 1 slightly favored over -1.

Using the formula above, the sum of 1,000 people playing has standard error of about \$32:

```
n <- 1000
sqrt(n) * 2 * sqrt(90)/19
#> [1] 31.6
```

As a result, when 1,000 people bet on red, the casino is expected to win $50 with a standard error of $32. It therefore seems like a safe bet. But we still haven't answered the question: how likely is it to lose money? Here the CLT will help.

Advanced note: Before continuing we should point out that exact probability calculations for the casino winnings can be performed with the binomial distribution. However, here we focus on the CLT, which can be generally applied to sums of random variables in a way that the binomial distribution can't.

14.6.1 Population SD versus the sample SD

The standard deviation of a list x (below we use heights as an example) is defined as the square root of the average of the squared differences:

```
library(dslabs)
x <- heights$height
m <- mean(x)
s <- sqrt(mean((x-m)^2))
```

Using mathematical notation we write:

$$\mu = \frac{1}{n}\sum_{i=1}^{n} x_i \sigma = \sqrt{\frac{1}{n}\sum_{i=1}^{n}(x_i - \mu)^2}$$

However, be aware that the sd function returns a slightly different result:

```
identical(s, sd(x))
#> [1] FALSE
s-sd(x)
#> [1] -0.00194
```

This is because the sd function R does not return the sd of the list, but rather uses a formula that estimates standard deviations of a population from a random sample $X_1, \ldots, X_N$ which, for reasons not discussed here, divide the sum of squares by the $N - 1$.

$$\bar{X} = \frac{1}{N}\sum_{i=1}^{N} X_i, \quad s = \sqrt{\frac{1}{N-1}\sum_{i=1}^{N}(X_i - \bar{X})^2}$$

You can see that this is the case by typing:

```
n <- length(x)
s-sd(x)*sqrt((n-1) / n)
#> [1] 0
```

For all the theory discussed here, you need to compute the actual standard deviation as defined:

```
sqrt(mean((x-m)^2))
```

So be careful when using the **sd** function in R. However, keep in mind that throughout the book we sometimes use the **sd** function when we really want the actual SD. This is because when the list size is big, these two are practically equivalent since $\sqrt{(N-1)/N} \approx 1$.

14.7 Central Limit Theorem

The Central Limit Theorem (CLT) tells us that when the number of draws, also called the *sample size*, is large, the probability distribution of the sum of the independent draws is approximately normal. Because sampling models are used for so many data generation processes, the CLT is considered one of the most important mathematical insights in history.

Previously, we discussed that if we know that the distribution of a list of numbers is approximated by the normal distribution, all we need to describe the list are the average and standard deviation. We also know that the same applies to probability distributions. If a random variable has a probability distribution that is approximated with the normal distribution, then all we need to describe the probability distribution are the average and standard deviation, referred to as the expected value and standard error.

We previously ran this Monte Carlo simulation:

```
n <- 1000
B <- 10000
roulette_winnings <- function(n){
  X <- sample(c(-1,1), n, replace = TRUE, prob=c(9/19, 10/19))
  sum(X)
}
S <- replicate(B, roulette_winnings(n))
```

The Central Limit Theorem (CLT) tells us that the sum S is approximated by a normal distribution. Using the formulas above, we know that the expected value and standard error are:

```
n * (20-18)/38
#> [1] 52.6
sqrt(n) * 2 * sqrt(90)/19
#> [1] 31.6
```

The theoretical values above match those obtained with the Monte Carlo simulation:

```
mean(S)
#> [1] 52.2
sd(S)
#> [1] 31.7
```

Using the CLT, we can skip the Monte Carlo simulation and instead compute the probability of the casino losing money using this approximation:

```
mu <- n * (20-18)/38
se <-  sqrt(n) * 2 * sqrt(90)/19
pnorm(0, mu, se)
#> [1] 0.0478
```

which is also in very good agreement with our Monte Carlo result:

```
mean(S < 0)
#> [1] 0.0458
```

14.7.1   How large is large in the Central Limit Theorem?

The CLT works when the number of draws is large. But large is a relative term. In many circumstances as few as 30 draws is enough to make the CLT useful. In some specific instances, as few as 10 is enough. However, these should not be considered general rules. Note, for example, that when the probability of success is very small, we need much larger sample sizes.

By way of illustration, let's consider the lottery. In the lottery, the chances of winning are less than 1 in a million. Thousands of people play so the number of draws is very large. Yet the number of winners, the sum of the draws, range between 0 and 4. This sum is certainly not well approximated by a normal distribution, so the CLT does not apply, even with the very large sample size. This is generally true when the probability of a success is very low. In these cases, the Poisson distribution is more appropriate.

You can examine the properties of the Poisson distribution using dpois and ppois. You can generate random variables following this distribution with rpois. However, we do not cover the theory here. You can learn about the Poisson distribution in any probability textbook and even Wikipedia[4]

14.8   Statistical properties of averages

There are several useful mathematical results that we used above and often employ when working with data. We list them below.

1. The expected value of the sum of random variables is the sum of each random variable's expected value. We can write it like this:

$$\mathrm{E}[X_1 + X_2 + \cdots + X_n] = \mathrm{E}[X_1] + \mathrm{E}[X_2] + \cdots + \mathrm{E}[X_n]$$

If the X are independent draws from the urn, then they all have the same expected value. Let's call it μ and thus:

---

[4]https://en.wikipedia.org/w/index.php?title=Poisson_distribution

$$E[X_1 + X_2 + \cdots + X_n] = n\mu$$

which is another way of writing the result we show above for the sum of draws.

2. The expected value of a non-random constant times a random variable is the non-random constant times the expected value of a random variable. This is easier to explain with symbols:

$$E[aX] = a \times E[X]$$

To see why this is intuitive, consider change of units. If we change the units of a random variable, say from dollars to cents, the expectation should change in the same way. A consequence of the above two facts is that the expected value of the average of independent draws from the same urn is the expected value of the urn, call it μ again:

$$E[(X_1 + X_2 + \cdots + X_n)/n] = E[X_1 + X_2 + \cdots + X_n]/n = n\mu/n = \mu$$

3. The square of the standard error of the sum of **independent** random variables is the sum of the square of the standard error of each random variable. This one is easier to understand in math form:

$$SE[X_1 + X_2 + \cdots + X_n] = \sqrt{SE[X_1]^2 + SE[X_2]^2 + \cdots + SE[X_n]^2}$$

The square of the standard error is referred to as the *variance* in statistical textbooks. Note that this particular property is not as intuitive as the previous three and more in depth explanations can be found in statistics textbooks.

4. The standard error of a non-random constant times a random variable is the non-random constant times the random variable's standard error. As with the expectation:

$$SE[aX] = a \times SE[X]$$

To see why this is intuitive, again think of units.

A consequence of 3 and 4 is that the standard error of the average of independent draws from the same urn is the standard deviation of the urn divided by the square root of n (the number of draws), call it σ:

$$
\begin{aligned}
SE[(X_1 + X_2 + \cdots + X_n)/n] &= SE[X_1 + X_2 + \cdots + X_n]/n \\
&= \sqrt{SE[X_1]^2 + SE[X_2]^2 + \cdots + SE[X_n]^2}/n \\
&= \sqrt{\sigma^2 + \sigma^2 + \cdots + \sigma^2}/n \\
&= \sqrt{n\sigma^2}/n \\
&= \sigma/\sqrt{n}
\end{aligned}
$$

5. If X is a normally distributed random variable, then if a and b are non-random constants, $aX + b$ is also a normally distributed random variable. All we are doing is changing the units of the random variable by multiplying by a, then shifting the center by b.

Note that statistical textbooks use the Greek letters μ and σ to denote the expected value and standard error, respectively. This is because μ is the Greek letter for m, the first letter of *mean*, which is another term used for expected value. Similarly, σ is the Greek letter for s, the first letter of standard error.

14.9   Law of large numbers

An important implication of the final result is that the standard error of the average becomes smaller and smaller as n grows larger. When n is very large, then the standard error is practically 0 and the average of the draws converges to the average of the urn. This is known in statistical textbooks as the law of large numbers or the law of averages.

14.9.1   Misinterpreting law of averages

The law of averages is sometimes misinterpreted. For example, if you toss a coin 5 times and see a head each time, you might hear someone argue that the next toss is probably a tail because of the law of averages: on average we should see 50% heads and 50% tails. A similar argument would be to say that red "is due" on the roulette wheel after seeing black come up five times in a row. These events are independent so the chance of a coin landing heads is 50% regardless of the previous 5. This is also the case for the roulette outcome. The law of averages applies only when the number of draws is very large and not in small samples. After a million tosses, you will definitely see about 50% heads regardless of the outcome of the first five tosses.

Another funny misuse of the law of averages is in sports when TV sportscasters predict a player is about to succeed because they have failed a few times in a row.

14.10   Exercises

1. In American Roulette you can also bet on green. There are 18 reds, 18 blacks and 2 greens (0 and 00). What are the chances the green comes out?

2. The payout for winning on green is $17 dollars. This means that if you bet a dollar and it lands on green, you get $17. Create a sampling model using sample to simulate the random variable X for your winnings. Hint: see the example below for how it should look like when betting on red.

```
x <- sample(c(1,-1), 1, prob = c(9/19, 10/19))
```

3. Compute the expected value of X.

4. Compute the standard error of X.

5. Now create a random variable S that is the sum of your winnings after betting on green

1000 times. Hint: change the argument `size` and `replace` in your answer to question 2. Start your code by setting the seed to 1 with `set.seed(1)`.

6. What is the expected value of S?

7. What is the standard error of S?

8. What is the probability that you end up winning money? Hint: use the CLT.

9. Create a Monte Carlo simulation that generates 1,000 outcomes of S. Compute the average and standard deviation of the resulting list to confirm the results of 6 and 7. Start your code by setting the seed to 1 with `set.seed(1)`.

10. Now check your answer to 8 using the Monte Carlo result.

11. The Monte Carlo result and the CLT approximation are close, but not that close. What could account for this?

    a. 1,000 simulations is not enough. If we do more, they match.
    b. The CLT does not work as well when the probability of success is small. In this case, it was 1/19. If we make the number of roulette plays bigger, they will match better.
    c. The difference is within rounding error.
    d. The CLT only works for averages.

12. Now create a random variable Y that is your average winnings per bet after playing off your winnings after betting on green 1,000 times.

13. What is the expected value of Y?

14. What is the standard error of Y?

15. What is the probability that you end up with winnings per game that are positive? Hint: use the CLT.

16. Create a Monte Carlo simulation that generates 2,500 outcomes of Y. Compute the average and standard deviation of the resulting list to confirm the results of 6 and 7. Start your code by setting the seed to 1 with `set.seed(1)`.

17. Now check your answer to 8 using the Monte Carlo result.

18. The Monte Carlo result and the CLT approximation are now much closer. What could account for this?

    a. We are now computing averages instead of sums.
    b. 2,500 Monte Carlo simulations is not better than 1,000.
    c. The CLT works better when the sample size is larger. We increased from 1,000 to 2,500.
    d. It is not closer. The difference is within rounding error.

14.11   Case study: The Big Short

14.11.1   Interest rates explained with chance model

More complex versions of the sampling models we have discussed are also used by banks to decide interest rates. Suppose you run a small bank that has a history of identifying potential homeowners that can be trusted to make payments. In fact, historically, in a given year, only 2% of your customers default, meaning that they don't pay back the money that you lent them. However, you are aware that if you simply loan money to everybody without interest, you will end up losing money due to this 2%. Although you know 2% of your clients will probably default, you don't know which ones. Yet by charging everybody just a bit extra in interest, you can make up the losses incurred due to that 2% and also cover your operating costs. You can also make a profit, but if you set the interest rates too high, your clients will go to another bank. We use all these facts and some probability theory to decide what interest rate you should charge.

Suppose your bank will give out 1,000 loans for $180,000 this year. Also, after adding up all costs, suppose your bank loses $200,000 per foreclosure. For simplicity, we assume this includes all operational costs. A sampling model for this scenario can be coded like this:

```
n <- 1000
loss_per_foreclosure <- -200000
p <- 0.02
defaults <- sample( c(0,1), n, prob=c(1-p, p), replace = TRUE)
sum(defaults * loss_per_foreclosure)
#> [1] -2800000
```

Note that the total loss defined by the final sum is a random variable. Every time you run the above code, you get a different answer. We can easily construct a Monte Carlo simulation to get an idea of the distribution of this random variable.

```
B <- 10000
losses <- replicate(B, {
    defaults <- sample( c(0,1), n, prob=c(1-p, p), replace = TRUE)
  sum(defaults * loss_per_foreclosure)
})
```

We don't really need a Monte Carlo simulation though. Using what we have learned, the CLT tells us that because our losses are a sum of independent draws, its distribution is approximately normal with expected value and standard errors given by:

```
n*(p*loss_per_foreclosure + (1-p)*0)
#> [1] -4e+06
sqrt(n)*abs(loss_per_foreclosure)*sqrt(p*(1-p))
#> [1] 885438
```

We can now set an interest rate to guarantee that, on average, we break even. Basically, we need to add a quantity x to each loan, which in this case are represented by draws, so that the expected value is 0. If we define l to be the loss per foreclosure, we need:

$$lp + x(1 - p) = 0$$

which implies x is

```
- loss_per_foreclosure*p/(1-p)
#> [1] 4082
```

or an interest rate of 0.023.

However, we still have a problem. Although this interest rate guarantees that on average we break even, there is a 50% chance that we lose money. If our bank loses money, we have to close it down. We therefore need to pick an interest rate that makes it unlikely for this to happen. At the same time, if the interest rate is too high, our clients will go to another bank so we must be willing to take some risks. So let's say that we want our chances of losing money to be 1 in 100, what does the x quantity need to be now? This one is a bit harder. We want the sum S to have:

$$\Pr(S < 0) = 0.01$$

We know that S is approximately normal. The expected value of S is

$$E[S] = \{lp + x(1 - p)\}n$$

with n the number of draws, which in this case represents loans. The standard error is

$$SD[S] = |x - l|\sqrt{np(1 - p)}.$$

Because x is positive and l negative $|x - l| = x - l$. Note that these are just an application of the formulas shown earlier, but using more compact symbols.

Now we are going to use a mathematical "trick" that is very common in statistics. We add and subtract the same quantities to both sides of the event $S < 0$ so that the probability does not change and we end up with a standard normal random variable on the left, which will then permit us to write down an equation with only x as an unknown. This "trick" is as follows:

If $\Pr(S < 0) = 0.01$ then

$$\Pr\left(\frac{S - E[S]}{SE[S]} < \frac{-E[S]}{SE[S]}\right)$$

And remember $E[S]$ and $SE[S]$ are the expected value and standard error of S, respectively. All we did above was add and divide by the same quantity on both sides. We did this because now the term on the left is a standard normal random variable, which we will rename Z. Now we fill in the blanks with the actual formula for expected value and standard error:

$$\Pr\left(Z < \frac{-\{lp + x(1 - p)\}n}{(x - l)\sqrt{np(1 - p)}}\right) = 0.01$$

It may look complicated, but remember that l, p and n are all known amounts, so eventually we will replace them with numbers.

Now because the Z is a normal random with expected value 0 and standard error 1, it means that the quantity on the right side of the $<$ sign must be equal to:

```
qnorm(0.01)
#> [1] -2.33
```

for the equation to hold true. Remember that $z = $`qnorm(0.01)` gives us the value of z for which:

$$\Pr(Z \leq z) = 0.01$$

So this means that the right side of the complicated equation must be $z=$`qnorm(0.01)`.

$$\frac{-\{lp + x(1-p)\}n}{(x-l)\sqrt{np(1-p)}} = z$$

The trick works because we end up with an expression containing x that we know has to be equal to a known quantity z. Solving for x is now simply algebra:

$$x = -l\frac{np - z\sqrt{np(1-p)}}{n(1-p) + z\sqrt{np(1-p)}}$$

which is:

```
l <- loss_per_foreclosure
z <- qnorm(0.01)
x <- -l*( n*p - z*sqrt(n*p*(1-p)))/ ( n*(1-p) + z*sqrt(n*p*(1-p)))
x
#> [1] 6249
```

Our interest rate now goes up to 0.035. This is still a very competitive interest rate. By choosing this interest rate, we now have an expected profit per loan of:

```
loss_per_foreclosure*p + x*(1-p)
#> [1] 2124
```

which is a total expected profit of about:

```
n*(loss_per_foreclosure*p + x*(1-p))
#> [1] 2124198
```

dollars!

We can run a Monte Carlo simulation to double check our theoretical approximations:

```
B <- 100000
profit <- replicate(B, {
    draws <- sample( c(x, loss_per_foreclosure), n,
                     prob=c(1-p, p), replace = TRUE)
    sum(draws)
})
mean(profit)
#> [1] 2121417
mean(profit<0)
#> [1] 0.0123
```

14.11.2   The Big Short

One of your employees points out that since the bank is making 2,124 dollars per loan, the bank should give out more loans! Why just n? You explain that finding those n clients was hard. You need a group that is predictable and that keeps the chances of defaults low. He then points out that even if the probability of default is higher, as long as our expected value is positive, you can minimize your chances of losses by increasing n and relying on the law of large numbers.

He claims that even if the default rate is twice as high, say 4%, if we set the rate just a bit higher than this value:

```
p <- 0.04
r <- (- loss_per_foreclosure*p/(1-p)) / 180000
r
#> [1] 0.0463
```

we will profit. At 5%, we are guaranteed a positive expected value of:

```
r <- 0.05
x <- r*180000
loss_per_foreclosure*p + x * (1-p)
#> [1] 640
```

and can minimize our chances of losing money by simply increasing n since:

$$\Pr(S < 0) = \Pr\left(Z < -\frac{\mathrm{E}[S]}{\mathrm{SE}[S]} \right)$$

with Z a standard normal random variable as shown earlier. If we define μ and σ to be the expected value and standard deviation of the urn, respectively (that is of a single loan), using the formulas above we have: $\mathrm{E}[S] = n\mu$ and $\mathrm{SE}[S] = \sqrt{n}\sigma$. So if we define z=qnorm(0.01), we have:

$$-\frac{n\mu}{\sqrt{n}\sigma} = -\frac{\sqrt{n}\mu}{\sigma} = z$$

which implies that if we let:

$$n \geq z^2\sigma^2/\mu^2$$

we are guaranteed to have a probability of less than 0.01. The implication is that, as long as μ is positive, we can find an n that minimizes the probability of a loss. This is a form of the law of large numbers: when n is large, our average earnings per loan converges to the expected earning μ.

With x fixed, now we can ask what n do we need for the probability to be 0.01? In our example, if we give out:

```
z <- qnorm(0.01)
n <- ceiling((z^2*(x-l)^2*p*(1-p))/(l*p + x*(1-p))^2)
n
#> [1] 22163
```

loans, the probability of losing is about 0.01 and we are expected to earn a total of

```
n*(loss_per_foreclosure*p + x * (1-p))
#> [1] 14184320
```

dollars! We can confirm this with a Monte Carlo simulation:

```
p <- 0.04
x <- 0.05*180000
profit <- replicate(B, {
    draws <- sample( c(x, loss_per_foreclosure), n,
                        prob=c(1-p, p), replace = TRUE)
    sum(draws)
})
mean(profit)
#> [1] 14207076
```

This seems like a no brainer. As a result, your colleague decides to leave your bank and start his own high-risk mortgage company. A few months later, your colleague's bank has gone bankrupt. A book is written and eventually a movie is made relating the mistake your friend, and many others, made. What happened?

Your colleague's scheme was mainly based on this mathematical formula:

$$\text{SE}[(X_1 + X_2 + \cdots + X_n)/n] = \sigma/\sqrt{n}$$

By making n large, we minimize the standard error of our per-loan profit. However, for this rule to hold, the Xs must be independent draws: one person defaulting must be independent of others defaulting. Note that in the case of averaging the **same** event over and over, an extreme example of events that are not independent, we get a standard error that is $\sqrt{n}$ times bigger:

$$\text{SE}[(X_1 + X_1 + \cdots + X_1)/n] = \text{SE}[nX_1/n] = \sigma > \sigma/\sqrt{n}$$

To construct a more realistic simulation than the original one your colleague ran, let's assume there is a global event that affects everybody with high-risk mortgages and changes their probability. We will assume that with 50-50 chance, all the probabilities go up or down slightly to somewhere between 0.03 and 0.05. But it happens to everybody at once, not just one person. These draws are no longer independent.

```
p <- 0.04
x <- 0.05*180000
profit <- replicate(B, {
    new_p <- 0.04 + sample(seq(-0.01, 0.01, length = 100), 1)
    draws <- sample( c(x, loss_per_foreclosure), n,
                          prob=c(1-new_p, new_p), replace = TRUE)
    sum(draws)
})
```

Note that our expected profit is still large:

```
mean(profit)
#> [1] 14093512
```

However, the probability of the bank having negative earnings shoots up to:

```
mean(profit<0)
#> [1] 0.349
```

Even scarier is that the probability of losing more than 10 million dollars is:

```
mean(profit < -10000000)
#> [1] 0.241
```

To understand how this happens look at the distribution:

```
data.frame(profit_in_millions=profit/10^6) %>%
  ggplot(aes(profit_in_millions)) +
  geom_histogram(color="black", binwidth = 5)
```

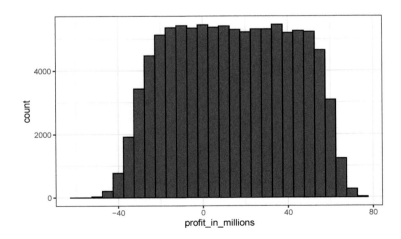

The theory completely breaks down and the random variable has much more variability than expected. The financial meltdown of 2007 was due, among other things, to financial "experts" assuming independence when there was none.

14.12   Exercises

1. Create a random variable S with the earnings of your bank if you give out 10,000 loans, the default rate is 0.3, and you lose \$200,000 in each foreclosure. Hint: use the code we showed in the previous section, but change the parameters.

2. Run a Monte Carlo simulation with 10,000 outcomes for S. Make a histogram of the results.

3. What is the expected value of S?

4. What is the standard error of S?

5. Suppose we give out loans for \$180,000. What should the interest rate be so that our expected value is 0?

6. (Harder) What should the interest rate be so that the chance of losing money is 1 in 20? In math notation, what should the interest rate be so that $\Pr(S < 0) = 0.05$?

7. If the bank wants to minimize the probabilities of losing money, which of the following does **not** make interest rates go up?

    a.   A smaller pool of loans.
    b.   A larger probability of default.
    c.   A smaller required probability of losing money.
    d.   The number of Monte Carlo simulations.

15

Statistical inference

In Chapter 16 we will describe, in some detail, how poll aggregators such as FiveThirtyEight use data to predict election outcomes. To understand how they do this, we first need to learn the basics of *Statistical Inference*, the part of statistics that helps distinguish patterns arising from signal from those arising from chance. Statistical inference is a broad topic and here we go over the very basics using polls as a motivating example. To describe the concepts, we complement the mathematical formulas with Monte Carlo simulations and R code.

15.1 Polls

Opinion polling has been conducted since the 19th century. The general goal is to describe the opinions held by a specific population on a given set of topics. In recent times, these polls have been pervasive during presidential elections. Polls are useful when interviewing every member of a particular population is logistically impossible. The general strategy is to interview a smaller group, chosen at random, and then infer the opinions of the entire population from the opinions of the smaller group. Statistical theory is used to justify the process. This theory is referred to as *inference* and it is the main topic of this chapter.

Perhaps the best known opinion polls are those conducted to determine which candidate is preferred by voters in a given election. Political strategists make extensive use of polls to decide, among other things, how to invest resources. For example, they may want to know in which geographical locations to focus their "get out the vote" efforts.

Elections are a particularly interesting case of opinion polls because the actual opinion of the entire population is revealed on election day. Of course, it costs millions of dollars to run an actual election which makes polling a cost effective strategy for those that want to forecast the results.

Although typically the results of these polls are kept private, similar polls are conducted by news organizations because results tend to be of interest to the general public and made public. We will eventually be looking at such data.

Real Clear Politics[1] is an example of a news aggregator that organizes and publishes poll results. For example, they present the following poll results reporting estimates of the popular vote for the 2016 presidential election[2]:

---

[1] http://www.realclearpolitics.com
[2] http://www.realclearpolitics.com/epolls/2016/president/us/general_election_trump_vs_clinton-5491.html

Poll	Date	Sample	MoE	Clinton	Trump	Spread
Final Results	–	–	–	48.2	46.1	Clinton +2.1
RCP Average	11/1 - 11/7	–	–	46.8	43.6	Clinton +3.2
Bloomberg	11/4 - 11/6	799 LV	3.5	46.0	43.0	Clinton +3
IBD	11/4 - 11/7	1107 LV	3.1	43.0	42.0	Clinton +1
Economist	11/4 - 11/7	3669 LV	–	49.0	45.0	Clinton +4
LA Times	11/1 - 11/7	2935 LV	4.5	44.0	47.0	Trump +3
ABC	11/3 - 11/6	2220 LV	2.5	49.0	46.0	Clinton +3
FOX News	11/3 - 11/6	1295 LV	2.5	48.0	44.0	Clinton +4
Monmouth	11/3 - 11/6	748 LV	3.6	50.0	44.0	Clinton +6
NBC News	11/3 - 11/5	1282 LV	2.7	48.0	43.0	Clinton +5
CBS News	11/2 - 11/6	1426 LV	3.0	47.0	43.0	Clinton +4
Reuters	11/2 - 11/6	2196 LV	2.3	44.0	39.0	Clinton +5

Although in the United States the popular vote does not determine the result of the presidential election, we will use it as an illustrative and simple example of how well polls work. Forecasting the election is a more complex process since it involves combining results from 50 states and DC and we describe it in Section 16.8.

Let's make some observations about the table above. First, note that different polls, all taken days before the election, report a different *spread*: the estimated difference between support for the two candidates. Notice also that the reported spreads hover around what ended up being the actual result: Clinton won the popular vote by 2.1%. We also see a column titled **MoE** which stands for *margin of error*.

In this section, we will show how the probability concepts we learned in the previous chapter can be applied to develop the statistical approaches that make polls an effective tool. We will learn the statistical concepts necessary to define *estimates* and *margins of errors*, and show how we can use these to forecast final results relatively well and also provide an estimate of the precision of our forecast. Once we learn this, we will be able to understand two concepts that are ubiquitous in data science: *confidence intervals* and *p-values*. Finally, to understand probabilistic statements about the probability of a candidate winning, we will have to learn about Bayesian modeling. In the final sections, we put it all together to recreate the simplified version of the FiveThirtyEight model and apply it to the 2016 election.

We start by connecting probability theory to the task of using polls to learn about a population.

15.1.1   The sampling model for polls

To help us understand the connection between polls and what we have learned, let's construct a similar situation to the one pollsters face. To mimic the challenge real pollsters face in terms of competing with other pollsters for media attention, we will use an urn full of beads to represent voters and pretend we are competing for a $25 dollar prize. The challenge is to guess the spread between the proportion of blue and red beads in this urn (in this case, a pickle jar):

Before making a prediction, you can take a sample (with replacement) from the urn. To mimic the fact that running polls is expensive, it costs you $0.10 per each bead you sample. Therefore, if your sample size is 250, and you win, you will break even since you will pay $25 to collect your $25 prize. Your entry into the competition can be an interval. If the interval you submit contains the true proportion, you get half what you paid and pass to the second phase of the competition. In the second phase, the entry with the smallest interval is selected as the winner.

The **dslabs** package includes a function that shows a random draw from this urn:

```
library(tidyverse)
library(dslabs)
take_poll(25)
```

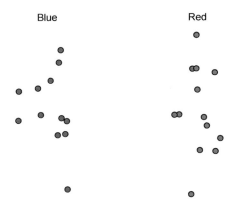

Think about how you would construct your interval based on the data shown above.

We have just described a simple sampling model for opinion polls. The beads inside the urn represent the individuals that will vote on election day. Those that will vote for the Republican candidate are represented with red beads and the Democrats with the blue beads. For simplicity, assume there are no other colors. That is, that there are just two parties: Republican and Democratic.

15.2   Populations, samples, parameters, and estimates

We want to predict the proportion of blue beads in the urn. Let's call this quantity p, which then tells us the proportion of red beads $1 - p$, and the spread $p - (1 - p)$, which simplifies to $2p - 1$.

In statistical textbooks, the beads in the urn are called the *population*. The proportion of blue beads in the population p is called a *parameter*. The 25 beads we see in the previous plot are called a *sample*. The task of statistical inference is to predict the parameter p using the observed data in the sample.

Can we do this with the 25 observations above? It is certainly informative. For example, given that we see 13 red and 12 blue beads, it is unlikely that $p > .9$ or $p < .1$. But are we ready to predict with certainty that there are more red beads than blue in the jar?

We want to construct an estimate of p using only the information we observe. An estimate should be thought of as a summary of the observed data that we think is informative about the parameter of interest. It seems intuitive to think that the proportion of blue beads in the sample 0.48 must be at least related to the actual proportion p. But do we simply predict p to be 0.48? First, remember that the sample proportion is a random variable. If we run the command `take_poll(25)` four times, we get a different answer each time, since the sample proportion is a random variable.

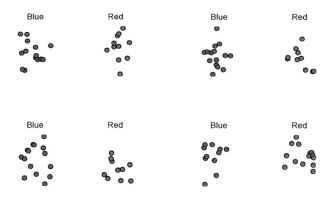

Note that in the four random samples shown above, the sample proportions range from 0.44 to 0.60. By describing the distribution of this random variable, we will be able to gain insights into how good this estimate is and how we can make it better.

15.2.1   The sample average

Conducting an opinion poll is being modeled as taking a random sample from an urn. We are proposing the use of the proportion of blue beads in our sample as an *estimate* of the parameter p. Once we have this estimate, we can easily report an estimate for the spread $2p - 1$, but for simplicity we will illustrate the concepts for estimating p. We will use our knowledge of probability to defend our use of the sample proportion and quantify how close we think it is from the population proportion p.

We start by defining the random variable X as: $X = 1$ if we pick a blue bead at random and $X = 0$ if it is red. This implies that the population is a list of 0s and 1s. If we sample N beads, then the average of the draws $X_1, \ldots, X_N$ is equivalent to the proportion of blue beads in our sample. This is because adding the Xs is equivalent to counting the blue beads and dividing this count by the total N is equivalent to computing a proportion. We use the symbol $\bar{X}$ to represent this average. In general, in statistics textbooks a bar on top of a symbol means the average. The theory we just learned about the sum of draws becomes useful because the average is a sum of draws multiplied by the constant $1/N$:

$$\bar{X} = 1/N \times \sum_{i=1}^{N} X_i$$

For simplicity, let's assume that the draws are independent: after we see each sampled bead, we return it to the urn. In this case, what do we know about the distribution of the sum of draws? First, we know that the expected value of the sum of draws is N times the average of the values in the urn. We know that the average of the 0s and 1s in the urn must be p, the proportion of blue beads.

Here we encounter an important difference with what we did in the Probability chapter: we don't know what is in the urn. We know there are blue and red beads, but we don't know how many of each. This is what we want to find out: we are trying to **estimate** p.

15.2.2 Parameters

Just like we use variables to define unknowns in systems of equations, in statistical inference we define *parameters* to define unknown parts of our models. In the urn model which we are using to mimic an opinion poll, we do not know the proportion of blue beads in the urn. We define the parameters p to represent this quantity. p is the average of the urn because if we take the average of the 1s (blue) and 0s (red), we get the proportion of blue beads. Since our main goal is figuring out what is p, we are going to *estimate this parameter*.

The ideas presented here on how we estimate parameters, and provide insights into how good these estimates are, extrapolate to many data science tasks. For example, we may want to determine the difference in health improvement between patients receiving treatment and a control group. We may ask, what are the health effects of smoking on a population? What are the differences in racial groups of fatal shootings by police? What is the rate of change in life expectancy in the US during the last 10 years? All these questions can be framed as a task of estimating a parameter from a sample.

15.2.3 Polling versus forecasting

Before we continue, let's make an important clarification related to the practical problem of forecasting the election. If a poll is conducted four months before the election, it is estimating the p for that moment and not for election day. The p for election night might be different since people's opinions fluctuate through time. The polls provided the night before the election tend to be the most accurate since opinions don't change that much in a day.

However, forecasters try to build tools that model how opinions vary across time and try to predict the election night results taking into consideration the fact that opinions fluctuate. We will describe some approaches for doing this in a later section.

15.2.4   Properties of our estimate: expected value and standard error

To understand how good our estimate is, we will describe the statistical properties of the random variable defined above: the sample proportion $\bar{X}$. Remember that $\bar{X}$ is the sum of independent draws so the rules we covered in the probability chapter apply.

Using what we have learned, the expected value of the sum $N\bar{X}$ is $N\times$ the average of the urn, p. So dividing by the non-random constant N gives us that the expected value of the average $\bar{X}$ is p. We can write it using our mathematical notation:

$$\mathrm{E}(\bar{X}) = p$$

We can also use what we learned to figure out the standard error: the standard error of the sum is $\sqrt{N}\times$ the standard deviation of the urn. Can we compute the standard error of the urn? We learned a formula that tells us that it is $(1 - 0)\sqrt{p(1 - p)} = \sqrt{p(1 - p)}$. Because we are dividing the sum by N, we arrive at the following formula for the standard error of the average:

$$\mathrm{SE}(\bar{X}) = \sqrt{p(1 - p)/N}$$

This result reveals the power of polls. The expected value of the sample proportion $\bar{X}$ is the parameter of interest p and we can make the standard error as small as we want by increasing N. The law of large numbers tells us that with a large enough poll, our estimate converges to p.

If we take a large enough poll to make our standard error about 1%, we will be quite certain about who will win. But how large does the poll have to be for the standard error to be this small?

One problem is that we do not know p, so we can't compute the standard error. However, for illustrative purposes, let's assume that $p = 0.51$ and make a plot of the standard error versus the sample size N:

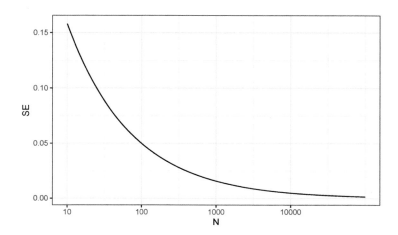

From the plot we see that we would need a poll of over 10,000 people to get the standard error that low. We rarely see polls of this size due in part to costs. From the Real Clear Politics table, we learn that the sample sizes in opinion polls range from 500-3,500 people. For a sample size of 1,000 and $p = 0.51$, the standard error is:

```
sqrt(p*(1-p))/sqrt(1000)
#> [1] 0.0158
```

or 1.5 percentage points. So even with large polls, for close elections, $\bar{X}$ can lead us astray if we don't realize it is a random variable. Nonetheless, we can actually say more about how close we get the p and we do that in Section 15.4.

15.3 Exercises

1. Suppose you poll a population in which a proportion p of voters are Democrats and $1 - p$ are Republicans. Your sample size is $N = 25$. Consider the random variable S which is the **total** number of Democrats in your sample. What is the expected value of this random variable? Hint: it's a function of p.

2. What is the standard error of S ? Hint: it's a function of p.

3. Consider the random variable S/N. This is equivalent to the sample average, which we have been denoting as $\bar{X}$. What is the expected value of the $\bar{X}$? Hint: it's a function of p.

4. What is the standard error of $\bar{X}$? Hint: it's a function of p.

5. Write a line of code that gives you the standard error se for the problem above for several values of p, specifically for p <- seq(0, 1, length = 100). Make a plot of se versus p.

6. Copy the code above and put it inside a for-loop to make the plot for $N = 25$, $N = 100$, and $N = 1000$.

7. If we are interested in the difference in proportions, $p - (1 - p)$, our estimate is $d = \bar{X} - (1 - \bar{X})$. Use the rules we learned about sums of random variables and scaled random variables to derive the expected value of d.

8. What is the standard error of d?

9. If the actual $p = .45$, it means the Republicans are winning by a relatively large margin since $d = -.1$, which is a 10% margin of victory. In this case, what is the standard error of $2\hat{X} - 1$ if we take a sample of $N = 25$?

10. Given the answer to 9, which of the following best describes your strategy of using a sample size of $N = 25$?

    a. The expected value of our estimate $2\bar{X} - 1$ is d, so our prediction will be right on.

    b. Our standard error is larger than the difference, so the chances of $2\bar{X} - 1$ being positive and throwing us off were not that small. We should pick a larger sample size.

    c. The difference is 10% and the standard error is about 0.2, therefore much smaller than the difference.

d. Because we don't know p, we have no way of knowing that making N larger would
    actually improve our standard error.

15.4   Central Limit Theorem in practice

The CLT tells us that the distribution function for a sum of draws is approximately normal.
We also learned that dividing a normally distributed random variable by a constant is also
a normally distributed variable. This implies that the distribution of $\bar{X}$ is approximately
normal.

In summary, we have that $\bar{X}$ has an approximately normal distribution with expected value
p and standard error $\sqrt{p(1-p)/N}$.

Now how does this help us? Suppose we want to know what is the probability that we are
within 1% from p. We are basically asking what is

$$\Pr(|\bar{X} - p| \leq .01)$$

which is the same as:

$$\Pr(\bar{X} \leq p + .01) - \Pr(\bar{X} \leq p - .01)$$

Can we answer this question? We can use the mathematical trick we learned in the previous
chapter. Subtract the expected value and divide by the standard error to get a standard
normal random variable, call it Z, on the left. Since p is the expected value and $\mathrm{SE}(\bar{X}) =$
$\sqrt{p(1-p)/N}$ is the standard error we get:

$$\Pr\left(Z \leq \frac{.01}{\mathrm{SE}(\bar{X})}\right) - \Pr\left(Z \leq -\frac{.01}{\mathrm{SE}(\bar{X})}\right)$$

One problem we have is that since we don't know p, we don't know $\mathrm{SE}(\bar{X})$. But it turns out
that the CLT still works if we estimate the standard error by using $\bar{X}$ in place of p. We say
that we *plug-in* the estimate. Our estimate of the standard error is therefore:

$$\hat{\mathrm{SE}}(\bar{X}) = \sqrt{\bar{X}(1 - \bar{X})/N}$$

In statistics textbooks, we use a little hat to denote estimates. The estimate can be constructed
using the observed data and N.

Now we continue with our calculation, but dividing by $\hat{\mathrm{SE}}(\bar{X}) = \sqrt{\bar{X}(1 - \bar{X})/N})$ instead.
In our first sample we had 12 blue and 13 red so $\bar{X} = 0.48$ and our estimate of standard
error is:

```
x_hat <- 0.48
se <- sqrt(x_hat*(1-x_hat)/25)
se
#> [1] 0.0999
```

And now we can answer the question of the probability of being close to p. The answer is:

```
pnorm(0.01/se) - pnorm(-0.01/se)
#> [1] 0.0797
```

Therefore, there is a small chance that we will be close. A poll of only $N = 25$ people is not really very useful, at least not for a close election.

Earlier we mentioned the *margin of error*. Now we can define it because it is simply two times the standard error, which we can now estimate. In our case it is:

```
1.96*se
#> [1] 0.196
```

Why do we multiply by 1.96? Because if you ask what is the probability that we are within 1.96 standard errors from p, we get:

$$\Pr\left(Z \leq 1.96\,\mathrm{SE}(\bar{X})/\mathrm{SE}(\bar{X})\right) - \Pr\left(Z \leq -1.96\,\mathrm{SE}(\bar{X})/\mathrm{SE}(\bar{X})\right)$$

which is:

$$\Pr\left(Z \leq 1.96\right) - \Pr\left(Z \leq -1.96\right)$$

which we know is about 95%:

```
pnorm(1.96)-pnorm(-1.96)
#> [1] 0.95
```

Hence, there is a 95% probability that $\bar{X}$ will be within $1.96 \times \hat{SE}(\bar{X})$, in our case within about 0.2, of p. Note that 95% is somewhat of an arbitrary choice and sometimes other percentages are used, but it is the most commonly used value to define margin of error. We often round 1.96 up to 2 for simplicity of presentation.

In summary, the CLT tells us that our poll based on a sample size of 25 is not very useful. We don't really learn much when the margin of error is this large. All we can really say is that the popular vote will not be won by a large margin. This is why pollsters tend to use larger sample sizes.

From the table above, we see that typical sample sizes range from 700 to 3500. To see how this gives us a much more practical result, notice that if we had obtained a $\bar{X}=0.48$ with a sample size of 2,000, our standard error $\hat{SE}(\bar{X})$ would have been 0.011. So our result is an estimate of 48% with a margin of error of 2%. In this case, the result is much more informative and would make us think that there are more red balls than blue. Keep in mind, however, that this is hypothetical. We did not take a poll of 2,000 since we don't want to ruin the competition.

15.4.1   A Monte Carlo simulation

Suppose we want to use a Monte Carlo simulation to corroborate the tools we have built using probability theory. To create the simulation, we would write code like this:

```
B <- 10000
N <- 1000
x_hat <- replicate(B, {
  x <- sample(c(0,1), size = N, replace = TRUE, prob = c(1-p, p))
  mean(x)
})
```

The problem is, of course, we don't know **p**. We could construct an urn like the one pictured above and run an analog (without a computer) simulation. It would take a long time, but you could take 10,000 samples, count the beads and keep track of the proportions of blue. We can use the function `take_poll(n=1000)` instead of drawing from an actual urn, but it would still take time to count the beads and enter the results.

One thing we therefore do to corroborate theoretical results is to pick one or several values of **p** and run the simulations. Let's set **p=0.45**. We can then simulate a poll:

```
p <- 0.45
N <- 1000

x <- sample(c(0,1), size = N, replace = TRUE, prob = c(1-p, p))
x_hat <- mean(x)
```

In this particular sample, our estimate is `x_hat`. We can use that code to do a Monte Carlo simulation:

```
B <- 10000
x_hat <- replicate(B, {
  x <- sample(c(0,1), size = N, replace = TRUE, prob = c(1-p, p))
  mean(x)
})
```

To review, the theory tells us that $\bar{X}$ is approximately normally distributed, has expected value $p = 0.45$ and standard error $\sqrt{p(1-p)/N} = 0.016$. The simulation confirms this:

```
mean(x_hat)
#> [1] 0.45
sd(x_hat)
#> [1] 0.0156
```

A histogram and qq-plot confirm that the normal approximation is accurate as well:

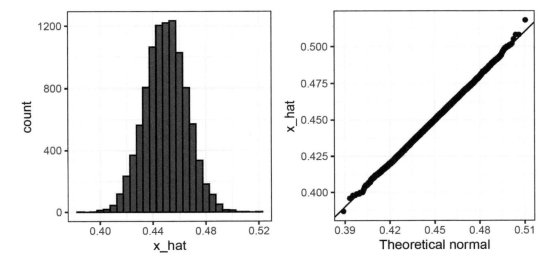

Of course, in real life we would never be able to run such an experiment because we don't know p. But we could run it for various values of p and N and see that the theory does indeed work well for most values. You can easily do this by re-running the code above after changing p and N.

15.4.2 The spread

The competition is to predict the spread, not the proportion p. However, because we are assuming there are only two parties, we know that the spread is $p - (1 - p) = 2p - 1$. As a result, everything we have done can easily be adapted to an estimate of $2p - 1$. Once we have our estimate $\bar{X}$ and $\hat{SE}(\bar{X})$, we estimate the spread with $2\bar{X} - 1$ and, since we are multiplying by 2, the standard error is $2\hat{SE}(\bar{X})$. Note that subtracting 1 does not add any variability so it does not affect the standard error.

For our 25 item sample above, our estimate p is .48 with margin of error .20 and our estimate of the spread is 0.04 with margin of error .40. Again, not a very useful sample size. However, the point is that once we have an estimate and standard error for p, we have it for the spread $2p - 1$.

15.4.3 Bias: why not run a very large poll?

For realistic values of p, say from 0.35 to 0.65, if we run a very large poll with 100,000 people, theory tells us that we would predict the election perfectly since the largest possible margin of error is around 0.3%:

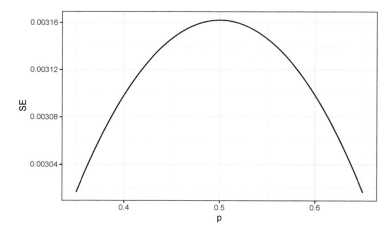

One reason is that running such a poll is very expensive. Another possibly more important reason is that theory has its limitations. Polling is much more complicated than picking beads from an urn. Some people might lie to pollsters and others might not have phones. But perhaps the most important way an actual poll differs from an urn model is that we actually don't know for sure who is in our population and who is not. How do we know who is going to vote? Are we reaching all possible voters? Hence, even if our margin of error is very small, it might not be exactly right that our expected value is p. We call this bias. Historically, we observe that polls are indeed biased, although not by that much. The typical bias appears to be about 1-2%. This makes election forecasting a bit more interesting and we will talk about how to model this in a later chapter.

15.5   Exercises

1. Write an *urn model* function that takes the proportion of Democrats p and the sample size N as arguments and returns the sample average if Democrats are 1s and Republicans are 0s. Call the function `take_sample`.

2. Now assume p <- 0.45 and that your sample size is $N = 100$. Take a sample 10,000 times and save the vector of `mean(X) - p` into an object called `errors`. Hint: use the function you wrote for exercise 1 to write this in one line of code.

3. The vector `errors` contains, for each simulated sample, the difference between the actual p and our estimate $\bar{X}$. We refer to this difference as the *error*. Compute the average and make a histogram of the errors generated in the Monte Carlo simulation and select which of the following best describes their distributions:

```
mean(errors)
hist(errors)
```

   a. The errors are all about 0.05.
   b. The errors are all about -0.05.
   c. The errors are symmetrically distributed around 0.
   d. The errors range from -1 to 1.

4. The error $\bar{X} - p$ is a random variable. In practice, the error is not observed because we do not know p. Here we observe it because we constructed the simulation. What is the average size of the error if we define the size by taking the absolute value $|\bar{X} - p|$?

5. The standard error is related to the typical **size** of the error we make when predicting. We say **size** because we just saw that the errors are centered around 0, so thus the average error value is 0. For mathematical reasons related to the Central Limit Theorem, we actually use the standard deviation of **errors** rather than the average of the absolute values to quantify the typical size. What is this standard deviation of the errors?

6. The theory we just learned tells us what this standard deviation is going to be because it is the standard error of $\bar{X}$. What does theory tell us is the standard error of $\bar{X}$ for a sample size of 100?

7. In practice, we don't know p, so we construct an estimate of the theoretical prediction based by plugging in $\bar{X}$ for p. Compute this estimate. Set the seed at 1 with `set.seed(1)`.

8. Note how close the standard error estimates obtained from the Monte Carlo simulation (exercise 5), the theoretical prediction (exercise 6), and the estimate of the theoretical prediction (exercise 7) are. The theory is working and it gives us a practical approach to knowing the typical error we will make if we predict p with $\bar{X}$. Another advantage that the theoretical result provides is that it gives an idea of how large a sample size is required to obtain the precision we need. Earlier we learned that the largest standard errors occur for $p = 0.5$. Create a plot of the largest standard error for N ranging from 100 to 5,000. Based on this plot, how large does the sample size have to be to have a standard error of about 1%?

   a. 100
   b. 500
   c. 2,500
   d. 4,000

9. For sample size $N = 100$, the central limit theorem tells us that the distribution of $\bar{X}$ is:

   a. practically equal to p.
   b. approximately normal with expected value p and standard error $\sqrt{p(1-p)/N}$.
   c. approximately normal with expected value $\bar{X}$ and standard error $\sqrt{\bar{X}(1-\bar{X})/N}$.
   d. not a random variable.

10. Based on the answer from exercise 8, the error $\bar{X} - p$ is:

   a. practically equal to 0.
   b. approximately normal with expected value 0 and standard error $\sqrt{p(1-p)/N}$.
   c. approximately normal with expected value p and standard error $\sqrt{p(1-p)/N}$.
   d. not a random variable.

11. To corroborate your answer to exercise 9, make a qq-plot of the **errors** you generated in exercise 2 to see if they follow a normal distribution.

12. If $p = 0.45$ and $N = 100$ as in exercise 2, use the CLT to estimate the probability that $\bar{X} > 0.5$. You can assume you know $p = 0.45$ for this calculation.

13. Assume you are in a practical situation and you don't know p. Take a sample of size

$N = 100$ and obtain a sample average of $\bar{X} = 0.51$. What is the CLT approximation for the probability that your error is equal to or larger than 0.01?

15.6 Confidence intervals

Confidence intervals are a very useful concept widely employed by data analysts. A version of these that are commonly seen come from the ggplot geometry geom_smooth. Here is an example using a temperature dataset available in R:

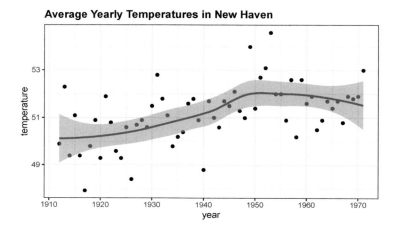

In the Machine Learning part we will learn how the curve is formed, but for now consider the shaded area around the curve. This is created using the concept of confidence intervals.

In our earlier competition, you were asked to give an interval. If the interval you submitted includes the p, you get half the money you spent on your "poll" back and pass to the next stage of the competition. One way to pass to the second round is to report a very large interval. For example, the interval $[0, 1]$ is guaranteed to include p. However, with an interval this big, we have no chance of winning the competition. Similarly, if you are an election forecaster and predict the spread will be between -100% and 100%, you will be ridiculed for stating the obvious. Even a smaller interval, such as saying the spread will be between -10 and 10%, will not be considered serious.

On the other hand, the smaller the interval we report, the smaller our chances are of winning the prize. Likewise, a bold pollster that reports very small intervals and misses the mark most of the time will not be considered a good pollster. We want to be somewhere in between.

We can use the statistical theory we have learned to compute the probability of any given interval including p. If we are asked to create an interval with, say, a 95% chance of including p, we can do that as well. These are called 95% confidence intervals.

When a pollster reports an estimate and a margin of error, they are, in a way, reporting a 95% confidence interval. Let's show how this works mathematically.

We want to know the probability that the interval $[\bar{X} - 2\hat{\text{SE}}(\bar{X}), \bar{X} - 2\hat{\text{SE}}(\bar{X})]$ contains the true proportion p. First, consider that the start and end of these intervals are random variables: every time we take a sample, they change. To illustrate this, run the Monte Carlo simulation above twice. We use the same parameters as above:

```
p <- 0.45
N <- 1000
```

And notice that the interval here:

```
x <- sample(c(0, 1), size = N, replace = TRUE, prob = c(1-p, p))
x_hat <- mean(x)
se_hat <- sqrt(x_hat * (1 - x_hat) / N)
c(x_hat - 1.96 * se_hat, x_hat + 1.96 * se_hat)
#> [1] 0.443 0.505
```

is different from this one:

```
x <- sample(c(0,1), size=N, replace=TRUE, prob=c(1-p, p))
x_hat <- mean(x)
se_hat <- sqrt(x_hat * (1 - x_hat) / N)
c(x_hat - 1.96 * se_hat, x_hat + 1.96 * se_hat)
#> [1] 0.417 0.479
```

Keep sampling and creating intervals and you will see the random variation.

To determine the probability that the interval includes p, we need to compute this:

$$\Pr\left(\bar{X} - 1.96\hat{\text{SE}}(\bar{X}) \leq p \leq \bar{X} + 1.96\hat{\text{SE}}(\bar{X})\right)$$

By subtracting and dividing the same quantities in all parts of the equation, we get that the above is equivalent to:

$$\Pr\left(-1.96 \leq \frac{\bar{X} - p}{\hat{\text{SE}}(\bar{X})} \leq 1.96\right)$$

The term in the middle is an approximately normal random variable with expected value 0 and standard error 1, which we have been denoting with Z, so we have:

$$\Pr\left(-1.96 \leq Z \leq 1.96\right)$$

which we can quickly compute using :

```
pnorm(1.96) - pnorm(-1.96)
#> [1] 0.95
```

proving that we have a 95% probability.

If we want to have a larger probability, say 99%, we need to multiply by whatever z satisfies the following:

$$\Pr\left(-z \le Z \le z\right) = 0.99$$

Using:

```
z <- qnorm(0.995)
z
#> [1] 2.58
```

will achieve this because by definition `pnorm(qnorm(0.995))` is 0.995 and by symmetry `pnorm(1-qnorm(0.995))` is 1 - 0.995. As a consequence, we have that:

```
pnorm(z) - pnorm(-z)
#> [1] 0.99
```

is 0.995 - 0.005 = 0.99. We can use this approach for any proportion p: we set `z = qnorm(1 - (1 - p)/2)` because $1 - (1 - p)/2 + (1 - p)/2 = p$.

So, for example, for $p = 0.95$, $1 - (1 - p)/2 = 0.975$ and we get the 1.96 we have been using:

```
qnorm(0.975)
#> [1] 1.96
```

15.6.1   A Monte Carlo simulation

We can run a Monte Carlo simulation to confirm that, in fact, a 95% confidence interval includes p 95% of the time.

```
N <- 1000
B <- 10000
inside <- replicate(B, {
  x <- sample(c(0,1), size = N, replace = TRUE, prob = c(1-p, p))
  x_hat <- mean(x)
  se_hat <- sqrt(x_hat * (1 - x_hat) / N)
  between(p, x_hat - 1.96 * se_hat, x_hat + 1.96 * se_hat)
})
mean(inside)
#> [1] 0.948
```

The following plot shows the first 100 confidence intervals. In this case, we created the simulation so the black line denotes the parameter we are trying to estimate:

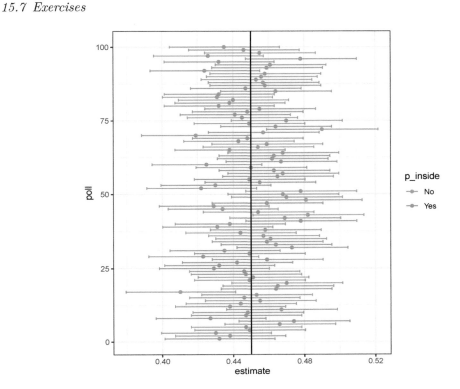

15.6.2 The correct language

When using the theory we described above, it is important to remember that it is the intervals that are random, not p. In the plot above, we can see the random intervals moving around and p, represented with the vertical line, staying in the same place. The proportion of blue in the urn p is not. So the 95% relates to the probability that this random interval falls on top of p. Saying the p has a 95% chance of being between this and that is technically an incorrect statement because p is not random.

15.7 Exercises

For these exercises, we will use actual polls from the 2016 election. You can load the data from the **dslabs** package.

```
library(dslabs)
data("polls_us_election_2016")
```

Specifically, we will use all the national polls that ended within one week before the election.

```
library(tidyverse)
polls <- polls_us_election_2016 %>%
   filter(enddate >= "2016-10-31" & state == "U.S.")
```

1. For the first poll, you can obtain the samples size and estimated Clinton percentage with:

```
N <- polls$samplesize[1]
x_hat <- polls$rawpoll_clinton[1]/100
```

Assume there are only two candidates and construct a 95% confidence interval for the election night proportion p.

2. Now use `dplyr` to add a confidence interval as two columns, call them `lower` and `upper`, to the object `poll`. Then use `select` to show the `pollster, enddate, x_hat,lower, upper` variables. Hint: define temporary columns `x_hat` and `se_hat`.

3. The final tally for the popular vote was Clinton 48.2% and Trump 46.1%. Add a column, call it `hit`, to the previous table stating if the confidence interval included the true proportion $p = 0.482$ or not.

4. For the table you just created, what proportion of confidence intervals included p?

5. If these confidence intervals are constructed correctly, and the theory holds up, what proportion should include p?

6. A much smaller proportion of the polls than expected produce confidence intervals containing p. If you look closely at the table, you will see that most polls that fail to include p are underestimating. The reason for this is undecided voters, individuals polled that do not yet know who they will vote for or do not want to say. Because, historically, undecideds divide evenly between the two main candidates on election day, it is more informative to estimate the spread or the difference between the proportion of two candidates d, which in this election was $0.482 - 0.461 = 0.021$. Assume that there are only two parties and that $d = 2p - 1$, redefine `polls` as below and re-do exercise 1, but for the difference.

```
polls <- polls_us_election_2016 %>%
  filter(enddate >= "2016-10-31" & state == "U.S.")  %>%
  mutate(d_hat = rawpoll_clinton / 100 - rawpoll_trump / 100)
```

7. Now repeat exercise 3, but for the difference.

8. Now repeat exercise 4, but for the difference.

9. Although the proportion of confidence intervals goes up substantially, it is still lower than 0.95. In the next chapter, we learn the reason for this. To motivate this, make a plot of the error, the difference between each poll's estimate and the actual $d = 0.021$. Stratify by pollster.

10. Redo the plot that you made for exercise 9, but only for pollsters that took five or more polls.

15.8 Power

Pollsters are not successful at providing correct confidence intervals, but rather at predicting who will win. When we took a 25 bead sample size, the confidence interval for the spread:

```
N <- 25
x_hat <- 0.48
(2 * x_hat - 1) + c(-1.96, 1.96) * 2 * sqrt(x_hat * (1 - x_hat) / N)
#> [1] -0.432  0.352
```

includes 0. If this were a poll and we were forced to make a declaration, we would have to say it was a "toss-up".

A problem with our poll results is that given the sample size and the value of p, we would have to sacrifice on the probability of an incorrect call to create an interval that does not include 0.

This does not mean that the election is close. It only means that we have a small sample size. In statistical textbooks this is called lack of *power*. In the context of polls, *power* is the probability of detecting spreads different from 0.

By increasing our sample size, we lower our standard error and therefore have a much better chance of detecting the direction of the spread.

15.9 p-values

p-values are ubiquitous in the scientific literature. They are related to confidence intervals so we introduce the concept here.

Let's consider the blue and red beads. Suppose that rather than wanting an estimate of the spread or the proportion of blue, I am interested only in the question: are there more blue beads or red beads? I want to know if the spread $2p - 1 > 0$.

Say we take a random sample of $N = 100$ and we observe 52 blue beads, which gives us $2\bar{X} - 1 = 0.04$. This seems to be pointing to the existence of more blue than red beads since 0.04 is larger than 0. However, as data scientists we need to be skeptical. We know there is chance involved in this process and we could get a 52 even when the actual spread is 0. We call the assumption that the spread is $2p - 1 = 0$ a *null hypothesis*. The null hypothesis is the skeptic's hypothesis. We have observed a random variable $2 * \bar{X} - 1 = 0.04$ and the p-value is the answer to the question: how likely is it to see a value this large, when the null hypothesis is true? So we write:

$$\Pr(|\bar{X} - 0.5| > 0.02)$$

assuming the $2p - 1 = 0$ or $p = 0.5$. Under the null hypothesis we know that:

$$\sqrt{N}\frac{\bar{X} - 0.5}{\sqrt{0.5(1 - 0.5)}}$$

is standard normal. We therefore can compute the probability above, which is the p-value.

$$\Pr\left(\sqrt{N}\frac{|\bar{X} - 0.5|}{\sqrt{0.5(1 - 0.5)}} > \sqrt{N}\frac{0.02}{\sqrt{0.5(1 - 0.5)}}\right)$$

```
N <- 100
z <- sqrt(N)*0.02/0.5
1 - (pnorm(z) - pnorm(-z))
#> [1] 0.689
```

In this case, there is actually a large chance of seeing 52 or larger under the null hypothesis.

Keep in mind that there is a close connection between p-values and confidence intervals. If a 95% confidence interval of the spread does not include 0, we know that the p-value must be smaller than 0.05.

To learn more about p-values, you can consult any statistics textbook. However, in general, we prefer reporting confidence intervals over p-values since it gives us an idea of the size of the estimate. If we just report the p-value we provide no information about the significance of the finding in the context of the problem.

15.10   Association tests

The statistical tests we have studied up to now leave out a substantial portion of data types. Specifically, we have not discussed inference for binary, categorical, and ordinal data. To give a very specific example, consider the following case study.

A 2014 PNAS paper[3] analyzed success rates from funding agencies in the Netherlands and concluded that their:

> results reveal gender bias favoring male applicants over female applicants in the prioritization of their "quality of researcher" (but not "quality of proposal") evaluations and success rates, as well as in the language use in instructional and evaluation materials.

The main evidence for this conclusion comes down to a comparison of the percentages. Table S1 in the paper includes the information we need. Here are the three columns showing the overall outcomes:

```
library(tidyverse)
library(dslabs)
data("research_funding_rates")
research_funding_rates %>% select(discipline, applications_total,
                                  success_rates_total) %>% head()
#>              discipline applications_total success_rates_total
#> 1   Chemical sciences                122                26.2
#> 2   Physical sciences                174                20.1
#> 3             Physics                 76                26.3
#> 4           Humanities                396                16.4
#> 5   Technical sciences               251                17.1
#> 6   Interdisciplinary               183                15.8
```

---

[3]http://www.pnas.org/content/112/40/12349.abstract

We have these values for each gender:

```
names(research_funding_rates)
#>  [1] "discipline"          "applications_total"  "applications_men"
#>  [4] "applications_women"  "awards_total"        "awards_men"
#>  [7] "awards_women"        "success_rates_total" "success_rates_men"
#> [10] "success_rates_women"
```

We can compute the totals that were successful and the totals that were not as follows:

```
totals <- research_funding_rates %>%
  select(-discipline) %>%
  summarize_all(sum) %>%
  summarize(yes_men = awards_men,
            no_men = applications_men - awards_men,
            yes_women = awards_women,
            no_women = applications_women - awards_women)
```

So we see that a larger percent of men than women received awards:

```
totals %>% summarize(percent_men = yes_men/(yes_men+no_men),
                     percent_women = yes_women/(yes_women+no_women))
#>    percent_men percent_women
#> 1      0.177        0.149
```

But could this be due just to random variability? Here we learn how to perform inference for this type of data.

15.10.1 Lady Tasting Tea

R.A. Fisher[4] was one of the first to formalize hypothesis testing. The "Lady Tasting Tea" is one of the most famous examples.

The story is as follows: an acquaintance of Fisher's claimed that she could tell if milk was added before or after tea was poured. Fisher was skeptical. He designed an experiment to test this claim. He gave her four pairs of cups of tea: one with milk poured first, the other after. The order was randomized. The null hypothesis here is that she is guessing. Fisher derived the distribution for the number of correct picks on the assumption that the choices were random and independent.

As an example, suppose she picked 3 out of 4 correctly. Do we believe she has a special ability? The basic question we ask is: if the tester is actually guessing, what are the chances that she gets 3 or more correct? Just as we have done before, we can compute a probability under the null hypothesis that she is guessing 4 of each. Under this null hypothesis, we can think of this particular example as picking 4 balls out of an urn with 4 blue (correct answer) and 4 red (incorrect answer) balls. Remember, she knows that there are four before tea and four after.

Under the null hypothesis that she is simply guessing, each ball has the same chance of being picked. We can then use combinations to figure out each probability. The probability of

---

[4]https://en.wikipedia.org/wiki/Ronald_Fisher

picking 3 is $\binom{4}{3}\binom{4}{1}/\binom{8}{4} = 16/70$. The probability of picking all 4 correct is $\binom{4}{4}\binom{4}{0}/\binom{8}{4} = 1/70$. Thus, the chance of observing a 3 or something more extreme, under the null hypothesis, is ≈ 0.24. This is the p-value. The procedure that produced this p-value is called *Fisher's exact test* and it uses the *hypergeometric distribution*.

15.10.2   Two-by-two tables

The data from the experiment is usually summarized by a table like this:

```
tab <- matrix(c(3,1,1,3),2,2)
rownames(tab)<-c("Poured Before","Poured After")
colnames(tab)<-c("Guessed before","Guessed after")
tab
#>                 Guessed before Guessed after
#> Poured Before                3             1
#> Poured After                 1             3
```

These are referred to as a two-by-two table. For each of the four combinations one can get with a pair of binary variables, they show the observed counts for each occurrence.

The function `fisher.test` performs the inference calculations above:

```
fisher.test(tab, alternative="greater")$p.value
#> [1] 0.243
```

15.10.3   Chi-square Test

Notice that, in a way, our funding rates example is similar to the Lady Tasting Tea. However, in the Lady Tasting Tea example, the number of blue and red beads is experimentally fixed and the number of answers given for each category is also fixed. This is because Fisher made sure there were four cups with milk poured before tea and four cups with milk poured after and the lady knew this, so the answers would also have to include four befores and four afters. If this is the case, the sum of the rows and the sum of the columns are fixed. This defines constraints on the possible ways we can fill the two by two table and also permits us to use the hypergeometric distribution. In general, this is not the case. Nonetheless, there is another approach, the Chi-squared test, which is described below.

Imagine we have 290, 1,345, 177, 1,011 applicants, some are men and some are women and some get funded, whereas others don't. We saw that the success rates for men and woman were:

```
totals %>% summarize(percent_men = yes_men/(yes_men+no_men),
                     percent_women = yes_women/(yes_women+no_women))
#>    percent_men percent_women
#> 1        0.177         0.149
```

respectively. Would we see this again if we randomly assign funding at the overall rate:

```
rate <- totals %>%
  summarize(percent_total =
                 (yes_men + yes_women)/
                 (yes_men + no_men +yes_women + no_women)) %>%
  pull(percent_total)
rate
#> [1] 0.165
```

The Chi-square test answers this question. The first step is to create the two-by-two data table:

```
two_by_two <- data.frame(awarded = c("no", "yes"),
                     men = c(totals$no_men, totals$yes_men),
                     women = c(totals$no_women, totals$yes_women))
two_by_two
#>    awarded  men women
#> 1       no 1345  1011
#> 2      yes  290   177
```

The general idea of the Chi-square test is to compare this two-by-two table to what you expect to see, which would be:

```
data.frame(awarded = c("no", "yes"),
       men = (totals$no_men + totals$yes_men) * c(1 - rate, rate),
       women = (totals$no_women + totals$yes_women) * c(1 - rate, rate))
#>    awarded  men women
#> 1       no 1365   991
#> 2      yes  270   197
```

We can see that more men than expected and fewer women than expected received funding. However, under the null hypothesis these observations are random variables. The Chi-square test tells us how likely it is to see a deviation this large or larger. This test uses an asymptotic result, similar to the CLT, related to the sums of independent binary outcomes. The R function chisq.test takes a two-by-two table and returns the results from the test:

```
chisq_test <- two_by_two %>% select(-awarded) %>% chisq.test()
```

We see that the p-value is 0.0509:

```
chisq_test$p.value
#> [1] 0.0509
```

15.10.4 The odds ratio

An informative summary statistic associated with two-by-two tables is the odds ratio. Define the two variables as $X = 1$ if you are a male and 0 otherwise, and $Y = 1$ if you are funded and 0 otherwise. The odds of getting funded if you are a man is defined:

$$\Pr(Y = 1 \mid X = 1)/\Pr(Y = 0 \mid X = 1)$$

and can be computed like this:

```
odds_men <- with(two_by_two, (men[2]/sum(men)) / (men[1]/sum(men)))
odds_men
#> [1] 0.216
```

And the odds of being funded if you are a woman is:

$$\Pr(Y = 1 \mid X = 0)/\Pr(Y = 0 \mid X = 0)$$

and can be computed like this:

```
odds_women <- with(two_by_two, (women[2]/sum(women)) / (women[1]/sum(women)))
odds_women
#> [1] 0.175
```

The odds ratio is the ratio for these two odds: how many times larger are the odds for men than for women?

```
odds_men / odds_women
#> [1] 1.23
```

We often see two-by-two tables written out as

	Men	Women
Awarded	a	b
Not Awarded	c	d

In this case, the odds ratio is $\frac{a/c}{b/d}$ which is equivalent to $(ad)/(bc)$

15.10.5 Confidence intervals for the odds ratio

Computing confidence intervals for the odds ratio is not mathematically straightforward. Unlike other statistics, for which we can derive useful approximations of their distributions, the odds ratio is not only a ratio, but a ratio of ratios. Therefore, there is no simple way of using, for example, the CLT.

However, statistical theory tells us that when all four entries of the two-by-two table are large enough, then the log of the odds ratio is approximately normal with standard error

$$\sqrt{1/a + 1/b + 1/c + 1/d}$$

This implies that a 95% confidence interval for the log odds ratio can be formed by:

$$\log\left(\frac{ad}{bc}\right) \pm 1.96\sqrt{1/a + 1/b + 1/c + 1/d}$$

By exponentiating these two numbers we can construct a confidence interval of the odds ratio.

Using R we can compute this confidence interval as follows:

```
log_or <- log(odds_men / odds_women)
se <- two_by_two %>% select(-awarded) %>%
  summarize(se = sqrt(sum(1/men) + sum(1/women))) %>%
  pull(se)
ci <- log_or + c(-1,1) * qnorm(0.975) * se
```

If we want to convert it back to the odds ratio scale, we can exponentiate:

```
exp(ci)
#> [1] 1.00 1.51
```

Note that 1 is not included in the confidence interval which must mean that the p-value is smaller than 0.05. We can confirm this using:

```
2*(1 - pnorm(log_or, 0, se))
#> [1] 0.0454
```

This is a slightly different p-value than that with the Chi-square test. This is because we are using a different asymptotic approximation to the null distribution. To learn more about inference and asymptotic theory for odds ratio, consult the *Generalized Linear Models* book by McCullagh and Nelder.

15.10.6 Small count correction

Note that the log odds ratio is not defined if any of the cells of the two-by-two table is 0. This is because if a, b, c, or d is 0, the $\log(\frac{ad}{bc})$ is either the log of 0 or has a 0 in the denominator. For this situation, it is common practice to avoid 0s by adding 0.5 to each cell. This is referred to as the *Haldane–Anscombe correction* and has been shown, both in practice and theory, to work well.

15.10.7 Large samples, small p-values

As mentioned earlier, reporting only p-values is not an appropriate way to report the results of data analysis. In scientific journals, for example, some studies seem to overemphasize p-values. Some of these studies have large sample sizes and report impressively small p-values. Yet when one looks closely at the results, we realize odds ratios are quite modest: barely bigger than 1. In this case the difference may not be *practically significant* or *scientifically significant.*

Note that the relationship between odds ratio and p-value is not one-to-one. It depends on the sample size. So a very small p-value does not necessarily mean a very large odds ratio. Notice what happens to the p-value if we multiply our two-by-two table by 10, which does not change the odds ratio:

```
two_by_two %>% select(-awarded) %>%
  mutate(men = men*10, women = women*10) %>%
  chisq.test() %>% .$p.value
#> [1] 2.63e-10
```

15.11   Exercises

1. A famous athlete has an impressive career, winning 70% of her 500 career matches. However, this athlete gets criticized because in important events, such as the Olympics, she has a losing record of 8 wins and 9 losses. Perform a Chi-square test to determine if this losing record can be simply due to chance as opposed to not performing well under pressure.

2. Why did we use the Chi-square test instead of Fisher's exact test in the previous exercise?

    a. It actually does not matter, since they give the exact same p-value.
    b. Fisher's exact and the Chi-square are different names for the same test.
    c. Because the sum of the rows and columns of the two-by-two table are not fixed so the hypergeometric distribution is not an appropriate assumption for the null hypothesis. For this reason, Fisher's exact test is rarely applicable with observational data.
    d. Because the Chi-square test runs faster.

3. Compute the odds ratio of "losing under pressure" along with a confidence interval.

4. Notice that the p-value is larger than 0.05 but the 95% confidence interval does not include 1. What explains this?

    a. We made a mistake in our code.
    b. These are not t-tests so the connection between p-value and confidence intervals does not apply.
    c. Different approximations are used for the p-value and the confidence interval calculation. If we had a larger sample size the match would be better.
    d. We should use the Fisher exact test to get confidence intervals.

5. Multiply the two-by-two table by 2 and see if the p-value and confidence retrieval are a better match.

16

Statistical models

"All models are wrong, but some are useful." –George E. P. Box

The day before the 2008 presidential election, Nate Silver's FiveThirtyEight stated that "Barack Obama appears poised for a decisive electoral victory". They went further and predicted that Obama would win the election with 349 electoral votes to 189, and the popular vote by a margin of 6.1%. FiveThirtyEight also attached a probabilistic statement to their prediction claiming that Obama had a 91% chance of winning the election. The predictions were quite accurate since, in the final results, Obama won the electoral college 365 to 173 and the popular vote by a 7.2% difference. Their performance in the 2008 election brought FiveThirtyEight to the attention of political pundits and TV personalities. Four years later, the week before the 2012 presidential election, FiveThirtyEight's Nate Silver was giving Obama a 90% chance of winning despite many of the experts thinking the final results would be closer. Political commentator Joe Scarborough said during his show[1]:

> Anybody that thinks that this race is anything but a toss-up right now is such an ideologue ... they're jokes.

To which Nate Silver responded via Twitter:

> If you think it's a toss-up, let's bet. If Obama wins, you donate $1,000 to the American Red Cross. If Romney wins, I do. Deal?

In 2016, Silver was not as certain and gave Hillary Clinton only a 71% of winning. In contrast, most other forecasters were almost certain she would win. She lost. But 71% is still more than 50%, so was Mr. Silver wrong? And what does probability mean in this context anyway? Are dice being tossed somewhere?

In this chapter we will demonstrate how *poll aggregators*, such as FiveThirtyEight, collected and combined data reported by different experts to produce improved predictions. We will introduce ideas behind the *statistical models*, also known as *probability models*, that were used by poll aggregators to improve election forecasts beyond the power of individual polls. In this chapter, we motivate the models, building on the statistical inference concepts we learned in Chapter 15. We start with relatively simple models, realizing that the actual data science exercise of forecasting elections involves rather complex ones, which we introduce towards the end of the chapter in Section 16.8.

---

[1] https://www.youtube.com/watch?v=TbKkjm-gheY

16.1   Poll aggregators

As we described earlier, a few weeks before the 2012 election Nate Silver was giving Obama a 90% chance of winning. How was Mr. Silver so confident? We will use a Monte Carlo simulation to illustrate the insight Mr. Silver had and others missed. To do this, we generate results for 12 polls taken the week before the election. We mimic sample sizes from actual polls and construct and report 95% confidence intervals for each of the 12 polls. We save the results from this simulation in a data frame and add a poll ID column.

```
library(tidyverse)
library(dslabs)
d <- 0.039
Ns <- c(1298, 533, 1342, 897, 774, 254, 812, 324, 1291, 1056, 2172, 516)
p <- (d + 1) / 2

polls <- map_df(Ns, function(N) {
  x <- sample(c(0,1), size=N, replace=TRUE, prob=c(1-p, p))
  x_hat <- mean(x)
  se_hat <- sqrt(x_hat * (1 - x_hat) / N)
  list(estimate = 2 * x_hat - 1,
    low = 2*(x_hat - 1.96*se_hat) - 1,
    high = 2*(x_hat + 1.96*se_hat) - 1,
    sample_size = N)
}) %>% mutate(poll = seq_along(Ns))
```

Here is a visualization showing the intervals the pollsters would have reported for the difference between Obama and Romney:

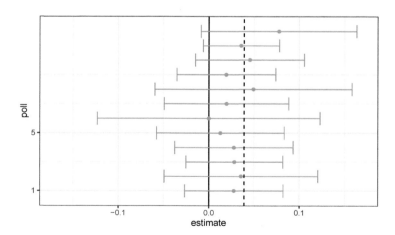

Not surprisingly, all 12 polls report confidence intervals that include the election night result (dashed line). However, all 12 polls also include 0 (solid black line) as well. Therefore, if asked individually for a prediction, the pollsters would have to say: it's a toss-up. Below we describe a key insight they are missing.

Poll aggregators, such as Nate Silver, realized that by combining the results of different polls you could greatly improve precision. By doing this, we are effectively conducting a poll with a huge sample size. We can therefore report a smaller 95% confidence interval and a more precise prediction.

Although as aggregators we do not have access to the raw poll data, we can use mathematics to reconstruct what we would have obtained had we made one large poll with:

```
sum(polls$sample_size)
#> [1] 11269
```

participants. Basically, we construct an estimate of the spread, let's call it d, with a weighted average in the following way:

```
d_hat <- polls %>%
  summarize(avg = sum(estimate*sample_size) / sum(sample_size)) %>%
  pull(avg)
```

Once we have an estimate of d, we can construct an estimate for the proportion voting for Obama, which we can then use to estimate the standard error. Once we do this, we see that our margin of error is 0.018.

Thus, we can predict that the spread will be 3.1 plus or minus 1.8, which not only includes the actual result we eventually observed on election night, but is quite far from including 0. Once we combine the 12 polls, we become quite certain that Obama will win the popular vote.

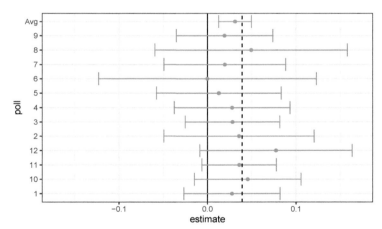

Of course, this was just a simulation to illustrate the idea. The actual data science exercise of forecasting elections is much more complicated and it involves modeling. Below we explain how pollsters fit multilevel models to the data and use this to forecast election results. In the 2008 and 2012 US presidential elections, Nate Silver used this approach to make an almost perfect prediction and silence the pundits.

Since the 2008 elections, other organizations have started their own election forecasting group that, like Nate Silver's, aggregates polling data and uses statistical models to make predictions. In 2016, forecasters underestimated Trump's chances of winning greatly. The day before the election the *New York Times* reported[2] the following probabilities for Hillary Clinton winning the presidency:

	NYT	538	HuffPost	PW	PEC	DK	Cook	Roth
Win Prob	85%	71%	98%	89%	>99%	92%	Lean Dem	Lean Dem

For example, the Princeton Election Consortium (PEC) gave Trump less than 1% chance of winning, while the Huffington Post gave him a 2% chance. In contrast, FiveThirtyEight had Trump's probability of winning at 29%, higher than tossing two coins and getting two heads. In fact, four days before the election FiveThirtyEight published an article titled *Trump Is Just A Normal Polling Error Behind Clinton*[3]. By understanding statistical models and how these forecasters use them, we will start to understand how this happened.

Although not nearly as interesting as predicting the electoral college, for illustrative purposes we will start by looking at predictions for the popular vote. FiveThirtyEight predicted a 3.6% advantage for Clinton[4], included the actual result of 2.1% (48.2% to 46.1%) in their interval, and was much more confident about Clinton winning the election, giving her an 81.4% chance. Their prediction was summarized with a chart like this:

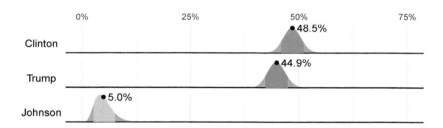

The colored areas represent values with an 80% chance of including the actual result, according to the FiveThirtyEight model.

We introduce actual data from the 2016 US presidential election to show how models are motivated and built to produce these predictions. To understand the "81.4% chance" statement we need to describe Bayesian statistics, which we do in Sections 16.4 and 16.8.1.

16.1.1   Poll data

We use public polling data organized by FiveThirtyEight for the 2016 presidential election. The data is included as part of the **dslabs** package:

```
data(polls_us_election_2016)
```

The table includes results for national polls, as well as state polls, taken during the year prior to the election. For this first example, we will filter the data to include national polls

---

[2]https://www.nytimes.com/interactive/2016/upshot/presidential-polls-forecast.html
[3]https://fivethirtyeight.com/features/trump-is-just-a-normal-polling-error-behind-clinton/
[4]https://projects.fivethirtyeight.com/2016-election-forecast/

conducted during the week before the election. We also remove polls that FiveThirtyEight has determined not to be reliable and graded with a "B" or less. Some polls have not been graded and we include those:

```
polls <- polls_us_election_2016 %>%
  filter(state == "U.S." & enddate >= "2016-10-31" &
         (grade %in% c("A+","A","A-","B+") | is.na(grade)))
```

We add a spread estimate:

```
polls <- polls %>%
  mutate(spread = rawpoll_clinton/100 - rawpoll_trump/100)
```

For this example, we will assume that there are only two parties and call p the proportion voting for Clinton and $1 - p$ the proportion voting for Trump. We are interested in the spread $2p - 1$. Let's call the spread d (for difference).

We have 49 estimates of the spread. The theory we learned tells us that these estimates are a random variable with a probability distribution that is approximately normal. The expected value is the election night spread d and the standard error is $2\sqrt{p(1-p)/N}$. Assuming the urn model we described earlier is a good one, we can use this information to construct a confidence interval based on the aggregated data. The estimated spread is:

```
d_hat <- polls %>%
  summarize(d_hat = sum(spread * samplesize) / sum(samplesize)) %>%
  pull(d_hat)
```

and the standard error is:

```
p_hat <- (d_hat+1)/2
moe <- 1.96 * 2 * sqrt(p_hat * (1 - p_hat) / sum(polls$samplesize))
moe
#> [1] 0.00662
```

So we report a spread of 1.43% with a margin of error of 0.66%. On election night, we discover that the actual percentage was 2.1%, which is outside a 95% confidence interval. What happened?

A histogram of the reported spreads shows a problem:

```
polls %>%
  ggplot(aes(spread)) +
  geom_histogram(color="black", binwidth = .01)
```

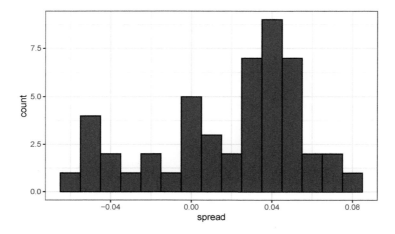

The data does not appear to be normally distributed and the standard error appears to be larger than 0.007. The theory is not quite working here.

16.1.2   Pollster bias

Notice that various pollsters are involved and some are taking several polls a week:

```
polls %>% group_by(pollster) %>% summarize(n())
#> # A tibble: 15 x 2
#>    pollster                                                         `n()`
#>    <fct>                                                            <int>
#> 1 ABC News/Washington Post                                           7
#> 2 Angus Reid Global                                                  1
#> 3 CBS News/New York Times                                            2
#> 4 Fox News/Anderson Robbins Research/Shaw & Company Research         2
#> 5 IBD/TIPP                                                           8
#> # ... with 10 more rows
```

Let's visualize the data for the pollsters that are regularly polling:

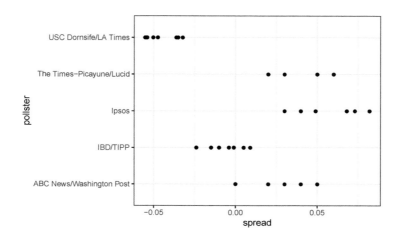

This plot reveals an unexpected result. First, consider that the standard error predicted by theory for each poll:

```
polls %>% group_by(pollster) %>%
  filter(n() >= 6) %>%
  summarize(se = 2 * sqrt(p_hat * (1-p_hat) / median(samplesize)))
#> # A tibble: 5 x 2
#>   pollster                     se
#>   <fct>                     <dbl>
#> 1 ABC News/Washington Post 0.0265
#> 2 IBD/TIPP                 0.0333
#> 3 Ipsos                    0.0225
#> 4 The Times-Picayune/Lucid 0.0196
#> 5 USC Dornsife/LA Times    0.0183
```

is between 0.018 and 0.033, which agrees with the within poll variation we see. However, there appears to be differences *across the polls*. Note, for example, how the USC Dornsife/LA Times pollster is predicting a 4% win for Trump, while Ipsos is predicting a win larger than 5% for Clinton. The theory we learned says nothing about different pollsters producing polls with different expected values. All the polls should have the same expected value. FiveThirtyEight refers to these differences as "house effects". We also call them *pollster bias*.

In the following section, rather than use the urn model theory, we are instead going to develop a data-driven model.

16.2 Data-driven models

For each pollster, let's collect their last reported result before the election:

```
one_poll_per_pollster <- polls %>% group_by(pollster) %>%
  filter(enddate == max(enddate)) %>%
  ungroup()
```

Here is a histogram of the data for these 15 pollsters:

```
qplot(spread, data = one_poll_per_pollster, binwidth = 0.01)
```

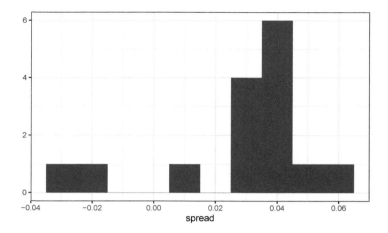

In the previous section, we saw that using the urn model theory to combine these results might not be appropriate due to the pollster effect. Instead, we will model this spread data directly.

The new model can also be thought of as an urn model, although the connection is not as direct. Rather than 0s (Republicans) and 1s (Democrats), our urn now contains poll results from all possible pollsters. We *assume* that the expected value of our urn is the actual spread $d = 2p - 1$.

Because instead of 0s and 1s, our urn contains continuous numbers between -1 and 1, the standard deviation of the urn is no longer $\sqrt{p(1-p)}$. Rather than voter sampling variability, the standard error now includes the pollster-to-pollster variability. Our new urn also includes the sampling variability from the polling. Regardless, this standard deviation is now an unknown parameter. In statistics textbooks, the Greek symbol σ is used to represent this parameter.

In summary, we have two unknown parameters: the expected value d and the standard deviation σ.

Our task is to estimate d. Because we model the observed values $X_1, \ldots X_N$ as a random sample from the urn, the CLT might still work in this situation because it is an average of independent random variables. For a large enough sample size N, the probability distribution of the sample average $\bar{X}$ is approximately normal with expected value μ and standard error $\sigma/\sqrt{N}$. If we are willing to consider $N = 15$ large enough, we can use this to construct confidence intervals.

A problem is that we don't know σ. But theory tells us that we can estimate the urn model σ with the *sample standard deviation* defined as $s = \sqrt{\sum_{i=1}^{N}(X_i - \bar{X})^2/(N-1)}$.

Unlike for the population standard deviation definition, we now divide by $N - 1$. This makes s a better estimate of σ. There is a mathematical explanation for this, which is explained in most statistics textbooks, but we don't cover it here.

The `sd` function in R computes the sample standard deviation:

```
sd(one_poll_per_pollster$spread)
#> [1] 0.0242
```

We are now ready to form a new confidence interval based on our new data-driven model:

```
results <- one_poll_per_pollster %>%
  summarize(avg = mean(spread),
            se = sd(spread) / sqrt(length(spread))) %>%
  mutate(start = avg - 1.96 * se,
         end = avg + 1.96 * se)
round(results * 100, 1)
#>    avg  se start end
#> 1 2.9 0.6   1.7 4.1
```

Our confidence interval is wider now since it incorporates the pollster variability. It does include the election night result of 2.1%. Also, note that it was small enough not to include 0, which means we were confident Clinton would win the popular vote.

Are we now ready to declare a probability of Clinton winning the popular vote? Not yet. In our model d is a fixed parameter so we can't talk about probabilities. To provide probabilities, we will need to learn about Bayesian statistics.

16.3   Exercises

We have been using urn models to motivate the use of probability models. Most data science applications are not related to data obtained from urns. More common are data that come from individuals. The reason probability plays a role here is because the data come from a random sample. The random sample is taken from a population and the urn serves as an analogy for the population.

Let's revisit the heights dataset. Suppose we consider the males in our course the population.

```
library(dslabs)
data(heights)
x <- heights %>% filter(sex == "Male") %>%
  pull(height)
```

1. Mathematically speaking, x is our population. Using the urn analogy, we have an urn with the values of x in it. What are the average and standard deviation of our population?

2. Call the population average computed above μ and the standard deviation σ. Now take a sample of size 50, with replacement, and construct an estimate for μ and σ.

3. What does the theory tell us about the sample average $\bar{X}$ and how it is related to μ?

    a. It is practically identical to μ.
    b. It is a random variable with expected value μ and standard error $\sigma/\sqrt{N}$.
    c. It is a random variable with expected value μ and standard error σ.
    d. Contains no information.

4. So how is this useful? We are going to use an oversimplified yet illustrative example. Suppose we want to know the average height of our male students, but we only get to measure 50 of the 708. We will use $\bar{X}$ as our estimate. We know from the answer to exercise

3 that the standard estimate of our error $\bar{X} - \mu$ is $\sigma/\sqrt{N}$. We want to compute this, but we don't know σ. Based on what is described in this section, show your estimate of σ.

5. Now that we have an estimate of σ, let's call our estimate s. Construct a 95% confidence interval for μ.

6. Now run a Monte Carlo simulation in which you compute 10,000 confidence intervals as you have just done. What proportion of these intervals include μ?

7. In this section, we talked about pollster bias. We used visualization to motivate the presence of such bias. Here we will give it a more rigorous treatment. We will consider two pollsters that conducted daily polls. We will look at national polls for the month before the election.

```
data(polls_us_election_2016)
polls <- polls_us_election_2016 %>%
   filter(pollster %in% c("Rasmussen Reports/Pulse Opinion Research",
                        "The Times-Picayune/Lucid") &
           enddate >= "2016-10-15" &
           state == "U.S.") %>%
   mutate(spread = rawpoll_clinton/100 - rawpoll_trump/100)
```

We want to answer the question: is there a poll bias? Make a plot showing the spreads for each poll.

8. The data does seem to suggest there is a difference. However, these data are subject to variability. Perhaps the differences we observe are due to chance.

The urn model theory says nothing about pollster effect. Under the urn model, both pollsters have the same expected value: the election day difference, that we call d.

To answer the question "is there an urn model?", we will model the observed data $Y_{i,j}$ in the following way:

$$Y_{i,j} = d + b_i + \varepsilon_{i,j}$$

with $i = 1, 2$ indexing the two pollsters, b_i the bias for pollster i and $\varepsilon_i j$ poll to poll chance variability. We assume the ε are independent from each other, have expected value 0 and standard deviation σ_i regardless of j.

Which of the following best represents our question?

    a. Is $\varepsilon_{i,j} = 0$?
    b. How close are the $Y_{i,j}$ to d?
    c. Is $b_1 \neq b_2$?
    d. Are $b_1 = 0$ and $b_2 = 0$?

9. In the right side of this model only $\varepsilon_{i,j}$ is a random variable. The other two are constants. What is the expected value of $Y_{1,j}$?

10. Suppose we define $\bar{Y}_1$ as the average of poll results from the first poll, $Y_{1,1}, \ldots, Y_{1,N_1}$ with N_1 the number of polls conducted by the first pollster:

```
polls %>%
  filter(pollster=="Rasmussen Reports/Pulse Opinion Research") %>%
  summarize(N_1 = n())
```

What is the expected values $\bar{Y}_1$?

11. What is the standard error of $\bar{Y}_1$?

12. Suppose we define $\bar{Y}_2$ as the average of poll results from the first poll, $Y_{2,1}, \ldots, Y_{2,N_2}$ with N_2 the number of polls conducted by the first pollster. What is the expected value $\bar{Y}_2$?

13. What is the standard error of $\bar{Y}_2$?

14. Using what we learned by answering the questions above, what is the expected value of $\bar{Y}_2 - \bar{Y}_1$?

15. Using what we learned by answering the questions above, what is the standard error of $\bar{Y}_2 - \bar{Y}_1$?

16. The answer to the question above depends on σ_1 and σ_2, which we don't know. We learned that we can estimate these with the sample standard deviation. Write code that computes these two estimates.

17. What does the CLT tell us about the distribution of $\bar{Y}_2 - \bar{Y}_1$?

   a. Nothing because this is not the average of a sample.
   b. Because the Y_{ij} are approximately normal, so are the averages.
   c. Note that $\bar{Y}_2$ and $\bar{Y}_1$ are sample averages, so if we assume N_2 and N_1 are large enough, each is approximately normal. The difference of normals is also normal.
   d. The data are not 0 or 1, so CLT does not apply.

18. We have constructed a random variable that has expected value $b_2 - b_1$, the pollster bias difference. If our model holds, then this random variable has an approximately normal distribution and we know its standard error. The standard error depends on σ_1 and σ_2, but we can plug the sample standard deviations we computed above. We started off by asking: is $b_2 - b_1$ different from 0? Use all the information we have learned above to construct a 95% confidence interval for the difference b_2 and b_1.

19. The confidence interval tells us there is relatively strong pollster effect resulting in a difference of about 5%. Random variability does not seem to explain it. We can compute a p-value to relay the fact that chance does not explain it. What is the p-value?

20. The statistic formed by dividing our estimate of $b_2 - b_1$ by its estimated standard error:

$$\frac{\bar{Y}_2 - \bar{Y}_1}{\sqrt{s_2^2/N_2 + s_1^2/N_1}}$$

is called the t-statistic. Now notice that we have more than two pollsters. We can also test for pollster effect using all pollsters, not just two. The idea is to compare the variability across polls to variability within polls. We can actually construct statistics to test for effects and approximate their distribution. The area of statistics that does this is called Analysis of Variance or ANOVA. We do not cover it here, but ANOVA provides a very useful set of tools to answer questions such as: is there a pollster effect?

For this exercise, create a new table:

```
polls <- polls_us_election_2016 %>%
  filter(enddate >= "2016-10-15" &
           state == "U.S.") %>%
  group_by(pollster) %>%
  filter(n() >= 5) %>%
  mutate(spread = rawpoll_clinton/100 - rawpoll_trump/100) %>%
  ungroup()
```

Compute the average and standard deviation for each pollster and examine the variability across the averages and how it compares to the variability within the pollsters, summarized by the standard deviation.

16.4   Bayesian statistics

What does it mean when an election forecaster tells us that a given candidate has a 90% chance of winning? In the context of the urn model, this would be equivalent to stating that the probability $p > 0.5$ is 90%. However, as we discussed earlier, in the urn model p is a fixed parameter and it does not make sense to talk about probability. With Bayesian statistics, we model p as random variable and thus a statement such as "90% chance of winning" is consistent with the approach.

Forecasters also use models to describe variability at different levels. For example, sampling variability, pollster to pollster variability, day to day variability, and election to election variability. One of the most successful approaches used for this are hierarchical models, which can be explained in the context of Bayesian statistics.

In this chapter we briefly describe Bayesian statistics. For an in-depth treatment of this topic we recommend one of the following textbooks:

- Berger JO (1985). Statistical Decision Theory and Bayesian Analysis, 2nd edition. Springer-Verlag.

- Lee PM (1989). Bayesian Statistics: An Introduction. Oxford.

16.4.1   Bayes theorem

We start by describing Bayes theorem. We do this using a hypothetical cystic fibrosis test as an example. Suppose a test for cystic fibrosis has an accuracy of 99%. We will use the following notation:

$$\text{Prob}(+ \mid D = 1) = 0.99, \text{Prob}(- \mid D = 0) = 0.99$$

with $+$ meaning a positive test and D representing if you actually have the disease (1) or not (0).

Suppose we select a random person and they test positive. What is the probability that they have the disease? We write this as $\text{Prob}(D = 1 \mid +)$? The cystic fibrosis rate is 1 in

3,900 which implies that $\mathrm{Prob}(D = 1) = 0.00025$. To answer this question, we will use Bayes theorem, which in general tells us that:

$$\Pr(A \mid B) = \frac{\Pr(B \mid A)\Pr(A)}{\Pr(B)}$$

This equation applied to our problem becomes:

$$\Pr(D = 1 \mid +) = \frac{P(+ \mid D = 1) \cdot P(D = 1)}{\Pr(+)}$$

$$= \frac{\Pr(+ \mid D = 1) \cdot P(D = 1)}{\Pr(+ \mid D = 1) \cdot P(D = 1) + \Pr(+ \mid D = 0)\Pr(D = 0)}$$

Plugging in the numbers we get:

$$\frac{0.99 \cdot 0.00025}{0.99 \cdot 0.00025 + 0.01 \cdot (.99975)} = 0.02$$

This says that despite the test having 0.99 accuracy, the probability of having the disease given a positive test is only 0.02. This may appear counter-intuitive to some, but the reason this is the case is because we have to factor in the very rare probability that a person, chosen at random, has the disease. To illustrate this, we run a Monte Carlo simulation.

16.5 Bayes theorem simulation

The following simulation is meant to help you visualize Bayes theorem. We start by randomly selecting 100,000 people from a population in which the disease in question has a 1 in 4,000 prevalence.

```
prev <- 0.00025
N <- 100000
outcome <- sample(c("Disease","Healthy"), N, replace = TRUE,
                  prob = c(prev, 1 - prev))
```

Note that there are very few people with the disease:

```
N_D <- sum(outcome == "Disease")
N_D
#> [1] 23
N_H <- sum(outcome == "Healthy")
N_H
#> [1] 99977
```

Also, there are many without the disease, which makes it more probable that we will see some false positives given that the test is not perfect. Now each person gets the test, which is correct 99% of the time:

```
accuracy <- 0.99
test <- vector("character", N)
test[outcome == "Disease"]  <- sample(c("+", "-"), N_D, replace = TRUE,
                                 prob = c(accuracy, 1 - accuracy))
test[outcome == "Healthy"]  <- sample(c("-", "+"), N_H, replace = TRUE,
                                 prob = c(accuracy, 1 - accuracy))
```

Because there are so many more controls than cases, even with a low false positive rate we get more controls than cases in the group that tested positive:

```
table(outcome, test)
#>            test
#> outcome         -      +
#>    Disease      0     23
#>    Healthy  99012    965
```

From this table, we see that the proportion of positive tests that have the disease is 23 out of 988. We can run this over and over again to see that, in fact, the probability converges to about 0.022.

16.5.1   Bayes in practice

José Iglesias is a professional baseball player. In April 2013, when he was starting his career, he was performing rather well:

Month	At Bats	H	AVG
April	20	9	.450

The batting average (AVG) statistic is one way of measuring success. Roughly speaking, it tells us the success rate when batting. An AVG of .450 means José has been successful 45% of the times he has batted (At Bats) which is rather high, historically speaking. Keep in mind that no one has finished a season with an AVG of .400 or more since Ted Williams did it in 1941! To illustrate the way hierarchical models are powerful, we will try to predict José's batting average at the end of the season. Note that in a typical season, players have about 500 at bats.

With the techniques we have learned up to now, referred to as *frequentist techniques*, the best we can do is provide a confidence interval. We can think of outcomes from hitting as a binomial with a success rate of p. So if the success rate is indeed .450, the standard error of just 20 at bats is:

$$\sqrt{\frac{.450(1 - .450)}{20}} = .111$$

This means that our confidence interval is $.450 - .222$ to $.450 + .222$ or .228 to .672.

This prediction has two problems. First, it is very large, so not very useful. Second, it is centered at .450, which implies that our best guess is that this new player will break Ted Williams' record.

If you follow baseball, this last statement will seem wrong and this is because you are implicitly using a hierarchical model that factors in information from years of following baseball. Here we show how we can quantify this intuition.

First, let's explore the distribution of batting averages for all players with more than 500 at bats during the previous three seasons:

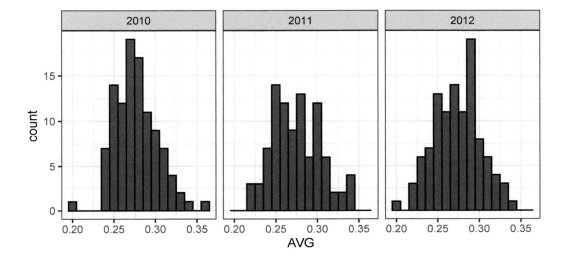

The average player had an `AVG` of .275 and the standard deviation of the population of players was 0.027. So we can see already that .450 would be quite an anomaly since it is over six standard deviations away from the mean.

So is José lucky or is he the best batter seen in the last 50 years? Perhaps it's a combination of both luck and talent. But how much of each? If we become convinced that he is lucky, we should trade him to a team that trusts the .450 observation and is maybe overestimating his potential.

16.6 Hierarchical models

The hierarchical model provides a mathematical description of how we came to see the observation of .450. First, we pick a player at random with an intrinsic ability summarized by, for example, p. Then we see 20 random outcomes with success probability p.

We use a model to represent two levels of variability in our data. First, each player is assigned a natural ability to hit. We will use the symbol p to represent this ability. You can think of p as the batting average you would converge to if this particular player batted over and over again.

Based on the plots we showed earlier, we assume that p has a normal distribution. With expected value .270 and standard error 0.027.

Now the second level of variability has to do with luck when batting. Regardless of how good the player is, sometimes you have bad luck and sometimes you have good luck. At each at bat, this player has a probability of success p. If we add up these successes and failures, then the CLT tells us that the observed average, call it Y, has a normal distribution with expected value p and standard error $\sqrt{p(1-p)/N}$ with N the number of at bats.

Statistical textbooks will write the model like this:

$$p \sim N(\mu, \tau^2)$$
$$Y \mid p \sim N(p, \sigma^2)$$

Here the $\sim$ symbol tells us the random variable on the left of the symbol follows the distribution on the right and $N(a, b^2)$ represents the normal distribution with mean a and standard deviation b. The $\mid$ is read as *conditioned on*, and it means that we are treating the random variable to the right of the symbol as known. We refer to the model as hierarchical because we need to know p, the first level, in order to model Y, the second level. In our example the first level describes randomness in assigning talent to a player and the second describes randomness in this particular player's performance once we have fixed the talent parameter. In a Bayesian framework, the first level is called a *prior distribution* and the second the *sampling distribution*. The data analysis we have conducted here suggests that we set $\mu = .270$, $\tau = 0.027$, and $\sigma^2 = p(1-p)/N$.

Now, let's use this model for José's data. Suppose we want to predict his innate ability in the form of his *true* batting average p. This would be the hierarchical model for our data:

$$p \sim N(.275, .027^2)$$
$$Y \mid p \sim N(p, .111^2)$$

We now are ready to compute a posterior distribution to summarize our prediction of p. The continuous version of Bayes' rule can be used here to derive the *posterior probability function*, which is the distribution of p assuming we observe $Y = y$. In our case, we can show that when we fix $Y = y$, p follows a normal distribution with expected value:

$$\begin{aligned}
\mathrm{E}(p \mid Y = y) &= B\mu + (1 - B)y \\
&= \mu + (1 - B)(y - \mu) \\
\text{with } B &= \frac{\sigma^2}{\sigma^2 + \tau^2}
\end{aligned}$$

This is a weighted average of the population average μ and the observed data y. The weight depends on the SD of the population τ and the SD of our observed data σ. This weighted average is sometimes referred to as *shrinking* because it *shrinks* estimates towards a prior mean. In the case of José Iglesias, we have:

$$\begin{aligned}
\mathrm{E}(p \mid Y = .450) &= B \times .275 + (1 - B) \times .450 \\
&= .275 + (1 - B)(.450 - .275) \\
B &= \frac{.111^2}{.111^2 + .027^2} = 0.944 \\
\mathrm{E}(p \mid Y = 450) &\approx .285
\end{aligned}$$

We do not show the derivation here, but the standard error can be shown to be:

$$\text{SE}(p \mid y)^2 = \frac{1}{1/\sigma^2 + 1/\tau^2} = \frac{1}{1/.111^2 + 1/.027^2} = 0.00069$$

and the standard deviation is therefore 0.026. So we started with a frequentist 95% confidence interval that ignored data from other players and summarized just José's data: $.450 \pm 0.220$. Then we used a Bayesian approach that incorporated data from other players and other years to obtain a posterior probability. This is actually referred to as an empirical Bayes approach because we used data to construct the prior. From the posterior, we can report what is called a 95% *credible interval* by reporting a region, centered at the mean, with a 95% chance of occurring. In our case, this turns out to be: $.285 \pm 0.052$.

The Bayesian credible interval suggests that if another team is impressed by the .450 observation, we should consider trading José as we are predicting he will be just slightly above average. Interestingly, the Red Sox traded José to the Detroit Tigers in July. Here are the José Iglesias batting averages for the next five months:

Month	At Bat	Hits	AVG
April	20	9	.450
May	26	11	.423
June	86	34	.395
July	83	17	.205
August	85	25	.294
September	50	10	.200
Total w/o April	330	97	.293

Although both intervals included the final batting average, the Bayesian credible interval provided a much more precise prediction. In particular, it predicted that he would not be as good during the remainder of the season.

16.7 Exercises

1. In 1999, in England, Sally Clark[5] was found guilty of the murder of two of her sons. Both infants were found dead in the morning, one in 1996 and another in 1998. In both cases, she claimed the cause of death was sudden infant death syndrome (SIDS). No evidence of physical harm was found on the two infants so the main piece of evidence against her was the testimony of Professor Sir Roy Meadow, who testified that the chances of two infants dying of SIDS was 1 in 73 million. He arrived at this figure by finding that the rate of SIDS was 1 in 8,500 and then calculating that the chance of two SIDS cases was $8,500 \times 8,500 \approx$ 73 million. Which of the following do you agree with?

    a. Sir Meadow assumed that the probability of the second son being affected by SIDS was independent of the first son being affected, thereby ignoring possible genetic

[5] https://en.wikipedia.org/wiki/Sally_Clark

causes. If genetics plays a role then: Pr(second case of SIDS | first case of SIDS) < Pr(first case of SIDS).

b. Nothing. The multiplication rule always applies in this way: $\Pr(A \text{ and } B) = \Pr(A)\Pr(B)$.

c. Sir Meadow is an expert and we should trust his calculations.

d. Numbers don't lie.

2. Let's assume that there is in fact a genetic component to SIDS and the probability of Pr(second case of SIDS | first case of SIDS) = 1/100, is much higher than 1 in 8,500. What is the probability of both of her sons dying of SIDS?

3. Many press reports stated that the expert claimed the probability of Sally Clark being innocent as 1 in 73 million. Perhaps the jury and judge also interpreted the testimony this way. This probability can be written as the probability of *a mother is a son-murdering psychopath* given that *two of her children are found dead with no evidence of physical harm.* According to Bayes' rule, what is this?

4. Assume that the chance of a son-murdering psychopath finding a way to kill her children, without leaving evidence of physical harm, is:

$$\Pr(A \mid B) = 0.50$$

with A = two of her children are found dead with no evidence of physical harm and B = a mother is a son-murdering psychopath = 0.50. Assume that the rate of son-murdering psychopaths mothers is 1 in 1,000,000. According to Bayes' theorem, what is the probability of $\Pr(B \mid A)$?

5/. After Sally Clark was found guilty, the Royal Statistical Society issued a statement saying that there was "no statistical basis" for the expert's claim. They expressed concern at the "misuse of statistics in the courts". Eventually, Sally Clark was acquitted in June 2003. What did the expert miss?

a. He made an arithmetic error.

b. He made two mistakes. First, he misused the multiplication rule and did not take into account how rare it is for a mother to murder her children. After using Bayes' rule, we found a probability closer to 0.5 than 1 in 73 million.

c. He mixed up the numerator and denominator of Bayes' rule.

d. He did not use R.

6. Florida is one of the most closely watched states in the U.S. election because it has many electoral votes, and the election is generally close, and Florida tends to be a swing state that can vote either way. Create the following table with the polls taken during the last two weeks:

```
library(tidyverse)
library(dslabs)
data(polls_us_election_2016)
polls <- polls_us_election_2016 %>%
   filter(state == "Florida" & enddate >= "2016-11-04" ) %>%
   mutate(spread = rawpoll_clinton/100 - rawpoll_trump/100)
```

Take the average spread of these polls. The CLT tells us this average is approximately normal.

Calculate an average and provide an estimate of the standard error. Save your results in an object called `results`.

7. Now assume a Bayesian model that sets the prior distribution for Florida's election night spread d to be Normal with expected value μ and standard deviation τ. What are the interpretations of μ and τ?

   a. μ and τ are arbitrary numbers that let us make probability statements about d.
   b. μ and τ summarize what we would predict for Florida before seeing any polls. Based on past elections, we would set μ close to 0 because both Republicans and Democrats have won, and τ at about 0.02, because these elections tend to be close.
   c. μ and τ summarize what we want to be true. We therefore set μ at 0.10 and τ at 0.01.
   d. The choice of prior has no effect on Bayesian analysis.

8. The CLT tells us that our estimate of the spread $\hat{d}$ has normal distribution with expected value d and standard deviation σ calculated in problem 6. Use the formulas we showed for the posterior distribution to calculate the expected value of the posterior distribution if we set $\mu = 0$ and $\tau = 0.01$.

9. Now compute the standard deviation of the posterior distribution.

10. Using the fact that the posterior distribution is normal, create an interval that has a 95% probability of occurring centered at the posterior expected value. Note that we call these credible intervals.

11. According to this analysis, what was the probability that Trump wins Florida?

12. Now use `sapply` function to change the prior variance from `seq(0.05, 0.05, len = 100)` and observe how the probability changes by making a plot.

16.8   Case study: election forecasting

In a previous section, we generated these data tables:

```
library(tidyverse)
library(dslabs)
polls <- polls_us_election_2016 %>%
   filter(state == "U.S." & enddate >= "2016-10-31" &
             (grade %in% c("A+","A","A-","B+") | is.na(grade))) %>%
   mutate(spread = rawpoll_clinton/100 - rawpoll_trump/100)

one_poll_per_pollster <- polls %>% group_by(pollster) %>%
   filter(enddate == max(enddate)) %>%
   ungroup()

results <- one_poll_per_pollster %>%
   summarize(avg = mean(spread), se = sd(spread)/sqrt(length(spread))) %>%
   mutate(start = avg - 1.96*se, end = avg + 1.96*se)
```

Below, we will use these for our forecasting.

16.8.1   Bayesian approach

Pollsters tend to make probabilistic statements about the results of the election. For example, "The chance that Obama wins the electoral college is 91%" is a probabilistic statement about a parameter which in previous sections we have denoted with d. We showed that for the 2016 election, FiveThirtyEight gave Clinton an 81.4% chance of winning the popular vote. To do this, they used the Bayesian approach we described.

We assume a hierarchical model similar to what we did to predict the performance of a baseball player. Statistical textbooks will write the model like this:

$$d \sim N(\mu, \tau^2) \text{ describes our best guess had we not seen any polling data}$$
$$\bar{X} \mid d \sim N(d, \sigma^2) \text{ describes randomness due to sampling and the pollster effect}$$

For our best guess, we note that before any poll data is available, we can use data sources other than polling data. A popular approach is to use what pollsters call *fundamentals*, which are based on properties about the current economy that historically appear to have an effect in favor or against the incumbent party. We won't use these here. Instead, we will use $\mu = 0$, which is interpreted as a model that simply does not provide any information on who will win. For the standard deviation, we will use recent historical data that shows the winner of the popular vote has an average spread of about 3.5%. Therefore, we set $\tau = 0.035$.

Now we can use the formulas for the posterior distribution for the parameter d: the probability of $d > 0$ given the observed poll data:

```
mu <- 0
tau <- 0.035
sigma <- results$se
Y <- results$avg
B <- sigma^2 / (sigma^2 + tau^2)

posterior_mean <- B*mu + (1-B)*Y
posterior_se <- sqrt( 1/ (1/sigma^2 + 1/tau^2))

posterior_mean
#> [1] 0.0281
posterior_se
#> [1] 0.00615
```

To make a probability statement, we use the fact that the posterior distribution is also normal. And we have a credible interval of:

```
posterior_mean + c(-1.96, 1.96)*posterior_se
#> [1] 0.0160 0.0401
```

The posterior probability $\Pr(d > 0 \mid \bar{X})$ can be computed like this:

```
1 - pnorm(0, posterior_mean, posterior_se)
#> [1] 1
```

This says we are 100% sure Clinton will win the popular vote, which seems too overconfident. Also, it is not in agreement with FiveThirtyEight's 81.4%. What explains this difference?

16.8.2 The general bias

After elections are over, one can look at the difference between pollster predictions and actual result. An important observation that our model does not take into account is that it is common to see a general bias that affects many pollsters in the same way making the observed data correlated. There is no good explanation for this, but we do observe it in historical data: in one election, the average of polls favors Democrats by 2%, then in the following election they favor Republicans by 1%, then in the next election there is no bias, then in the following one Republicans are favored by 3%, and so on. In 2016, the polls were biased in favor of the Democrats by 1-2%.

Although we know this bias term affects our polls, we have no way of knowing what this bias is until election night. So we can't correct our polls accordingly. What we can do is include a term in our model that accounts for this variability.

16.8.3 Mathematical representations of models

Suppose we are collecting data from one pollster and we assume there is no general bias. The pollster collects several polls with a sample size of N, so we observe several measurements of the spread $X_1, \ldots, X_J$. The theory tells us that these random variables have expected value d and standard error $2\sqrt{p(1-p)/N}$. Let's start by using the following model to describe the observed variability:

$$X_j = d + \varepsilon_j.$$

We use the index j to represent the different polls and we define ε_j to be a random variable that explains the poll-to-poll variability introduced by sampling error. To do this, we assume its average is 0 and standard error is $2\sqrt{p(1-p)/N}$. If d is 2.1 and the sample size for these polls is 2,000, we can simulate $J = 6$ data points from this model like this:

```
set.seed(3)
J <- 6
N <- 2000
d <- .021
p <- (d + 1)/2
X <- d + rnorm(J, 0, 2 * sqrt(p * (1 - p) / N))
```

Now suppose we have $J = 6$ data points from $I = 5$ different pollsters. To represent this we now need two indexes, one for pollster and one for the polls each pollster takes. We use X_{ij} with i representing the pollster and j representing the j-th poll from that pollster. If we apply the same model, we write:

$$X_{i,j} = d + \varepsilon_{i,j}$$

To simulate data, we now have to loop through the pollsters:

```
I <- 5
J <- 6
N <- 2000
X <- sapply(1:I, function(i){
  d + rnorm(J, 0, 2 * sqrt(p * (1 - p) / N))
})
```

The simulated data does not really seem to capture the features of the actual data:

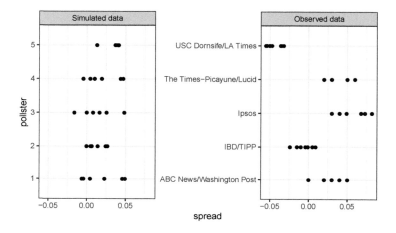

The model above does not account for pollster-to-pollster variability. To fix this, we add a new term for the pollster effect. We will use h_i to represent the house effect of the i-th pollster. The model is now augmented to:

$$X_{i,j} = d + h_i + \varepsilon_{i,j}$$

To simulate data from a specific pollster, we now need to draw an h_i and then add the εs. Here is how we would do it for one specific pollster. We assume σ_h is 0.025:

```
I <- 5
J <- 6
N <- 2000
d <- .021
p <- (d + 1) / 2
h <- rnorm(I, 0, 0.025)
X <- sapply(1:I, function(i){
  d + h[i] + rnorm(J, 0, 2 * sqrt(p * (1 - p) / N))
})
```

The simulated data now looks more like the actual data:

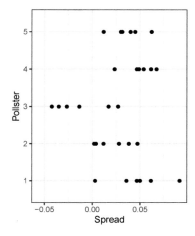

Note that h_i is common to all the observed spreads from a specific pollster. Different pollsters have a different h_i, which explains why we can see the groups of points shift up and down from pollster to pollster.

Now, in the model above, we assume the average house effect is 0. We think that for every pollster biased in favor of our party, there is another one in favor of the other and assume the standard deviation is σ_h. But historically we see that every election has a general bias affecting all polls. We can observe this with the 2016 data, but if we collect historical data, we see that the average of polls misses by more than models like the one above predict. To see this, we would take the average of polls for each election year and compare it to the actual value. If we did this, we would see a difference with a standard deviation of between 2-3%. To incorporate this into the model, we can add another term to account for this variability:

$$X_{i,j} = d + b + h_i + \varepsilon_{i,j}.$$

Here b is a random variable that accounts for the election-to-election variability. This random variable changes from election to election, but for any given election, it is the same for all pollsters and polls within on election. This is why it does not have indexes. This implies that all the random variables $X_{i,j}$ for an election year are correlated since they all have b in common.

One way to interpret b is as the difference between the average of all polls from all pollsters and the actual result of the election. Because we don't know the actual result until after the election, we can't estimate b until after the election. However, we can estimate b from previous elections and study the distribution of these values. Based on this approach we assume that, across election years, b has expected value 0 and the standard error is about $\sigma_b = 0.025$.

An implication of adding this term to the model is that the standard deviation for $X_{i,j}$ is actually higher than what we earlier called σ, which combines the pollster variability and the sample in variability, and was estimated with:

```
sd(one_poll_per_pollster$spread)
#> [1] 0.0242
```

This estimate does not include the variability introduced by b. Note that because

$$\bar{X} = d + b + \frac{1}{N} \sum_{i=1}^{N} X_i,$$

the standard deviation of $\bar{X}$ is:

$$\sqrt{\sigma^2/N + \sigma_b^2}.$$

Since the same b is in every measurement, the average does not reduce the variability introduced by the b term. This is an important point: it does not matter how many polls you take, this bias does not get reduced.

If we redo the Bayesian calculation taking this variability into account, we get a result much closer to FiveThirtyEight's:

```
mu <- 0
tau <- 0.035
sigma <- sqrt(results$se^2 + .025^2)
Y <- results$avg
B <- sigma^2 / (sigma^2 + tau^2)

posterior_mean <- B*mu + (1-B)*Y
posterior_se <- sqrt( 1/ (1/sigma^2 + 1/tau^2))

1 - pnorm(0, posterior_mean, posterior_se)
#> [1] 0.817
```

16.8.4   Predicting the electoral college

Up to now we have focused on the popular vote. But in the United States, elections are not decided by the popular vote but rather by what is known as the electoral college. Each state gets a number of electoral votes that depends, in a somewhat complex way, on the population size of the state. Here are the top 5 states ranked by electoral votes in 2016.

```
results_us_election_2016 %>% top_n(5, electoral_votes)
#>            state electoral_votes clinton trump others
#> 1    California              55    61.7  31.6    6.7
#> 2         Texas              38    43.2  52.2    4.5
#> 3       Florida              29    47.8  49.0    3.2
#> 4      New York              29    59.0  36.5    4.5
#> 5      Illinois              20    55.8  38.8    5.4
#> 6  Pennsylvania              20    47.9  48.6    3.6
```

With some minor exceptions we don't discuss, the electoral votes are won all or nothing. For example, if you win California by just 1 vote, you still get all 55 of its electoral votes. This means that by winning a few big states by a large margin, but losing many small states by small margins, you can win the popular vote and yet lose the electoral college. This happened in 1876, 1888, 2000, and 2016. The idea behind this is to avoid a few large states having the power to dominate the presidential election. Nonetheless, many people in the US consider the electoral college unfair and would like to see it abolished.

We are now ready to predict the electoral college result for 2016. We start by aggregating results from a poll taken during the last week before the election. We use the `str_detect`, a function we introduce later in Section 24.1, to remove polls that are not for entire states.

```
results <- polls_us_election_2016 %>%
  filter(state!="U.S." &
           !str_detect(state, "CD") &
           enddate >="2016-10-31" &
           (grade %in% c("A+","A","A-","B+") | is.na(grade))) %>%
  mutate(spread = rawpoll_clinton/100 - rawpoll_trump/100) %>%
  group_by(state) %>%
  summarize(avg = mean(spread), sd = sd(spread), n = n()) %>%
  mutate(state = as.character(state))
```

Here are the five closest races according to the polls:

```
results %>% arrange(abs(avg))
#> # A tibble: 47 x 4
#>    state             avg      sd      n
#>    <chr>           <dbl>   <dbl>  <int>
#> 1 Florida         0.00356 0.0163     7
#> 2 North Carolina -0.00730 0.0306     9
#> 3 Ohio           -0.0104  0.0252     6
#> 4 Nevada          0.0169  0.0441     7
#> 5 Iowa           -0.0197  0.0437     3
#> # ... with 42 more rows
```

We now introduce the command `left_join` that will let us easily add the number of electoral votes for each state from the dataset `us_electoral_votes_2016`. We will describe this function in detail in the Wrangling chapter. Here, we simply say that the function combines the two datasets so that the information from the second argument is added to the information in the first:

```
results <- left_join(results, results_us_election_2016, by = "state")
```

Notice that some states have no polls because the winner is pretty much known:

```
results_us_election_2016 %>% filter(!state %in% results$state) %>%
  pull(state)
#> [1] "Rhode Island"         "Alaska"           "Wyoming"
#> [4] "District of Columbia"
```

No polls were conducted in DC, Rhode Island, Alaska, and Wyoming because Democrats are sure to win in the first two and Republicans in the last two.

Because we can't estimate the standard deviation for states with just one poll, we will estimate it as the median of the standard deviations estimated for states with more than one poll:

```
results <- results %>%
  mutate(sd = ifelse(is.na(sd), median(results$sd, na.rm = TRUE), sd))
```

To make probabilistic arguments, we will use a Monte Carlo simulation. For each state, we apply the Bayesian approach to generate an election day d. We could construct the priors for each state based on recent history. However, to keep it simple, we assign a prior to each state that assumes we know nothing about what will happen. Since from election year to election year the results from a specific state don't change that much, we will assign a standard deviation of 2% or $\tau = 0.02$. For now, we will assume, incorrectly, that the poll results from each state are independent. The code for the Bayesian calculation under these assumptions looks like this:

```
#> # A tibble: 47 x 12
#>   state      avg     sd       n electoral_votes clinton trump others
#>   <chr>    <dbl>   <dbl> <int>           <int>   <dbl> <dbl>  <dbl>
#> 1 Alab~  -0.149  2.53e-2     3               9    34.4  62.1    3.6
#> 2 Ariz~  -0.0326 2.70e-2     9              11    45.1  48.7    6.2
#> 3 Arka~  -0.151  9.90e-4     2               6    33.7  60.6    5.8
#> 4 Cali~   0.260  3.87e-2     5              55    61.7  31.6    6.7
#> 5 Colo~   0.0452 2.95e-2     7               9    48.2  43.3    8.6
#> # ... with 42 more rows, and 4 more variables: sigma <dbl>, B <dbl>,
#> #   posterior_mean <dbl>, posterior_se <dbl>
```

The estimates based on posterior do move the estimates towards 0, although the states with many polls are influenced less. This is expected as the more poll data we collect, the more we trust those results:

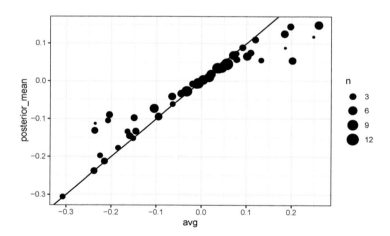

Now we repeat this 10,000 times and generate an outcome from the posterior. In each iteration, we keep track of the total number of electoral votes for Clinton. Remember that Trump gets 270 minus the votes for Clinton. Also note that the reason we add 7 in the code is to account for Rhode Island and D.C.:

```
B <- 10000
mu <- 0
tau <- 0.02
clinton_EV <- replicate(B, {
  results %>% mutate(sigma = sd/sqrt(n),
                     B = sigma^2 / (sigma^2 + tau^2),
                     posterior_mean = B * mu + (1 - B) * avg,
                     posterior_se = sqrt(1 / (1/sigma^2 + 1/tau^2)),
                     result = rnorm(length(posterior_mean),
                                    posterior_mean, posterior_se),
                     clinton = ifelse(result > 0, electoral_votes, 0)) %>%
    summarize(clinton = sum(clinton)) %>%
    pull(clinton) + 7
})

mean(clinton_EV > 269)
#> [1] 0.998
```

This model gives Clinton over 99% chance of winning. A similar prediction was made by the Princeton Election Consortium. We now know it was quite off. What happened?

The model above ignores the general bias and assumes the results from different states are independent. After the election, we realized that the general bias in 2016 was not that big: it was between 1 and 2%. But because the election was close in several big states and these states had a large number of polls, pollsters that ignored the general bias greatly underestimated the standard error. Using the notation we introduce, they assumed the standard error was $\sqrt{\sigma^2/N}$ which with large N is quite smaller than the more accurate estimate $\sqrt{\sigma^2/N + \sigma_b^2}$. FiveThirtyEight, which models the general bias in a rather sophisticated way, reported a closer result. We can simulate the results now with a bias term. For the state level, the general bias can be larger so we set it at $\sigma_b = 0.03$:

```
tau <- 0.02
bias_sd <- 0.03
clinton_EV_2 <- replicate(1000, {
  results %>% mutate(sigma = sqrt(sd^2/n  + bias_sd^2),
                     B = sigma^2 / (sigma^2 + tau^2),
                     posterior_mean = B*mu + (1-B)*avg,
                     posterior_se = sqrt( 1/ (1/sigma^2 + 1/tau^2)),
                     result = rnorm(length(posterior_mean),
                                    posterior_mean, posterior_se),
                     clinton = ifelse(result>0, electoral_votes, 0)) %>%
    summarize(clinton = sum(clinton) + 7) %>%
    pull(clinton)
})
mean(clinton_EV_2 > 269)
#> [1] 0.848
```

This gives us a much more sensible estimate. Looking at the outcomes of the simulation, we see how the bias term adds variability to the final results.

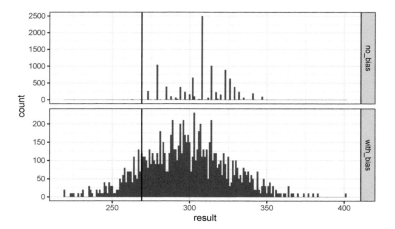

FiveThirtyEight includes many other features we do not include here. One is that they model variability with distributions that have high probabilities for extreme events compared to the normal. One way we could do this is by changing the distribution used in the simulation from a normal distribution to a t-distribution. FiveThirtyEight predicted a probability of 71%.

16.8.5 Forecasting

Forecasters like to make predictions well before the election. The predictions are adapted as new polls come out. However, an important question forecasters must ask is: how informative are polls taken several weeks before the election about the actual election? Here we study the variability of poll results across time.

To make sure the variability we observe is not due to pollster effects, let's study data from one pollster:

```
one_pollster <- polls_us_election_2016 %>%
  filter(pollster == "Ipsos" & state == "U.S.") %>%
  mutate(spread = rawpoll_clinton/100 - rawpoll_trump/100)
```

Since there is no pollster effect, then perhaps the theoretical standard error matches the data-derived standard deviation. We compute both here:

```
se <- one_pollster %>%
  summarize(empirical = sd(spread),
            theoretical = 2 * sqrt(mean(spread) * (1 - mean(spread)) /
                                   min(samplesize)))
se
#>    empirical theoretical
#> 1     0.0403      0.0326
```

But the empirical standard deviation is higher than the highest possible theoretical estimate. Furthermore, the spread data does not look normal as the theory would predict:

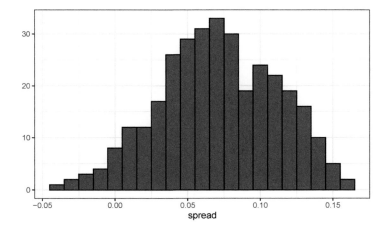

The models we have described include pollster-to-pollster variability and sampling error. But this plot is for one pollster and the variability we see is certainly not explained by sampling error. Where is the extra variability coming from? The following plots make a strong case that it comes from time fluctuations not accounted for by the theory that assumes p is fixed:

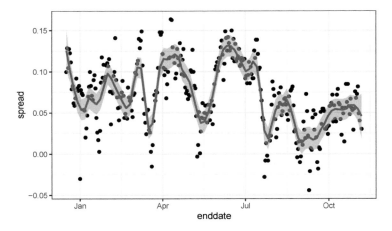

Some of the peaks and valleys we see coincide with events such as the party conventions, which tend to give the candidate a boost. We can see the peaks and valleys are consistent across several pollsters:

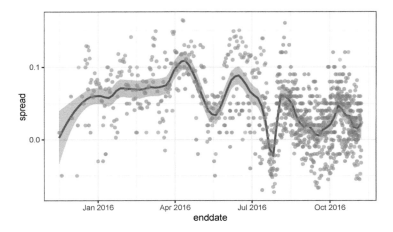

This implies that, if we are going to forecast, our model must include a term to accounts for the time effect. We need to write a model including a bias term for time:

$$Y_{i,j,t} = d + b + h_j + b_t + \varepsilon_{i,j,t}$$

The standard deviation of b_t would depend on t since the closer we get to election day, the closer to 0 this bias term should be.

Pollsters also try to estimate trends from these data and incorporate these into their predictions. We can model the time trend with a function $f(t)$ and rewrite the model like this: The blue lines in the plots above:

$$Y_{i,j,t} = d + b + h_j + b_t + f(t) + \varepsilon_{i,jt,}$$

We usually see the estimated $f(t)$ not for the difference, but for the actual percentages for each candidate like this:

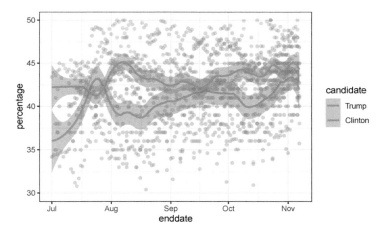

Once a model like the one above is selected, we can use historical and present data to estimate all the necessary parameters to make predictions. There is a variety of methods for estimating trends $f(t)$ which we discuss in the Machine Learning part.

16.9 Exercises

1. Create this table:

```
library(tidyverse)
library(dslabs)
data("polls_us_election_2016")
polls <- polls_us_election_2016 %>%
   filter(state != "U.S." & enddate >= "2016-10-31") %>%
   mutate(spread = rawpoll_clinton/100 - rawpoll_trump/100)
```

Now for each poll use the CLT to create a 95% confidence interval for the spread reported by each poll. Call the resulting object cis with columns lower and upper for the limits of the confidence intervals. Use the `select` function to keep the columns `state`, `startdate`, `end date`, `pollster`, `grade`, `spread`, `lower`, `upper`.

2. You can add the final result to the `cis` table you just created using the `right_join` function like this:

```
add <- results_us_election_2016 %>%
   mutate(actual_spread = clinton/100 - trump/100) %>%
   select(state, actual_spread)
cis <- cis %>%
   mutate(state = as.character(state)) %>%
   left_join(add, by = "state")
```

Now determine how often the 95% confidence interval includes the actual result.

3. Repeat this, but show the proportion of hits for each pollster. Show only pollsters with more than 5 polls and order them from best to worst. Show the number of polls conducted by each pollster and the FiveThirtyEight grade of each pollster. Hint: use `n=n()`, `grade = grade[1]` in the call to summarize.

4. Repeat exercise 3, but instead of pollster, stratify by state. Note that here we can't show grades.

5. Make a barplot based on the result of exercise 4. Use `coord_flip`.

6. Add two columns to the `cis` table by computing, for each poll, the difference between the predicted spread and the actual spread, and define a column `hit` that is true if the signs are the same. Hint: use the function `sign`. Call the object `resids`.

7. Create a plot like in exercise 5, but for the proportion of times the sign of the spread agreed.

8. In exercise 7, we see that for most states the polls had it right 100% of the time. For only 9 states did the polls miss more than 25% of the time. In particular, notice that in Wisconsin every single poll got it wrong. In Pennsylvania and Michigan more than 90% of the polls had the signs wrong. Make a histogram of the errors. What is the median of these errors?

9. We see that at the state level, the median error was 3% in favor of Clinton. The distribution is not centered at 0, but at 0.03. This is the general bias we described in the section above.

Create a boxplot to see if the bias was general to all states or it affected some states differently. Use `filter(grade %in% c("A+","A","A-","B+") | is.na(grade)))` to only include pollsters with high grades.

10. Some of these states only have a few polls. Repeat exercise 9, but only include states with 5 good polls or more. Hint: use `group_by`, `filter` then `ungroup`. You will see that the West (Washington, New Mexico, California) underestimated Hillary's performance, while the Midwest (Michigan, Pennsylvania, Wisconsin, Ohio, Missouri) overestimated it. In our simulation, we did not model this behavior since we added general bias, rather than a regional bias. Note that some pollsters may now be modeling correlation between similar states and estimating this correlation from historical data. To learn more about this, you can learn about random effects and mixed models.

16.10   The t-distribution

Above we made use of the CLT with a sample size of 15. Because we are estimating a second parameters σ, further variability is introduced into our confidence interval which results in intervals that are too small. For very large sample sizes this extra variability is negligible, but, in general, for values smaller than 30 we need to be cautious about using the CLT.

However, if the data in the urn is known to follow a normal distribution, then we actually have mathematical theory that tells us how much bigger we need to make the intervals to account for the estimation of σ. Using this theory, we can construct confidence intervals for any N. But again, this works only if **the data in the urn is known to follow a normal distribution**. So for the 0, 1 data of our previous urn model, this theory definitely does not apply.

The statistic on which confidence intervals for d are based is

$$Z = \frac{\bar{X} - d}{\sigma/\sqrt{N}}$$

CLT tells us that Z is approximately normally distributed with expected value 0 and standard error 1. But in practice we don't know σ so we use:

$$Z = \frac{\bar{X} - d}{s/\sqrt{N}}$$

By substituting σ with s we introduce some variability. The theory tells us that Z follows a t-distribution with $N - 1$ *degrees of freedom*. The degrees of freedom is a parameter that controls the variability via fatter tails:

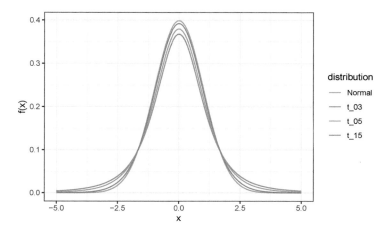

If we are willing to assume the pollster effect data is normally distributed, based on the sample data $X_1, \ldots, X_N$,

```
one_poll_per_pollster %>%
  ggplot(aes(sample=spread)) + stat_qq()
```

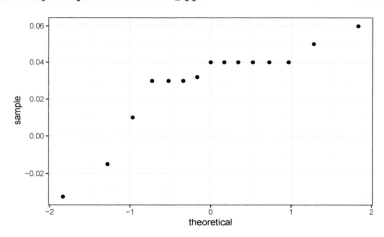

then Z follows a t-distribution with $N-1$ degrees of freedom. So perhaps a better confidence interval for d is:

```
z <- qt(0.975,  nrow(one_poll_per_pollster)-1)
one_poll_per_pollster %>%
  summarize(avg = mean(spread), moe = z*sd(spread)/sqrt(length(spread))) %>%
  mutate(start = avg - moe, end = avg + moe)
#> # A tibble: 1 x 4
#>      avg     moe   start     end
#>    <dbl>   <dbl>   <dbl>   <dbl>
#> 1 0.0290  0.0134  0.0156  0.0424
```

A bit larger than the one using normal is

```
qt(0.975, 14)
#> [1] 2.14
```

is bigger than

```
qnorm(0.975)
#> [1] 1.96
```

The t-distribution can also be used to model errors in bigger deviations that are more likely
than with the normal distribution, as seen in the densities we previously saw. Fivethirtyeight
uses the t-distribution to generate errors that better model the deviations we see in election
data. For example, in Wisconsin the average of six polls was 7% in favor of Clinton with
a standard deviation of 1%, but Trump won by 0.7%. Even after taking into account the
overall bias, this 7.7% residual is more in line with t-distributed data than the normal
distribution.

```
data("polls_us_election_2016")
polls_us_election_2016 %>%
  filter(state =="Wisconsin" &
           enddate >="2016-10-31" &
           (grade %in% c("A+","A","A-","B+") | is.na(grade))) %>%
  mutate(spread = rawpoll_clinton/100 - rawpoll_trump/100) %>%
  mutate(state = as.character(state)) %>%
  left_join(results_us_election_2016, by = "state") %>%
  mutate(actual = clinton/100 - trump/100) %>%
  summarize(actual = first(actual), avg = mean(spread),
            sd = sd(spread), n = n()) %>%
  select(actual, avg, sd, n)
#>    actual    avg      sd n
#> 1 -0.007 0.0711 0.0104 6
```

17

Regression

Up to this point, this book has focused mainly on single variables. However, in data science applications, it is very common to be interested in the relationship between two or more variables. For instance, in Chapter 18 we will use a data-driven approach that examines the relationship between player statistics and success to guide the building of a baseball team with a limited budget. Before delving into this more complex example, we introduce necessary concepts needed to understand regression using a simpler illustration. We actually use the dataset from which regression was born.

The example is from genetics. Francis Galton[1] studied the variation and heredity of human traits. Among many other traits, Galton collected and studied height data from families to try to understand heredity. While doing this, he developed the concepts of correlation and regression, as well as a connection to pairs of data that follow a normal distribution. Of course, at the time this data was collected our knowledge of genetics was quite limited compared to what we know today. A very specific question Galton tried to answer was: how well can we predict a child's height based on the parents' height? The technique he developed to answer this question, regression, can also be applied to our baseball question. Regression can be applied in many other circumstances as well.

Historical note: Galton made important contributions to statistics and genetics, but he was also one of the first proponents of eugenics, a scientifically flawed philosophical movement favored by many biologists of Galton's time but with horrific historical consequences. You can read more about it here: https://pged.org/history-eugenics-and-genetics/.

17.1 Case study: is height hereditary?

We have access to Galton's family height data through the **HistData** package. This data contains heights on several dozen families: mothers, fathers, daughters, and sons. To imitate Galton's analysis, we will create a dataset with the heights of fathers and a randomly selected son of each family:

```
library(tidyverse)
library(HistData)
data("GaltonFamilies")

set.seed(1983)
galton_heights <- GaltonFamilies %>%
  filter(gender == "male") %>%
```

---

[1]https://en.wikipedia.org/wiki/Francis_Galton

```
group_by(family) %>%
sample_n(1) %>%
ungroup() %>%
select(father, childHeight) %>%
rename(son = childHeight)
```

In the exercises, we will look at other relationships including mothers and daughters.

Suppose we were asked to summarize the father and son data. Since both distributions are well approximated by the normal distribution, we could use the two averages and two standard deviations as summaries:

```
galton_heights %>%
  summarize(mean(father), sd(father), mean(son), sd(son))
#> # A tibble: 1 x 4
#>   `mean(father)` `sd(father)` `mean(son)` `sd(son)`
#>            <dbl>        <dbl>       <dbl>     <dbl>
#> 1           69.1         2.55        69.2      2.71
```

However, this summary fails to describe an important characteristic of the data: the trend that the taller the father, the taller the son.

```
galton_heights %>% ggplot(aes(father, son)) +
  geom_point(alpha = 0.5)
```

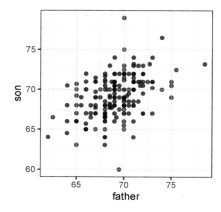

We will learn that the correlation coefficient is an informative summary of how two variables move together and then see how this can be used to predict one variable using the other.

17.2   The correlation coefficient

The correlation coefficient is defined for a list of pairs $(x_1, y_1), \ldots, (x_n, y_n)$ as the average of the product of the standardized values:

$$\rho = \frac{1}{n} \sum_{i=1}^{n} \left(\frac{x_i - \mu_x}{\sigma_x} \right) \left(\frac{y_i - \mu_y}{\sigma_y} \right)$$

with μ_x, μ_y the averages of $x_1, \ldots, x_n$ and $y_1, \ldots, y_n$, respectively, and σ_x, σ_y the standard deviations. The Greek letter ρ is commonly used in statistics books to denote the correlation. The Greek letter for r, ρ, because it is the first letter of regression. Soon we learn about the connection between correlation and regression. We can represent the formula above with R code using:

```
rho <- mean(scale(x) * scale(y))
```

To understand why this equation does in fact summarize how two variables move together, consider the i-th entry of x is $\left(\frac{x_i - \mu_x}{\sigma_x} \right)$ SDs away from the average. Similarly, the y_i that is paired with x_i, is $\left(\frac{y_1 - \mu_y}{\sigma_y} \right)$ SDs away from the average y. If x and y are unrelated, the product $\left(\frac{x_i - \mu_x}{\sigma_x} \right) \left(\frac{y_i - \mu_y}{\sigma_y} \right)$ will be positive ($+ \times +$ and $- \times -$) as often as negative ($+ \times -$ and $- \times +$) and will average out to about 0. This correlation is the average and therefore unrelated variables will have 0 correlation. If instead the quantities vary together, then we are averaging mostly positive products ($+ \times +$ and $- \times -$) and we get a positive correlation. If they vary in opposite directions, we get a negative correlation.

The correlation coefficient is always between -1 and 1. We can show this mathematically: consider that we can't have higher correlation than when we compare a list to itself (perfect correlation) and in this case the correlation is:

$$\rho = \frac{1}{n} \sum_{i=1}^{n} \left(\frac{x_i - \mu_x}{\sigma_x} \right)^2 = \frac{1}{\sigma_x^2} \frac{1}{n} \sum_{i=1}^{n} (x_i - \mu_x)^2 = \frac{1}{\sigma_x^2} \sigma_x^2 = 1$$

A similar derivation, but with x and its exact opposite, proves the correlation has to be bigger or equal to -1.

For other pairs, the correlation is in between -1 and 1. The correlation between father and son's heights is about 0.5:

```
galton_heights %>% summarize(r = cor(father, son)) %>% pull(r)
#> [1] 0.433
```

To see what data looks like for different values of ρ, here are six examples of pairs with correlations ranging from -0.9 to 0.99:

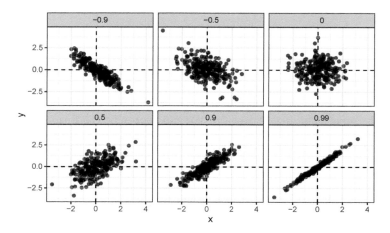

17.2.1   Sample correlation is a random variable

Before we continue connecting correlation to regression, let's remind ourselves about random variability.

In most data science applications, we observe data that includes random variation. For example, in many cases, we do not observe data for the entire population of interest but rather for a random sample. As with the average and standard deviation, the *sample correlation* is the most commonly used estimate of the population correlation. This implies that the correlation we compute and use as a summary is a random variable.

By way of illustration, let's assume that the 179 pairs of fathers and sons is our entire population. A less fortunate geneticist can only afford measurements from a random sample of 25 pairs. The sample correlation can be computed with:

```
R <- sample_n(galton_heights, 25, replace = TRUE) %>%
  summarize(r = cor(father, son)) %>% pull(r)
```

R is a random variable. We can run a Monte Carlo simulation to see its distribution:

```
B <- 1000
N <- 25
R <- replicate(B, {
  sample_n(galton_heights, N, replace = TRUE) %>%
    summarize(r=cor(father, son)) %>%
    pull(r)
})
qplot(R, geom = "histogram", binwidth = 0.05, color = I("black"))
```

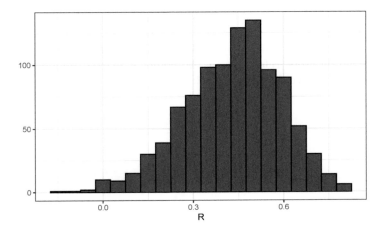

We see that the expected value of R is the population correlation:

```
mean(R)
#> [1] 0.431
```

and that it has a relatively high standard error relative to the range of values R can take:

```
sd(R)
#> [1] 0.161
```

So, when interpreting correlations, remember that correlations derived from samples are estimates containing uncertainty.

Also, note that because the sample correlation is an average of independent draws, the central limit actually applies. Therefore, for large enough N, the distribution of R is approximately normal with expected value ρ. The standard deviation, which is somewhat complex to derive, is $\sqrt{\frac{1-r^2}{N-2}}$.

In our example, $N = 25$ does not seem to be large enough to make the approximation a good one:

```
ggplot(aes(sample=R), data = data.frame(R)) +
  stat_qq() +
  geom_abline(intercept = mean(R), slope = sqrt((1-mean(R)^2)/(N-2)))
```

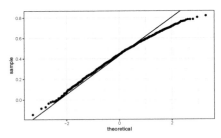

If you increase N, you will see the distribution converging to normal.

17.2.2   Correlation is not always a useful summary

Correlation is not always a good summary of the relationship between two variables. The following four artificial datasets, referred to as Anscombe's quartet, famously illustrate this point. All these pairs have a correlation of 0.82:

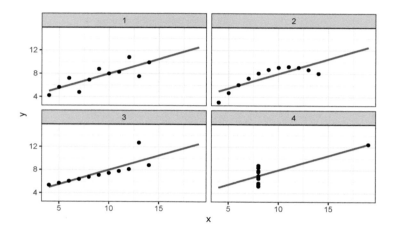

Correlation is only meaningful in a particular context. To help us understand when it is that correlation is meaningful as a summary statistic, we will return to the example of predicting a son's height using his father's height. This will help motivate and define linear regression. We start by demonstrating how correlation can be useful for prediction.

17.3   Conditional expectations

Suppose we are asked to guess the height of a randomly selected son and we don't know his father's height. Because the distribution of sons' heights is approximately normal, we know the average height, 69.2, is the value with the highest proportion and would be the prediction with the highest chance of minimizing the error. But what if we are told that the father is taller than average, say 72 inches tall, do we still guess 69.2 for the son?

It turns out that if we were able to collect data from a very large number of fathers that are 72 inches, the distribution of their sons' heights would be normally distributed. This implies that the average of the distribution computed on this subset would be our best prediction.

In general, we call this approach *conditioning*. The general idea is that we stratify a population into groups and compute summaries in each group. Conditioning is therefore related to the concept of stratification described in Section 8.13. To provide a mathematical description of conditioning, consider we have a population of pairs of values $(x_1, y_1), \ldots, (x_n, y_n)$, for example all father and son heights in England. In the previous chapter we learned that if you take a random pair (X, Y), the expected value and best predictor of Y is $E(Y) = \mu_y$, the population average $1/n \sum_{i=1}^{n} y_i$. However, we are no longer interested in the general population, instead we are interested in only the subset of a population with a specific x_i

value, 72 inches in our example. This subset of the population, is also a population and thus the same principles and properties we have learned apply. The y_i in the subpopulation have a distribution, referred to as the *conditional distribution*, and this distribution has an expected value referred to as the *conditional expectation*. In our example, the conditional expectation is the average height of all sons in England with fathers that are 72 inches. The statistical notation for the conditional expectation is

$$E(Y \mid X = x)$$

with x representing the fixed value that defines that subset, for example 72 inches. Similarly, we denote the standard deviation of the strata with

$$SD(Y \mid X = x) = \sqrt{\text{Var}(Y \mid X = x)}$$

Because the conditional expectation $E(Y \mid X = x)$ is the best predictor for the random variable Y for an individual in the strata defined by $X = x$, many data science challenges reduce to estimating this quantity. The conditional standard deviation quantifies the precision of the prediction.

In the example we have been considering, we are interested in computing the average son height *conditioned* on the father being 72 inches tall. We want to estimate $E(Y|X = 72)$ using the sample collected by Galton. We previously learned that the sample average is the preferred approach to estimating the population average. However, a challenge when using this approach to estimating conditional expectations is that for continuous data we don't have many data points matching exactly one value in our sample. For example, we have only:

```
sum(galton_heights$father == 72)
#> [1] 8
```

fathers that are exactly 72-inches. If we change the number to 72.5, we get even fewer data points:

```
sum(galton_heights$father == 72.5)
#> [1] 1
```

A practical way to improve these estimates of the conditional expectations, is to define strata of with similar values of x. In our example, we can round father heights to the nearest inch and assume that they are all 72 inches. If we do this, we end up with the following prediction for the son of a father that is 72 inches tall:

```
conditional_avg <- galton_heights %>%
  filter(round(father) == 72) %>%
  summarize(avg = mean(son)) %>%
  pull(avg)
conditional_avg
#> [1] 70.5
```

Note that a 72-inch father is taller than average – specifically, 72 - 69.1/2.5 = 1.1 standard deviations taller than the average father. Our prediction 70.5 is also taller than average, but only 0.49 standard deviations larger than the average son. The sons of 72-inch fathers have *regressed* some to the average height. We notice that the reduction in how many SDs taller is about 0.5, which happens to be the correlation. As we will see in a later section, this is not a coincidence.

If we want to make a prediction of any height, not just 72, we could apply the same approach to each strata. Stratification followed by boxplots lets us see the distribution of each group:

```
galton_heights %>% mutate(father_strata = factor(round(father))) %>%
  ggplot(aes(father_strata, son)) +
  geom_boxplot() +
  geom_point()
```

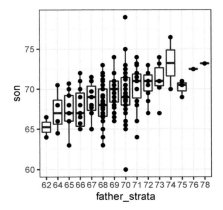

Not surprisingly, the centers of the groups are increasing with height. Furthermore, these centers appear to follow a linear relationship. Below we plot the averages of each group. If we take into account that these averages are random variables with standard errors, the data is consistent with these points following a straight line:

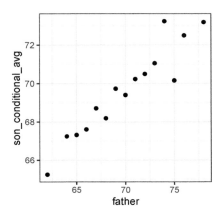

The fact that these conditional averages follow a line is not a coincidence. In the next section, we explain that the line these averages follow is what we call the *regression line*, which improves the precision of our estimates. However, it is not always appropriate to estimate conditional expectations with the regression line so we also describe Galton's theoretical justification for using the regression line.

17.4 The regression line

If we are predicting a random variable Y knowing the value of another $X = x$ using a regression line, then we predict that for every standard deviation, σ_X, that x increases above the average μ_X, Y increase ρ standard deviations σ_Y above the average μ_Y with ρ the correlation between X and Y. The formula for the regression is therefore:

$$\left(\frac{Y - \mu_Y}{\sigma_Y}\right) = \rho\left(\frac{x - \mu_X}{\sigma_X}\right)$$

We can rewrite it like this:

$$Y = \mu_Y + \rho\left(\frac{x - \mu_X}{\sigma_X}\right)\sigma_Y$$

If there is perfect correlation, the regression line predicts an increase that is the same number of SDs. If there is 0 correlation, then we don't use x at all for the prediction and simply predict the average μ_Y. For values between 0 and 1, the prediction is somewhere in between. If the correlation is negative, we predict a reduction instead of an increase.

Note that if the correlation is positive and lower than 1, our prediction is closer, in standard units, to the average height than the value used to predict, x, is to the average of the xs. This is why we call it *regression*: the son regresses to the average height. In fact, the title of Galton's paper was: *Regression toward mediocrity in hereditary stature*. To add regression lines to plots, we will need the above formula in the form:

$$y = b + mx \text{ with slope } m = \rho\frac{\sigma_y}{\sigma_x} \text{ and intercept } b = \mu_y - m\mu_x$$

Here we add the regression line to the original data:

```
mu_x <- mean(galton_heights$father)
mu_y <- mean(galton_heights$son)
s_x <- sd(galton_heights$father)
s_y <- sd(galton_heights$son)
r <- cor(galton_heights$father, galton_heights$son)

galton_heights %>%
  ggplot(aes(father, son)) +
  geom_point(alpha = 0.5) +
  geom_abline(slope = r * s_y/s_x, intercept = mu_y - r * s_y/s_x * mu_x)
```

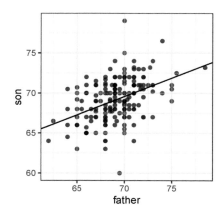

The regression formula implies that if we first standardize the variables, that is subtract the average and divide by the standard deviation, then the regression line has intercept 0 and slope equal to the correlation ρ. You can make same plot, but using standard units like this:

```
galton_heights %>%
  ggplot(aes(scale(father), scale(son))) +
  geom_point(alpha = 0.5) +
  geom_abline(intercept = 0, slope = r)
```

17.4.1   Regression improves precision

Let's compare the two approaches to prediction that we have presented:

1. Round fathers' heights to closest inch, stratify, and then take the average.
2. Compute the regression line and use it to predict.

We use a Monte Carlo simulation sampling $N = 50$ families:

```
B <- 1000
N <- 50

set.seed(1983)
conditional_avg <- replicate(B, {
  dat <- sample_n(galton_heights, N)
  dat %>% filter(round(father) == 72) %>%
    summarize(avg = mean(son)) %>%
    pull(avg)
})

regression_prediction <- replicate(B, {
  dat <- sample_n(galton_heights, N)
  mu_x <- mean(dat$father)
  mu_y <- mean(dat$son)
  s_x <- sd(dat$father)
  s_y <- sd(dat$son)
```

```
  r <- cor(dat$father, dat$son)
  mu_y + r*(72 - mu_x)/s_x*s_y
})
```

Although the expected value of these two random variables is about the same:

```
mean(conditional_avg, na.rm = TRUE)
#> [1] 70.5
mean(regression_prediction)
#> [1] 70.5
```

The standard error for the regression prediction is substantially smaller:

```
sd(conditional_avg, na.rm = TRUE)
#> [1] 0.964
sd(regression_prediction)
#> [1] 0.452
```

The regression line is therefore much more stable than the conditional mean. There is an intuitive reason for this. The conditional average is computed on a relatively small subset: the fathers that are about 72 inches tall. In fact, in some of the permutations we have no data, which is why we use `na.rm=TRUE`. The regression always uses all the data.

So why not always use the regression for prediction? Because it is not always appropriate. For example, Anscombe provided cases for which the data does not have a linear relationship. So are we justified in using the regression line to predict? Galton answered this in the positive for height data. The justification, which we include in the next section, is somewhat more advanced than the rest of the chapter.

17.4.2 Bivariate normal distribution (advanced)

Correlation and the regression slope are a widely used summary statistic, but they are often misused or misinterpreted. Anscombe's examples provide over-simplified cases of dataset in which summarizing with correlation would be a mistake. But there are many more real-life examples.

The main way we motivate the use of correlation involves what is called the *bivariate normal distribution.*

When a pair of random variables is approximated by the bivariate normal distribution, scatterplots look like ovals. As we saw in Section 17.2, they can be thin (high correlation) or circle-shaped (no correlation).

A more technical way to define the bivariate normal distribution is the following: if X is a normally distributed random variable, Y is also a normally distributed random variable, and the conditional distribution of Y for any $X = x$ is approximately normal, then the pair is approximately bivariate normal.

If we think the height data is well approximated by the bivariate normal distribution, then we should see the normal approximation hold for each strata. Here we stratify the son heights by the standardized father heights and see that the assumption appears to hold:

```
galton_heights %>%
  mutate(z_father = round((father - mean(father)) / sd(father))) %>%
  filter(z_father %in% -2:2) %>%
  ggplot() +
  stat_qq(aes(sample = son)) +
  facet_wrap( ~ z_father)
```

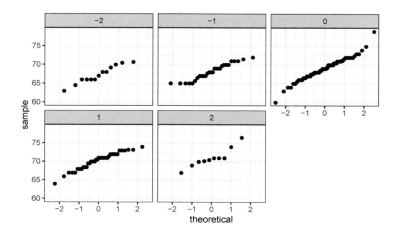

Now we come back to defining correlation. Galton used mathematical statistics to demonstrate that, when two variables follow a bivariate normal distribution, computing the regression line is equivalent to computing conditional expectations. We don't show the derivation here, but we can show that under this assumption, for any given value of x, the expected value of the Y in pairs for which $X = x$ is:

$$E(Y|X = x) = \mu_Y + \rho \frac{X - \mu_X}{\sigma_X} \sigma_Y$$

This is the regression line, with slope

$$\rho \frac{\sigma_Y}{\sigma_X}$$

and intercept $\mu_y - m\mu_X$. It is equivalent to the regression equation we showed earlier which can be written like this:

$$\frac{E(Y \mid X = x) - \mu_Y}{\sigma_Y} = \rho \frac{x - \mu_X}{\sigma_X}$$

This implies that, if our data is approximately bivariate, the regression line gives the conditional probability. Therefore, we can obtain a much more stable estimate of the conditional expectation by finding the regression line and using it to predict.

In summary, if our data is approximately bivariate, then the conditional expectation, the best prediction of Y given we know the value of X, is given by the regression line.

17.4.3   Variance explained

The bivariate normal theory also tells us that the standard deviation of the *conditional* distribution described above is:

$$SD(Y \mid X = x) = \sigma_Y \sqrt{1 - \rho^2}$$

To see why this is intuitive, notice that without conditioning, $SD(Y) = \sigma_Y$, we are looking at the variability of all the sons. But once we condition, we are only looking at the variability of the sons with a tall, 72-inch, father. This group will all tend to be somewhat tall so the standard deviation is reduced.

Specifically, it is reduced to $\sqrt{1 - \rho^2} = \sqrt{1 - 0.25} = 0.86$ of what it was originally. We could say that the fathers' heights "explains" 14% of the sons' heights variability.

The statement "X explains such and such percent of the variability" is commonly used in academic papers. In this case, this percent actually refers to the variance (the SD squared). So if the data is bivariate normal, the variance is reduced by $1 - \rho^2$, so we say that X explains $1 - (1 - \rho^2) = \rho^2$ (the correlation squared) of the variance.

But it is important to remember that the "variance explained" statement only makes sense when the data is approximated by a bivariate normal distribution.

17.4.4   Warning: there are two regression lines

We computed a regression line to predict the son's height from father's height. We used these calculations:

```
mu_x <- mean(galton_heights$father)
mu_y <- mean(galton_heights$son)
s_x <- sd(galton_heights$father)
s_y <- sd(galton_heights$son)
r <- cor(galton_heights$father, galton_heights$son)
m_1 <-  r * s_y / s_x
b_1 <- mu_y - m_1*mu_x
```

which gives us the function $E(Y \mid X = x) = 37.3 + 0.46\, x$.

What if we want to predict the father's height based on the son's? It is important to know that this is not determined by computing the inverse function: $x = \{E(Y \mid X = x) - 37.3\} / 0.5$.

We need to compute $E(X \mid Y = y)$. Since the data is approximately bivariate normal, the theory described above tells us that this conditional expectation will follow a line with slope and intercept:

```
m_2 <-  r * s_x / s_y
b_2 <- mu_x - m_2 * mu_y
```

So we get $E(X \mid Y = y) = 40.9 + 0.41y$. Again we see regression to the average: the prediction for the father is closer to the father average than the son heights y is to the son average.

Here is a plot showing the two regression lines, with blue for the predicting son heights with father heights and red for predicting father heights with son heights:

```
galton_heights %>%
  ggplot(aes(father, son)) +
  geom_point(alpha = 0.5) +
  geom_abline(intercept = b_1, slope = m_1, col = "blue") +
  geom_abline(intercept = -b_2/m_2, slope = 1/m_2, col = "red")
```

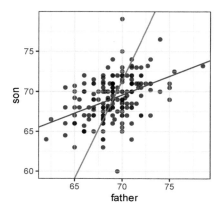

17.5   Exercises

1. Load the `GaltonFamilies` data from the **HistData**. The children in each family are listed by gender and then by height. Create a dataset called `galton_heights` by picking a male and female at random.

2. Make a scatterplot for heights between mothers and daughters, mothers and sons, fathers and daughters, and fathers and sons.

3. Compute the correlation in heights between mothers and daughters, mothers and sons, fathers and daughters, and fathers and sons.

18

Linear models

Since Galton's original development, regression has become one of the most widely used tools in data science. One reason has to do with the fact that regression permits us to find relationships between two variables taking into account the effects of other variables that affect both. This has been particularly popular in fields where randomized experiments are hard to run, such as economics and epidemiology.

When we are not able to randomly assign each individual to a treatment or control group, confounding is particularly prevalent. For example, consider estimating the effect of eating fast foods on life expectancy using data collected from a random sample of people in a jurisdiction. Fast food consumers are more likely to be smokers, drinkers, and have lower incomes. Therefore, a naive regression model may lead to an overestimate of the negative health effect of fast food. So how do we account for confounding in practice? In this chapter we learn how linear models can help with such situations and can be used to describe how one or more variables affect an outcome variable.

18.1 Case study: Moneyball

Moneyball: The Art of Winning an Unfair Game is a book by Michael Lewis about the Oakland Athletics (A's) baseball team and its general manager, the person tasked with building the team, Billy Beane.

Traditionally, baseball teams use *scouts* to help them decide what players to hire. These scouts evaluate players by observing them perform. Scouts tend to favor athletic players with observable physical abilities. For this reason, scouts tend to agree on who the best players are and, as a result, these players tend to be in high demand. This in turn drives up their salaries.

From 1989 to 1991, the A's had one of the highest payrolls in baseball. They were able to buy the best players and, during that time, they were one of the best teams. However, in 1995 the A's team owner changed and the new management cut the budget drastically, leaving then general manager, Sandy Alderson, with one of the lowest payrolls in baseball. He could no longer afford the most sought-after players. Alderson began using a statistical approach to find inefficiencies in the market. Alderson was a mentor to Billy Beane, who succeeded him in 1998 and fully embraced data science, as opposed to scouts, as a method for finding low-cost players that data predicted would help the team win. Today, this strategy has been adapted by most baseball teams. As we will see, regression plays a large role in this approach.

As motivation for this chapter, we will pretend it is 2002 and try to build a baseball team with a limited budget, just like the A's had to do. To appreciate what you are up against, note that in 2002 the Yankees' payroll of \$125,928,583 more than tripled the Oakland A's \$39,679,746:

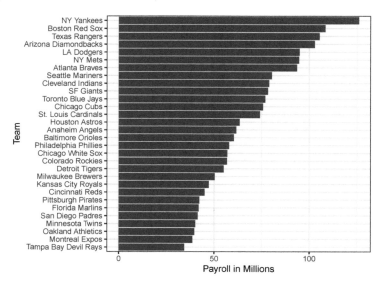

18.1.1   Sabermetics

Statistics have been used in baseball since its beginnings. The dataset we will be using, included in the **Lahman** library, goes back to the 19th century. For example, a summary statistics we will describe soon, the *batting average*, has been used for decades to summarize a batter's success. Other statistics[1] such as home runs (HR), runs batted in (RBI), and stolen bases (SB) are reported for each player in the game summaries included in the sports section of newspapers, with players rewarded for high numbers. Although summary statistics such as these were widely used in baseball, data analysis per se was not. These statistics were arbitrarily decided on without much thought as to whether they actually predicted anything or were related to helping a team win.

This changed with Bill James[2]. In the late 1970s, this aspiring writer and baseball fan started publishing articles describing more in-depth analysis of baseball data. He named the approach of using data to predict what outcomes best predicted if a team would win *sabermetrics*[3]. Until Billy Beane made sabermetrics the center of his baseball operation, Bill James' work was mostly ignored by the baseball world. Currently, sabermetrics popularity is no longer limited to just baseball; other sports have started to use this approach as well.

In this chapter, to simplify the exercise, we will focus on scoring runs and ignore the two other important aspects of the game: pitching and fielding. We will see how regression analysis can help develop strategies to build a competitive baseball team with a constrained budget. The approach can be divided into two separate data analyses. In the first, we determine which recorded player-specific statistics predict runs. In the second, we examine if players were undervalued based on what our first analysis predicts.

---

[1]http://mlb.mlb.com/stats/league_leaders.jsp
[2]https://en.wikipedia.org/wiki/Bill_James
[3]https://en.wikipedia.org/wiki/Sabermetrics

18.1.2 Baseball basics

To see how regression will help us find undervalued players, we actually don't need to understand all the details about the game of baseball, which has over 100 rules. Here, we distill the sport to the basic knowledge one needs to know how to effectively attack the data science problem.

The goal of a baseball game is to score more runs (points) than the other team. Each team has 9 batters that have an opportunity to hit a ball with a bat in a predetermined order. After the 9th batter has had their turn, the first batter bats again, then the second, and so on. Each time a batter has an opportunity to bat, we call it a plate appearance (PA). At each PA, the other team's *pitcher* throws the ball and the batter tries to hit it. The PA ends with an binary outcome: the batter either makes an *out* (failure) and returns to the bench or the batter doesn't (success) and can run around the bases, and potentially score a run (reach all 4 bases). Each team gets nine tries, referred to as *innings*, to score runs and each inning ends after three outs (three failures).

Here is a video showing a success: https://www.youtube.com/watch?v=HL-XjMCPfio. And here is one showing a failure: https://www.youtube.com/watch?v=NeloljCx-1g. In these videos, we see how luck is involved in the process. When at bat, the batter wants to hit the ball hard. If the batter hits it hard enough, it is a HR, the best possible outcome as the batter gets at least one automatic run. But sometimes, due to chance, the batter hits the ball very hard and a defender catches it, resulting in an out. In contrast, sometimes the batter hits the ball softly, but it lands just in the right place. The fact that there is chance involved hints at why probability models will be involved.

Now there are several ways to succeed. Understanding this distinction will be important for our analysis. When the batter hits the ball, the batter wants to pass as many *bases* as possible. There are four bases with the fourth one called *home plate*. Home plate is where batters start by trying to hit, so the bases form a cycle.

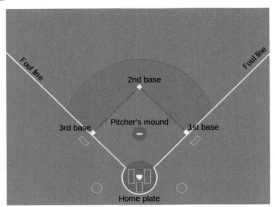

(Courtesy of Cburnett[4]. CC BY-SA 3.0 license[5].)

A batter who *goes around the bases* and arrives home, scores a run.

We are simplifying a bit, but there are five ways a batter can succeed, that is, not make an out:

---

[4]https://en.wikipedia.org/wiki/User:Cburnett
[5]https://creativecommons.org/licenses/by-sa/3.0/deed.en

- Bases on balls (BB) - the pitcher fails to throw the ball through a predefined area considered to be hittable (the strikezone), so the batter is permitted to go to first base.
- Single - Batter hits the ball and gets to first base.
- Double (2B) - Batter hits the ball and gets to second base.
- Triple (3B) - Batter hits the ball and gets to third base.
- Home Run (HR) - Batter hits the ball and goes all the way home and scores a run.

Here is an example of a HR: https://www.youtube.com/watch?v=xYxSZJ9GZ-w. If a batter gets to a base, the batter still has a chance of getting home and scoring a run if the next batter hits successfully. While the batter is *on base*, the batter can also try to steal a base (SB). If a batter runs fast enough, the batter can try to go from one base to the next without the other team tagging the runner. [Here] is an example of a stolen base: https://www.youtube.com/watch?v=JSE5kfxkzfk.

All these events are kept track of during the season and are available to us through the **Lahman** package. Now we will start discussing how data analysis can help us decide how to use these statistics to evaluate players.

18.1.3 No awards for BB

Historically, the *batting average* has been considered the most important offensive statistic. To define this average, we define a *hit* (H) and an *at bat* (AB). Singles, doubles, triples, and home runs are hits. The fifth way to be successful, BB, is not a hit. An AB is the number of times you either get a hit or make an out; BBs are excluded. The batting average is simply H/AB and is considered the main measure of a success rate. Today this success rate ranges from 20% to 38%. We refer to the batting average in thousands so, for example, if your success rate is 28%, we call it *batting 280*.

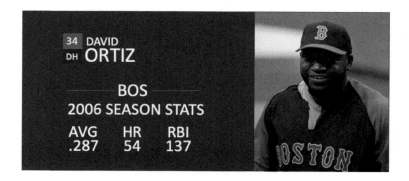

(Picture courtesy of Keith Allison[6]. CC BY-SA 2.0 license[7].)

One of Bill James' first important insights is that the batting average ignores BB, but a BB is a success. He proposed we use the *on base percentage* (OBP) instead of batting average. He defined OBP as (H+BB)/(AB+BB) which is simply the proportion of plate appearances that don't result in an out, a very intuitive measure. He noted that a player that gets many more BB than the average player might not be recognized if the batter does

---

[6]https://www.flickr.com/people/27003603@N00
[7]https://creativecommons.org/licenses/by-sa/2.0

not excel in batting average. But is this player not helping produce runs? No award is given to the player with the most BB. However, bad habits are hard to break and baseball did not immediately adopt OBP as an important statistic. In contrast, total stolen bases were considered important and an award[8] given to the player with the most. But players with high totals of SB also made more outs as they did not always succeed. Does a player with high SB total help produce runs? Can we use data science to determine if it's better to pay for players with high BB or SB?

18.1.4 Base on balls or stolen bases?

One of the challenges in this analysis is that it is not obvious how to determine if a player produces runs because so much depends on his teammates. We do keep track of the number of runs scored by a player. However, remember that if a player X bats right before someone who hits many HRs, batter X will score many runs. But these runs don't necessarily happen if we hire player X but not his HR hitting teammate. However, we can examine team-level statistics. How do teams with many SB compare to teams with few? How about BB? We have data! Let's examine some.

Let's start with an obvious one: HRs. Do teams that hit more home runs score more runs? We examine data from 1961 to 2001. The visualization of choice when exploring the relationship between two variables, such as HRs and wins, is a scatterplot:

```
library(Lahman)
```

```
Teams %>% filter(yearID %in% 1961:2001) %>%
  mutate(HR_per_game = HR / G, R_per_game = R / G) %>%
  ggplot(aes(HR_per_game, R_per_game)) +
  geom_point(alpha = 0.5)
```

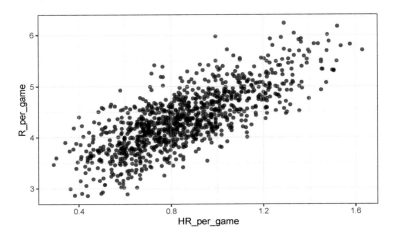

The plot shows a strong association: teams with more HRs tend to score more runs. Now let's examine the relationship between stolen bases and runs:

---

[8]http://www.baseball-almanac.com/awards/lou_brock_award.shtml

```
Teams %>% filter(yearID %in% 1961:2001) %>%
  mutate(SB_per_game = SB / G, R_per_game = R / G) %>%
  ggplot(aes(SB_per_game, R_per_game)) +
  geom_point(alpha = 0.5)
```

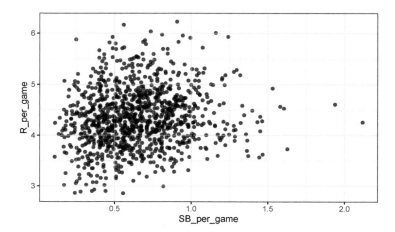

Here the relationship is not as clear. Finally, let's examine the relationship between BB and runs:

```
Teams %>% filter(yearID %in% 1961:2001) %>%
  mutate(BB_per_game = BB/G, R_per_game = R/G) %>%
  ggplot(aes(BB_per_game, R_per_game)) +
  geom_point(alpha = 0.5)
```

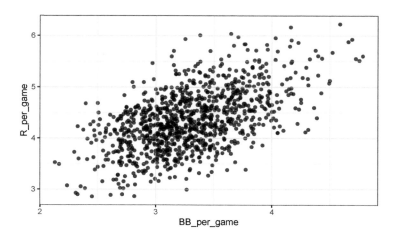

Here again we see a clear association. But does this mean that increasing a team's BBs **causes** an increase in runs? One of the most important lessons you learn in this book is that **association is not causation.**

In fact, it looks like BBs and HRs are also associated:

```
Teams %>% filter(yearID %in% 1961:2001 ) %>%
  mutate(HR_per_game = HR/G, BB_per_game = BB/G) %>%
  ggplot(aes(HR_per_game, BB_per_game)) +
  geom_point(alpha = 0.5)
```

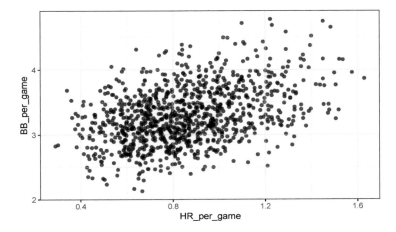

We know that HRs cause runs because, as the name "home run" implies, when a player hits a HR they are guaranteed at least one run. Could it be that HRs also cause BB and this makes it appear as if BB cause runs? When this happens we say there is *confounding*, an important concept we will learn more about throughout this chapter.

Linear regression will help us parse all this out and quantify the associations. This will then help us determine what players to recruit. Specifically, we will try to predict things like how many more runs will a team score if we increase the number of BBs, but keep the HRs fixed? Regression will help us answer questions like this one.

18.1.5 Regression applied to baseball statistics

Can we use regression with these data? First, notice that the HR and Run data appear to be bivariate normal. We save the plot into the object p as we will use it again later.

```
library(Lahman)
p <- Teams %>% filter(yearID %in% 1961:2001 ) %>%
  mutate(HR_per_game = HR/G, R_per_game = R/G) %>%
  ggplot(aes(HR_per_game, R_per_game)) +
  geom_point(alpha = 0.5)
p
```

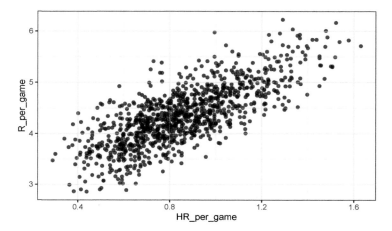

The qq-plots confirm that the normal approximation is useful here:

```
Teams %>% filter(yearID %in% 1961:2001 ) %>%
  mutate(z_HR = round((HR - mean(HR))/sd(HR)),
         R_per_game = R/G) %>%
  filter(z_HR %in% -2:3) %>%
  ggplot() +
  stat_qq(aes(sample=R_per_game)) +
  facet_wrap(~z_HR)
```

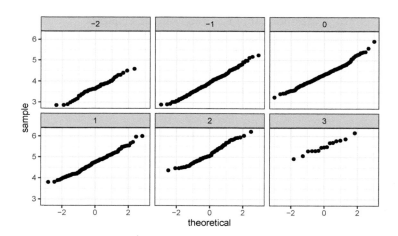

Now we are ready to use linear regression to predict the number of runs a team will score if we know how many home runs the team hits. All we need to do is compute the five summary statistics:

```
summary_stats <- Teams %>%
  filter(yearID %in% 1961:2001 ) %>%
  mutate(HR_per_game = HR/G, R_per_game = R/G) %>%
```

```
summarize(avg_HR = mean(HR_per_game),
          s_HR = sd(HR_per_game),
          avg_R = mean(R_per_game),
          s_R = sd(R_per_game),
          r = cor(HR_per_game, R_per_game))
summary_stats
#>    avg_HR  s_HR avg_R   s_R     r
#> 1   0.855 0.243  4.36 0.589 0.762
```

and use the formulas given above to create the regression lines:

```
reg_line <- summary_stats %>% summarize(slope = r*s_R/s_HR,
                       intercept = avg_R - slope*avg_HR)

p + geom_abline(intercept = reg_line$intercept, slope = reg_line$slope)
```

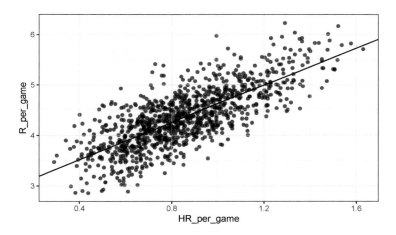

Soon we will learn R functions, such as `lm`, that make fitting regression lines much easier. Another example is the **ggplot2** function `geom_smooth` which computes and adds a regression line to plot along with confidence intervals, which we also learn about later. We use the argument `method = "lm"` which stands for *linear model*, the title of an upcoming section. So we can simplify the code above like this:

```
p + geom_smooth(method = "lm")
```

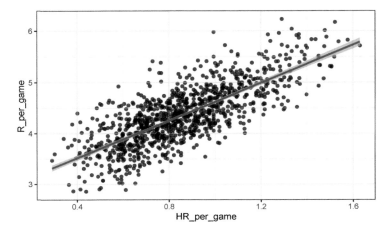

In the example above, the slope is 1.845. So this tells us that teams that hit 1 more HR per game than the average team, score 1.845 more runs per game than the average team. Given that the most common final score is a difference of a run, this can certainly lead to a large increase in wins. Not surprisingly, HR hitters are very expensive. Because we are working on a budget, we will need to find some other way to increase wins. So in the next section we move our attention to BB.

18.2   Confounding

Previously, we noted a strong relationship between Runs and BB. If we find the regression line for predicting runs from bases on balls, we a get slope of:

```
library(tidyverse)
library(Lahman)
get_slope <- function(x, y) cor(x, y) / (sd(x) * sd(y))

bb_slope <- Teams %>%
  filter(yearID %in% 1961:2001 ) %>%
  mutate(BB_per_game = BB/G, R_per_game = R/G) %>%
  summarize(slope = get_slope(R_per_game, BB_per_game))

bb_slope
#>    slope
#> 1   2.12
```

So does this mean that if we go and hire low salary players with many BB, and who therefore increase the number of walks per game by 2, our team will score 4.2 more runs per game?

We are again reminded that association is not causation. The data does provide strong evidence that a team with two more BB per game than the average team, scores 4.2 runs per game. But this does not mean that BB are the cause.

Note that if we compute the regression line slope for singles we get:

```
singles_slope <- Teams %>%
  filter(yearID %in% 1961:2001 ) %>%
  mutate(Singles_per_game = (H-HR-X2B-X3B)/G, R_per_game = R/G) %>%
  summarize(slope = get_slope(R_per_game, Singles_per_game))

singles_slope
#>    slope
#> 1    1.3
```

which is a lower value than what we obtain for BB.

Also, notice that a single gets you to first base just like a BB. Those that know about baseball will tell you that with a single, runners on base have a better chance of scoring than with a BB. So how can BB be more predictive of runs? The reason this happen is because of confounding. Here we show the correlation between HR, BB, and singles:

```
Teams %>%
  filter(yearID %in% 1961:2001 ) %>%
  mutate(Singles = (H-HR-X2B-X3B)/G, BB = BB/G, HR = HR/G) %>%
  summarize(cor(BB, HR), cor(Singles, HR), cor(BB, Singles))
#>   cor(BB, HR) cor(Singles, HR) cor(BB, Singles)
#> 1     0.404          -0.174          -0.056
```

It turns out that pitchers, afraid of HRs, will sometimes avoid throwing strikes to HR hitters. As a result, HR hitters tend to have more BBs and a team with many HRs will also have more BBs. Although it may appear that BBs cause runs, it is actually the HRs that cause most of these runs. We say that BBs are *confounded* with HRs. Nonetheless, could it be that BBs still help? To find out, we somehow have to adjust for the HR effect. Regression can help with this as well.

18.2.1 Understanding confounding through stratification

A first approach is to keep HRs fixed at a certain value and then examine the relationship between BB and runs. As we did when we stratified fathers by rounding to the closest inch, here we can stratify HR per game to the closest ten. We filter out the strata with few points to avoid highly variable estimates:

```
dat <- Teams %>% filter(yearID %in% 1961:2001) %>%
  mutate(HR_strata = round(HR/G, 1),
         BB_per_game = BB / G,
         R_per_game = R / G) %>%
  filter(HR_strata >= 0.4 & HR_strata <=1.2)
```

and then make a scatterplot for each strata:

```
dat %>%
  ggplot(aes(BB_per_game, R_per_game)) +
  geom_point(alpha = 0.5) +
  geom_smooth(method = "lm") +
  facet_wrap( ~ HR_strata)
```

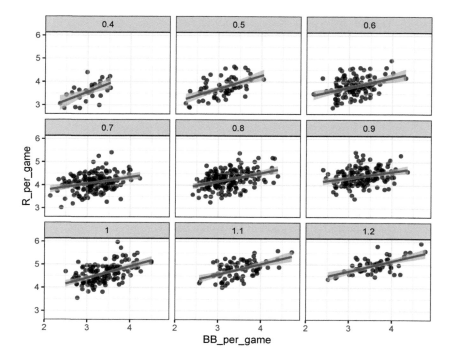

Remember that the regression slope for predicting runs with BB was 2.1. Once we stratify by HR, these slopes are substantially reduced:

```
dat %>%
  group_by(HR_strata) %>%
  summarize(slope = get_slope(BB_per_game, R_per_game))
#> # A tibble: 9 x 2
#>   HR_strata slope
#>       <dbl> <dbl>
#> 1       0.4  4.19
#> 2       0.5  3.30
#> 3       0.6  2.55
#> 4       0.7  2.03
#> 5       0.8  2.47
#> # ... with 4 more rows
```

The slopes are reduced, but they are not 0, which indicates that BBs are helpful for producing runs, just not as much as previously thought. In fact, the values above are closer to the slope we obtained from singles, 1.3, which is more consistent with our intuition. Since both singles and BB get us to first base, they should have about the same predictive power.

Although our understanding of the application tells us that HR cause BB but not the other way around, we can still check if stratifying by BB makes the effect of BB go down. To do this, we use the same code except that we swap HR and BBs to get this plot:

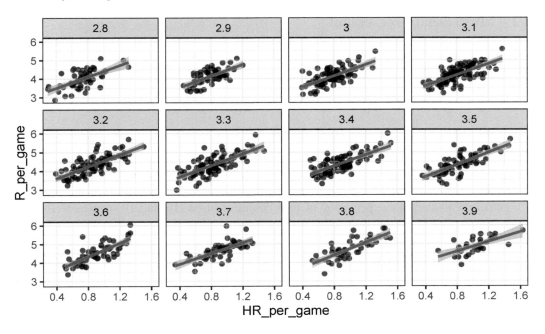

In this case, the slopes do not change much from the original:

```
dat %>% group_by(BB_strata) %>%
    summarize(slope = get_slope(HR_per_game, R_per_game))
#> # A tibble: 12 x 2
#>   BB_strata slope
#>       <dbl> <dbl>
#> 1       2.8  6.47
#> 2       2.9  8.87
#> 3       3    7.40
#> 4       3.1  7.34
#> 5       3.2  5.89
#> # ... with 7 more rows
```

They are reduced a bit, which is consistent with the fact that BB do in fact cause some runs.

```
hr_slope <- Teams %>%
  filter(yearID %in% 1961:2001 ) %>%
  mutate(HR_per_game = HR/G, R_per_game = R/G) %>%
  summarize(slope = get_slope(R_per_game, HR_per_game))

hr_slope
#>    slope
#> 1  5.33
```

Regardless, it seems that if we stratify by HR, we have bivariate distributions for runs versus BB. Similarly, if we stratify by BB, we have approximate bivariate normal distributions for HR versus runs.

18.2.2   Multivariate regression

It is somewhat complex to be computing regression lines for each strata. We are essentially fitting models like this:

$$E[R \mid BB = x_1, \, HR = x_2] = \beta_0 + \beta_1(x_2)x_1 + \beta_2(x_1)x_2$$

with the slopes for x_1 changing for different values of x_2 and vice versa. But is there an easier approach?

If we take random variability into account, the slopes in the strata don't appear to change much. If these slopes are in fact the same, this implies that $\beta_1(x_2)$ and $\beta_2(x_1)$ are constants. This in turn implies that the expectation of runs conditioned on HR and BB can be written like this:

$$E[R \mid BB = x_1, \, HR = x_2] = \beta_0 + \beta_1 x_1 + \beta_2 x_2$$

This model suggests that if the number of HR is fixed at x_2, we observe a linear relationship between runs and BB with an intercept of $\beta_0 + \beta_2 x_2$. Our exploratory data analysis suggested this. The model also suggests that as the number of HR grows, the intercept growth is linear as well and determined by $\beta_1 x_1$.

In this analysis, referred to as *multivariate regression*, you will often hear people say that the BB slope β_1 is *adjusted* for the HR effect. If the model is correct then confounding has been accounted for. But how do we estimate β_1 and β_2 from the data? For this, we learn about linear models and least squares estimates.

18.3   Least squares estimates

We have described how if data is bivariate normal then the conditional expectations follow the regression line. The fact that the conditional expectation is a line is not an extra assumption but rather a derived result. However, in practice it is common to explicitly write down a model that describes the relationship between two or more variables using a *linear model*.

We note that "linear" here does not refer to lines exclusively, but rather to the fact that the conditional expectation is a linear combination of known quantities. In mathematics, when we multiply each variable by a constant and then add them together, we say we formed a linear combination of the variables. For example, $3x - 4y + 5z$ is a linear combination of x, y, and z. We can also add a constant so $2 + 3x - 4y + 5z$ is also linear combination of x, y, and z.

So $\beta_0 + \beta_1 x_1 + \beta_2 x_2$, is a linear combination of x_1 and x_2. The simplest linear model is a constant β_0; the second simplest is a line $\beta_0 + \beta_1 x$. If we were to specify a linear model for Galton's data, we would denote the N observed father heights with $x_1, \ldots, x_n$, then we model the N son heights we are trying to predict with:

$$Y_i = \beta_0 + \beta_1 x_i + \varepsilon_i, \, i = 1, \ldots, N.$$

Here x_i is the father's height, which is fixed (not random) due to the conditioning, and Y_i is the random son's height that we want to predict. We further assume that ε_i are independent from each other, have expected value 0 and the standard deviation, call it σ, does not depend on i.

In the above model, we know the x_i, but to have a useful model for prediction, we need β_0 and β_1. We estimate these from the data. Once we do this, we can predict son's heights for any father's height x. We show how to do this in the next section.

Note that if we further assume that the ε is normally distributed, then this model is exactly the same one we derived earlier by assuming bivariate normal data. A somewhat nuanced difference is that in the first approach we assumed the data was bivariate normal and that the linear model was derived, not assumed. In practice, linear models are just assumed without necessarily assuming normality: the distribution of the εs is not specified. Nevertheless, if your data is bivariate normal, the above linear model holds. If your data is not bivariate normal, then you will need to have other ways of justifying the model.

18.3.1 Interpreting linear models

One reason linear models are popular is that they are interpretable. In the case of Galton's data, we can interpret the data like this: due to inherited genes, the son's height prediction grows by β_1 for each inch we increase the father's height x. Because not all sons with fathers of height x are of equal height, we need the term ε, which explains the remaining variability. This remaining variability includes the mother's genetic effect, environmental factors, and other biological randomness.

Given how we wrote the model above, the intercept β_0 is not very interpretable as it is the predicted height of a son with a father with no height. Due to regression to the mean, the prediction will usually be a bit larger than 0. To make the slope parameter more interpretable, we can rewrite the model slightly as:

$$Y_i = \beta_0 + \beta_1(x_i - \bar{x}) + \varepsilon_i, \; i = 1, \ldots, N$$

with $\bar{x} = 1/N \sum_{i=1}^{N} x_i$ the average of the x. In this case β_0 represents the height when $x_i = \bar{x}$, which is the height of the son of an average father.

18.3.2 Least Squares Estimates (LSE)

For linear models to be useful, we have to estimate the unknown βs. The standard approach in science is to find the values that minimize the distance of the fitted model to the data. The following is called the least squares (LS) equation and we will see it often in this chapter. For Galton's data, we would write:

$$RSS = \sum_{i=1}^{n} \{y_i - (\beta_0 + \beta_1 x_i)\}^2$$

This quantity is called the residual sum of squares (RSS). Once we find the values that minimize the RSS, we will call the values the least squares estimates (LSE) and denote them with $\hat{\beta}_0$ and $\hat{\beta}_1$. Let's demonstrate this with the previously defined dataset:

```
library(HistData)
data("GaltonFamilies")
set.seed(1983)
galton_heights <- GaltonFamilies %>%
  filter(gender == "male") %>%
  group_by(family) %>%
  sample_n(1) %>%
  ungroup() %>%
  select(father, childHeight) %>%
  rename(son = childHeight)
```

Let's write a function that computes the RSS for any pair of values β_0 and β_1.

```
rss <- function(beta0, beta1, data){
  resid <- galton_heights$son - (beta0+beta1*galton_heights$father)
  return(sum(resid^2))
}
```

So for any pair of values, we get an RSS. Here is a plot of the RSS as a function of β_1 when we keep the β_0 fixed at 25.

```
beta1 = seq(0, 1, len=nrow(galton_heights))
results <- data.frame(beta1 = beta1,
                      rss = sapply(beta1, rss, beta0 = 25))
results %>% ggplot(aes(beta1, rss)) + geom_line() +
  geom_line(aes(beta1, rss))
```

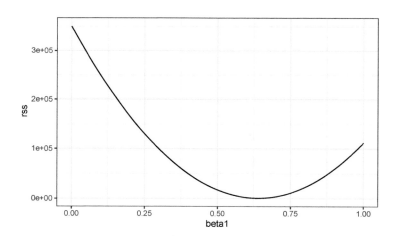

We can see a clear minimum for β_1 at around 0.65. However, this minimum for β_1 is for when $\beta_0 = 25$, a value we arbitrarily picked. We don't know if (25, 0.65) is the pair that minimizes the equation across all possible pairs.

Trial and error is not going to work in this case. We could search for a minimum within a fine grid of β_0 and β_1 values, but this is unnecessarily time-consuming since we can use calculus: take the partial derivatives, set them to 0 and solve for β_1 and β_2. Of course, if we have many parameters, these equations can get rather complex. But there are functions in R that do these calculations for us. We will learn these next. To learn the mathematics behind this, you can consult a book on linear models.

18.3.3 The `lm` function

In R, we can obtain the least squares estimates using the `lm` function. To fit the model:

$$Y_i = \beta_0 + \beta_1 x_i + \varepsilon_i$$

with Y_i the son's height and x_i the father's height, we can use this code to obtain the least squares estimates.

```
fit <- lm(son ~ father, data = galton_heights)
fit$coef
#> (Intercept)        father
#>      37.288         0.461
```

The most common way we use `lm` is by using the character ~ to let `lm` know which is the variable we are predicting (left of ~) and which we are using to predict (right of ~). The intercept is added automatically to the model that will be fit.

The object `fit` includes more information about the fit. We can use the function `summary` to extract more of this information (not shown):

```
summary(fit)
#>
#> Call:
#> lm(formula = son ~ father, data = galton_heights)
#>
#> Residuals:
#>     Min      1Q Median      3Q     Max
#> -9.354  -1.566 -0.008   1.726   9.415
#>
#> Coefficients:
#>              Estimate Std. Error t value Pr(>|t|)
#> (Intercept)  37.2876     4.9862    7.48  3.4e-12 ***
#> father        0.4614     0.0721    6.40  1.4e-09 ***
#> ---
#> Signif. codes:  0 '***' 0.001 '**' 0.01 '*' 0.05 '.' 0.1 ' ' 1
#>  ·
#> Residual standard error: 2.45 on 177 degrees of freedom
#> Multiple R-squared:  0.188,   Adjusted R-squared:  0.183
#> F-statistic: 40.9 on 1 and 177 DF,  p-value: 1.36e-09
```

To understand some of the information included in this summary we need to remember that the LSE are random variables. Mathematical statistics gives us some ideas of the distribution of these random variables

18.3.4   LSE are random variables

The LSE is derived from the data $y_1, \ldots, y_N$, which are a realization of random variables $Y_1, \ldots, Y_N$. This implies that our estimates are random variables. To see this, we can run a Monte Carlo simulation in which we assume the son and father height data defines a population, take a random sample of size $N = 50$, and compute the regression slope coefficient for each one:

```
B <- 1000
N <- 50
lse <- replicate(B, {
  sample_n(galton_heights, N, replace = TRUE) %>%
    lm(son ~ father, data = .) %>%
    .$coef
})
lse <- data.frame(beta_0 = lse[1,], beta_1 = lse[2,])
```

We can see the variability of the estimates by plotting their distributions:

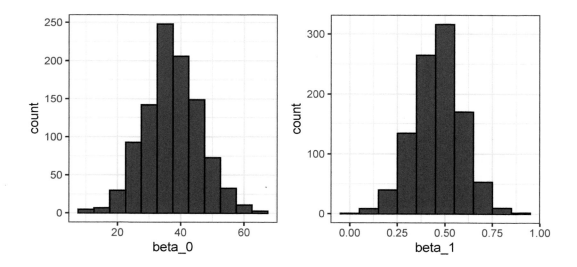

The reason these look normal is because the central limit theorem applies here as well: for large enough N, the least squares estimates will be approximately normal with expected value β_0 and β_1, respectively. The standard errors are a bit complicated to compute, but mathematical theory does allow us to compute them and they are included in the summary provided by the lm function. Here it is for one of our simulated data sets:

```
sample_n(galton_heights, N, replace = TRUE) %>%
  lm(son ~ father, data = .) %>%
  summary %>% .$coef
#>             Estimate Std. Error t value Pr(>|t|)
#> (Intercept)    19.28     11.656    1.65 1.05e-01
#> father          0.72      0.169    4.25 9.79e-05
```

You can see that the standard errors estimates reported by the `summary` are close to the standard errors from the simulation:

```
lse %>% summarize(se_0 = sd(beta_0), se_1 = sd(beta_1))
#>     se_0  se_1
#> 1  8.84  0.128
```

The `summary` function also reports t-statistics (`t value`) and p-values (`Pr(>|t|)`). The t-statistic is not actually based on the central limit theorem but rather on the assumption that the εs follow a normal distribution. Under this assumption, mathematical theory tells us that the LSE divided by their standard error, $\hat{\beta}_0/\hat{\text{SE}}(\hat{\beta}_0)$ and $\hat{\beta}_1/\hat{\text{SE}}(\hat{\beta}_1)$, follow a t-distribution with $N - p$ degrees of freedom, with p the number of parameters in our model. In the case of height $p = 2$, the two p-values are testing the null hypothesis that $\beta_0 = 0$ and $\beta_1 = 0$, respectively.

Remember that, as we described in Section 16.10 for large enough N, the CLT works and the t-distribution becomes almost the same as the normal distribution. Also, notice that we can construct confidence intervals, but we will soon learn about **broom**, an add-on package that makes this easy.

Although we do not show examples in this book, hypothesis testing with regression models is commonly used in epidemiology and economics to make statements such as "the effect of A on B was statistically significant after adjusting for X, Y, and Z". However, several assumptions have to hold for these statements to be true.

18.3.5 Predicted values are random variables

Once we fit our model, we can obtain prediction of Y by plugging in the estimates into the regression model. For example, if the father's height is x, then our prediction $\hat{Y}$ for the son's height will be:

$$\hat{Y} = \hat{\beta}_0 + \hat{\beta}_1 x$$

When we plot $\hat{Y}$ versus x, we see the regression line.

Keep in mind that the prediction $\hat{Y}$ is also a random variable and mathematical theory tells us what the standard errors are. If we assume the errors are normal, or have a large enough sample size, we can use theory to construct confidence intervals as well. In fact, the **ggplot2** layer `geom_smooth(method = "lm")` that we previously used plots $\hat{Y}$ and surrounds it by confidence intervals:

```
galton_heights %>% ggplot(aes(son, father)) +
  geom_point() +
  geom_smooth(method = "lm")
```

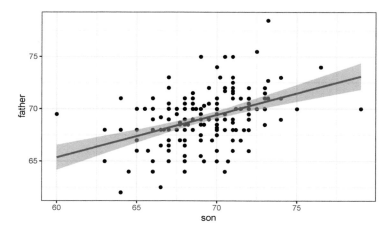

The R function `predict` takes an `lm` object as input and returns the prediction. If requested, the standard errors and other information from which we can construct confidence intervals is provided:

```
fit <- galton_heights %>% lm(son ~ father, data = .)

y_hat <- predict(fit, se.fit = TRUE)

names(y_hat)
#> [1] "fit"            "se.fit"         "df"            "residual.scale"
```

18.4   Exercises

We have shown how BB and singles have similar predictive power for scoring runs. Another way to compare the usefulness of these baseball metrics is by assessing how stable they are across the years. Since we have to pick players based on their previous performances, we will prefer metrics that are more stable. In these exercises, we will compare the stability of singles and BBs.

1. Before we get started, we want to generate two tables. One for 2002 and another for the average of 1999-2001 seasons. We want to define per plate appearance statistics. Here is how we create the 2017 table. Keeping only players with more than 100 plate appearances.

```
library(Lahman)
dat <- Batting %>% filter(yearID == 2002) %>%
  mutate(pa = AB + BB,
         singles = (H - X2B - X3B - HR) / pa, bb = BB / pa) %>%
  filter(pa >= 100) %>%
  select(playerID, singles, bb)
```

Now compute a similar table but with rates computed over 1999-2001.

2. In Section 22.1 we learn about the `inner_join`, which you can use to have the 2001 data and averages in the same table:

```
dat <- inner_join(dat, avg, by = "playerID")
```

Compute the correlation between 2002 and the previous seasons for singles and BB.

3. Note that the correlation is higher for BB. To quickly get an idea of the uncertainty associated with this correlation estimate, we will fit a linear model and compute confidence intervals for the slope coefficient. However, first make scatterplots to confirm that fitting a linear model is appropriate.

4. Now fit a linear model for each metric and use the `confint` function to compare the estimates.

18.5 Linear regression in the tidyverse

To see how we use the `lm` function in a more complex analysis, let's go back to the baseball example. In a previous example, we estimated regression lines to predict runs for BB in different HR strata. We first constructed a data frame similar to this:

```
dat <- Teams %>% filter(yearID %in% 1961:2001) %>%
  mutate(HR = round(HR/G, 1),
         BB = BB/G,
         R = R/G) %>%
  select(HR, BB, R) %>%
  filter(HR >= 0.4 & HR<=1.2)
```

Since we didn't know the `lm` function, to compute the regression line in each strata, we used the formula directly like this:

```
get_slope <- function(x, y) cor(x, y) / (sd(x) * sd(y))
dat %>%
  group_by(HR) %>%
  summarize(slope = get_slope(BB, R))
```

We argued that the slopes are similar and that the differences were perhaps due to random variation. To provide a more rigorous defense of the slopes being the same, which led to our multivariate model, we could compute confidence intervals for each slope. We have not learned the formula for this, but the `lm` function provides enough information to construct them.

First, note that if we try to use the `lm` function to get the estimated slope like this:

```
dat %>%
  group_by(HR) %>%
  lm(R ~ BB, data = .) %>% .$coef
#> (Intercept)          BB
#>       2.198       0.638
```

we don't get the result we want. The lm function ignores the group_by. This is expected because lm is not part of the **tidyverse** and does not know how to handle the outcome of a grouped tibble.

The **tidyverse** functions know how to interpret grouped tibbles. Furthermore, to facilitate stringing commands through the pipe %>%, **tidyverse** functions consistently return data frames, since this assures that the output of a function is accepted as the input of another. But most R functions do not recognize grouped tibbles nor do they return data frames. The lm function is an example. The do functions serves as a bridge between R functions, such as lm, and the **tidyverse**. The do function understands grouped tibbles and always returns a data frame.

So, let's try to use the do function to fit a regression line to each HR strata:

```
dat %>%
  group_by(HR) %>%
  do(fit = lm(R ~ BB, data = .))
#> Source: local data frame [9 x 2]
#> Groups: <by row>
#>
#> # A tibble: 9 x 2
#>       HR fit
#> * <dbl> <list>
#> 1   0.4 <lm>
#> 2   0.5 <lm>
#> 3   0.6 <lm>
#> 4   0.7 <lm>
#> 5   0.8 <lm>
#> # ... with 4 more rows
```

Notice that we did in fact fit a regression line to each strata. The do function will create a data frame with the first column being the strata value and a column named fit (we chose the name, but it can be anything). The column will contain the result of the lm call. Therefore, the returned tibble has a column with lm objects, which is not very useful.

Also, if we do not name a column (note above we named it fit), then do will return the actual output of lm, not a data frame, and this will result in an error since do is expecting a data frame as output.

```
dat %>%
  group_by(HR) %>%
  do(lm(R ~ BB, data = .))
```

Error: Results 1, 2, 3, 4, 5, ... must be data frames, not lm

For a useful data frame to be constructed, the output of the function must be a data frame too. We could build a function that returns only what we want in the form of a data frame:

```
get_slope <- function(data){
  fit <- lm(R ~ BB, data = data)
  data.frame(slope = fit$coefficients[2],
             se = summary(fit)$coefficient[2,2])
}
```

And then use **do without** naming the output, since we are already getting a data frame:

```
dat %>%
  group_by(HR) %>%
  do(get_slope(.))
#> # A tibble: 9 x 3
#> # Groups:    HR [9]
#>       HR slope      se
#>    <dbl> <dbl>   <dbl>
#> 1    0.4 0.734  0.208
#> 2    0.5 0.566  0.110
#> 3    0.6 0.412  0.0974
#> 4    0.7 0.285  0.0705
#> 5    0.8 0.365  0.0653
#> # ... with 4 more rows
```

If we name the output, then we get something we do not want, a column containing data frames:

```
dat %>%
  group_by(HR) %>%
  do(slope = get_slope(.))
#> Source: local data frame [9 x 2]
#> Groups: <by row>
#>
#> # A tibble: 9 x 2
#>       HR slope
#> * <dbl> <list>
#> 1    0.4 <df[,2] [1 x 2]>
#> 2    0.5 <df[,2] [1 x 2]>
#> 3    0.6 <df[,2] [1 x 2]>
#> 4    0.7 <df[,2] [1 x 2]>
#> 5    0.8 <df[,2] [1 x 2]>
#> # ... with 4 more rows
```

This is not very useful, so let's cover one last feature of **do**. If the data frame being returned has more than one row, these will be concatenated appropriately. Here is an example in which we return both estimated parameters:

```
get_lse <- function(data){
  fit <- lm(R ~ BB, data = data)
  data.frame(term = names(fit$coefficients),
    slope = fit$coefficients,
    se = summary(fit)$coefficient[,2])
}

dat %>%
  group_by(HR) %>%
  do(get_lse(.))
#> # A tibble: 18 x 4
#> # Groups:    HR [9]
```

```
#>       HR term         slope    se
#>    <dbl> <fct>         <dbl> <dbl>
#> 1   0.4 (Intercept)  1.36  0.631
#> 2   0.4 BB           0.734 0.208
#> 3   0.5 (Intercept)  2.01  0.344
#> 4   0.5 BB           0.566 0.110
#> 5   0.6 (Intercept)  2.53  0.305
#> # ... with 13 more rows
```

If you think this is all a bit too complicated, you are not alone. To simplify things, we introduce the **broom** package which was designed to facilitate the use of model fitting functions, such as `lm`, with the **tidyverse**.

18.5.1   The broom package

Our original task was to provide an estimate and confidence interval for the slope estimates of each strata. The **broom** package will make this quite easy.

The **broom** package has three main functions, all of which extract information from the object returned by `lm` and return it in a **tidyverse** friendly data frame. These functions are `tidy`, `glance`, and `augment`. The `tidy` function returns estimates and related information as a data frame:

```
library(broom)
fit <- lm(R ~ BB, data = dat)
tidy(fit)
#> # A tibble: 2 x 5
#>   term         estimate std.error statistic  p.value
#>   <chr>           <dbl>     <dbl>     <dbl>    <dbl>
#> 1 (Intercept)      2.20     0.113     19.4 1.12e-70
#> 2 BB               0.638    0.0344    18.5 1.35e-65
```

We can add other important summaries, such as confidence intervals:

```
tidy(fit, conf.int = TRUE)
#> # A tibble: 2 x 7
#>   term        estimate std.error statistic  p.value conf.low conf.high
#>   <chr>          <dbl>     <dbl>     <dbl>    <dbl>    <dbl>     <dbl>
#> 1 (Intercept)     2.20     0.113     19.4 1.12e-70     1.98      2.42
#> 2 BB              0.638    0.0344    18.5 1.35e-65     0.570     0.705
```

Because the outcome is a data frame, we can immediately use it with `do` to string together the commands that produce the table we are after. Because a data frame is returned, we can filter and select the rows and columns we want, which facilitates working with **ggplot2**:

```
dat %>%
  group_by(HR) %>%
  do(tidy(lm(R ~ BB, data = .), conf.int = TRUE)) %>%
  filter(term == "BB") %>%
  select(HR, estimate, conf.low, conf.high) %>%
```

```
ggplot(aes(HR, y = estimate, ymin = conf.low, ymax = conf.high)) +
geom_errorbar() +
geom_point()
```

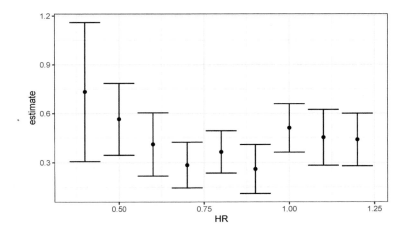

Now we return to discussing our original task of determining if slopes changed. The plot we just made, using do and tidy, shows that the confidence intervals overlap, which provides a nice visual confirmation that our assumption that the slope does not change is safe.

The other functions provided by **broom**, glance, and augment, relate to model-specific and observation-specific outcomes, respectively. Here, we can see the model fit summaries glance returns:

```
glance(fit)
#> # A tibble: 1 x 11
#>   r.squared adj.r.squared sigma statistic  p.value    df logLik   AIC
#>       <dbl>         <dbl> <dbl>     <dbl>    <dbl> <int>  <dbl> <dbl>
#> 1     0.266         0.265 0.454      343. 1.35e-65     2  -596. 1199.
#> # ... with 3 more variables: BIC <dbl>, deviance <dbl>,
#> #   df.residual <int>
```

You can learn more about these summaries in any regression text book.

We will see an example of augment in the next section.

18.6   Exercises

1. In a previous section, we computed the correlation between mothers and daughters, mothers and sons, fathers and daughters, and fathers and sons, and noticed that the highest correlation is between fathers and sons and the lowest is between mothers and sons. We can compute these correlations using:

```
data("GaltonFamilies")
set.seed(1)
galton_heights <- GaltonFamilies %>%
  group_by(family, gender) %>%
  sample_n(1) %>%
  ungroup()

cors <- galton_heights %>%
  gather(parent, parentHeight, father:mother) %>%
  mutate(child = ifelse(gender == "female", "daughter", "son")) %>%
  unite(pair, c("parent", "child")) %>%
  group_by(pair) %>%
  summarize(cor = cor(parentHeight, childHeight))
```

Are these differences statistically significant? To answer this, we will compute the slopes of the regression line along with their standard errors. Start by using `lm` and the **broom** package to compute the slopes LSE and the standard errors.

2. Repeat the exercise above, but compute a confidence interval as well.

3. Plot the confidence intervals and notice that they overlap, which implies that the data is consistent with the inheritance of height being independent of sex.

4. Because we are selecting children at random, we can actually do something like a permutation test here. Repeat the computation of correlations 100 times taking a different sample each time. Hint: use similar code to what we used with simulations.

5. Fit a linear regression model to obtain the effects of BB and HR on Runs (at the team level) in 1971. Use the `tidy` function in the **broom** package to obtain the results in a data frame.

6. Now let's repeat the above for each year since 1961 and make a plot. Use `do` and the **broom** package to fit this model for every year since 1961.

7. Use the results of the previous exercise to plot the estimated effects of BB on runs.

8. **Advanced**. Write a function that takes R, HR, and BB as arguments and fits two linear models: R ~ BB and R~BB+HR. Then use the `do` function to obtain the BB for both models for each year since 1961. Then plot these against each other as a function of time.

18.7   Case study: Moneyball (continued)

In trying to answer how well BBs predict runs, data exploration led us to a model:

$$E[R \mid BB = x_1, HR = x_2] = \beta_0 + \beta_1 x_1 + \beta_2 x_2$$

Here, the data is approximately normal and conditional distributions were also normal. Thus, we are justified in using a linear model:

$$Y_i = \beta_0 + \beta_1 x_{i,1} + \beta_2 x_{i,2} + \varepsilon_i$$

with Y_i runs per game for team i, $x_{i,1}$ walks per game, and $x_{i,2}$. To use `lm` here, we need to let the function know we have two predictor variables. So we use the `+` symbol as follows:

```
fit <- Teams %>%
  filter(yearID %in% 1961:2001) %>%
  mutate(BB = BB/G, HR = HR/G,  R = R/G) %>%
  lm(R ~ BB + HR, data = .)
```

We can use `tidy` to see a nice summary:

```
tidy(fit, conf.int = TRUE)
#> # A tibble: 3 x 7
#>   term         estimate std.error statistic   p.value conf.low conf.high
#>   <chr>           <dbl>     <dbl>     <dbl>     <dbl>    <dbl>     <dbl>
#> 1 (Intercept)      1.74    0.0824      21.2  7.62e- 83     1.58      1.91
#> 2 BB               0.387   0.0270      14.3  1.20e- 42     0.334     0.440
#> 3 HR               1.56    0.0490      31.9  1.78e-155     1.47      1.66
```

When we fit the model with only one variable, the estimated slopes were 2.12261805015402 and 5.32531443555834 for BB and HR, respectively. Note that when fitting the multivariate model both go down, with the BB effect decreasing much more.

Now we want to construct a metric to pick players, we need to consider singles, doubles, and triples as well. Can we build a model that predicts runs based on all these outcomes?

We now are going to take somewhat of a "leap of faith" and assume that these five variables are jointly normal. This means that if we pick any one of them, and hold the other four fixed, the relationship with the outcome is linear and the slope does not depend on the four values held constant. If this is true, then a linear model for our data is:

$$Y_i = \beta_0 + \beta_1 x_{i,1} + \beta_2 x_{i,2} + \beta_3 x_{i,3} + \beta_4 x_{i,4} + \beta_5 x_{i,5} + \varepsilon_i$$

with $x_{i,1}, x_{i,2}, x_{i,3}, x_{i,4}, x_{i,5}$ representing BB, singles, doubles, triples, and HR respectively.

Using `lm`, we can quickly find the LSE for the parameters using:

```
fit <- Teams %>%
  filter(yearID %in% 1961:2001) %>%
  mutate(BB = BB / G,
         singles = (H - X2B - X3B - HR) / G,
         doubles = X2B / G,
         triples = X3B / G,
         HR = HR / G,
         R = R / G) %>%
  lm(R ~ BB + singles + doubles + triples + HR, data = .)
```

We can see the coefficients using `tidy`:

```
coefs <- tidy(fit, conf.int = TRUE)

coefs
```

```
#> # A tibble: 6 x 7
#>   term         estimate std.error statistic   p.value conf.low conf.high
#>   <chr>           <dbl>     <dbl>     <dbl>     <dbl>    <dbl>     <dbl>
#> 1 (Intercept)     -2.77    0.0862     -32.1 4.76e-157    -2.94     -2.60
#> 2 BB               0.371   0.0117      31.6 1.87e-153     0.348     0.394
#> 3 singles          0.519   0.0127      40.8 8.67e-217     0.494     0.544
#> 4 doubles          0.771   0.0226      34.1 8.44e-171     0.727     0.816
#> 5 triples          1.24    0.0768      16.1 2.12e- 52     1.09      1.39
#> # ... with 1 more row
```

To see how well our metric actually predicts runs, we can predict the number of runs for each team in 2002 using the function `predict`, then make a plot:

```
Teams %>%
  filter(yearID %in% 2002) %>%
  mutate(BB = BB/G,
         singles = (H-X2B-X3B-HR)/G,
         doubles = X2B/G,
         triples =X3B/G,
         HR=HR/G,
         R=R/G)  %>%
  mutate(R_hat = predict(fit, newdata = .)) %>%
  ggplot(aes(R_hat, R, label = teamID)) +
  geom_point() +
  geom_text(nudge_x=0.1, cex = 2) +
  geom_abline()
```

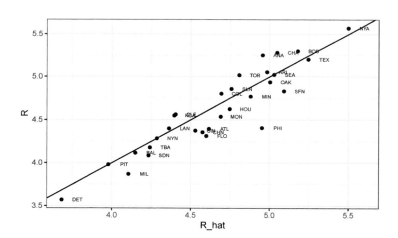

Our model does quite a good job as demonstrated by the fact that points from the observed versus predicted plot fall close to the identity line.

So instead of using batting average, or just number of HR, as a measure of picking players, we can use our fitted model to form a metric that relates more directly to run production. Specifically, to define a metric for player A, we imagine a team made up of players just like player A and use our fitted regression model to predict how many runs this team would produce. The formula would look like this: -2.769 + 0.371 × BB + 0.519 × singles + 0.771 × doubles + 1.24 × triples + 1.443 × HR.

To define a player-specific metric, we have a bit more work to do. A challenge here is that we derived the metric for teams, based on team-level summary statistics. For example, the HR value that is entered into the equation is HR per game for the entire team. If we compute the HR per game for a player, it will be much lower since the total is accumulated by 9 batters. Furthermore, if a player only plays part of the game and gets fewer opportunities than average, it is still considered a game played. For players, a rate that takes into account opportunities is the per-plate-appearance rate.

To make the per-game team rate comparable to the per-plate-appearance player rate, we compute the average number of team plate appearances per game:

```
pa_per_game <- Batting %>% filter(yearID == 2002) %>%
  group_by(teamID) %>%
  summarize(pa_per_game = sum(AB+BB)/max(G)) %>%
  pull(pa_per_game) %>%
  mean
```

We compute the per-plate-appearance rates for players available in 2002 on data from 1997-2001. To avoid small sample artifacts, we filter players with less than 200 plate appearances per year. Here is the entire calculation in one line:

```
players <- Batting %>% filter(yearID %in% 1997:2001) %>%
  group_by(playerID) %>%
  mutate(PA = BB + AB) %>%
  summarize(G = sum(PA)/pa_per_game,
    BB = sum(BB)/G,
    singles = sum(H-X2B-X3B-HR)/G,
    doubles = sum(X2B)/G,
    triples = sum(X3B)/G,
    HR = sum(HR)/G,
    AVG = sum(H)/sum(AB),
    PA = sum(PA)) %>%
  filter(PA >= 1000) %>%
  select(-G) %>%
  mutate(R_hat = predict(fit, newdata = .))
```

The player-specific predicted runs computed here can be interpreted as the number of runs we predict a team will score if all batters are exactly like that player. The distribution shows that there is wide variability across players:

```
qplot(R_hat, data = players, binwidth = 0.5, color = I("black"))
```

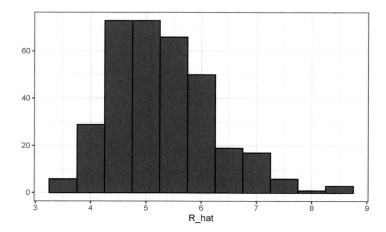

18.7.1   Adding salary and position information

To actually build the team, we will need to know their salaries as well as their defensive position. For this, we join the `players` data frame we just created with the player information data frame included in some of the other Lahman data tables. We will learn more about the join function we learned in Section 22.1.

Start by adding the 2002 salary of each player:

```
players <- Salaries %>%
  filter(yearID == 2002) %>%
  select(playerID, salary) %>%
  right_join(players, by="playerID")
```

Next, we add their defensive position. This is a somewhat complicated task because players play more than one position each year. The **Lahman** package table `Appearances` tells how many games each player played in each position, so we can pick the position that was most played using `which.max` on each row. We use `apply` to do this. However, because some players are traded, they appear more than once on the table, so we first sum their appearances across teams. Here, we pick the one position the player most played using the `top_n` function. To make sure we only pick one position, in the case of ties, we pick the first row of the resulting data frame. We also remove the `OF` position which stands for outfielder, a generalization of three positions: left field (LF), center field (CF), and right field (RF). We also remove pitchers since they don't bat in the league in which the A's play.

```
position_names <-
  paste0("G_", c("p","c","1b","2b","3b","ss","lf","cf","rf", "dh"))

tmp <- Appearances %>%
  filter(yearID == 2002) %>%
  group_by(playerID) %>%
  summarize_at(position_names, sum) %>%
  ungroup()

pos <- tmp %>%
```

```
  select(position_names) %>%
  apply(., 1, which.max)

players <- tibble(playerID = tmp$playerID, POS = position_names[pos]) %>%
  mutate(POS = str_to_upper(str_remove(POS, "G_"))) %>%
  filter(POS != "P") %>%
  right_join(players, by="playerID") %>%
  filter(!is.na(POS)  & !is.na(salary))
```

Finally, we add their first and last name:

```
players <- Master %>%
  select(playerID, nameFirst, nameLast, debut) %>%
  mutate(debut = as.Date(debut)) %>%
  right_join(players, by="playerID")
```

If you are a baseball fan, you will recognize the top 10 players:

```
players %>% select(nameFirst, nameLast, POS, salary, R_hat) %>%
  arrange(desc(R_hat)) %>% top_n(10)
#> Selecting by R_hat
#>    nameFirst nameLast POS   salary R_hat
#> 1      Barry    Bonds  LF 15000000  8.44
#> 2      Larry   Walker  RF 12666667  8.34
#> 3       Todd   Helton  1B  5000000  7.76
#> 4      Manny  Ramirez  LF 15462727  7.71
#> 5     Sammy     Sosa  RF 15000000  7.56
#> 6       Jeff  Bagwell  1B 11000000  7.41
#> 7       Mike   Piazza   C 10571429  7.34
#> 8      Jason   Giambi  1B 10428571  7.26
#> 9      Edgar Martinez  DH  7086668  7.26
#> 10       Jim    Thome  1B  8000000  7.23
```

18.7.2   Picking nine players

On average, players with a higher metric have higher salaries:

```
players %>% ggplot(aes(salary, R_hat, color = POS)) +
  geom_point() +
  scale_x_log10()
```

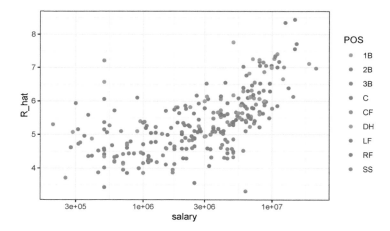

We can search for good deals by looking at players who produce many more runs than others with similar salaries. We can use this table to decide what players to pick and keep our total salary below the 40 million dollars Billy Beane had to work with. This can be done using what computer scientists call linear programming. This is not something we teach, but here are the position players selected with this approach:

nameFirst	nameLast	POS	salary	R_hat
Todd	Helton	1B	5000000	7.76
Mike	Piazza	C	10571429	7.34
Edgar	Martinez	DH	7086668	7.26
Jim	Edmonds	CF	7333333	6.55
Jeff	Kent	2B	6000000	6.39
Phil	Nevin	3B	2600000	6.16
Matt	Stairs	RF	500000	6.06
Henry	Rodriguez	LF	300000	5.94
John	Valentin	SS	550000	5.27

We see that all these players have above average BB and most have above average HR rates, while the same is not true for singles. Here is a table with statistics standardized across players so that, for example, above average HR hitters have values above 0.

nameLast	BB	singles	doubles	triples	HR	AVG	R_hat
Helton	0.909	-0.215	2.649	-0.311	1.522	2.670	2.532
Piazza	0.328	0.423	0.204	-1.418	1.825	2.199	2.089
Martinez	2.135	-0.005	1.265	-1.224	0.808	2.203	2.000
Edmonds	1.071	-0.558	0.791	-1.152	0.973	0.854	1.256
Kent	0.232	-0.732	2.011	0.448	0.766	0.787	1.087
Nevin	0.307	-0.905	0.479	-1.191	1.193	0.105	0.848
Stairs	1.100	-1.513	-0.046	-1.129	1.121	-0.561	0.741
Rodriguez	0.201	-1.596	0.332	-0.782	1.320	-0.672	0.610
Valentin	0.180	-0.929	1.794	-0.435	-0.045	-0.472	-0.089

18.8 The regression fallacy

Wikipedia defines the *sophomore slump* as:

> A sophomore slump or sophomore jinx or sophomore jitters refers to an instance in which a second, or sophomore, effort fails to live up to the standards of the first effort. It is commonly used to refer to the apathy of students (second year of high school, college or university), the performance of athletes (second season of play), singers/bands (second album), television shows (second seasons) and movies (sequels/prequels).

In Major League Baseball, the rookie of the year (ROY) award is given to the first-year player who is judged to have performed the best. The *sophomore slump* phrase is used to describe the observation that ROY award winners don't do as well during their second year. For example, this Fox Sports article[9] asks "Will MLB's tremendous rookie class of 2015 suffer a sophomore slump?".

Does the data confirm the existence of a sophomore slump? Let's take a look. Examining the data for batting average, we see that this observation holds true for the top performing ROYs:

nameFirst	nameLast	rookie_year	rookie	sophomore
Willie	McCovey	1959	0.354	0.238
Ichiro	Suzuki	2001	0.350	0.321
Al	Bumbry	1973	0.337	0.233
Fred	Lynn	1975	0.331	0.314
Albert	Pujols	2001	0.329	0.314

In fact, the proportion of players that have a lower batting average their sophomore year is 0.686.

So is it "jitters" or "jinx"? To answer this question, let's turn our attention to all players that played the 2013 and 2014 seasons and batted more than 130 times (minimum to win Rookie of the Year).

---

[9] http://www.foxsports.com/mlb/story/kris-bryant-carlos-correa-rookies-of-year-award-matt-duffy-francisco-lindor-kang-sano-120715

The same pattern arises when we look at the top performers: batting averages go down for most of the top performers.

nameFirst	nameLast	2013	2014
Miguel	Cabrera	0.348	0.313
Hanley	Ramirez	0.345	0.283
Michael	Cuddyer	0.331	0.332
Scooter	Gennett	0.324	0.289
Joe	Mauer	0.324	0.277

But these are not rookies! Also, look at what happens to the worst performers of 2013:

nameFirst	nameLast	2013	2014
Danny	Espinosa	0.158	0.219
Dan	Uggla	0.179	0.149
Jeff	Mathis	0.181	0.200
Melvin	Upton	0.184	0.208
Adam	Rosales	0.190	0.262

Their batting averages mostly go up! Is this some sort of reverse sophomore slump? It is not. There is no such thing as the sophomore slump. This is all explained with a simple statistical fact: the correlation for performance in two separate years is high, but not perfect:

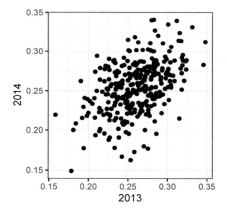

The correlation is 0.46 and the data look very much like a bivariate normal distribution, which means we predict a 2014 batting average Y for any given player that had a 2013 batting average X with:

$$\frac{Y - .255}{.032} = 0.46 \left(\frac{X - .261}{.023} \right)$$

Because the correlation is not perfect, regression tells us that, on average, expect high performers from 2013 to do a bit worse in 2014. It's not a jinx; it's just due to chance. The ROY are selected from the top values of X so it is expected that Y will regress to the mean.

18.9   Measurement error models

Up to now, all our linear regression examples have been applied to two or more random variables. We assume the pairs are bivariate normal and use this to motivate a linear model. This approach covers most real-life examples of linear regression. The other major application comes from measurement errors models. In these applications, it is common to have a non-random covariate, such as time, and randomness is introduced from measurement error rather than sampling or natural variability.

To understand these models, imagine you are Galileo in the 16th century trying to describe the velocity of a falling object. An assistant climbs the Tower of Pisa and drops a ball, while several other assistants record the position at different times. Let's simulate some data using the equations we know today and adding some measurement error. The **dslabs** function `rfalling_object` generates these simulations:

```
library(dslabs)
falling_object <- rfalling_object()
```

The assistants hand the data to Galileo and this is what he sees:

```
falling_object %>%
  ggplot(aes(time, observed_distance)) +
  geom_point() +
  ylab("Distance in meters") +
  xlab("Time in seconds")
```

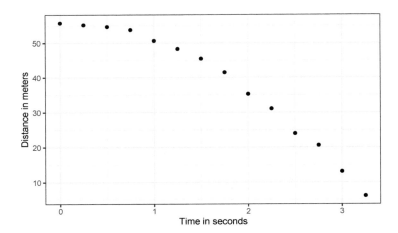

Galileo does not know the exact equation, but by looking at the plot above, he deduces that the position should follow a parabola, which we can write like this:

$$f(x) = \beta_0 + \beta_1 x + \beta_2 x^2$$

The data does not fall exactly on a parabola. Galileo knows this is due to measurement error.

His helpers make mistakes when measuring the distance. To account for this, he models the data with:

$$Y_i = \beta_0 + \beta_1 x_i + \beta_2 x_i^2 + \varepsilon_i, i = 1, \ldots, n$$

with Y_i representing distance in meters, x_i representing time in seconds, and ε accounting for measurement error. The measurement error is assumed to be random, independent from each other, and having the same distribution for each i. We also assume that there is no bias, which means the expected value $E[\varepsilon] = 0$.

Note that this is a linear model because it is a linear combination of known quantities (x and x^2 are known) and unknown parameters (the βs are unknown parameters to Galileo). Unlike our previous examples, here x is a fixed quantity; we are not conditioning.

To pose a new physical theory and start making predictions about other falling objects, Galileo needs actual numbers, rather than unknown parameters. Using LSE seems like a reasonable approach. How do we find the LSE?

LSE calculations do not require the errors to be approximately normal. The `lm` function will find the β s that will minimize the residual sum of squares:

```
fit <- falling_object %>%
  mutate(time_sq = time^2) %>%
  lm(observed_distance~time+time_sq, data=.)
tidy(fit)
#> # A tibble: 3 x 5
#>   term          estimate std.error statistic  p.value
#>   <chr>            <dbl>     <dbl>     <dbl>     <dbl>
#> 1 (Intercept)      56.1     0.592      94.9  2.23e-17
#> 2 time            -0.786    0.845     -0.930 3.72e- 1
#> 3 time_sq         -4.53     0.251     -18.1  1.58e- 9
```

Let's check if the estimated parabola fits the data. The **broom** function `augment` lets us do this easily:

```
augment(fit) %>%
  ggplot() +
  geom_point(aes(time, observed_distance)) +
  geom_line(aes(time, .fitted), col = "blue")
```

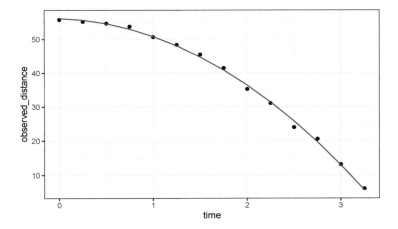

Thanks to my high school physics teacher, I know that the equation for the trajectory of a falling object is:

$$d = h_0 + v_0 t - 0.5 \times 9.8t^2$$

with h_0 and v_0 the starting height and velocity, respectively. The data we simulated above followed this equation and added measurement error to simulate n observations for dropping the ball ($v_0 = 0$) from the tower of Pisa ($h_0 = 55.86$).

These are consistent with the parameter estimates:

```
tidy(fit, conf.int = TRUE)
#> # A tibble: 3 x 7
#>    term          estimate std.error statistic  p.value conf.low conf.high
#>    <chr>            <dbl>     <dbl>     <dbl>     <dbl>    <dbl>     <dbl>
#> 1 (Intercept)     56.1      0.592     94.9    2.23e-17    54.8     57.4
#> 2 time            -0.786    0.845     -0.930  3.72e- 1    -2.65     1.07
#> 3 time_sq         -4.53     0.251    -18.1    1.58e- 9    -5.08    -3.98
```

The Tower of Pisa height is within the confidence interval for β_0, the initial velocity 0 is in the confidence interval for β_1 (note the p-value is larger than 0.05), and the acceleration constant is in a confidence interval for $-2 \times \beta_2$.

18.10   Exercises

Since the 1980s, sabermetricians have used a summary statistic different from batting average to evaluate players. They realized walks were important and that doubles, triples, and HRs, should be weighed more than singles. As a result, they proposed the following metric:

$$\frac{BB}{PA} + \frac{Singles + 2Doubles + 3Triples + 4HR}{AB}$$

They called this on-base-percentage plus slugging percentage (OPS). Although the sabermetricians probably did not use regression, here we show how this metric is close to what one gets with regression.

1. Compute the OPS for each team in the 2001 season. Then plot Runs per game versus OPS.

2. For every year since 1961, compute the correlation between runs per game and OPS; then plot these correlations as a function of year.

3. Note that we can rewrite OPS as a weighted average of BBs, singles, doubles, triples, and HRs. We know that the weights for doubles, triples, and HRs are 2, 3, and 4 times that of singles. But what about BB? What is the weight for BB relative to singles? Hint: the weight for BB relative to singles will be a function of AB and PA.

4. Note that the weight for BB, $\frac{AB}{PA}$, will change from team to team. To see how variable it is, compute and plot this quantity for each team for each year since 1961. Then plot it again, but instead of computing it for every team, compute and plot the ratio for the entire year. Then, once you are convinced that there is not much of a time or team trend, report the overall average.

5. So now we know that the formula for OPS is proportional to $0.91 \times BB + \text{singles} + 2 \times \text{doubles} + 3 \times \text{triples} + 4 \times HR$. Let's see how these coefficients compare to those obtained with regression. Fit a regression model to the data after 1961, as done earlier: using per game statistics for each year for each team. After fitting this model, report the coefficients as weights relative to the coefficient for singles.

6. We see that our linear regression model coefficients follow the same general trend as those used by OPS, but with slightly less weight for metrics other than singles. For each team in years after 1961, compute the OPS, the predicted runs with the regression model and compute the correlation between the two as well as the correlation with runs per game.

7. We see that using the regression approach predicts runs slightly better than OPS, but not that much. However, note that we have been computing OPS and predicting runs for teams when these measures are used to evaluate players. Let's show that OPS is quite similar to what one obtains with regression at the player level. For the 1961 season and after, compute the OPS and the predicted runs from our model for each player and plot them. Use the PA per game correction we used in the previous chapter:

8. What players have show the largest difference between their rank by predicted runs and OPS?

19

Association is not causation

Association is not causation is perhaps the most important lesson one learns in a statistics class. *Correlation is not causation* is another way to say this. Throughout the Statistics part of the book, we have described tools useful for quantifying associations between variables. However, we must be careful not to over interpret these associations.

There are many reasons that a variable X can be correlated with a variable Y without having any direct effect on Y. Here we examine three common ways that can lead to misinterpreting data.

19.1 Spurious correlation

The following comical example underscores that correlation is not causation. It shows a very strong correlation between divorce rates and margarine consumption.

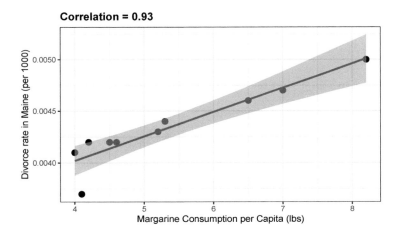

Does this mean that margarine causes divorces? Or do divorces cause people to eat more margarine? Of course the answer to both these questions is no. This is just an example of what we call a *spurious correlation*.

You can see many more absurd examples on the Spurious Correlations website[1].

The cases presented in the spurious correlation site are all instances of what is generally

---

[1] http://tylervigen.com/spurious-correlations

called *data dredging*, *data fishing*, or *data snooping*. It's basically a form of what in the US they call *cherry picking*. An example of data dredging would be if you look through many results produced by a random process and pick the one that shows a relationship that supports a theory you want to defend.

A Monte Carlo simulation can be used to show how data dredging can result in finding high correlations among uncorrelated variables. We will save the results of our simulation into a tibble:

```
N <- 25
g <- 1000000
sim_data <- tibble(group = rep(1:g, each=N),
                   x = rnorm(N * g),
                   y = rnorm(N * g))
```

The first column denotes group. We created groups and for each one we generated a pair of independent vectors, X and Y, with 25 observations each, stored in the second and third columns. Because we constructed the simulation, we know that X and Y are not correlated.

Next, we compute the correlation between X and Y for each group and look at the max:

```
res <- sim_data %>%
  group_by(group) %>%
  summarize(r = cor(x, y)) %>%
  arrange(desc(r))
res
#> # A tibble: 1,000,000 x 2
#>     group       r
#>     <int>   <dbl>
#> 1   19673   0.808
#> 2  901379   0.799
#> 3  297494   0.780
#> 4  633789   0.768
#> 5  168119   0.764
#> # ... with 1e+06 more rows
```

We see a maximum correlation of 0.808 and if you just plot the data from the group achieving this correlation, it shows a convincing plot that X and Y are in fact correlated:

```
sim_data %>% filter(group == res$group[which.max(res$r)]) %>%
  ggplot(aes(x, y)) +
  geom_point() +
  geom_smooth(method = "lm")
```

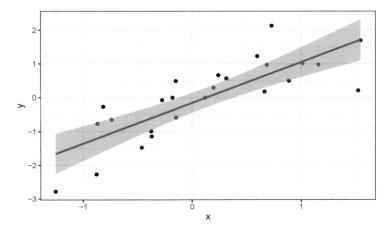

Remember that the correlation summary is a random variable. Here is the distribution generated by the Monte Carlo simulation:

```
res %>% ggplot(aes(x=r)) + geom_histogram(binwidth = 0.1, color = "black")
```

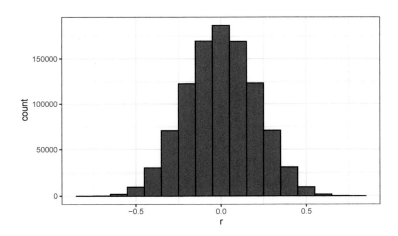

It's just a mathematical fact that if we observe random correlations that are expected to be 0, but have a standard error of 0.204, the largest one will be close to 1.

If we performed regression on this group and interpreted the p-value, we would incorrectly claim this was a statistically significant relation:

```
library(broom)
sim_data %>%
  filter(group == res$group[which.max(res$r)]) %>%
  do(tidy(lm(y ~ x, data = .))) %>%
  filter(term == "x")
#> # A tibble: 1 x 5
#>   term  estimate std.error statistic   p.value
#>   <chr>    <dbl>     <dbl>     <dbl>     <dbl>
#> 1 x         1.20     0.183      6.57 0.00000104
```

This particular form of data dredging is referred to as *p-hacking.* P-hacking is a topic of much discussion because it is a problem in scientific publications. Because publishers tend to reward statistically significant results over negative results, there is an incentive to report significant results. In epidemiology and the social sciences, for example, researchers may look for associations between an adverse outcome and several exposures and report only the one exposure that resulted in a small p-value. Furthermore, they might try fitting several different models to account for confounding and pick the one that yields the smallest p-value. In experimental disciplines, an experiment might be repeated more than once, yet only the results of the one experiment with a small p-value reported. This does not necessarily happen due to unethical behavior, but rather as a result of statistical ignorance or wishful thinking. In advanced statistics courses, you can learn methods to adjust for these multiple comparisons.

19.2   Outliers

Suppose we take measurements from two independent outcomes, X and Y, and we standardize the measurements. However, imagine we make a mistake and forget to standardize entry 23. We can simulate such data using:

```
set.seed(1985)
x <- rnorm(100,100,1)
y <- rnorm(100,84,1)
x[-23] <- scale(x[-23])
y[-23] <- scale(y[-23])
```

The data look like this:

```
qplot(x, y)
```

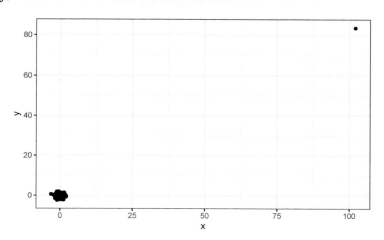

Not surprisingly, the correlation is very high:

```
cor(x,y)
#> [1] 0.988
```

But this is driven by the one outlier. If we remove this outlier, the correlation is greatly reduced to almost 0, which is what it should be:

```
cor(x[-23], y[-23])
#> [1] -0.0442
```

In Section 11 we described alternatives to the average and standard deviation that are robust to outliers. There is also an alternative to the sample correlation for estimating the population correlation that is robust to outliers. It is called *Spearman correlation*. The idea is simple: compute the correlation on the ranks of the values. Here is a plot of the ranks plotted against each other:

```
qplot(rank(x), rank(y))
```

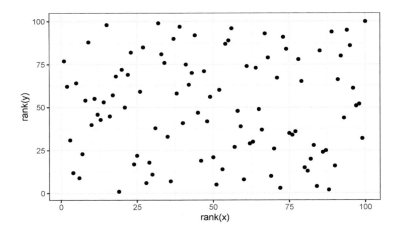

The outlier is no longer associated with a very large value and the correlation comes way down:

```
cor(rank(x), rank(y))
#> [1] 0.00251
```

Spearman correlation can also be calculated like this:

```
cor(x, y, method = "spearman")
#> [1] 0.00251
```

There are also methods for robust fitting of linear models which you can learn about in, for instance, this book: Robust Statistics: Edition 2 by Peter J. Huber & Elvezio M. Ronchetti.

19.3   Reversing cause and effect

Another way association is confused with causation is when the cause and effect are reversed. An example of this is claiming that tutoring makes students perform worse because they test lower than peers that are not tutored. In this case, the tutoring is not causing the low test scores, but the other way around.

A form of this claim actually made it into an op-ed in the New York Times titled Parental Involvement Is Overrated[2]. Consider this quote from the article:

> When we examined whether regular help with homework had a positive impact on children's academic performance, we were quite startled by what we found. Regardless of a family's social class, racial or ethnic background, or a child's grade level, consistent homework help almost never improved test scores or grades... Even more surprising to us was that when parents regularly helped with homework, kids usually performed worse.

A very likely possibility is that the children needing regular parental help, receive this help because they don't perform well in school.

We can easily construct an example of cause and effect reversal using the father and son height data. If we fit the model:

$$X_i = \beta_0 + \beta_1 y_i + \varepsilon_i, i = 1, \ldots, N$$

to the father and son height data, with X_i the father height and y_i the son height, we do get a statistically significant result:

```
library(HistData)
data("GaltonFamilies")
GaltonFamilies %>%
  filter(childNum == 1 & gender == "male") %>%
  select(father, childHeight) %>%
  rename(son = childHeight) %>%
  do(tidy(lm(father ~ son, data = .)))
#> # A tibble: 2 x 5
#>   term         estimate std.error statistic  p.value
#>   <chr>          <dbl>     <dbl>     <dbl>    <dbl>
#> 1 (Intercept)    34.0      4.57      7.44  4.31e-12
#> 2 son             0.499    0.0648    7.70  9.47e-13
```

The model fits the data very well. If we look at the mathematical formulation of the model above, it could easily be incorrectly interpreted so as to suggest that the son being tall caused the father to be tall. But given what we know about genetics and biology, we know it's the other way around. The model is technically correct. The estimates and p-values were obtained correctly as well. What is wrong here is the interpretation.

---

[2]https://opinionator.blogs.nytimes.com/2014/04/12/parental-involvement-is-overrated

19.4 Confounders

Confounders are perhaps the most common reason that leads to associations begin misinterpreted.

If X and Y are correlated, we call Z a *confounder* if changes in Z causes changes in both X and Y. Earlier, when studying baseball data, we saw how Home Runs was a confounder that resulted in a higher correlation than expected when studying the relationship between Bases on Balls and Runs. In some cases, we can use linear models to account for confounders. However, this is not always the case.

Incorrect interpretation due to confounders is ubiquitous in the lay press and they are often hard to detect. Here, we present a widely used example related to college admissions.

19.4.1 Example: UC Berkeley admissions

Admission data from six U.C. Berkeley majors, from 1973, showed that more men were being admitted than women: 44% men were admitted compared to 30% women. PJ Bickel, EA Hammel, and JW O'Connell. Science (1975). We can load the data and a statistical test, which clearly rejects the hypothesis that gender and admission are independent:

```
data(admissions)
admissions %>% group_by(gender) %>%
  summarize(total_admitted = round(sum(admitted / 100 * applicants)),
            not_admitted = sum(applicants) - sum(total_admitted)) %>%
  select(-gender) %>%
  do(tidy(chisq.test(.))) %>% .$p.value
#> [1] 1.06e-21
```

But closer inspection shows a paradoxical result. Here are the percent admissions by major:

```
admissions %>% select(major, gender, admitted) %>%
  spread(gender, admitted) %>%
  mutate(women_minus_men = women - men)
#>    major men women women_minus_men
#> 1      A  62    82              20
#> 2      B  63    68               5
#> 3      C  37    34              -3
#> 4      D  33    35               2
#> 5      E  28    24              -4
#> 6      F   6     7               1
```

Four out of the six majors favor women. More importantly, all the differences are much smaller than the 14.2 difference that we see when examining the totals.

The paradox is that analyzing the totals suggests a dependence between admission and gender, but when the data is grouped by major, this dependence seems to disappear. What's going on? This actually can happen if an uncounted confounder is driving most of the variability.

So let's define three variables: X is 1 for men and 0 for women, Y is 1 for those admitted and 0 otherwise, and Z quantifies the selectivity of the major. A gender bias claim would be based on the fact that $\Pr(Y = 1 | X = x)$ is higher for $x = 1$ than $x = 0$. However, Z is an important confounder to consider. Clearly Z is associated with Y, as the more selective a major, the lower $\Pr(Y = 1 | Z = z)$. But is major selectivity Z associated with gender X?

One way to see this is to plot the total percent admitted to a major versus the percent of women that made up the applicants:

```
admissions %>%
  group_by(major) %>%
  summarize(major_selectivity = sum(admitted * applicants)/sum(applicants),
            percent_women_applicants = sum(applicants * (gender=="women")) /
                                       sum(applicants) * 100) %>%
  ggplot(aes(major_selectivity, percent_women_applicants, label = major)) +
  geom_text()
```

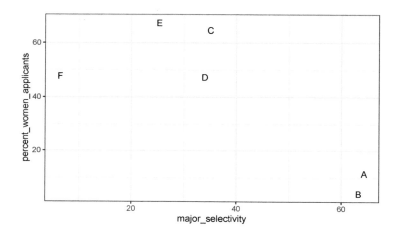

There seems to be association. The plot suggests that women were much more likely to apply to the two "hard" majors: gender and major's selectivity are confounded. Compare, for example, major B and major E. Major E is much harder to enter than major B and over 60% of applicants to major E were women, while less than 30% of the applicants of major B were women.

19.4.2   Confounding explained graphically

The following plot shows the number of applicants that were admitted and those that were not by:

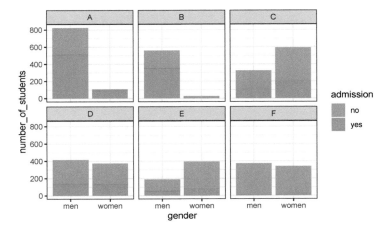

It also breaks down the acceptances by major. This breakdown allows us to see that the majority of accepted men came from two majors: A and B. It also lets us see that few women applied to these majors.

19.4.3   Average after stratifying

In this plot, we can see that if we condition or stratify by major, and then look at differences, we control for the confounder and this effect goes away:

```
admissions %>%
  ggplot(aes(major, admitted, col = gender, size = applicants)) +
  geom_point()
```

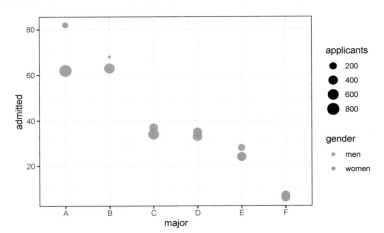

Now we see that major by major, there is not much difference. The size of the dot represents the number of applicants, and explains the paradox: we see large red dots and small blue dots for the easiest majors, A and B.

If we average the difference by major, we find that the percent is actually 3.5% higher for women.

```
admissions %>% group_by(gender) %>% summarize(average = mean(admitted))
#> # A tibble: 2 x 2
#>   gender average
#>   <chr>    <dbl>
#> 1 men       38.2
#> 2 women     41.7
```

19.5   Simpson's paradox

The case we have just covered is an example of Simpson's paradox. It is called a paradox because we see the sign of the correlation flip when comparing the entire publication and specific strata. As an illustrative example, suppose you have three random variables X, Y, and Z and that we observe realizations of these. Here is a plot of simulated observations for X and Y along with the sample correlation:

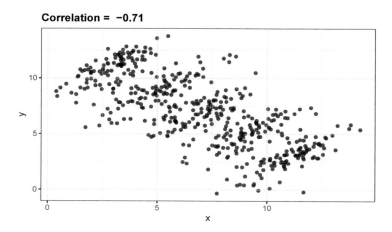

You can see that X and Y are negatively correlated. However, once we stratify by Z (shown in different colors below) another pattern emerges:

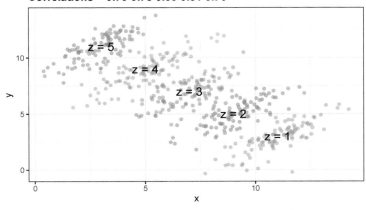

Correlations = 0.79 0.75 0.69 0.81 0.76

It is really Z that is negatively correlated with X. If we stratify by Z, the X and Y are actually positively correlated as seen in the plot above.

19.6 Exercises

For the next set of exercises, we examine the data from a 2014 PNAS paper[3] that analyzed success rates from funding agencies in the Netherlands and concluded:

> Our results reveal gender bias favoring male applicants over female applicants in the prioritization of their "quality of researcher" (but not "quality of proposal") evaluations and success rates, as well as in the language used in instructional and evaluation materials.

A response[4] was published a few months later titled *No evidence that gender contributes to personal research funding success in The Netherlands: A reaction to Van der Lee and Ellemers* which concluded:

> However, the overall gender effect borders on statistical significance, despite the large sample. Moreover, their conclusion could be a prime example of Simpson's paradox; if a higher percentage of women apply for grants in more competitive scientific disciplines (i.e., with low application success rates for both men and women), then an analysis across all disciplines could incorrectly show "evidence" of gender inequality.

Who is right here? The original paper or the response? Here, you will examine the data and come to your own conclusion.

1. The main evidence for the conclusion of the original paper comes down to a comparison of the percentages. Table S1 in the paper includes the information we need:

```
library(dslabs)
data("research_funding_rates")
research_funding_rates
```

---

[3] http://www.pnas.org/content/112/40/12349.abstract
[4] http://www.pnas.org/content/112/51/E7036.extract

Construct the two-by-two table used for the conclusion about differences in awards by gender.

2. Compute the difference in percentage from the two-by-two table.

3. In the previous exercise, we noticed that the success rate is lower for women. But is it significant? Compute a p-value using a Chi-square test.

4. We see that the p-value is about 0.05. So there appears to be some evidence of an association. But can we infer causation here? Is gender bias causing this observed difference? The response to the original paper claims that what we see here is similar to the UC Berkeley admissions example. Specifically they state that this "could be a prime example of Simpson's paradox; if a higher percentage of women apply for grants in more competitive scientific disciplines, then an analysis across all disciplines could incorrectly show 'evidence' of gender inequality." To settle this dispute, create a dataset with number of applications, awards, and success rate for each gender. Re-order the disciplines by their overall success rate. Hint: use the `reorder` function to re-order the disciplines in a first step, then use `gather`, `separate`, and `spread` to create the desired table.

5. To check if this is a case of Simpson's paradox, plot the success rates versus disciplines, which have been ordered by overall success, with colors to denote the genders and size to denote the number of applications.

6. We definitely do not see the same level of confounding as in the UC Berkeley example. It is hard to say there is a confounder here. However, we do see that, based on the observed rates, some fields favor men and others favor women and we do see that the two fields with the largest difference favoring men are also the fields with the most applications. But, unlike the UC Berkeley example, women are not more likely to apply for the harder subjects. So perhaps some of the selection committees are biased and others are not.

But, before we conclude this, we must check if these differences are any different than what we get by chance. Are any of the differences seen above statistically significant? Keep in mind that even when there is no bias, we will see differences due to random variability in the review process as well as random variability across candidates. Perform a Chi-square test for each discipline. Hint: define a function that receives the total of a two-by-two table and returns a data frame with the p-value. Use the 0.5 correction. Then use the do function.

7. For the medical sciences, there appears to be a statistically significant difference. But is this a spurious correlation? We performed 9 tests. Reporting only the one case with a p-value less than 0.05 might be considered an example of cherry picking. Repeat the exercise above, but instead of a p-value, compute a log odds ratio divided by their standard error. Then use qq-plot to see how much these log odds ratios deviate from the normal distribution we would expect: a standard normal distribution.

Part IV

Data Wrangling

20

Introduction to data wrangling

The datasets used in this book have been made available to you as R objects, specifically as data frames. The US murders data, the reported heights data, and the Gapminder data were all data frames. These datasets come included in the **dslabs** package and we loaded them using the `data` function. Furthermore, we have made the data available in what is referred to as `tidy` form. The tidyverse packages and functions assume that the data is `tidy` and this assumption is a big part of the reason these packages work so well together.

However, very rarely in a data science project is data easily available as part of a package. We did quite a bit of work "behind the scenes" to get the original raw data into the *tidy* tables you worked with. Much more typical is for the data to be in a file, a database, or extracted from a document, including web pages, tweets, or PDFs. In these cases, the first step is to import the data into R and, when using the **tidyverse**, tidy the data. This initial step in the data analysis process usually involves several, often complicated, steps to convert data from its raw form to the *tidy* form that greatly facilitates the rest of the analysis. We refer to this process as `data wrangling`.

Here we cover several common steps of the data wrangling process including tidying data, string processing, html parsing, working with dates and times, and text mining. Rarely are all these wrangling steps necessary in a single analysis, but data scientists will likely face them all at some point. Some of the examples we use to demonstrate data wrangling techniques are based on the work we did to convert raw data into the tidy datasets provided by the **dslabs** package and used in the book as examples.

21

Reshaping data

As we have seen through the book, having data in *tidy* format is what makes the tidyverse flow. After the first step in the data analysis process, importing data, a common next step is to reshape the data into a form that facilitates the rest of the analysis. The **tidyr** package includes several functions that are useful for tidying data.

We will use the fertility wide format dataset described in Section 4.1 as an example in this section.

```
library(tidyverse)
library(dslabs)
path <- system.file("extdata", package="dslabs")
filename <- file.path(path, "fertility-two-countries-example.csv")
wide_data <- read_csv(filename)
```

21.1 gather

One of the most used functions in the **tidyr** package is `gather`, which is useful for converting wide data into tidy data.

As with most tidyverse functions, the `gather` function's first argument is the data frame that will be converted. Here we want to reshape the `wide_data` dataset so that each row represents a fertility observation, which implies we need three columns to store the year, country, and the observed value. In its current form, data from different years are in different columns with the year values stored in the column names. Through the second and third argument we will tell `gather` the column names we want to assign to the columns containing the current column names and observations, respectively. In this case a good choice for these two arguments would be `year` and `fertility`. Note that nowhere in the data file does it tell us this is fertility data. Instead, we deciphered this from the file name. Through the fourth argument we specify the columns containing observed values; these are the columns that will be *gathered*. The default is to gather all columns so, in most cases, we have to specify the columns. In our example we want columns 1960, 1961 up to 2015.

The code to gather the fertility data therefore looks like this:

```
new_tidy_data <- gather(wide_data, year, fertility, `1960`:`2015`)
```

We can also use the pipe like this:

```
new_tidy_data <- wide_data %>% gather(year, fertility, `1960`:`2015`)
```

We can see that the data have been converted to tidy format with columns `year` and `fertility`:

```
head(new_tidy_data)
#> # A tibble: 6 x 3
#>     country       year   fertility
#>     <chr>         <chr>     <dbl>
#> 1 Germany         1960       2.41
#> 2 South Korea     1960       6.16
#> 3 Germany         1961       2.44
#> 4 South Korea     1961       5.99
#> 5 Germany         1962       2.47
#> # ... with 1 more row
```

and that each year resulted in two rows since we have two countries and this column was not gathered. A somewhat quicker way to write this code is to specify which column will **not** be gathered, rather than all the columns that will be gathered:

```
new_tidy_data <- wide_data %>%
  gather(year, fertility, -country)
```

The `new_tidy_data` object looks like the original `tidy_data` we defined this way

```
data("gapminder")
tidy_data <- gapminder %>%
  filter(country %in% c("South Korea", "Germany") & !is.na(fertility)) %>%
  select(country, year, fertility)
```

with just one minor difference. Can you spot it? Look at the data type of the year column:

```
class(tidy_data$year)
#> [1] "integer"
class(new_tidy_data$year)
#> [1] "character"
```

The `gather` function assumes that column names are characters. So we need a bit more wrangling before we are ready to make a plot. We need to convert the year column to be numbers. The `gather` function includes the `convert` argument for this purpose:

```
new_tidy_data <- wide_data %>%
  gather(year, fertility, -country, convert = TRUE)
class(new_tidy_data$year)
#> [1] "integer"
```

Note that we could have also used the `mutate` and `as.numeric`.

Now that the data is tidy, we can use this relatively simple ggplot code:

```
new_tidy_data%>%ggplot(aes(year, fertility, color = country)) + geom_point()
```

21.2 spread

As we will see in later examples, it is sometimes useful for data wrangling purposes to convert tidy data into wide data. We often use this as an intermediate step in tidying up data. The spread function is basically the inverse of `gather`. The first argument is for the data, but since we are using the pipe, we don't show it. The second argument tells spread which variable will be used as the column names. The third argument specifies which variable to use to fill out the cells:

```
new_wide_data <- new_tidy_data %>% spread(year, fertility)
select(new_wide_data, country, `1960`:`1967`)
#> # A tibble: 2 x 9
#>   country     `1960` `1961` `1962` `1963` `1964` `1965` `1966` `1967`
#>   <chr>       <dbl>  <dbl>  <dbl>  <dbl>  <dbl>  <dbl>  <dbl>  <dbl>
#> 1 Germany     2.41   2.44   2.47   2.49   2.49   2.48   2.44   2.37
#> 2 South Korea 6.16   5.99   5.79   5.57   5.36   5.16   4.99   4.85
```

The following diagram can help remind you how these two functions work:

tidyr::**gather(cases, "year", "n", 2:4)**
Gather columns into rows.

tidyr::**spread(pollution, size, amount)**
Spread rows into columns.

(Image courtesy of RStudio[1]. CC-BY-4.0 license[2]. Cropped from original.)

21.3 separate

The data wrangling shown above was simple compared to what is usually required. In our example spreadsheet files, we include an illustration that is slightly more complicated. It contains two variables: life expectancy and fertility. However, the way it is stored is not tidy and, as we will explain, not optimal.

```
path <- system.file("extdata", package = "dslabs")
```

---

[1] https://github.com/rstudio/cheatsheets
[2] https://github.com/rstudio/cheatsheets/blob/master/LICENSE

```
filename <- "life-expectancy-and-fertility-two-countries-example.csv"
filename <- file.path(path, filename)

raw_dat <- read_csv(filename)
select(raw_dat, 1:5)
#> # A tibble: 2 x 5
#>   country `1960_fertility` `1960_life_expe~ `1961_fertility`
#>   <chr>              <dbl>            <dbl>            <dbl>
#> 1 Germany             2.41             69.3             2.44
#> 2 South ~             6.16             53.0             5.99
#> # ... with 1 more variable: `1961_life_expectancy` <dbl>
```

First, note that the data is in wide format. Second, notice that this table includes values for two variables, fertility and life expectancy, with the column name encoding which column represents which variable. Encoding information in the column names is not recommended but, unfortunately, it is quite common. We will put our wrangling skills to work to extract this information and store it in a tidy fashion.

We can start the data wrangling with the **gather** function, but we should no longer use the column name **year** for the new column since it also contains the variable type. We will call it **key**, the default, for now:

```
dat <- raw_dat %>% gather(key, value, -country)
head(dat)
#> # A tibble: 6 x 3
#>   country     key                   value
#>   <chr>       <chr>                 <dbl>
#> 1 Germany     1960_fertility         2.41
#> 2 South Korea 1960_fertility         6.16
#> 3 Germany     1960_life_expectancy  69.3
#> 4 South Korea 1960_life_expectancy  53.0
#> 5 Germany     1961_fertility         2.44
#> # ... with 1 more row
```

The result is not exactly what we refer to as tidy since each observation is associated with two, not one, rows. We want to have the values from the two variables, fertility and life expectancy, in two separate columns. The first challenge to achieve this is to separate the **key** column into the year and the variable type. Notice that the entries in this column separate the year from the variable name with an underscore:

```
dat$key[1:5]
#> [1] "1960_fertility"       "1960_fertility"       "1960_life_expectancy"
#> [4] "1960_life_expectancy" "1961_fertility"
```

Encoding multiple variables in a column name is such a common problem that the **readr** package includes a function to separate these columns into two or more. Apart from the data, the **separate** function takes three arguments: the name of the column to be separated, the names to be used for the new columns, and the character that separates the variables. So, a first attempt at this is:

```
dat %>% separate(key, c("year", "variable_name"), "_")
```

Because _ is the default separator assumed by **separate**, we do not have to include it in the code:

```
dat %>% separate(key, c("year", "variable_name"))
#> Warning: Expected 2 pieces. Additional pieces discarded in 112 rows [3,
#> 4, 7, 8, 11, 12, 15, 16, 19, 20, 23, 24, 27, 28, 31, 32, 35, 36, 39,
#> 40, ...].
#> # A tibble: 224 x 4
#>   country     year  variable_name value
#>   <chr>       <chr> <chr>         <dbl>
#> 1 Germany     1960  fertility      2.41
#> 2 South Korea 1960  fertility      6.16
#> 3 Germany     1960  life          69.3
#> 4 South Korea 1960  life          53.0
#> 5 Germany     1961  fertility      2.44
#> # ... with 219 more rows
```

The function does separate the values, but we run into a new problem. We receive the warning `Too many values at 112 locations:` and that the `life_expectancy` variable is truncated to `life`. This is because the _ is used to separate `life` and `expectancy`, not just year and variable name! We could add a third column to catch this and let the **separate** function know which column to *fill in* with missing values, `NA`, when there is no third value. Here we tell it to fill the column on the right:

```
var_names <- c("year", "first_variable_name", "second_variable_name")
dat %>% separate(key, var_names, fill = "right")
#> # A tibble: 224 x 5
#>   country     year  first_variable_name second_variable_name value
#>   <chr>       <chr> <chr>               <chr>                <dbl>
#> 1 Germany     1960  fertility           <NA>                  2.41
#> 2 South Korea 1960  fertility           <NA>                  6.16
#> 3 Germany     1960  life                expectancy           69.3
#> 4 South Korea 1960  life                expectancy           53.0
#> 5 Germany     1961  fertility           <NA>                  2.44
#> # ... with 219 more rows
```

However, if we read the **separate** help file, we find that a better approach is to merge the last two variables when there is an extra separation:

```
dat %>% separate(key, c("year", "variable_name"), extra = "merge")
#> # A tibble: 224 x 4
#>   country     year  variable_name   value
#>   <chr>       <chr> <chr>           <dbl>
#> 1 Germany     1960  fertility        2.41
#> 2 South Korea 1960  fertility        6.16
#> 3 Germany     1960  life_expectancy 69.3
#> 4 South Korea 1960  life_expectancy 53.0
#> 5 Germany     1961  fertility        2.44
#> # ... with 219 more rows
```

This achieves the separation we wanted. However, we are not done yet. We need to create a column for each variable. As we learned, the spread function can do this:

```
dat %>%
  separate(key, c("year", "variable_name"), extra = "merge") %>%
  spread(variable_name, value)
#> # A tibble: 112 x 4
#>    country year  fertility life_expectancy
#>    <chr>   <chr>     <dbl>           <dbl>
#> 1 Germany 1960       2.41            69.3
#> 2 Germany 1961       2.44            69.8
#> 3 Germany 1962       2.47            70.0
#> 4 Germany 1963       2.49            70.1
#> 5 Germany 1964       2.49            70.7
#> # ... with 107 more rows
```

The data is now in tidy format with one row for each observation with three variables: year, fertility, and life expectancy.

21.4 unite

It is sometimes useful to do the inverse of separate, unite two columns into one. To demonstrate how to use unite, we show code that, although *not* the optimal approach, serves as an illustration. Suppose that we did not know about extra and used this command to separate:

```
dat %>%
  separate(key, var_names, fill = "right")
#> # A tibble: 224 x 5
#>    country     year  first_variable_name second_variable_name value
#>    <chr>       <chr> <chr>               <chr>                <dbl>
#> 1 Germany     1960  fertility           <NA>                  2.41
#> 2 South Korea 1960  fertility           <NA>                  6.16
#> 3 Germany     1960  life                expectancy           69.3
#> 4 South Korea 1960  life                expectancy           53.0
#> 5 Germany     1961  fertility           <NA>                  2.44
#> # ... with 219 more rows
```

We can achieve the same final result by uniting the second and third columns, then spreading the columns and renaming fertility_NA to fertility:

```
dat %>%
  separate(key, var_names, fill = "right") %>%
  unite(variable_name, first_variable_name, second_variable_name) %>%
  spread(variable_name, value) %>%
  rename(fertility = fertility_NA)
#> # A tibble: 112 x 4
```

```
#>    country year   fertility life_expectancy
#>    <chr>    <chr>     <dbl>           <dbl>
#> 1 Germany 1960       2.41            69.3
#> 2 Germany 1961       2.44            69.8
#> 3 Germany 1962       2.47            70.0
#> 4 Germany 1963       2.49            70.1
#> 5 Germany 1964       2.49            70.7
#> # ... with 107 more rows
```

21.5 Exercises

1. Run the following command to define the `co2_wide` object:

```
co2_wide <- data.frame(matrix(co2, ncol = 12, byrow = TRUE)) %>%
  setNames(1:12) %>%
  mutate(year = as.character(1959:1997))
```

Use the gather function to wrangle this into a tidy dataset. Call the column with the CO2 measurements co2 and call the month column `month`. Call the resulting object `co2_tidy`.

2. Plot CO2 versus month with a different curve for each year using this code:

```
co2_tidy %>% ggplot(aes(month, co2, color = year)) + geom_line()
```

If the expected plot is not made, it is probably because `co2_tidy$month` is not numeric:

```
class(co2_tidy$month)
```

Rewrite the call to gather using an argument that assures the month column will be numeric. Then make the plot.

3. What do we learn from this plot?

   a. CO2 measures increase monotonically from 1959 to 1997.
   b. CO2 measures are higher in the summer and the yearly average increased from 1959 to 1997.
   c. CO2 measures appear constant and random variability explains the differences.
   d. CO2 measures do not have a seasonal trend.

4. Now load the `admissions` data set, which contains admission information for men and women across six majors and keep only the admitted percentage column:

```
load(admissions)
dat <- admissions %>% select(-applicants)
```

If we think of an observation as a major, and that each observation has two variables (men admitted percentage and women admitted percentage) then this is not tidy. Use the `spread` function to wrangle into tidy shape: one row for each major.

5. Now we will try a more advanced wrangling challenge. We want to wrangle the admissions data so that for each major we have 4 observations: `admitted_men`, `admitted_women`, `applicants_men` and `applicants_women`. The *trick* we perform here is actually quite common: first gather to generate an intermediate data frame and then spread to obtain the tidy data we want. We will go step by step in this and the next two exercises.

Use the gather function to create a `tmp` data.frame with a column containing the type of observation `admitted` or `applicants`. Call the new columns `key` and `value`.

6. Now you have an object `tmp` with columns `major`, `gender`, `key` and `value`. Note that if you combine the key and gender, we get the column names we want: `admitted_men`, `admitted_women`, `applicants_men` and `applicants_women`. Use the function `unite` to create a new column called `column_name`.

7. Now use the `spread` function to generate the tidy data with four variables for each major.

8. Now use the pipe to write a line of code that turns `admissions` to the table produced in the previous exercise.

22

Joining tables

The information we need for a given analysis may not be just in one table. For example, when forecasting elections we used the function `left_join` to combine the information from two tables. Here we use a simpler example to illustrate the general challenge of combining tables.

Suppose we want to explore the relationship between population size for US states and electoral votes. We have the population size in this table:

```
library(tidyverse)
library(dslabs)
data(murders)
head(murders)
#>        state abb region population total
#> 1    Alabama  AL  South     4779736   135
#> 2     Alaska  AK   West      710231    19
#> 3    Arizona  AZ   West     6392017   232
#> 4   Arkansas  AR  South     2915918    93
#> 5 California  CA   West    37253956  1257
#> 6   Colorado  CO   West     5029196    65
```

and electoral votes in this one:

```
data(polls_us_election_2016)
head(results_us_election_2016)
#>          state electoral_votes clinton trump others
#> 1   California              55    61.7  31.6    6.7
#> 2       Texas              38    43.2  52.2    4.5
#> 3     Florida              29    47.8  49.0    3.2
#> 4    New York              29    59.0  36.5    4.5
#> 5    Illinois              20    55.8  38.8    5.4
#> 6 Pennsylvania             20    47.9  48.6    3.6
```

Just concatenating these two tables together will not work since the order of the states is not the same.

```
identical(results_us_election_2016$state, murders$state)
#> [1] FALSE
```

The *join* functions, described below, are designed to handle this challenge.

22.1   Joins

The *join* functions in the **dplyr** package make sure that the tables are combined so that matching rows are together. If you know SQL, you will see that the approach and syntax is very similar. The general idea is that one needs to identify one or more columns that will serve to match the two tables. Then a new table with the combined information is returned. Notice what happens if we join the two tables above by state using `left_join` (we will remove the `others` column and rename `electoral_votes` so that the tables fit on the page):

```
tab <- left_join(murders, results_us_election_2016, by = "state") %>%
  select(-others) %>% rename(ev = electoral_votes)
head(tab)
#>          state abb region population total ev clinton trump
#> 1     Alabama  AL  South     4779736   135  9    34.4  62.1
#> 2      Alaska  AK   West      710231    19  3    36.6  51.3
#> 3     Arizona  AZ   West     6392017   232 11    45.1  48.7
#> 4    Arkansas  AR  South     2915918    93  6    33.7  60.6
#> 5  California  CA   West    37253956  1257 55    61.7  31.6
#> 6    Colorado  CO   West     5029196    65  9    48.2  43.3
```

The data has been successfully joined and we can now, for example, make a plot to explore the relationship:

```
library(ggrepel)
tab %>% ggplot(aes(population/10^6, ev, label = abb)) +
  geom_point() +
  geom_text_repel() +
  scale_x_continuous(trans = "log2") +
  scale_y_continuous(trans = "log2") +
  geom_smooth(method = "lm", se = FALSE)
```

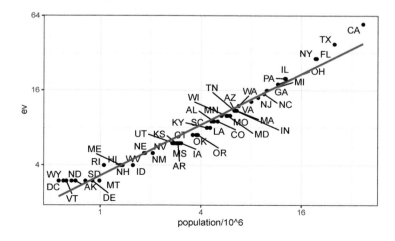

We see the relationship is close to linear with about 2 electoral votes for every million persons, but with very small states getting higher ratios.

In practice, it is not always the case that each row in one table has a matching row in the other. For this reason, we have several versions of join. To illustrate this challenge, we will take subsets of the tables above. We create the tables **tab1** and **tab2** so that they have some states in common but not all:

```
tab_1 <- slice(murders, 1:6) %>% select(state, population)
tab_1
#>         state population
#> 1     Alabama    4779736
#> 2      Alaska     710231
#> 3     Arizona    6392017
#> 4    Arkansas    2915918
#> 5  California   37253956
#> 6    Colorado    5029196
tab_2 <- results_us_election_2016 %>%
   filter(state%in%c("Alabama", "Alaska", "Arizona",
                     "California", "Connecticut", "Delaware")) %>%
   select(state, electoral_votes) %>% rename(ev = electoral_votes)
tab_2
#>         state ev
#> 1  California 55
#> 2     Arizona 11
#> 3     Alabama  9
#> 4 Connecticut  7
#> 5      Alaska  3
#> 6    Delaware  3
```

We will use these two tables as examples in the next sections.

22.1.1   Left join

Suppose we want a table like **tab_1**, but adding electoral votes to whatever states we have available. For this, we use **left_join** with **tab_1** as the first argument. We specify which column to use to match with the **by** argument.

```
left_join(tab_1, tab_2, by = "state")
#>         state population ev
#> 1     Alabama    4779736  9
#> 2      Alaska     710231  3
#> 3     Arizona    6392017 11
#> 4    Arkansas    2915918 NA
#> 5  California   37253956 55
#> 6    Colorado    5029196 NA
```

Note that NAs are added to the two states not appearing in **tab_2**. Also, notice that this function, as well as all the other joins, can receive the first arguments through the pipe:

```
tab_1 %>% left_join(tab_2, by = "state")
```

22.1.2   Right join

If instead of a table with the same rows as first table, we want one with the same rows as second table, we can use `right_join`:

```
tab_1 %>% right_join(tab_2, by = "state")
#>           state population ev
#> 1   California   37253956 55
#> 2      Arizona    6392017 11
#> 3      Alabama    4779736  9
#> 4  Connecticut         NA  7
#> 5       Alaska     710231  3
#> 6     Delaware         NA  3
```

Now the NAs are in the column coming from `tab_1`.

22.1.3   Inner join

If we want to keep only the rows that have information in both tables, we use `inner_join`. You can think of this as an intersection:

```
inner_join(tab_1, tab_2, by = "state")
#>           state population ev
#> 1      Alabama    4779736  9
#> 2       Alaska     710231  3
#> 3      Arizona    6392017 11
#> 4   California   37253956 55
```

22.1.4   Full join

If we want to keep all the rows and fill the missing parts with NAs, we can use `full_join`. You can think of this as a union:

```
full_join(tab_1, tab_2, by = "state")
#>           state population ev
#> 1      Alabama    4779736  9
#> 2       Alaska     710231  3
#> 3      Arizona    6392017 11
#> 4     Arkansas    2915918 NA
#> 5   California   37253956 55
#> 6     Colorado    5029196 NA
#> 7  Connecticut         NA  7
#> 8     Delaware         NA  3
```

22.1.5 Semi join

The `semi_join` function lets us keep the part of first table for which we have information in the second. It does not add the columns of the second:

```
semi_join(tab_1, tab_2, by = "state")
#>         state population
#> 1     Alabama    4779736
#> 2      Alaska     710231
#> 3     Arizona    6392017
#> 4  California   37253956
```

22.1.6 Anti join

The function `anti_join` is the opposite of `semi_join`. It keeps the elements of the first table for which there is no information in the second:

```
anti_join(tab_1, tab_2, by = "state")
#>        state population
#> 1 Arkansas     2915918
#> 2 Colorado     5029196
```

The following diagram summarizes the above joins:

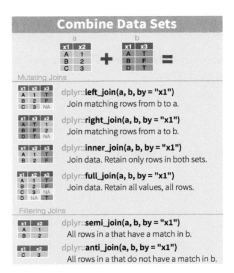

(Image courtesy of RStudio[1]. CC-BY-4.0 license[2]. Cropped from original.)

22.2   Binding

Although we have yet to use it in this book, another common way in which datasets are combined is by *binding* them. Unlike the join function, the binding functions do not try to match by a variable, but instead simply combine datasets. If the datasets don't match by the appropriate dimensions, one obtains an error.

22.2.1   Binding columns

The **dplyr** function *bind_cols* binds two objects by making them columns in a tibble. For example, we quickly want to make a data frame consisting of numbers we can use.

```
bind_cols(a = 1:3, b = 4:6)
#> # A tibble: 3 x 2
#>       a     b
#>   <int> <int>
#> 1     1     4
#> 2     2     5
#> 3     3     6
```

This function requires that we assign names to the columns. Here we chose a and b.

Note that there is an R-base function cbind with the exact same functionality. An important difference is that cbind can create different types of objects, while bind_cols always produces a data frame.

bind_cols can also bind two different data frames. For example, here we break up the tab data frame and then bind them back together:

```
tab_1 <- tab[, 1:3]
tab_2 <- tab[, 4:6]
tab_3 <- tab[, 7:8]
new_tab <- bind_cols(tab_1, tab_2, tab_3)
head(new_tab)
#>         state abb region population total ev clinton trump
#> 1     Alabama  AL  South     4779736   135  9    34.4  62.1
#> 2      Alaska  AK   West      710231    19  3    36.6  51.3
#> 3     Arizona  AZ   West     6392017   232 11    45.1  48.7
#> 4    Arkansas  AR  South     2915918    93  6    33.7  60.6
#> 5  California  CA   West    37253956  1257 55    61.7  31.6
#> 6    Colorado  CO   West     5029196    65  9    48.2  43.3
```

22.2.2   Binding by rows

The bind_rows function is similar to bind_cols, but binds rows instead of columns:

```
tab_1 <- tab[1:2,]
tab_2 <- tab[3:4,]
bind_rows(tab_1, tab_2)
#>      state abb region population total ev clinton trump
#> 1  Alabama  AL  South    4779736   135  9    34.4  62.1
#> 2   Alaska  AK   West     710231    19  3    36.6  51.3
#> 3  Arizona  AZ   West    6392017   232 11    45.1  48.7
#> 4 Arkansas  AR  South    2915918    93  6    33.7  60.6
```

This is based on an R-base function `rbind`.

22.3 Set operators

Another set of commands useful for combining datasets are the set operators. When applied to vectors, these behave as their names suggest. Examples are `intersect`, `union`, `setdiff`, and `setequal`. However, if the **tidyverse**, or more specifically **dplyr**, is loaded, these functions can be used on data frames as opposed to just on vectors.

22.3.1 Intersect

You can take intersections of vectors of any type, such as numeric:

```
intersect(1:10, 6:15)
#> [1]  6  7  8  9 10
```

or characters:

```
intersect(c("a","b","c"), c("b","c","d"))
#> [1] "b" "c"
```

The **dplyr** package includes an `intersect` function that can be applied to tables with the same column names. This function returns the rows in common between two tables. To make sure we use the **dplyr** version of `intersect` rather than the base package version, we can use `dplyr::intersect` like this:

```
tab_1 <- tab[1:5,]
tab_2 <- tab[3:7,]
dplyr::intersect(tab_1, tab_2)
#>        state abb region population total ev clinton trump
#> 1    Arizona  AZ   West    6392017   232 11    45.1  48.7
#> 2   Arkansas  AR  South    2915918    93  6    33.7  60.6
#> 3 California  CA   West   37253956  1257 55    61.7  31.6
```

22.3.2   Union

Similarly *union* takes the union of vectors. For example:

```
union(1:10, 6:15)
#>  [1]  1  2  3  4  5  6  7  8  9 10 11 12 13 14 15
union(c("a","b","c"), c("b","c","d"))
#> [1] "a" "b" "c" "d"
```

The **dplyr** package includes a version of **union** that combines all the rows of two tables with the same column names.

```
tab_1 <- tab[1:5,]
tab_2 <- tab[3:7,]
dplyr::union(tab_1, tab_2)
#>           state abb     region population total ev clinton trump
#> 1       Alabama  AL      South    4779736   135  9    34.4  62.1
#> 2        Alaska  AK       West     710231    19  3    36.6  51.3
#> 3       Arizona  AZ       West    6392017   232 11    45.1  48.7
#> 4      Arkansas  AR      South    2915918    93  6    33.7  60.6
#> 5    California  CA       West   37253956  1257 55    61.7  31.6
#> 6      Colorado  CO       West    5029196    65  9    48.2  43.3
#> 7   Connecticut  CT  Northeast    3574097    97  7    54.6  40.9
```

22.3.3   setdiff

The set difference between a first and second argument can be obtained with `setdiff`. Unlike `intersect` and `union`, this function is not symmetric:

```
setdiff(1:10, 6:15)
#> [1] 1 2 3 4 5
setdiff(6:15, 1:10)
#> [1] 11 12 13 14 15
```

As with the functions shown above, **dplyr** has a version for data frames:

```
tab_1 <- tab[1:5,]
tab_2 <- tab[3:7,]
dplyr::setdiff(tab_1, tab_2)
#>     state abb region population total ev clinton trump
#> 1 Alabama  AL  South    4779736   135  9    34.4  62.1
#> 2  Alaska  AK   West     710231    19  3    36.6  51.3
```

22.3.4   setequal

Finally, the function `setequal` tells us if two sets are the same, regardless of order. So notice that:

```
setequal(1:5, 1:6)
#> [1] FALSE
```

but:

```
setequal(1:5, 5:1)
#> [1] TRUE
```

When applied to data frames that are not equal, regardless of order, the **dplyr** version provides a useful message letting us know how the sets are different:

```
dplyr::setequal(tab_1, tab_2)
#> [1] FALSE
```

22.4   Exercises

1. Install and load the **Lahman** library. This database includes data related to baseball teams. It includes summary statistics about how the players performed on offense and defense for several years. It also includes personal information about the players.

The `Batting` data frame contains the offensive statistics for all players for many years. You can see, for example, the top 10 hitters by running this code:

```
library(Lahman)

top <- Batting %>%
  filter(yearID == 2016) %>%
  arrange(desc(HR)) %>%
  slice(1:10)

top %>% as_tibble()
```

But who are these players? We see an ID, but not the names. The player names are in this table

```
Master %>% as_tibble()
```

We can see column names `nameFirst` and `nameLast`. Use the `left_join` function to create a table of the top home run hitters. The table should have `playerID`, first name, last name, and number of home runs (HR). Rewrite the object `top` with this new table.

2. Now use the `Salaries` data frame to add each player's salary to the table you created in exercise 1. Note that salaries are different every year so make sure to filter for the year 2016, then use `right_join`. This time show first name, last name, team, HR, and salary.

3. In a previous exercise, we created a tidy version of the `co2` dataset:

```
co2_wide <- data.frame(matrix(co2, ncol = 12, byrow = TRUE)) %>%
  setNames(1:12) %>%
  mutate(year = 1959:1997) %>%
  gather(month, co2, -year, convert = TRUE)
```

We want to see if the monthly trend is changing so we are going to remove the year effects and then plot the results. We will first compute the year averages. Use the `group_by` and `summarize` to compute the average co2 for each year. Save in an object called `yearly_avg`.

4. Now use the `left_join` function to add the yearly average to the `co2_wide` dataset. Then compute the residuals: observed co2 measure - yearly average.

5. Make a plot of the seasonal trends by year but only after removing the year effect.

23

Web scraping

The data we need to answer a question is not always in a spreadsheet ready for us to read. For example, the US murders dataset we used in the R Basics chapter originally comes from this Wikipedia page:

```
url <- paste0("https://en.wikipedia.org/w/index.php?title=",
"Gun_violence_in_the_United_States_by_state",
"&direction=prev&oldid=810166167")
```

You can see the data table when you visit the webpage:

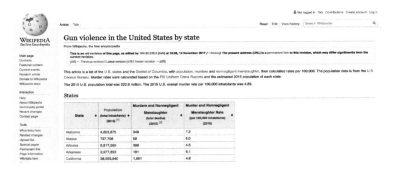

(Web page courtesy of Wikipedia[1]. CC-BY-SA-3.0 license[2]. Screenshot of part of the page.)

Unfortunately, there is no link to a data file. To make the data frame that is loaded when we type `data(murders)`, we had to do some *web scraping*.

Web scraping, or *web harvesting*, is the term we use to describe the process of extracting data from a website. The reason we can do this is because the information used by a browser to render webpages is received as a text file from a server. The text is code written in hyper text markup language (HTML). Every browser has a way to show the html source code for a page, each one different. On Chrome, you can use Control-U on a PC and command+alt+U on a Mac. You will see something like this:

---

[1]https://en.wikipedia.org/w/index.php?title=Gun_violence_in_the_United_States_by_state&direction=prev&oldid=810166167
[2]https://en.wikipedia.org/wiki/Wikipedia:Text_of_Creative_Commons_Attribution-ShareAlike_3.0_Unported_License

```
1 <!DOCTYPE html>
2 <html class="client-nojs" lang="en" dir="ltr">
3 <head>
4 <meta charset="UTF-8"/>
5 <title>Gun violence in the United States by state - Wikipedia</title>
6 <script>document.documentElement.className="client-js";RLCONF=
{"wgCanonicalNamespace":"","wgCanonicalSpecialPageName":!1,"wgNamespaceNumber":0,"wgPageName":"Gun_violence_in_the_United_States_by_state","wgTitle":"Gun violence in the United
States by state","wgCurRevisionId":919733895,"wgRevisionId":910166087,"wgArticleId":8071382,"wgIsArticle":!0,"wgIsRedirect":!1,"wgAction":"view","wgUserName":null,"wgUserGroups":
["*"],"wgCategories":["States of the United States law-related lists","Gun violence in the United
States"],"wgBreakFrames":!1,"wgPageContentLanguage":"en","wgPageContentModel":"wikitext","wgSeparatorTransformTable":["",""],"wgDigitTransformTable":
["",""],"wgDefaultDateFormat":"dmy","wgMonthNames":
["","January","February","March","April","May","June","July","August","September","October","November","December"],"wgMonthNamesShort":
["","Jan","Feb","Mar","Apr","May","Jun","Jul","Aug","Sep","Oct","Nov","Dec"],"wgRelevantPageName":"Gun_violence_in_the_United_States_by_state",
7 "wgRelevantArticleId":8071382,"wgRequestId":"XaknjwpAIDEAAC36JdMAAADP","wgCSPNonce":!1,"wgIsProbablyEditable":!0,"wgRelevantPageIsProbablyEditable":!0,"wgRestrictionEdit":
[],"wgRestrictionMove":[],"wgMediaViewerOnClick":!0,"wgMediaViewerEnabledByDefault":!0,"wgPopupsReferencePreviews":!1,"wgPopupsConflictsWithNavPopupGadget":!1,"wgVisualEditor":
{"pageLanguageCode":"en","pageLanguageDir":"ltr","pageVariantFallbacks":"en"},"wgMFDisplayWikibaseDescriptions":
{"search":!0,"nearby":!0,"watchlist":!0,"tagline":!1},"wgWMESchemaEditAttemptStepOversample":!1,"wgULSCurrentAutonym":"English","wgNoticeProject":"wikipedia","wgWikibaseItemId":"Q
5618453","wgCentralAuthMobileDomain":!1,"wgEditSubmitButtonLabelPublish":!0};RLSTATE=
{"ext.globalCssJs.user.styles":"ready","site.styles":"ready","noscript":"ready","user.styles":"ready","ext.globalCssJs.user":"ready","user":"ready","user.options":"ready","user.to
kens":"loading","ext.cite.styles":"ready",
8 "mediawiki.legacy.shared":"ready","mediawiki.legacy.commonPrint":"ready","jquery.tablesorter.styles":"ready","jquery.makeCollapsible.styles":"ready","wikibase.client.init":"ready"
,"ext.visualEditor.desktopArticleTarget.noscript":"ready","ext.uls.interlanguage":"ready","ext.wikimediaBadges":"ready","ext.3d.styles":"ready","mediawiki.skinning.interface":"rea
dy","skins.vector.styles":"ready"};RLPAGEMODULES=["ext.cite.ux-
enhancements","ext.cite.tracking","site","mediawiki.page.startup","mediawiki.page.ready","jquery.tablesorter","jquery.makeCollapsible","mediawiki.searchSuggest","ext.gadget.teahou
se","ext.gadget.ReferenceTooltips","ext.gadget.watchlist-notice","ext.gadget.DRN-wizard","ext.gadget.charinsert","ext.gadget.refToolbar","ext.gadget.extra-toolbar-
buttons","ext.gadget.switcher","ext.centralauth.centralautologin","mmv.head","mmv.bootstrap.autostart","ext.popups","ext.visualEditor.desktopArticleTarget.init","ext.visualEditor.
targetLoader","ext.eventLogging","ext.wikimediaEvents",
9 "ext.navigationTiming","ext.uls.compactlinks","ext.uls.interface","ext.cx.eventlogging.campaigns","ext.quicksurveys.init","ext.centralNotice.geoIP","ext.centralNotice.startUp","sk
ins.vector.js"];</script>
10 <script>(RLQ=window.RLQ||[]).push(function(){mw.loader.implement("user.tokens@tffin",function($,jQuery,require,module)
{/*@nomin*/mw.user.tokens.set({"patrolToken":"+\\","watchToken":"+\\","csrfToken":"+\\"});
11 });});</script>
12 <link rel="stylesheet" href="/w/load.php?
lang=en&amp;modules=ext.3d.styles%7Cext.cite.styles%7Cext.uls.interlanguage%7Cext.visualEditor.desktopArticleTarget.noscript%7Cext.wikimediaBadges%7Cjquery.makeCollapsible.styles%
7C!jquery.tablesorter.styles%7Cmediawiki.legacy.commonPrint%2Cshared%7Cmediawiki.skinning.interface%7Cskins.vector.styles%7Cwikibase.client.init&amp;only=styles&amp;skin=vector"/>
13 <script async="" src="/w/load.php?lang=en&amp;modules=startup&amp;only=scripts&amp;raw=1&amp;skin=vector"></script>
14 <meta name="ResourceLoaderDynamicStyles" content=""/>
15 <link rel="stylesheet" href="/w/load.php?lang=en&amp;modules=site.styles&amp;only=styles&amp;skin=vector"/>
16 <meta name="generator" content="MediaWiki 1.35.0-wmf.1"/>
17 <meta name="referrer" content="origin"/>
18 <meta name="referrer" content="origin-when-crossorigin"/>
19 <meta name="referrer" content="origin-when-cross-origin"/>
20 <meta name="robots" content="noindex,nofollow"/>
21 <meta property="og:image" content="https://upload.wikimedia.org/wikipedia/commons/thumb/4/4d/Gun_Ownership_Related_to_Gun_Violence_by_State_%28United_States%29.svg/1200px-
Gun_Ownership_Related_to_Gun_Violence_by_State_%28United_States%29.svg.png"/>
22 <link rel="alternate" type="application/x-wiki" title="Edit this page" href="/w/index.php?title=Gun_violence_in_the_United_States_by_state&amp;action=edit"/>
23 <link rel="edit" title="Edit this page" href="/w/index.php?title=Gun_violence_in_the_United_States_by_state&amp;action=edit"/>
24 <link rel="apple-touch-icon" href="/static/apple-touch/wikipedia.png"/>
25 <link rel="shortcut icon" href="/static/favicon/wikipedia.ico"/>
26 <link rel="search" type="application/opensearchdescription+xml" href="/w/opensearch_desc.php" title="Wikipedia (en)"/>
27 <link rel="EditURI" type="application/rsd+xml" href="//en.wikipedia.org/w/api.php?action=rsd"/>
28 <link rel="license" href="//creativecommons.org/licenses/by-sa/3.0/"/>
29 <link rel="alternate" type="application/atom+xml" title="Wikipedia Atom feed" href="/w/index.php?title=Special:RecentChanges&amp;feed=atom"/>
30 <link rel="canonical" href="https://en.wikipedia.org/wiki/Gun_violence_in_the_United_States_by_state"/>
```

23.1 HTML

Because this code is accessible, we can download the HTML file, import it into R, and then write programs to extract the information we need from the page. However, once we look at HTML code, this might seem like a daunting task. But we will show you some convenient tools to facilitate the process. To get an idea of how it works, here are a few lines of code from the Wikipedia page that provides the US murders data:

```
<table class="wikitable sortable">
<tr>
<th>State</th>
<th><a href="/wiki/List_of_U.S._states_and_territories_by_population"
title="List of U.S. states and territories by population">Population</a><br />
<small>(total inhabitants)</small><br />
<small>(2015)</small> <sup id="cite_ref-1" class="reference">
<a href="#cite_note-1">[1]</a></sup></th>
<th>Murders and Nonnegligent
<p>Manslaughter<br />
<small>(total deaths)</small><br />
<small>(2015)</small> <sup id="cite_ref-2" class="reference">
<a href="#cite_note-2">[2]</a></sup></p>
</th>
<th>Murder and Nonnegligent
<p>Manslaughter Rate<br />
<small>(per 100,000 inhabitants)</small><br />
<small>(2015)</small></p>
</th>
</tr>
```

```
<tr>
<td><a href="/wiki/Alabama" title="Alabama">Alabama</a></td>
<td>4,853,875</td>
<td>348</td>
<td>7.2</td>
</tr>
<tr>
<td><a href="/wiki/Alaska" title="Alaska">Alaska</a></td>
<td>737,709</td>
<td>59</td>
<td>8.0</td>
</tr>
<tr>
```

You can actually see the data, except data values are surrounded by html code such as `<td>`. We can also see a pattern of how it is stored. If you know HTML, you can write programs that leverage knowledge of these patterns to extract what we want. We also take advantage of a language widely used to make webpages look "pretty" called Cascading Style Sheets (CSS). We say more about this in Section 23.3.

Although we provide tools that make it possible to scrape data without knowing HTML, as a data scientist it is quite useful to learn some HTML and CSS. Not only does this improve your scraping skills, but it might come in handy if you are creating a webpage to showcase your work. There are plenty of online courses and tutorials for learning these. Two examples are Codeacademy[3] and W3schools[4].

23.2 The rvest package

The **tidyverse** provides a web harvesting package called **rvest**. The first step using this package is to import the webpage into R. The package makes this quite simple:

```
library(tidyverse)
library(rvest)
h <- read_html(url)
```

Note that the entire Murders in the US Wikipedia webpage is now contained in **h**. The class of this object is:

```
class(h)
#> [1] "xml_document" "xml_node"
```

The **rvest** package is actually more general; it handles XML documents. XML is a general markup language (that's what the ML stands for) that can be used to represent any kind

---

[3]https://www.codecademy.com/learn/learn-html
[4]https://www.w3schools.com/

of data. HTML is a specific type of XML specifically developed for representing webpages. Here we focus on HTML documents.

Now, how do we extract the table from the object h? If we print h, we don't really see much:

```
h
#> {xml_document}
#> <html class="client-nojs" lang="en" dir="ltr">
#> [1] <head>\n<meta http-equiv="Content-Type" content="text/html; chars ...
#> [2] <body class="mediawiki ltr sitedir-ltr mw-hide-empty-elt ns-0 ns- ...
```

We can see all the code that defines the downloaded webpage using the `html_text` function like this:

```
html_text(h)
```

We don't show the output here because it includes thousands of characters, but if we look at it, we can see the data we are after are stored in an HTML table: you can see this in this line of the HTML code above `<table class="wikitable sortable">`. The different parts of an HTML document, often defined with a message in between < and > are referred to as *nodes*. The **rvest** package includes functions to extract nodes of an HTML document: `html_nodes` extracts all nodes of different types and `html_node` extracts the first one. To extract the tables from the html code we use:

```
tab <- h %>% html_nodes("table")
```

Now, instead of the entire webpage, we just have the html code for the tables in the page:

```
tab
#> {xml_nodeset (2)}
#> [1] <table class="wikitable sortable"><tbody>\n<tr>\n<th>State\n</th>...
#> [2] <table class="nowraplinks hlist mw-collapsible mw-collapsed navbo...
```

The table we are interested is the first one:

```
tab[[1]]
#> {xml_node}
#> <table class="wikitable sortable">
#> [1] <tbody>\n<tr>\n<th>State\n</th>\n<th>\n<a href="/wiki/List_of_U.S...
```

This is clearly not a tidy dataset, not even a data frame. In the code above, you can definitely see a pattern and writing code to extract just the data is very doable. In fact, **rvest** includes a function just for converting HTML tables into data frames:

```
tab <- tab[[1]] %>% html_table
class(tab)
#> [1] "data.frame"
```

We are now much closer to having a usable data table:

```
tab <- tab %>% setNames(c("state", "population", "total", "murder_rate"))
head(tab)
#> state population total murder_rate
#> 1 Alabama 4,853,875 348 7.2
#> 2 Alaska 737,709 59 8.0
#> 3 Arizona 6,817,565 309 4.5
#> 4 Arkansas 2,977,853 181 6.1
#> 5 California 38,993,940 1,861 4.8
#> 6 Colorado 5,448,819 176 3.2
```

We still have some wrangling to do. For example, we need to remove the commas and turn characters into numbers. Before continuing with this, we will learn a more general approach to extracting information from web sites.

23.3 CSS selectors

The default look of a webpage made with the most basic HTML is quite unattractive. The aesthetically pleasing pages we see today are made using CSS to define the look and style of webpages. The fact that all pages for a company have the same style usually results from their use of the same CSS file to define the style. The general way these CSS files work is by defining how each of the elements of a webpage will look. The title, headings, itemized lists, tables, and links, for example, each receive their own style including font, color, size, and distance from the margin. CSS does this by leveraging patterns used to define these elements, referred to as *selectors*. An example of such a pattern, which we used above, is `table`, but there are many, many more.

If we want to grab data from a webpage and we happen to know a selector that is unique to the part of the page containing this data, we can use the `html_nodes` function. However, knowing which selector can be quite complicated. In fact, the complexity of webpages has been increasing as they become more sophisticated. For some of the more advanced ones, it seems almost impossible to find the nodes that define a particular piece of data. However, selector gadgets actually make this possible.

SelectorGadget[5] is piece of software that allows you to interactively determine what CSS selector you need to extract specific components from the webpage. If you plan on scraping data other than tables from html pages, we highly recommend you install it. A Chrome extension is available which permits you to turn on the gadget and then, as you click through the page, it highlights parts and shows you the selector you need to extract these parts. There are various demos of how to do this including **rvest** author Hadley Wickham's vignette[6] and other tutorials based on the vignette[7] [8].

---

[5] http://selectorgadget.com/

[6] https://cran.r-project.org/web/packages/rvest/vignettes/selectorgadget.html

[7] https://stat4701.github.io/edav/2015/04/02/rvest_tutorial/

[8] https://www.analyticsvidhya.com/blog/2017/03/beginners-guide-on-web-scraping-in-r-using-rvest-with-hands-on-knowledge/

23.4 JSON

Sharing data on the internet has become more and more common. Unfortunately, providers use different formats, which makes it harder for data scientists to wrangle data into R. Yet there are some standards that are also becoming more common. Currently, a format that is widely being adopted is the JavaScript Object Notation or JSON. Because this format is very general, it is nothing like a spreadsheet. This JSON file looks more like the code you use to define a list. Here is an example of information stored in a JSON format:

```
#>
#> Attaching package: 'jsonlite'
#> The following object is masked from 'package:purrr':
#>
#>     flatten
#> [
#>   {
#>     "name": "Miguel",
#>     "student_id": 1,
#>     "exam_1": 85,
#>     "exam_2": 86
#>   },
#>   {
#>     "name": "Sofia",
#>     "student_id": 2,
#>     "exam_1": 94,
#>     "exam_2": 93
#>   },
#>   {
#>     "name": "Aya",
#>     "student_id": 3,
#>     "exam_1": 87,
#>     "exam_2": 88
#>   },
#>   {
#>     "name": "Cheng",
#>     "student_id": 4,
#>     "exam_1": 90,
#>     "exam_2": 91
#>   }
#> ]
```

The file above actually represents a data frame. To read it, we can use the function `fromJSON` from the **jsonlite** package. Note that JSON files are often made available via the internet. Several organizations provide a JSON API or a web service that you can connect directly to and obtain data. Here is an example:

```
library(jsonlite)
citi_bike <- fromJSON("http://citibikenyc.com/stations/json")
```

This downloads a list. The first argument tells you when you downloaded it:

```
citi_bike$executionTime
#> [1] "2019-09-16 04:54:16 PM"
```

and the second is a data table:

```
citi_bike$stationBeanList %>% as_tibble()
#> # A tibble: 844 x 18
#>        id stationName availableDocks totalDocks latitude longitude
#>     <int> <chr>                <int>      <int>    <dbl>     <dbl>
#> 1     281 Grand Army~             36         66     40.8     -74.0
#> 2     285 Broadway &~             51         53     40.7     -74.0
#> 3     304 Broadway &~             22         33     40.7     -74.0
#> 4     305 E 58 St & ~             17         33     40.8     -74.0
#> 5     337 Old Slip &~             29         37     40.7     -74.0
#> # ... with 839 more rows, and 12 more variables: statusValue <chr>,
#> #   statusKey <int>, availableBikes <int>, stAddress1 <chr>,
#> #   stAddress2 <chr>, city <chr>, postalCode <chr>, location <chr>,
#> #   altitude <chr>, testStation <lgl>, lastCommunicationTime <chr>,
#> #   landMark <chr>
```

You can learn much more by examining tutorials and help files from the **jsonlite** package. This package is intended for relatively simple tasks such as converging data into tables. For more flexibility, we recommend `rjson`.

23.5 Exercises

1. Visit the following web page: https://web.archive.org/web/20181024132313/http://www.stevetheump.com/Payrolls.htm

Notice there are several tables. Say we are interested in comparing the payrolls of teams across the years. The next few exercises take us through the steps needed to do this.

Start by applying what you learned to read in the website into an object called h.

2. Note that, although not very useful, we can actually see the content of the page by typing:

```
html_text(h)
```

The next step is to extract the tables. For this, we can use the `html_nodes` function. We learned that tables in html are associated with the `table` node. Use the `html_nodes` function and the `table` node to extract the first table. Store it in an object `nodes`.

3. The `html_nodes` function returns a list of objects of class `xml_node`. We can see the content of each one using, for example, the `html_text` function. You can see the content for an arbitrarily picked component like this:

```
html_text(nodes[[8]])
```

If the content of this object is an html table, we can use the `html_table` function to convert it to a data frame. Use the `html_table` function to convert the 8th entry of `nodes` into a table.

4. Repeat the above for the first 4 components of `nodes`. Which of the following are payroll tables:

    a. All of them.
    b. 1
    c. 2
    d. 2-4

5. Repeat the above for the first **last** 3 components of `nodes`. Which of the following is true:

    a. The last entry in `nodes` shows the average across all teams through time, not payroll per team.
    b. All three are payroll per team tables.
    c. All three are like the first entry, not a payroll table.
    d. All of the above.

6. We have learned that the first and last entries of `nodes` are not payroll tables. Redefine `nodes` so that these two are removed.

7. We saw in the previous analysis that the first table node is not actually a table. This happens sometimes in html because tables are used to make text look a certain way, as opposed to storing numeric values. Remove the first component and then use `sapply` and `html_table` to convert each node in `nodes` into a table. Note that in this case, `sapply` will return a list of tables. You can also use `lapply` to assure that a list is applied.

8. Look through the resulting tables. Are they all the same? Could we just join them with `bind_rows`?

9. Create two tables, call them `tab_1` and `tab_2` using entries 10 and 19.

10. Use a `full_join` function to combine these two tables. Before you do this you will have to fix the missing header problem. You will also need to make the names match.

11. After joining the tables, you see several NAs. This is because some teams are in one table and not the other. Use the `anti_join` function to get a better idea of why this is happening.

12. We see see that one of the problems is that Yankees are listed as both *N.Y. Yankees* and *NY Yankees*. In the next section, we will learn efficient approaches to fixing problems like this. Here we can do it "by hand" as follows:

```
tab_1 <- tab_1 %>%
  mutate(Team = ifelse(Team == "N.Y. Yankees", "NY Yankees", Team))
```

Now join the tables and show only Oakland and the Yankees and the payroll columns.

13. Advanced: extract the titles of the movies that won Best Picture from this website: https://m.imdb.com/chart/bestpicture/

24

String processing

One of the most common data wrangling challenges involves extracting numeric data contained in character strings and converting them into the numeric representations required to make plots, compute summaries, or fit models in R. Also common is processing unorganized text into meaningful variable names or categorical variables. Many of the string processing challenges a data scientist faces are unique and often unexpected. It is therefore quite ambitious to write a comprehensive section on this topic. Here we use a series of case studies that help us demonstrate how string processing is a necessary step for many data wrangling challenges. Specifically, we describe the process of converting the not yet shown original *raw* data from which we extracted the `murders`, `heights`, and `research_funding_rates` example into the data frames we have studied in this book.

By going over these case studies, we will cover some of the most common tasks in string processing including extracting numbers from strings, removing unwanted characters from text, finding and replacing characters, extracting specific parts of strings, converting free form text to more uniform formats, and splitting strings into multiple values.

Base R includes functions to perform all these tasks. However, they don't follow a unifying convention, which makes them a bit hard to memorize and use. The **stringr** package basically repackages this functionality, but uses a more consistent approach of naming functions and ordering their arguments. For example, in **stringr**, all the string processing functions start with `str_`. This means that if you type `str_` and hit tab, R will auto-complete and show all the available functions. As a result, we don't necessarily have to memorize all the function names. Another advantage is that in the functions in this package the string being processed is always the first argument, which means we can more easily use the pipe. Therefore, we will start by describing how to use the functions in the **stringr** package.

Most of the examples will come from the second case study which deals with self-reported heights by students and most of the chapter is dedicated to learning regular expressions (regex), and functions in the **stringr** package.

24.1 The stringr package

```
library(tidyverse)
library(stringr)
```

In general, string processing tasks can be divided into **detecting**, **locating**, **extracting**, or **replacing** patterns in strings. We will see several examples. The table below includes the

functions available to you in the **stringr** package. We split them by task. We also include
the R-base equivalent when available.

All these functions take a character vector as first argument. Also, for each function,
operations are vectorized: the operation gets applied to each string in the vector.

Finally, note that in this table we mention *groups*. These will be explained in Section 24.5.9.

stringr	Task	Description	R-base
str_detect	Detect	Is the pattern in the string?	grepl
str_which	Detect	Returns the index of entries that contain the pattern.	grep
str_subset	Detect	Returns the subset of strings that contain the pattern.	grep with value = TRUE
str_locate	Locate	Returns positions of first occurrence of pattern in a string.	regexpr
str_locate_all	Locate	Returns position of all occurrences of pattern in a string.	gregexpr
str_view	Locate	Show the first part of the string that matches pattern.	
str_view_all	Locate	Show me all the parts of the string that match the pattern.	
str_extract	Extract	Extract the first part of the string that matches the pattern.	
str_extract_all	Extract	Extract all parts of the string that match the pattern.	
str_match	Extract	Extract first part of the string that matches the groups and the patterns defined by the groups.	
str_match_all	Extract	Extract all parts of the string that matches the groups and the patterns defined by the groups.	
str_sub	Extract	Extract a substring.	substring
str_split	Extract	Split a string into a list with parts separated by pattern.	strsplit
str_split_fixed	Extract	Split a string into a matrix with parts separated by pattern.	strsplit with fixed = TRUE
str_count	Describe	Count number of times a pattern appears in a string.	
str_length	Describe	Number of character in string.	nchar
str_replace	Replace	Replace first part of a string matching a pattern with another.	
str_replace_all	Replace	Replace all parts of a string matching a pattern with another.	gsub
str_to_upper	Replace	Change all characters to upper case.	toupper
str_to_lower	Replace	Change all characters to lower case.	tolower
str_to_title	Replace	Change first character to upper and rest to lower.	
str_replace_na	Replace	Replace all NAs to a new value.	

stringr	Task	Description	R-base
str_trim	Replace	Remove white space from start and end of string.	
str_c	Manipulate	Join multiple strings.	paste0
str_conv	Manipulate	Change the encoding of the string.	
str_sort	Manipulate	Sort the vector in alphabetical order.	sort
str_order	Manipulate	Index needed to order the vector in alphabetical order.	order
str_trunc	Manipulate	Truncate a string to a fixed size.	
str_pad	Manipulate	Add white space to string to make it a fixed size.	
str_dup	Manipulate	Repeat a string.	rep then paste
str_wrap	Manipulate	Wrap things into formatted paragraphs.	
str_interp	Manipulate	String interpolation.	sprintf

24.2 Case study 1: US murders data

In this section we introduce some of the more simple string processing challenges with the following datasets as an example:

```
library(rvest)
url <- paste0("https://en.wikipedia.org/w/index.php?title=",
              "Gun_violence_in_the_United_States_by_state",
              "&direction=prev&oldid=810166167")
murders_raw <- read_html(url) %>%
  html_node("table") %>%
  html_table() %>%
  setNames(c("state", "population", "total", "murder_rate"))
```

The code above shows the first step in constructing the dataset

```
library(dslabs)
data(murders)
```

from the raw data, which was extracted from a Wikipedia page.

In general, string processing involves a string and a pattern. In R, we usually store strings in a character vector such as murders$population. The first three strings in this vector defined by the population variable are:

```
murders_raw$population[1:3]
#> [1] "4,853,875" "737,709"   "6,817,565"
```

The usual coercion does not work here:

```
as.numeric(murders_raw$population[1:3])
#> Warning: NAs introduced by coercion
#> [1] NA NA NA
```

This is because of the commas ,. The string processing we want to do here is remove the **pattern**, ,, from the **strings** in murders_raw$population and then coerce to numbers. We can use the str_detect function to see that two of the three columns have commas in the entries:

```
commas <- function(x) any(str_detect(x, ","))
murders_raw %>% summarize_all(commas)
#>   state population total murder_rate
#> 1 FALSE       TRUE  TRUE       FALSE
```

We can then use the str_replace_all function to remove them:

```
test_1 <- str_replace_all(murders_raw$population, ",", "")
test_1 <- as.numeric(test_1)
```

We can then use mutate_all to apply this operation to each column, since it won't affect the columns without commas.

It turns out that this operation is so common that readr includes the function parse_number specifically meant to remove non-numeric characters before coercing:

```
test_2 <- parse_number(murders_raw$population)
identical(test_1, test_2)
#> [1] TRUE
```

So we can obtain our desired table using:

```
murders_new <- murders_raw %>% mutate_at(2:3, parse_number)
head(murders_new)
#>        state population total murder_rate
#> 1    Alabama    4853875   348         7.2
#> 2     Alaska     737709    59         8.0
#> 3    Arizona    6817565   309         4.5
#> 4   Arkansas    2977853   181         6.1
#> 5 California   38993940  1861         4.8
#> 6   Colorado    5448819   176         3.2
```

This case is relatively simple compared to the string processing challenges that we typically face in data science. The next example is a rather complex one and it provides several challenges that will permit us to learn many string processing techniques.

24.3   Case study 2: self-reported heights

The **dslabs** package includes the raw data from which the heights dataset was obtained. You can load it like this:

```
data(reported_heights)
```

These heights were obtained using a web form in which students were asked to enter their heights. They could enter anything, but the instructions asked for *height in inches*, a number. We compiled 1,095 submissions, but unfortunately the column vector with the reported heights had several non-numeric entries and as a result became a character vector:

```
class(reported_heights$height)
#> [1] "character"
```

If we try to parse it into numbers, we get a warning:

```
x <- as.numeric(reported_heights$height)
#> Warning: NAs introduced by coercion
```

Although most values appear to be height in inches as requested:

```
head(x)
#> [1] 75 70 68 74 61 65
```

we do end up with many NAs:

```
sum(is.na(x))
#> [1] 81
```

We can see some of the entries that are not successfully converted by using `filter` to keep only the entries resulting in NAs:

```
reported_heights %>%
  mutate(new_height = as.numeric(height)) %>%
  filter(is.na(new_height)) %>%
  head(n=10)
#>               time_stamp    sex                     height new_height
#> 1  2014-09-02 15:16:28    Male                        5' 4"         NA
#> 2  2014-09-02 15:16:37 Female                        165cm         NA
#> 3  2014-09-02 15:16:52    Male                         5'7         NA
#> 4  2014-09-02 15:16:56    Male                        >9000         NA
#> 5  2014-09-02 15:16:56    Male                        5'7"         NA
#> 6  2014-09-02 15:17:09 Female                        5'3"         NA
#> 7  2014-09-02 15:18:00    Male 5 feet and 8.11 inches         NA
#> 8  2014-09-02 15:19:48    Male                        5'11         NA
#> 9  2014-09-04 00:46:45    Male                        5'9''         NA
#> 10 2014-09-04 10:29:44    Male                       5'10''         NA
```

We immediately see what is happening. Some of the students did not report their heights in inches as requested. We could discard these data and continue. However, many of the entries follow patterns that, in principle, we can easily convert to inches. For example, in the output above, we see various cases that use the format x'y'' with x and y representing feet and inches, respectively. Each one of these cases can be read and converted to inches by a human, for example 5'4'' is 5*12 + 4 = 64. So we could fix all the problematic entries *by hand*. However, humans are prone to making mistakes, so an automated approach is preferable. Also, because we plan on continuing to collect data, it will be convenient to write code that automatically does this.

A first step in this type of task is to survey the problematic entries and try to define specific patterns followed by a large groups of entries. The larger these groups, the more entries we can fix with a single programmatic approach. We want to find patterns that can be accurately described with a rule, such as "a digit, followed by a feet symbol, followed by one or two digits, followed by an inches symbol".

To look for such patterns, it helps to remove the entries that are consistent with being in inches and to view only the problematic entries. We thus write a function to automatically do this. We keep entries that either result in NAs when applying as.numeric or are outside a range of plausible heights. We permit a range that covers about 99.9999% of the adult population. We also use suppressWarnings to avoid the warning message we know as.numeric will gives us.

```
not_inches <- function(x, smallest = 50, tallest = 84){
  inches <- suppressWarnings(as.numeric(x))
  ind <- is.na(inches) | inches < smallest | inches > tallest
  ind
}
```

We apply this function and find the number of problematic entries:

```
problems <- reported_heights %>%
  filter(not_inches(height)) %>%
  pull(height)
length(problems)
#> [1] 292
```

We can now view all the cases by simply printing them. We don't do that here because there are length(problems), but after surveying them carefully, we see that three patterns can be used to define three large groups within these exceptions.

1. A pattern of the form x'y or x' y'' or x'y" with x and y representing feet and inches, respectively. Here are ten examples:

```
#> 5' 4" 5'7 5'7" 5'3" 5'11 5'9'' 5'10'' 5' 10 5'5" 5'2"
```

2. A pattern of the form x.y or x,y with x feet and y inches. Here are ten examples:

```
#> 5.3 5.5 6.5 5.8 5.6 5,3 5.9 6,8 5.5 6.2
```

3. Entries that were reported in centimeters rather than inches. Here are ten examples:

```
#> 150 175 177 178 163 175 178 165 165 180
```

Once we see these large groups following specific patterns, we can develop a plan of attack. Remember that there is rarely just one way to perform these tasks. Here we pick one that helps us teach several useful techniques. But surely there is a more efficient way of performing the task.

Plan of attack: we will convert entries fitting the first two patterns into a standardized one. We will then leverage the standardization to extract the feet and inches and convert to inches. We will then define a procedure for identifying entries that are in centimeters and convert them to inches. After applying these steps, we will then check again to see what entries were not fixed and see if we can tweak our approach to be more comprehensive.

At the end, we hope to have a script that makes web-based data collection methods robust to the most common user mistakes.

To achieve our goal, we will use a technique that enables us to accurately detect patterns and extract the parts we want: *regular expressions* (regex). But first, we quickly describe how to *escape* the function of certain characters so that they can be included in strings.

24.4 How to *escape* when defining strings

To define strings in R, we can use either double quotes:

```
s <- "Hello!"
```

or single quotes:

```
s <- 'Hello!'
```

Make sure you choose the correct single quote since using the back quote will give you an error:

```
s <- `Hello`
```

```
Error: object 'Hello' not found
```

Now, what happens if the string we want to define includes double quotes? For example, if we want to write 10 inches like this 10"? In this case you can't use:

```
s <- "10""
```

because this is just the string 10 followed by a double quote. If you type this into R, you get an error because you have an *unclosed* double quote. To avoid this, we can use the single quotes:

```
s <- '10"'
```

If we print out s we see that the double quotes are *escaped* with the backslash \.

```
s
#> [1] "10\""
```

In fact, escaping with the backslash provides a way to define the string while still using the double quotes to define strings:

```
s <- "10\""
```

In R, the function `cat` lets us see what the string actually looks like:

```
cat(s)
#> 10"
```

Now, what if we want our string to be 5 feet written like this 5'? In this case, we can use the double quotes:

```
s <- "5'"
cat(s)
#> 5'
```

So we've learned how to write 5 feet and 10 inches separately, but what if we want to write them together to represent *5 feet and 10 inches* like this 5'10"? In this case, neither the single nor double quotes will work. This:

```
s <- '5'10"'
```

closes the string after 5 and this:

```
s <- "5'10""
```

closes the string after 10. Keep in mind that if we type one of the above code snippets into R, it will get stuck waiting for you to close the open quote and you will have to exit the execution with the *esc* button.

In this situation, we need to escape the function of the quotes with the backslash \. You can escape either character like this:

```
s <- '5\'10"'
cat(s)
#> 5'10"
```

or like this:

```
s <- "5'10\""
cat(s)
#> 5'10"
```

Escaping characters is something we often have to use when processing strings.

24.5 Regular expressions

A regular expression (regex) is a way to describe specific patterns of characters of text. They can be used to determine if a given string matches the pattern. A set of rules has been defined to do this efficiently and precisely and here we show some examples. We can learn more about these rules by reading a detailed tutorials[1] [2]. This RStudio cheat sheet[3] is also very useful.

The patterns supplied to the **stringr** functions can be a regex rather than a standard string. We will learn how this works through a series of examples.

Throughout this section you will see that we create strings to test out our regex. To do this, we define patterns that we know should match and also patterns that we know should not. We will call them `yes` and `no`, respectively. This permits us to check for the two types of errors: failing to match and incorrectly matching.

24.5.1 Strings are a regexp

Technically any string is a regex, perhaps the simplest example is a single character. So the comma , used in the next code example is a simple example of searching with regex.

```
pattern <- ","
str_detect(murders_raw$total, pattern)
```

We suppress the output which is logical vector telling us which entries have commas.

Above, we noted that an entry included a `cm`. This is also a simple example of a regex. We can show all the entries that used `cm` like this:

```
str_subset(reported_heights$height, "cm")
#> [1] "165cm"  "170 cm"
```

24.5.2 Special characters

Now let's consider a slightly more complicated example. Which of the following strings contain the pattern `cm` or `inches`?

---

[1] https://www.regular-expressions.info/tutorial.html
[2] http://r4ds.had.co.nz/strings.html#matching-patterns-with-regular-expressions
[3] https://www.rstudio.com/wp-content/uploads/2016/09/RegExCheatsheet.pdf

```
yes <- c("180 cm", "70 inches")
no <- c("180", "70''")
s <- c(yes, no)
```

```
str_detect(s, "cm") | str_detect(s, "inches")
#> [1]   TRUE   TRUE FALSE FALSE
```

However, we don't need to do this. The main feature that distinguishes the regex *language* from plain strings is that we can use special characters. These are characters with a meaning. We start by introducing | which means *or*. So if we want to know if either cm or inches appears in the strings, we can use the regex cm|inches:

```
str_detect(s, "cm|inches")
#> [1]   TRUE   TRUE FALSE FALSE
```

and obtain the correct answer.

Another special character that will be useful for identifying feet and inches values is \d which means any digit: 0, 1, 2, 3, 4, 5, 6, 7, 8, 9. The backslash is used to distinguish it from the character d. In R, we have to *escape* the backslash \ so we actually have to use \\d to represent digits. Here is an example:

```
yes <- c("5", "6", "5'10", "5 feet", "4'11")
no <- c("", ".", "Five", "six")
s <- c(yes, no)
pattern <- "\\d"
str_detect(s, pattern)
#> [1]   TRUE   TRUE   TRUE   TRUE   TRUE FALSE FALSE FALSE FALSE
```

We take this opportunity to introduce the **str_view** function, which is helpful for troubleshooting as it shows us the first match for each string:

```
str_view(s, pattern)
```

> 5
>
> 6
>
> 5'10
>
> 5 feet
>
> 4'11
>
> .
>
> Five
>
> six

and str_view_all shows us all the matches, so 3'2 has two matches and 5'10 has three.

```
str_view_all(s, pattern)
```

```
5

6

5'10

5 feet

4'11

.

Five

six
```

There are many other special characters. We will learn some others below, but you can see most or all of them in the cheat sheet[4] mentioned earlier.

24.5.3 Character classes

Character classes are used to define a series of characters that can be matched. We define character classes with square brackets []. So, for example, if we want the pattern to match only if we have a 5 or a 6, we use the regex [56]:

```
str_view(s, "[56]")
```

```
5

6

5'10

5 feet

4'11

.

Five

six
```

Suppose we want to match values between 4 and 7. A common way to define character classes is with ranges. So, for example, [0-9] is equivalent to \\d. The pattern we want is therefore [4-7].

```
yes <- as.character(4:7)
no <- as.character(1:3)
s <- c(yes, no)
str_detect(s, "[4-7]")
#> [1]  TRUE  TRUE  TRUE  TRUE FALSE FALSE FALSE
```

However, it is important to know that in regex everything is a character; there are no numbers. So 4 is the character 4 not the number four. Notice, for example, that [1-20]

---

[4]https://www.rstudio.com/wp-content/uploads/2016/09/RegExCheatsheet.pdf

does **not** mean 1 through 20, it means the characters 1 through 2 or the character 0. So [1-20] simply means the character class composed of 0, 1, and 2.

Keep in mind that characters do have an order and the digits do follow the numeric order. So 0 comes before 1 which comes before 2 and so on. For the same reason, we can define lower case letters as [a-z], upper case letters as [A-Z], and [a-zA-z] as both.

24.5.4   Anchors

What if we want a match when we have exactly 1 digit? This will be useful in our case study since feet are never more than 1 digit so a restriction will help us. One way to do this with regex is by using *anchors*, which let us define patterns that must start or end at a specific place. The two most common anchors are ^ and $ which represent the beginning and end of a string, respectively. So the pattern ^\\d$ is read as "start of the string followed by one digit followed by end of string".

This pattern now only detects the strings with exactly one digit:

```
pattern <- "^\\d$"
yes <- c("1", "5", "9")
no <- c("12", "123", " 1", "a4", "b")
s <- c(yes, no)
str_view_all(s, pattern)
```

| 1 |
| 5 |
| 9 |
| 12 |
| 123 |
| 1 |
| a4 |
| b |

The 1 does not match because it does not start with the digit but rather with a space, which is actually not easy to see.

24.5.5   Quantifiers

For the inches part, we can have one or two digits. This can be specified in regex with *quantifiers*. This is done by following the pattern with curly brackets containing the number of times the previous entry can be repeated. We use an example to illustrate. The pattern for one or two digits is:

```
pattern <- "^\\d{1,2}$"
yes <- c("1", "5", "9", "12")
no <- c("123", "a4", "b")
str_view(c(yes, no), pattern)
```

```
1

5

9

12

123

a4

b
```

In this case, 123 does **not** match, but 12 does. So to look for our feet and inches pattern, we can add the symbols for feet ' and inches " after the digits.

With what we have learned, we can now construct an example for the pattern x'y\" with x feet and y inches.

```
pattern <- "^[4-7]'\\d{1,2}\"$"
```

The pattern is now getting complex, but you can look at it carefully and break it down:

- ^ = start of the string
- [4-7] = one digit, either 4,5,6 or 7
- ' = feet symbol
- \\d{1,2} = one or two digits
- \" = inches symbol
- $ = end of the string

Let's test it out:

```
yes <- c("5'7\"",  "6'2\"",   "5'12\"")
no <- c("6,2\"",  "6.2\"","I am 5'11\"", "3'2\"", "64")
str_detect(yes, pattern)
#> [1] TRUE TRUE TRUE
str_detect(no, pattern)
#> [1] FALSE FALSE FALSE FALSE FALSE
```

For now, we are permitting the inches to be 12 or larger. We will add a restriction later as the regex for this is a bit more complex than we are ready to show.

24.5.6 White space \s

Another problem we have are spaces. For example, our pattern does not match 5' 4" because there is a space between ' and 4 which our pattern does not permit. Spaces are characters and R does not ignore them:

```
identical("Hi", "Hi ")
#> [1] FALSE
```

In regex, \s represents white space. To find patterns like 5' 4, we can change our pattern to:

```
pattern_2 <- "^[4-7]'\\s\\d{1,2}\"$"
str_subset(problems, pattern_2)
#> [1] "5' 4\""  "5' 11\""  "5' 7\""
```

However, this will not match the patterns with no space. So do we need more than one regex pattern? It turns out we can use a quantifier for this as well.

24.5.7   Quantifiers: *, ?, +

We want the pattern to permit spaces but not require them. Even if there are several spaces, like in this example 5'   4, we still want it to match. There is a quantifier for exactly this purpose. In regex, the character * means zero or more instances of the previous character. Here is an example:

```
yes <- c("AB", "A1B", "A11B", "A111B", "A1111B")
no <- c("A2B", "A21B")
str_detect(yes, "A1*B")
#> [1] TRUE TRUE TRUE TRUE TRUE
str_detect(no, "A1*B")
#> [1] FALSE FALSE
```

The above matches the first string which has zero 1s and all the strings with one or more 1. We can then improve our pattern by adding the * after the space character \s.

There are two other similar quantifiers. For none or once, we can use ?, and for one or more, we can use +. You can see how they differ with this example:

```
data.frame(string = c("AB", "A1B", "A11B", "A111B", "A1111B"),
           none_or_more = str_detect(yes, "A1*B"),
           nore_or_once = str_detect(yes, "A1?B"),
           once_or_more = str_detect(yes, "A1+B"))
#>    string none_or_more nore_or_once once_or_more
#> 1      AB        TRUE        TRUE       FALSE
#> 2     A1B        TRUE        TRUE        TRUE
#> 3    A11B        TRUE       FALSE        TRUE
#> 4   A111B        TRUE       FALSE        TRUE
#> 5  A1111B        TRUE       FALSE        TRUE
```

We will actually use all three in our reported heights example, but we will see these in a later section.

24.5.8   Not

To specify patterns that we do **not** want to detect, we can use the ^ symbol but only **inside** square brackets. Remember that outside the square bracket ^ means the start of the string. So, for example, if we want to detect digits that are preceded by anything except a letter we can do the following:

```
pattern <- "[^a-zA-Z]\\d"
yes <- c(".3", "+2", "-0","*4")
no <- c("A3", "B2", "C0", "E4")
str_detect(yes, pattern)
#> [1] TRUE TRUE TRUE TRUE
str_detect(no, pattern)
#> [1] FALSE FALSE FALSE FALSE
```

Another way to generate a pattern that searches for *everything except* is to use the upper case of the special character. For example \\D means anything other than a digit, \\S means anything except a space, and so on.

24.5.9 Groups

Groups are a powerful aspect of regex that permits the extraction of values. Groups are defined using parentheses. They don't affect the pattern matching per se. Instead, it permits tools to identify specific parts of the pattern so we can extract them.

We want to change heights written like 5.6 to 5'6.

To avoid changing patterns such as 70.2, we will require that the first digit be between 4 and 7 [4-7] and that the second be none or more digits \\d*. Let's start by defining a simple pattern that matches this:

```
pattern_without_groups <- "^[4-7],\\d*$"
```

We want to extract the digits so we can then form the new version using a period. These are our two groups, so we encapsulate them with parentheses:

```
pattern_with_groups <-  "^([4-7]),(\\d*)$"
```

We encapsulate the part of the pattern that matches the parts we want to keep for later use. Adding groups does not affect the detection, since it only signals that we want to save what is captured by the groups. Note that both patterns return the same result when using str_detect:

```
yes <- c("5,9", "5,11", "6,", "6,1")
no <- c("5'9", ",", "2,8", "6.1.1")
s <- c(yes, no)
str_detect(s, pattern_without_groups)
#> [1]  TRUE   TRUE   TRUE   TRUE FALSE FALSE FALSE FALSE
str_detect(s, pattern_with_groups)
#> [1]  TRUE   TRUE   TRUE   TRUE FALSE FALSE FALSE FALSE
```

Once we define groups, we can use the function str_match to extract the values these groups define:

```
str_match(s, pattern_with_groups)
#>         [,1]     [,2] [,3]
#> [1,] "5,9"    "5"  "9"
#> [2,] "5,11"   "5"  "11"
#> [3,] "6,"     "6"  ""
#> [4,] "6,1"    "6"  "1"
#> [5,] NA       NA   NA
#> [6,] NA       NA   NA
#> [7,] NA       NA   NA
#> [8,] NA       NA   NA
```

Notice that the second and third columns contain feet and inches, respectively. The first column is the part of the string matching the pattern. If no match occurred, we see an `NA`.

Now we can understand the difference between the functions `str_extract` and `str_match`: `str_extract` extracts only strings that match a pattern, not the values defined by groups:

```
str_extract(s, pattern_with_groups)
#> [1] "5,9"  "5,11" "6,"   "6,1"  NA     NA     NA     NA
```

24.6   Search and replace with regex

Earlier we defined the object `problems` containing the strings that do not appear to be in inches. We can see that not too many of our problematic strings match the pattern:

```
pattern <- "^[4-7]'\\d{1,2}\"$"
sum(str_detect(problems, pattern))
#> [1] 14
```

To see why this is, we show some examples that expose why we don't have more matches:

```
problems[c(2, 10, 11, 12, 15)] %>% str_view(pattern)
```

```
5' 4"
5'7"
5'3"
5 feet and 8.11 inches
5.5
```

An initial problem we see immediately is that some students wrote out the words "feet" and "inches". We can see the entries that did this with the `str_subset` function:

```
str_subset(problems, "inches")
#> [1] "5 feet and 8.11 inches" "Five foot eight inches"
#> [3] "5 feet 7inches"         "5ft 9 inches"
#> [5] "5 ft 9 inches"          "5 feet 6 inches"
```

We also see that some entries used two single quotes '' instead of a double quote ".

```
str_subset(problems, "'''")
#> [1] "5'9''"   "5'10''"  "5'10''"  "5'3''"   "5'7''"   "5'6''"
#> [7] "5'7.5''" "5'7.5''" "5'10''"  "5'11''"  "5'10''"  "5'5''"
```

To correct this, we can replace the different ways of representing inches and feet with a uniform symbol. We will use ' for feet, whereas for inches we will simply not use a symbol since some entries were of the form x'y. Now, if we no longer use the inches symbol, we have to change our pattern accordingly:

```
pattern <- "^[4-7]'\\d{1,2}$"
```

If we do this replacement before the matching, we get many more matches:

```
problems %>%
  str_replace("feet|ft|foot", "'") %>% # replace feet, ft, foot with '
  str_replace("inches|in|''|\"", "") %>% # remove all inches symbols
  str_detect(pattern) %>%
  sum()
#> [1] 48
```

However, we still have many cases to go.

Note that in the code above, we leveraged the **stringr** consistency and used the pipe.

For now, we improve our pattern by adding \\s* in front of and after the feet symbol ' to permit space between the feet symbol and the numbers. Now we match a few more entries:

```
pattern <- "^[4-7]\\s*'\\s*\\d{1,2}$"
problems %>%
  str_replace("feet|ft|foot", "'") %>% # replace feet, ft, foot with '
  str_replace("inches|in|''|\"", "") %>% # remove all inches symbols
  str_detect(pattern) %>%
  sum
#> [1] 53
```

We might be tempted to avoid doing this by removing all the spaces with str_replace_all. However, when doing such an operation we need to make sure that it does not have unintended effects. In our reported heights examples, this will be a problem because some entries are of the form x y with space separating the feet from the inches. If we remove all spaces, we will incorrectly turn x y into xy which implies that a 6 1 would become 61 inches instead of 73 inches.

The second large type of problematic entries were of the form x.y, x,y and x y. We want to change all these to our common format x'y. But we can't just do a search and replace because we would change values such as 70.5 into 70'5. Our strategy will therefore be to search for a very specific pattern that assures us feet and inches are being provided and then, for those that match, replace appropriately.

24.6.1   Search and replace using groups

Another powerful aspect of groups is that you can refer to the extracted values in a regex when searching and replacing.

The regex special character for the i-th group is \\i. So \\1 is the value extracted from the first group, \\2 the value from the second and so on. As a simple example, note that the following code will replace a comma with period, but only if it is between two digits:

```
pattern_with_groups <-  "^([4-7]),(\\d*)$"
yes <- c("5,9", "5,11", "6,", "6,1")
no <- c("5'9", ",", "2,8", "6.1.1")
s <- c(yes, no)
str_replace(s, pattern_with_groups, "\\1'\\2")
#> [1] "5'9"    "5'11"   "6'"     "6'1"    "5'9"    ","      "2,8"    "6.1.1"
```

We can use this to convert cases in our reported heights.

We are now ready to define a pattern that helps us convert all the x.y, x,y and x y to our preferred format. We need to adapt pattern_with_groups to be a bit more flexible and capture all the cases.

```
pattern_with_groups <-"^([4-7])\\s*[,\\.\\s+]\\s*(\\d*)$"
```

Let's break this one down:

* ^ = start of the string
* [4-7] = one digit, either 4, 5, 6, or 7
* \\s* = none or more white space
* [,\\.\\s+] = feet symbol is either ,, . or at least one space
* \\s* = none or more white space
* \\d* = none or more digits
* $ = end of the string

We can see that it appears to be working:

```
str_subset(problems, pattern_with_groups) %>% head()
#> [1] "5.3"  "5.25" "5.5"  "6.5"  "5.8"  "5.6"
```

and will be able to perform the search and replace:

```
str_subset(problems, pattern_with_groups) %>%
  str_replace(pattern_with_groups, "\\1'\\2") %>% head
#> [1] "5'3"  "5'25" "5'5"  "6'5"  "5'8"  "5'6"
```

Again, we will deal with the inches-larger-than-twelve challenge later.

24.7 Testing and improving

Developing the right regex on the first try is often difficult. Trial and error is a common approach to finding the regex pattern that satisfies all desired conditions. In the previous sections, we have developed a powerful string processing technique that can help us catch many of the problematic entries. Here we will test our approach, search for further problems, and tweak our approach for possible improvements. Let's write a function that captures all the entries that can't be converted into numbers remembering that some are in centimeters (we will deal with those later):

```
not_inches_or_cm <- function(x, smallest = 50, tallest = 84){
  inches <- suppressWarnings(as.numeric(x))
  ind <- !is.na(inches) &
    ((inches >= smallest & inches <= tallest) |
       (inches/2.54 >= smallest & inches/2.54 <= tallest))
  !ind
}

problems <- reported_heights %>%
  filter(not_inches_or_cm(height)) %>%
  pull(height)
length(problems)
#> [1] 200
```

Let's see what proportion of these fit our pattern after the processing steps we developed above:

```
converted <- problems %>%
  str_replace("feet|foot|ft", "'") %>% # convert feet symbols to '
  str_replace("inches|in|''|\"", "") %>%  # remove inches symbols
  str_replace("^([4-7])\\s*[,\\.\\s+]\\s*(\\d*)$", "\\1'\\2")# change format

pattern <- "^[4-7]\\s*'\\s*\\d{1,2}$"
index <- str_detect(converted, pattern)
mean(index)
#> [1] 0.615
```

Note how we leveraged the pipe, one of the advantages of using **stringr**. This last piece of code shows that we have matched well over half of the strings. Let's examine the remaining cases:

```
converted[!index]
#>  [1] "6"       "165cm"    "511"            "6"
#>  [5] "2"       ">9000"    "5 ' and 8.11 "  "11111"
#>  [9] "6"       "103.2"    "19"             "5"
#> [13] "300"     "6'"       "6"              "Five ' eight "
#> [17] "7"       "214"      "6"              "0.7"
#> [21] "6"       "2'33"     "612"            "1,70"
```

```
#> [25]  "87"          "5'7.5"        "5'7.5"        "111"
#> [29]  "5' 7.78"     "12"           "6"            "yyy"
#> [33]  "89"          "34"           "25"           "6"
#> [37]  "6"           "22"           "684"          "6"
#> [41]  "1"           "1"            "6*12"         "87"
#> [45]  "6"           "1.6"          "120"          "120"
#> [49]  "23"          "1.7"          "6"            "5"
#> [53]  "69"          "5' 9 "        "5 ' 9 "       "6"
#> [57]  "6"           "86"           "708,661"      "5 ' 6 "
#> [61]  "6"           "649,606"      "10000"        "1"
#> [65]  "728,346"     "0"            "6"            "6"
#> [69]  "6"           "100"          "88"           "6"
#> [73]  "170 cm"      "7,283,465"    "5"            "5"
#> [77]  "34"
```

Four clear patterns arise:

1. Many students measuring exactly 5 or 6 feet did not enter any inches, for example 6', and our pattern requires that inches be included.
2. Some students measuring exactly 5 or 6 feet entered just that number.
3. Some of the inches were entered with decimal points. For example 5'7.5''. Our pattern only looks for two digits.
4. Some entries have spaces at the end, for example 5 ' 9.

Although not as common, we also see the following problems:

5. Some entries are in meters and some of these use European decimals: 1.6, 1,70.
6. Two students added cm.
7. A student spelled out the numbers: Five foot eight inches.

It is not necessarily clear that it is worth writing code to handle these last three cases since they might be rare enough. However, some of them provide us with an opportunity to learn a few more regex techniques, so we will build a fix.

For case 1, if we add a '0 after the first digit, for example, convert all 6 to 6'0, then our previously defined pattern will match. This can be done using groups:

```
yes <- c("5", "6", "5")
no <- c("5'", "5''", "5'4")
s <- c(yes, no)
str_replace(s, "^([4-7])$", "\\1'0")
#> [1] "5'0" "6'0" "5'0" "5'"  "5''" "5'4"
```

The pattern says it has to start (^) with a digit between 4 and 7 and end there ($). The parenthesis defines the group that we pass as \\1 to generate the replacement regex string.

We can adapt this code slightly to handle the case 2 as well, which covers the entry 5'. Note 5' is left untouched. This is because the extra ' makes the pattern not match since we have to end with a 5 or 6. We want to permit the 5 or 6 to be followed by 0 or 1 feet sign. So we can simply add '{0,1} after the ' to do this. However, we can use the none or once special character ?. As we saw above, this is different from * which is none or more. We now see that the fourth case is also converted:

```
str_replace(s, "^([56])'?$", "\\1'0")
#> [1] "5'0" "6'0" "5'0" "5'0" "5''" "5'4"
```

Here we only permit 5 and 6, but not 4 and 7. This is because 5 and 6 feet tall is quite common, so we assume those that typed 5 or 6 really meant 60 or 72 inches. However, 4 and 7 feet tall are so rare that, although we accept 84 as a valid entry, we assume 7 was entered in error.

We can use quantifiers to deal with **case 3**. These entries are not matched because the inches include decimals and our pattern does not permit this. We need to allow the second group to include decimals not just digits. This means we must permit zero or one period . then zero or more digits. So we will be using both ? and *. Also remember that, for this particular case, the period needs to be escaped since it is a special character (it means any character except line break). Here is a simple example of how we can use *.

So we can adapt our pattern, currently `^[4-7]\\s*'\\s*\\d{1,2}$` to permit a decimal at the end:

```
pattern <- "^[4-7]\\s*'\\s*(\\d+\\.?\\d*)$"
```

Case 4, meters using commas, we can approach similarly to how we converted the x.y to x'y. A difference is that we require that the first digit be 1 or 2:

```
yes <- c("1,7", "1, 8", "2, " )
no <- c("5,8", "5,3,2", "1.7")
s <- c(yes, no)
str_replace(s, "^([12])\\s*,\\s*(\\d*)$", "\\1\\.\\2")
#> [1] "1.7"   "1.8"   "2."    "5,8"   "5,3,2" "1.7"
```

We will later check if the entries are meters using their numeric values. We will come back to the case study after introducing two widely used functions in string processing that will come in handy when developing our final solution for the self-reported heights.

24.8   Trimming

In general, spaces at the start or end of the string are uninformative. These can be particularly deceptive because sometimes they can be hard to see:

```
s <- "Hi "
cat(s)
#> Hi
identical(s, "Hi")
#> [1] FALSE
```

This is a general enough problem that there is a function dedicated to removing them: `str_trim`.

```
str_trim("5 ' 9 ")
#> [1] "5 ' 9"
```

_____

24.9   Changing lettercase

Notice that regex is case sensitive. Often we want to match a word regardless of case. One approach to doing this is to first change everything to lower case and then proceeding ignoring case. As an example, note that one of the entries writes out numbers as words `Five foot eight inches`. Although not efficient, we could add 13 extra `str_replace` calls to convert `zero` to 0, `one` to 1, and so on. To avoid having to write two separate operations for `Zero` and `zero`, `One` and `one`, etc., we can use the `str_to_lower` function to make all works lower case first:

```
s <- c("Five feet eight inches")
str_to_lower(s)
#> [1] "five feet eight inches"
```

Other related functions are `str_to_upper` and `str_to_title`. We are now ready to define a procedure that converts all the problematic cases to inches.

_____

24.10   Case study 2: self-reported heights (continued)

We now put all of what we have learned together into a function that takes a string vector and tries to convert as many strings as possible to one format. We write a function that puts together what we have done above.

```
convert_format <- function(s){
  s %>%
    str_replace("feet|foot|ft", "'") %>%
    str_replace_all("inches|in|''|\"|cm|and", "") %>%
    str_replace("^([4-7])\\s*[,\\.\\s+]\\s*(\\d*)$", "\\1'\\2") %>%
    str_replace("^([56])'?$", "\\1'0") %>%
    str_replace("^([12])\\s*,\\s*(\\d*)$", "\\1\\.\\2") %>%
    str_trim()
}
```

We can also write a function that converts words to numbers:

```
library(english)
words_to_numbers <- function(s){
  s <- str_to_lower(s)
  for(i in 0:11)
    s <- str_replace_all(s, words(i), as.character(i))
  s
}
```

Note that we can perform the above operation more efficiently with the function `recode`, which we learn about in Section 24.13. Now we can see which problematic entries remain:

```
converted <- problems %>% words_to_numbers() %>% convert_format()
remaining_problems <- converted[not_inches_or_cm(converted)]
pattern <- "^[4-7]\\s*'\\s*\\d+\\.?\\d*$"
index <- str_detect(remaining_problems, pattern)
remaining_problems[!index]
#>  [1] "511"       "2"         ">9000"     "11111"     "103.2"
#>  [6] "19"        "300"       "7"         "214"       "0.7"
#> [11] "2'33"      "612"       "1.70"      "87"        "111"
#> [16] "12"        "yyy"       "89"        "34"        "25"
#> [21] "22"        "684"       "1"         "1"         "6*12"
#> [26] "87"        "1.6"       "120"       "120"       "23"
#> [31] "1.7"       "86"        "708,661"   "649,606"   "10000"
#> [36] "1"         "728,346"   "0"         "100"       "88"
#> [41] "7,283,465" "34"
```

apart from the cases reported as meters, which we will fix below, they all seem to be cases that are impossible to fix.

24.10.1   The `extract` function

The `extract` function is a useful **tidyverse** function for string processing that we will use in our final solution, so we introduce it here. In a previous section, we constructed a regex that lets us identify which elements of a character vector match the feet and inches pattern. However, we want to do more. We want to extract and save the feet and number values so that we can convert them to inches when appropriate.

If we have a simpler case like this:

```
s <- c("5'10", "6'1")
tab <- data.frame(x = s)
```

In Section 21.3 we learned about the `separate` function, which can be used to achieve our current goal:

```
tab %>% separate(x, c("feet", "inches"), sep = "'")
#>   feet inches
#> 1    5     10
#> 2    6      1
```

The `extract` function from the **tidyr** package lets us use regex groups to extract the desired values. Here is the equivalent to the code above using `separate` but using `extract`:

```
library(tidyr)
tab %>% extract(x, c("feet", "inches"), regex = "(\\d)'(\\d{1,2})")
#>   feet inches
#> 1    5     10
#> 2    6      1
```

So why do we even need the new function `extract`? We have seen how small changes can throw off exact pattern matching. Groups in regex give us more flexibility. For example, if we define:

```
s <- c("5'10", "6'1\"","5'8inches")
tab <- data.frame(x = s)
```

and we only want the numbers, `separate` fails:

```
tab %>% separate(x, c("feet","inches"), sep = "'", fill = "right")
#>   feet  inches
#> 1    5      10
#> 2    6      1"
#> 3    5 8inches
```

However, we can use `extract`. The regex here is a bit more complicated since we have to permit ' with spaces and `feet`. We also do not want the " included in the value, so we do not include that in the group:

```
tab %>% extract(x, c("feet", "inches"), regex = "(\\d)'(\\d{1,2})")
#>   feet inches
#> 1    5     10
#> 2    6      1
#> 3    5      8
```

24.10.2 Putting it all together

We are now ready to put it all together and wrangle our reported heights data to try to recover as many heights as possible. The code is complex, but we will break it down into parts.

We start by cleaning up the `height` column so that the heights are closer to a feet'inches format. We added an original heights column so we can compare before and after.

Now we are ready to wrangle our reported heights dataset:

```
pattern <- "^([4-7])\\s*'\\s*(\\d+\\.?\\d*)$"

smallest <- 50
tallest <- 84
new_heights <- reported_heights %>%
  mutate(original = height,
```

```
            height = words_to_numbers(height) %>% convert_format()) %>%
    extract(height, c("feet", "inches"), regex = pattern, remove = FALSE) %>%
    mutate_at(c("height", "feet", "inches"), as.numeric) %>%
    mutate(guess = 12 * feet + inches) %>%
    mutate(height = case_when(
      is.na(height) ~ as.numeric(NA),
      between(height, smallest, tallest) ~ height,   #inches
      between(height/2.54, smallest, tallest) ~ height/2.54, #cm
      between(height*100/2.54, smallest, tallest) ~ height*100/2.54, #meters
      TRUE ~ as.numeric(NA))) %>%
    mutate(height = ifelse(is.na(height) &
                                inches < 12 & between(guess, smallest, tallest),
                          guess, height)) %>%
    select(-guess)
```

We can check all the entries we converted by typing:

```
new_heights %>%
  filter(not_inches(original)) %>%
  select(original, height) %>%
  arrange(height) %>%
  View()
```

A final observation is that if we look at the shortest students in our course:

```
new_heights %>% arrange(height) %>% head(n=7)
#>               time_stamp    sex height feet inches original
#> 1 2017-07-04 01:30:25   Male   50.0   NA     NA       50
#> 2 2017-09-07 10:40:35   Male   50.0   NA     NA       50
#> 3 2014-09-02 15:18:30 Female   51.0   NA     NA       51
#> 4 2016-06-05 14:07:20 Female   52.0   NA     NA       52
#> 5 2016-06-05 14:07:38 Female   52.0   NA     NA       52
#> 6 2014-09-23 03:39:56 Female   53.0   NA     NA       53
#> 7 2015-01-07 08:57:29   Male   53.8   NA     NA    53.77
```

We see heights of 53, 54, and 55. In the originals, we also have 51 and 52. These short heights are rare and it is likely that the students actually meant 5'1, 5'2, 5'3, 5'4, and 5'5. Because we are not completely sure, we will leave them as reported. The object `new_heights` contains our final solution for this case study.

24.11   String splitting

Another very common data wrangling operation is string splitting. To illustrate how this comes up, we start with an illustrative example. Suppose we did not have the function `read_csv` or `read.csv` available to us. We instead have to read a csv file using the base R function `readLines` like this:

```
filename <- system.file("extdata/murders.csv", package = "dslabs")
lines <- readLines(filename)
```

This function reads-in the data line-by-line to create a vector of strings. In this case, one string for each row in the spreadsheet. The first six lines are:

```
lines %>% head()
#> [1] "state,abb,region,population,total"
#> [2] "Alabama,AL,South,4779736,135"
#> [3] "Alaska,AK,West,710231,19"
#> [4] "Arizona,AZ,West,6392017,232"
#> [5] "Arkansas,AR,South,2915918,93"
#> [6] "California,CA,West,37253956,1257"
```

We want to extract the values that are separated by a comma for each string in the vector. The command `str_split` does exactly this:

```
x <- str_split(lines, ",")
x %>% head(2)
#> [[1]]
#> [1] "state"      "abb"         "region"     "population" "total"
#>
#> [[2]]
#> [1] "Alabama" "AL"         "South"    "4779736" "135"
```

Note that the first entry has the column names, so we can separate that out:

```
col_names <- x[[1]]
x <- x[-1]
```

To convert our list into a data frame, we can use a shortcut provided by the `map` functions in the **purrr** package. The map function applies the same function to each element in a list. So if we want to extract the first entry of each element in x, we can write:

```
library(purrr)
map(x, function(y) y[1]) %>% head(2)
#> [[1]]
#> [1] "Alabama"
#>
#> [[2]]
#> [1] "Alaska"
```

However, because this is such a common task, **purrr** provides a shortcut. If the second argument receives an integer instead of a function, it assumes we want that entry. So the code above can be written more efficiently like this:

```
map(x, 1)
```

To force `map` to return a character vector instead of a list, we can use `map_chr`. Similarly, `map_int` returns integers. So to create our data frame, we can use:

```
dat <- tibble(map_chr(x, 1),
              map_chr(x, 2),
              map_chr(x, 3),
              map_chr(x, 4),
              map_chr(x, 5)) %>%
  mutate_all(parse_guess) %>%
  setNames(col_names)
dat %>% head
#> # A tibble: 6 x 5
#>   state      abb   region population total
#>   <chr>      <chr> <chr>       <dbl> <dbl>
#> 1 Alabama    AL    South     4779736   135
#> 2 Alaska     AK    West       710231    19
#> 3 Arizona    AZ    West      6392017   232
#> 4 Arkansas   AR    South     2915918    93
#> 5 California CA    West     37253956  1257
#> # ... with 1 more row
```

If you learn more about the **purrr** package, you will learn that you perform the above with the following, more efficient, code:

```
dat <- x %>%
  transpose() %>%
  map( ~ parse_guess(unlist(.))) %>%
  setNames(col_names) %>%
  as_tibble()
```

It turns out that we can avoid all the work shown above after the call to `str_split`. Specifically, if we know that the data we are extracting can be represented as a table, we can use the argument `simplify=TRUE` and `str_split` returns a matrix instead of a list:

```
x <- str_split(lines, ",", simplify = TRUE)
col_names <- x[1,]
x <- x[-1,]
colnames(x) <- col_names
x %>% as_tibble() %>%
  mutate_all(parse_guess) %>%
  head(5)
#> # A tibble: 5 x 5
#>   state      abb   region population total
#>   <chr>      <chr> <chr>       <dbl> <dbl>
#> 1 Alabama    AL    South     4779736   135
#> 2 Alaska     AK    West       710231    19
#> 3 Arizona    AZ    West      6392017   232
#> 4 Arkansas   AR    South     2915918    93
#> 5 California CA    West     37253956  1257
```

24.12 Case study 3: extracting tables from a PDF

One of the datasets provided in **dslabs** shows scientific funding rates by gender in the Netherlands:

```
library(dslabs)
data("research_funding_rates")
research_funding_rates %>%
  select("discipline", "success_rates_men", "success_rates_women")
#>            discipline success_rates_men success_rates_women
#> 1   Chemical sciences              26.5                25.6
#> 2   Physical sciences              19.3                23.1
#> 3             Physics              26.9                22.2
#> 4          Humanities              14.3                19.3
#> 5  Technical sciences              15.9                21.0
#> 6   Interdisciplinary              11.4                21.8
#> 7 Earth/life sciences              24.4                14.3
#> 8     Social sciences              15.3                11.5
#> 9    Medical sciences              18.8                11.2
```

The data comes from a paper published in the Proceedings of the National Academy of Science (PNAS)[5], a widely read scientific journal. However, the data is not provided in a spreadsheet; it is in a table in a PDF document. Here is a screenshot of the table:

Table S1. Numbers of applications and awarded grants, along with success rates for male and female applicants, by scientific discipline

Discipline	Applications, n			Awards, n			Success rates, %		
	Total	Men	Women	Total	Men	Women	Total	Men	Women
Total	2,823	1,635	1,188	467	290	177	16.5	17.7a	14.9b
Chemical sciences	122	83	39	32	22	10	26.2	26.5a	25.6a
Physical sciences	174	135	39	35	26	9	20.1	19.3a	23.1a
Physics	76	67	9	20	18	2	26.3	26.9a	22.2a
Humanities	396	230	166	65	33	32	16.4	14.3a	19.3a
Technical sciences	251	189	62	43	30	13	17.1	15.9a	21.0a
Interdisciplinary	183	105	78	29	12	17	15.8	11.4a	21.8a
Earth/life sciences	282	156	126	56	38	18	19.9	24.4a	14.3b
Social sciences	834	425	409	112	65	47	13.4	15.3a	11.5a
Medical sciences	505	245	260	75	46	29	14.9	18.8a	11.2b

Success rates for male and female applicants with different subscripts differ reliably from one another ($P < 0.05$).

(Source: Romy van der Lee and Naomi Ellemers, PNAS 2015 112 (40) 12349-12353[6].)

We could extract the numbers by hand, but this could lead to human error. Instead, we can try to wrangle the data using R. We start by downloading the pdf document, then importing into R:

```
library("pdftools")
temp_file <- tempfile()
url <- paste0("http://www.pnas.org/content/suppl/2015/09/16/",
              "1510159112.DCSupplemental/pnas.201510159SI.pdf")
```

---

[5] http://www.pnas.org/content/112/40/12349.abstract
[6] http://www.pnas.org/content/112/40/12349

```
download.file(url, temp_file)
txt <- pdf_text(temp_file)
file.remove(temp_file)
```

If we examine the object text, we notice that it is a character vector with an entry for each page. So we keep the page we want:

```
raw_data_research_funding_rates <- txt[2]
```

The steps above can actually be skipped because we include this raw data in the **dslabs** package as well:

```
data("raw_data_research_funding_rates")
```

Examining the object `raw_data_research_funding_rates` we see that it is a long string and each line on the page, including the table rows, are separated by the symbol for newline: \n. We therefore can create a list with the lines of the text as elements as follows:

```
tab <- str_split(raw_data_research_funding_rates, "\n")
```

Because we start off with just one element in the string, we end up with a list with just one entry.

```
tab <- tab[[1]]
```

By examining `tab` we see that the information for the column names is the third and fourth entries:

```
the_names_1 <- tab[3]
the_names_2 <- tab[4]
```

The first of these rows looks like this:

```
#>                                                             Applications, n
#>                              Awards, n            Success rates, %
```

We want to create one vector with one name for each column. Using some of the functions we have just learned, we do this. Let's start with `the_names_1`, shown above. We want to remove the leading space and anything following the comma. We use regex for the latter. Then we can obtain the elements by splitting strings separated by space. We want to split only when there are 2 or more spaces to avoid splitting `Success rates`. So we use the regex \\s{2,}

```
the_names_1 <- the_names_1 %>%
  str_trim() %>%
  str_replace_all(",\\s.", "") %>%
  str_split("\\s{2,}", simplify = TRUE)
the_names_1
#>      [,1]           [,2]     [,3]
#> [1,] "Applications" "Awards" "Success rates"
```

Now we will look at `the_names_2`:

```
#>                         Discipline              Total     Men      Women
#> n          Total      Men        Women          Total     Men      Women
```

Here we want to trim the leading space and then split by space as we did for the first line:

```
the_names_2 <- the_names_2 %>%
  str_trim() %>%
  str_split("\\s+", simplify = TRUE)
the_names_2
#>     [,1]          [,2]      [,3]   [,4]      [,5]      [,6]   [,7]      [,8]
#> [1,] "Discipline" "Total"  "Men"  "Women"  "Total"  "Men"  "Women"  "Total"
#>     [,9]   [,10]
#> [1,] "Men"  "Women"
```

We can then join these to generate one name for each column:

```
tmp_names <- str_c(rep(the_names_1, each = 3), the_names_2[-1], sep = "_")
the_names <- c(the_names_2[1], tmp_names) %>%
  str_to_lower() %>%
  str_replace_all("\\s", "_")
the_names
#>  [1] "discipline"           "applications_total"   "applications_men"
#>  [4] "applications_women"   "awards_total"         "awards_men"
#>  [7] "awards_women"         "success_rates_total"  "success_rates_men"
#> [10] "success_rates_women"
```

Now we are ready to get the actual data. By examining the `tab` object, we notice that the information is in lines 6 through 14. We can use `str_split` again to achieve our goal:

```
new_research_funding_rates <- tab[6:14] %>%
  str_trim %>%
  str_split("\\s{2,}", simplify = TRUE) %>%
  data.frame(stringsAsFactors = FALSE) %>%
  setNames(the_names) %>%
  mutate_at(-1, parse_number)
new_research_funding_rates %>% as_tibble()
#> # A tibble: 9 x 10
#>   discipline applications_to~ applications_men applications_wo~
#>   <chr>                 <dbl>            <dbl>            <dbl>
#> 1 Chemical ~              122               83               39
#> 2 Physical ~              174              135               39
#> 3 Physics                  76               67                9
#> 4 Humanities              396              230              166
#> 5 Technical~              251              189               62
#> # ... with 4 more rows, and 6 more variables: awards_total <dbl>,
#> #   awards_men <dbl>, awards_women <dbl>, success_rates_total <dbl>,
#> #   success_rates_men <dbl>, success_rates_women <dbl>
```

We can see that the objects are identical:

```
identical(research_funding_rates, new_research_funding_rates)
#> [1] TRUE
```

24.13   Recoding

Another common operation involving strings is recoding the names of categorical variables. Let's say you have really long names for your levels and you will be displaying them in plots, you might want to use shorter versions of these names. For example, in character vectors with country names, you might want to change "United States of America" to "USA" and "United Kingdom" to UK, and so on. We can do this with **case_when**, although the **tidyverse** offers an option that is specifically designed for this task: the **recode** function.

Here is an example that shows how to rename countries with long names:

```
library(dslabs)
data("gapminder")
```

Suppose we want to show life expectancy time series by country for the Caribbean:

```
gapminder %>%
  filter(region == "Caribbean") %>%
  ggplot(aes(year, life_expectancy, color = country)) +
  geom_line()
```

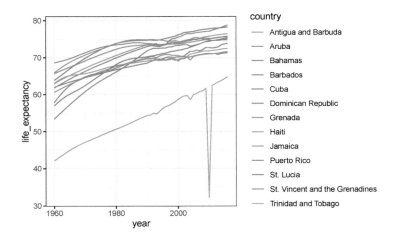

The plot is what we want, but much of the space is wasted to accommodate some of the long country names. We have four countries with names longer than 12 characters. These names appear once for each year in the Gapminder dataset. Once we pick nicknames, we need to change them all consistently. The **recode** function can be used to do this:

```
gapminder %>% filter(region=="Caribbean") %>%
  mutate(country = recode(country,
                          `Antigua and Barbuda` = "Barbuda",
                          `Dominican Republic` = "DR",
                          `St. Vincent and the Grenadines` = "St. Vincent",
                          `Trinidad and Tobago` = "Trinidad")) %>%
  ggplot(aes(year, life_expectancy, color = country)) +
  geom_line()
```

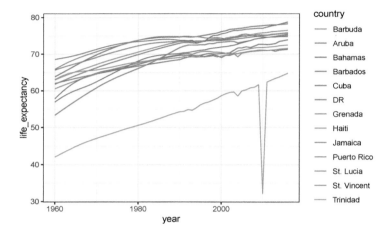

There are other similar functions in other R packages, such as `recode_factor` and `fct_recoder` in the **forcats** package.

24.14   Exercises

1. Complete all lessons and exercises in the https://regexone.com/ online interactive tutorial.

2. In the `extdata` directory of the **dslabs** package, you will find a PDF file containing daily mortality data for Puerto Rico from Jan 1, 2015 to May 31, 2018. You can find the file like this:

```
fn <- system.file("extdata", "RD-Mortality-Report_2015-18-180531.pdf",
                  package="dslabs")
```

Find and open the file or open it directly from RStudio. On a Mac, you can type:

```
system2("open", args = fn)
```

and on Windows, you can type:

```
system("cmd.exe", input = paste("start", fn))
```

Which of the following best describes this file:

    a. It is a table. Extracting the data will be easy.
    b. It is a report written in prose. Extracting the data will be impossible.
    c. It is a report combining graphs and tables. Extracting the data seems possible.
    d. It shows graphs of the data. Extracting the data will be difficult.

3. We are going to create a tidy dataset with each row representing one observation. The variables in this dataset will be year, month, day, and deaths. Start by installing and loading the **pdftools** package:

```
install.packages("pdftools")
library(pdftools)
```

Now read-in `fn` using the `pdf_text` function and store the results in an object called `txt`. Which of the following best describes what you see in `txt`?

    a. A table with the mortality data.
    b. A character string of length 12. Each entry represents the text in each page. The mortality data is in there somewhere.
    c. A character string with one entry containing all the information in the PDF file.
    d. An html document.

4. Extract the ninth page of the PDF file from the object `txt`, then use the `str_split` from the **stringr** package so that you have each line in a different entry. Call this string vector `s`. Then look at the result and choose the one that best describes what you see.

    a. It is an empty string.
    b. I can see the figure shown in page 1.
    c. It is a tidy table.
    d. I can see the table! But there is a bunch of other stuff we need to get rid of.

5. What kind of object is `s` and how many entries does it have?

6. We see that the output is a list with one component. Redefine `s` to be the first entry of the list. What kind of object is `s` and how many entries does it have?

7. When inspecting the string we obtained above, we see a common problem: white space before and after the other characters. Trimming is a common first step in string processing. These extra spaces will eventually make splitting the strings hard so we start by removing them. We learned about the command `str_trim` that removes spaces at the start or end of the strings. Use this function to trim `s`.

8. We want to extract the numbers from the strings stored in `s`. However, there are many non-numeric characters that will get in the way. We can remove these, but before doing this we want to preserve the string with the column header, which includes the month abbreviation. Use the `str_which` function to find the rows with a header. Save these results to `header_index`. Hint: find the first string that matches the pattern 2015 using the `str_which` function.

9. Now we are going to define two objects: `month` will store the month and `header` will store the column names. Identify which row contains the header of the table. Save the content of the row into an object called `header`, then use `str_split` to help define the two objects we

need. Hints: the separator here is one or more spaces. Also, consider using the `simplify` argument.

10. Notice that towards the end of the page you see a *totals* row followed by rows with other summary statistics. Create an object called `tail_index` with the index of the *totals* entry.

11. Because our PDF page includes graphs with numbers, some of our rows have just one number (from the y-axis of the plot). Use the `str_count` function to create an object `n` with the number of numbers in each each row. Hint: you can write a regex for number like this `\\d+`.

12. We are now ready to remove entries from rows that we know we don't need. The entry `header_index` and everything before it should be removed. Entries for which n is 1 should also be removed, and the entry `tail_index` and everything that comes after it should be removed as well.

13. Now we are ready to remove all the non-numeric entries. Do this using regex and the `str_remove_all` function. Hint: remember that in regex, using the upper case version of a special character usually means the opposite. So `\\D` means "not a digit". Remember you also want to keep spaces.

14. To convert the strings into a table, use the `str_split_fixed` function. Convert `s` into a data matrix with just the day and death count data. Hints: note that the separator is one or more spaces. Make the argument `n` a value that limits the number of columns to the values in the 4 columns and the last column captures all the extra stuff. Then keep only the first four columns.

15. Now you are almost ready to finish. Add column names to the matrix, including one called `day`. Also, add a column with the month. Call the resulting object `dat`. Finally, make sure the day is an integer not a character. Hint: use only the first five columns.

16. Now finish it up by tidying `tab` with the gather function.

17. Make a plot of deaths versus day with color to denote year. Exclude 2018 since we do not have data for the entire year.

18. Now that we have wrangled this data step-by-step, put it all together in one R chunk, using the pipe as much as possible. Hint: first define the indexes, then write one line of code that does all the string processing.

19. Advanced: let's return to the MLB Payroll example from the web scraping section. Use what you have learned in the web scraping and string processing chapters to extract the payroll for the New York Yankees, Boston Red Sox, and Oakland A's and plot them as a function of time.

25

Parsing dates and times

25.1 The date data type

We have described three main types of vectors: numeric, character, and logical. In data science projects, we very often encounter variables that are dates. Although we can represent a date with a string, for example `November 2, 2017`, once we pick a reference day, referred to as the *epoch*, they can be converted to numbers by calculating the number of days since the epoch. Computer languages usually use January 1, 1970, as the epoch. So, for example, January 2, 2017 is day 1, December 31, 1969 is day -1, and November 2, 2017, is day 17,204.

Now how should we represent dates and times when analyzing data in R? We could just use days since the epoch, but then it is almost impossible to interpret. If I tell you it's November 2, 2017, you know what this means immediately. If I tell you it's day 17,204, you will be quite confused. Similar problems arise with times and even more complications can appear due to time zones.

For this reason, R defines a data type just for dates and times. We saw an example in the polls data:

```
library(tidyverse)
library(dslabs)
data("polls_us_election_2016")
polls_us_election_2016$startdate %>% head
#> [1] "2016-11-03" "2016-11-01" "2016-11-02" "2016-11-04" "2016-11-03"
#> [6] "2016-11-03"
```

These look like strings, but they are not:

```
class(polls_us_election_2016$startdate)
#> [1] "Date"
```

Look at what happens when we convert them to numbers:

```
as.numeric(polls_us_election_2016$startdate) %>% head
#> [1] 17108 17106 17107 17109 17108 17108
```

It turns them into days since the epoch. The `as.Date` function can convert a character into a date. So to see that the epoch is day 0 we can type

```
as.Date("1970-01-01") %>% as.numeric
#> [1] 0
```

Plotting functions, such as those in ggplot, are aware of the date format. This means that, for example, a scatterplot can use the numeric representation to decide on the position of the point, but include the string in the labels:

```
polls_us_election_2016 %>% filter(pollster == "Ipsos" & state =="U.S.") %>%
    ggplot(aes(startdate, rawpoll_trump)) +
    geom_line()
```

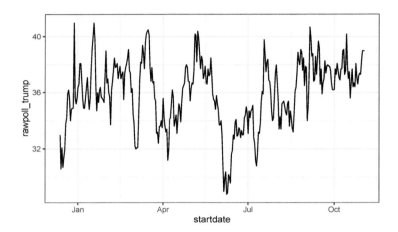

Note in particular that the month names are displayed, a very convenient feature.

25.2   The lubridate package

The **tidyverse** includes functionality for dealing with dates through the **lubridate** package.

```
library(lubridate)
```

We will take a random sample of dates to show some of the useful things one can do:

```
set.seed(2002)
dates <- sample(polls_us_election_2016$startdate, 10) %>% sort
dates
#>  [1] "2016-05-31" "2016-08-08" "2016-08-19" "2016-09-22" "2016-09-27"
#>  [6] "2016-10-12" "2016-10-24" "2016-10-26" "2016-10-29" "2016-10-30"
```

The functions year, month and day extract those values:

```
tibble(date = dates,
       month = month(dates),
       day = day(dates),
       year = year(dates))
#> # A tibble: 10 x 4
#>    date        month   day  year
#>    <date>      <dbl> <int> <dbl>
#> 1 2016-05-31      5    31  2016
#> 2 2016-08-08      8     8  2016
#> 3 2016-08-19      8    19  2016
#> 4 2016-09-22      9    22  2016
#> 5 2016-09-27      9    27  2016
#> # ... with 5 more rows
```

We can also extract the month labels:

```
month(dates, label = TRUE)
#>  [1] May Aug Aug Sep Sep Oct Oct Oct Oct Oct
#> 12 Levels: Jan < Feb < Mar < Apr < May < Jun < Jul < Aug < ... < Dec
```

Another useful set of functions are the *parsers* that convert strings into dates. The function ymd assumes the dates are in the format YYYY-MM-DD and tries to parse as well as possible.

```
x <- c(20090101, "2009-01-02", "2009 01 03", "2009-1-4",
       "2009-1, 5", "Created on 2009 1 6", "200901 !!! 07")
ymd(x)
#> [1] "2009-01-01" "2009-01-02" "2009-01-03" "2009-01-04" "2009-01-05"
#> [6] "2009-01-06" "2009-01-07"
```

A further complication comes from the fact that dates often come in different formats in which the order of year, month, and day are different. The preferred format is to show year (with all four digits), month (two digits), and then day, or what is called the ISO 8601. Specifically we use YYYY-MM-DD so that if we order the string, it will be ordered by date. You can see the function ymd returns them in this format.

But, what if you encounter dates such as "09/01/02"? This could be September 1, 2002 or January 2, 2009 or January 9, 2002. In these cases, examining the entire vector of dates will help you determine what format it is by process of elimination. Once you know, you can use the many parses provided by **lubridate**.

For example, if the string is:

```
x <- "09/01/02"
```

The ymd function assumes the first entry is the year, the second is the month, and the third is the day, so it converts it to:

```
ymd(x)
#> [1] "2009-01-02"
```

The mdy function assumes the first entry is the month, then the day, then the year:

```
mdy(x)
#> [1] "2002-09-01"
```

The *lubridate* package provides a function for every possibility:

```
ydm(x)
#> [1] "2009-02-01"
myd(x)
#> [1] "2001-09-02"
dmy(x)
#> [1] "2002-01-09"
dym(x)
#> [1] "2001-02-09"
```

The **lubridate** package is also useful for dealing with times. In R base, you can get the current time typing `Sys.time()`. The **lubridate** package provides a slightly more advanced function, `now`, that permits you to define the time zone:

```
now()
#> [1] "2019-09-16 16:54:07 EDT"
now("GMT")
#> [1] "2019-09-16 20:54:07 GMT"
```

You can see all the available time zones with `OlsonNames()` function.

We can also extract hours, minutes, and seconds:

```
now() %>% hour()
#> [1] 16
now() %>% minute()
#> [1] 54
now() %>% second()
#> [1] 7.67
```

The package also includes a function to parse strings into times as well as parsers for time objects that include dates:

```
x <- c("12:34:56")
hms(x)
#> [1] "12H 34M 56S"
x <- "Nov/2/2012 12:34:56"
mdy_hms(x)
#> [1] "2012-11-02 12:34:56 UTC"
```

This package has many other useful functions. We describe two of these here that we find particularly useful.

The `make_date` function can be used to quickly create a date object. It takes three arguments: year, month, day, hour, minute, seconds, and time zone defaulting to the epoch values on UTC time. So create an date object representing, for example, July 6, 2019 we write:

```
make_date(1970, 7, 6)
#> [1] "1970-07-06"
```

To make a vector of January 1 for the 80s we write:

```
make_date(1980:1989)
#>  [1] "1980-01-01" "1981-01-01" "1982-01-01" "1983-01-01" "1984-01-01"
#>  [6] "1985-01-01" "1986-01-01" "1987-01-01" "1988-01-01" "1989-01-01"
```

Another very useful function is the `round_date`. It can be used to *round* dates to nearest year, quarter, month, week, day, hour, minutes, or seconds. So if we want to group all the polls by week of the year we can do the following:

```
polls_us_election_2016 %>%
  mutate(week = round_date(startdate, "week")) %>%
  group_by(week) %>%
  summarize(margin = mean(rawpoll_clinton - rawpoll_trump)) %>%
  qplot(week, margin, data = .)
```

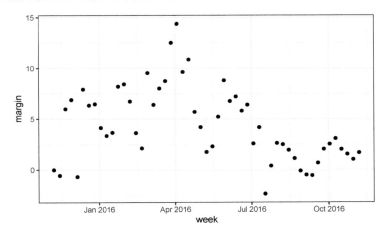

25.3 Exercises

In the previous exercise section, we wrangled data from a PDF file containing vital statistics from Puerto Rico. We did this for the month of September. Below we include code that does it for all 12 months.

```
library(tidyverse)
library(purrr)
library(pdftools)

fn <- system.file("extdata", "RD-Mortality-Report_2015-18-180531.pdf",
                  package="dslabs")
tab <- map_df(str_split(pdf_text(fn), "\n"), function(s){
```

```
  s <- str_trim(s)
  header_index <- str_which(s, "2015")[1]
  tmp <- str_split(s[header_index], "\\s+", simplify = TRUE)
  month <- tmp[1]
  header <- tmp[-1]
  tail_index  <- str_which(s, "Total")
  n <- str_count(s, "\\d+")
  out <- c(1:header_index, which(n==1), which(n>=28), tail_index:length(s))
  s[-out] %>% str_remove_all("[^\\d\\s]") %>% str_trim() %>%
    str_split_fixed("\\s+", n = 6) %>% .[,1:5] %>% as_tibble() %>%
    setNames(c("day", header)) %>%
    mutate(month = month, day = as.numeric(day)) %>%
    gather(year, deaths, -c(day, month)) %>%
    mutate(deaths = as.numeric(deaths))
})
```

1. We want to make a plot of death counts versus date. A first step is to convert the month variable from characters to numbers. Note that the month abbreviations are in Spanglish. Use the `recode` function to convert months to numbers and redefine `tab`.

2. Create a new column `date` with the date for each observation. Hint: use the `make_date` function.

3. Plot deaths versus date.

4. Note that after May 31, 2018, the deaths are all 0. The data is probably not entered yet. We also see a drop off starting around May 1. Redefine `tab` to exclude observations taken on or after May 1, 2018. Then, remake the plot.

5. Remake the plot above but this time plot deaths against the day of the year, for example, Jan 12, 2016 and Jan 12, 2017 are both day 12. Use color to denote the different years. Hint: use the **lubridate** function `yday`.

6. Remake the plot above but, this time, use two different colors for before and after September 20, 2017.

7. Advanced: remake the plot above, but this time show the month in the x-axis. Hint: create a variable with the date for a given year. Then use the `scale_x_date` function to show just the months.

8. Remake the deaths versus day but with weekly averages. Hint: use the function `round_date`.

9. Remake the plot but with monthly averages. Hint: use the function `round_date` again.

26

Text mining

With the exception of labels used to represent categorical data, we have focused on numerical data. But in many applications, data starts as text. Well-known examples are spam filtering, cyber-crime prevention, counter-terrorism and sentiment analysis. In all these cases, the raw data is composed of free form text. Our task is to extract insights from these data. In this section, we learn how to generate useful numerical summaries from text data to which we can apply some of the powerful data visualization and analysis techniques we have learned.

26.1 Case study: Trump tweets

During the 2016 US presidential election, then candidate Donald J. Trump used his twitter account as a way to communicate with potential voters. On August 6, 2016, Todd Vaziri tweeted[1] about Trump that "Every non-hyperbolic tweet is from iPhone (his staff). Every hyperbolic tweet is from Android (from him)." Data scientist David Robinson conducted an analysis[2] to determine if data supported this assertion. Here, we go through David's analysis to learn some of the basics of text mining. To learn more about text mining in R, we recommend the Text Mining with R book[3] by Julia Silge and David Robinson.

We will use the following libraries:

```
library(tidyverse)
library(lubridate)
library(scales)
```

In general, we can extract data directly from Twitter using the **rtweet** package. However, in this case, a group has already compiled data for us and made it available at http://www.trumptwitterarchive.com. We can get the data from their JSON API using a script like this:

```
url <- 'http://www.trumptwitterarchive.com/data/realdonaldtrump/%s.json'
trump_tweets <- map(2009:2017, ~sprintf(url, .x)) %>%
  map_df(jsonlite::fromJSON, simplifyDataFrame = TRUE) %>%
  filter(!is_retweet & !str_detect(text, '^"')) %>%
  mutate(created_at = parse_date_time(created_at,
                                orders = "a b! d! H!:M!:S! z!* Y!",
                                tz="EST"))
```

---

[1]https://twitter.com/tvaziri/status/762005541388378112/photo/1
[2]http://varianceexplained.org/r/trump-tweets/
[3]https://www.tidytextmining.com/

For convenience, we include the result of the code above in the **dslabs** package:

```
library(dslabs)
data("trump_tweets")
```

You can see the data frame with information about the tweets by typing

```
head(trump_tweets)
```

with the following variables included:

```
names(trump_tweets)
#> [1] "source"            "id_str"
#> [3] "text"              "created_at"
#> [5] "retweet_count"     "in_reply_to_user_id_str"
#> [7] "favorite_count"    "is_retweet"
```

The help file ?trump_tweets provides details on what each variable represents. The tweets are represented by the text variable:

```
trump_tweets$text[16413] %>% str_wrap(width = options()$width) %>% cat
#> Great to be back in Iowa! #TBT with @JerryJrFalwell joining me in
#> Davenport- this past winter. #MAGA https://t.co/A5IFOQHnic
```

and the source variable tells us which device was used to compose and upload each tweet:

```
trump_tweets %>% count(source) %>% arrange(desc(n)) %>% head(5)
#> # A tibble: 5 x 2
#>   source                  n
#>   <chr>               <int>
#> 1 Twitter Web Client  10718
#> 2 Twitter for Android  4652
#> 3 Twitter for iPhone   3962
#> 4 TweetDeck             468
#> 5 TwitLonger Beta       288
```

We are interested in what happened during the campaign, so for this analysis we will focus on what was tweeted between the day Trump announced his campaign and election day. We define the following table containing just the tweets from that time period. Note that we use extract to remove the Twitter for part of the source and filter out retweets.

```
campaign_tweets <- trump_tweets %>%
  extract(source, "source", "Twitter for (.*)") %>%
  filter(source %in% c("Android", "iPhone") &
           created_at >= ymd("2015-06-17") &
           created_at < ymd("2016-11-08")) %>%
  filter(!is_retweet) %>%
  arrange(created_at)
```

We can now use data visualization to explore the possibility that two different groups were tweeting from these devices. For each tweet, we will extract the hour, East Coast time (EST), it was tweeted and then compute the proportion of tweets tweeted at each hour for each device:

```
ds_theme_set()
campaign_tweets %>%
  mutate(hour = hour(with_tz(created_at, "EST"))) %>%
  count(source, hour) %>%
  group_by(source) %>%
  mutate(percent = n / sum(n)) %>%
  ungroup %>%
  ggplot(aes(hour, percent, color = source)) +
  geom_line() +
  geom_point() +
  scale_y_continuous(labels = percent_format()) +
  labs(x = "Hour of day (EST)", y = "% of tweets", color = "")
```

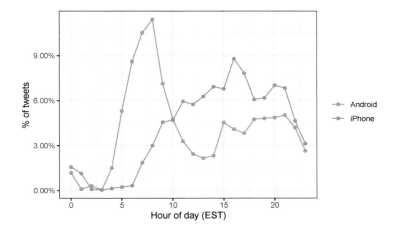

We notice a big peak for the Android in the early hours of the morning, between 6 and 8 AM. There seems to be a clear difference in these patterns. We will therefore assume that two different entities are using these two devices.

We will now study how the tweets differ when we compare Android to iPhone. To do this, we introduce the **tidytext** package.

26.2 Text as data

The **tidytext** package helps us convert free form text into a tidy table. Having the data in this format greatly facilitates data visualization and the use of statistical techniques.

```
library(tidytext)
```

The main function needed to achieve this is **unnest_tokens**. A *token* refers to a unit that we are considering to be a data point. The most common *token* will be words, but they can also be single characters, ngrams, sentences, lines, or a pattern defined by a regex. The functions will take a vector of strings and extract the tokens so that each one gets a row in the new table. Here is a simple example:

```
poem <- c("Roses are red,", "Violets are blue,",
          "Sugar is sweet,", "And so are you.")
example <- tibble(line = c(1, 2, 3, 4),
                        text = poem)
example
#> # A tibble: 4 x 2
#>    line text
#>   <dbl> <chr>
#> 1     1 Roses are red,
#> 2     2 Violets are blue,
#> 3     3 Sugar is sweet,
#> 4     4 And so are you.
example %>% unnest_tokens(word, text)
#> # A tibble: 13 x 2
#>    line word
#>   <dbl> <chr>
#> 1     1 roses
#> 2     1 are
#> 3     1 red
#> 4     2 violets
#> 5     2 are
#> # ... with 8 more rows
```

Now let's look at an example from the tweets. We will look at tweet number 3008 because it will later permit us to illustrate a couple of points:

```
i <- 3008
campaign_tweets$text[i] %>% str_wrap(width = 65) %>% cat()
#> Great to be back in Iowa! #TBT with @JerryJrFalwell joining me in
#> Davenport- this past winter. #MAGA https://t.co/A5IFOQHnic
campaign_tweets[i,] %>%
  unnest_tokens(word, text) %>%
  pull(word)
#>  [1] "great"            "to"         "be"      "back"
#>  [5] "in"               "iowa"       "tbt"     "with"
#>  [9] "jerryjrfalwell"   "joining"    "me"      "in"
#> [13] "davenport"        "this"       "past"    "winter"
#> [17] "maga"             "https"      "t.co"    "a5if0qhnic"
```

Note that the function tries to convert tokens into words. To do this, however, it strips characters that are important in the context of twitter. Namely, the function removes all the # and @. A *token* in the context of Twitter is not the same as in the context of spoken or written English. For this reason, instead of using the default, words, we use the **tweets** token includes patterns that start with @ and #:

```
campaign_tweets[i,] %>%
  unnest_tokens(word, text, token = "tweets") %>%
  pull(word)
#>  [1] "great"                "to"
#>  [3] "be"                   "back"
#>  [5] "in"                   "iowa"
#>  [7] "#tbt"                 "with"
#>  [9] "@jerryjrfalwell"      "joining"
#> [11] "me"                   "in"
#> [13] "davenport"            "this"
#> [15] "past"                 "winter"
#> [17] "#maga"                "https://t.co/a5ifOqhnic"
```

Another minor adjustment we want to make is to remove the links to pictures:

```
links <- "https://t.co/[A-Za-z\\d]+|&amp;"
campaign_tweets[i,] %>%
  mutate(text = str_replace_all(text, links, "")) %>%
  unnest_tokens(word, text, token = "tweets") %>%
  pull(word)
#>  [1] "great"          "to"        "be"
#>  [4] "back"           "in"        "iowa"
#>  [7] "#tbt"           "with"      "@jerryjrfalwell"
#> [10] "joining"        "me"        "in"
#> [13] "davenport"      "this"      "past"
#> [16] "winter"         "#maga"
```

Now we are now ready to extract the words for all our tweets.

```
tweet_words <- campaign_tweets %>%
  mutate(text = str_replace_all(text, links, "")) %>%
  unnest_tokens(word, text, token = "tweets")
```

And we can now answer questions such as "what are the most commonly used words?":

```
tweet_words %>%
  count(word) %>%
  arrange(desc(n))
#> # A tibble: 6,620 x 2
#>   word      n
#>   <chr> <int>
#> 1 the    2329
#> 2 to     1410
#> 3 and    1239
#> 4 in     1185
#> 5 i      1143
#> # ... with 6,615 more rows
```

It is not surprising that these are the top words. The top words are not informative. The *tidytext* package has a database of these commonly used words, referred to as *stop words*, in text mining:

```
stop_words
#> # A tibble: 1,149 x 2
#>    word   lexicon
#>    <chr>  <chr>
#> 1 a       SMART
#> 2 a's     SMART
#> 3 able    SMART
#> 4 about   SMART
#> 5 above   SMART
#> # ... with 1,144 more rows
```

If we filter out rows representing stop words with `filter(!word %in% stop_words$word)`:

```
tweet_words <- campaign_tweets %>%
  mutate(text = str_replace_all(text, links, ""))  %>%
  unnest_tokens(word, text, token = "tweets") %>%
  filter(!word %in% stop_words$word )
```

we end up with a much more informative set of top 10 tweeted words:

```
tweet_words %>%
  count(word) %>%
  top_n(10, n) %>%
  mutate(word = reorder(word, n)) %>%
  arrange(desc(n))
#> # A tibble: 10 x 2
#>    word                      n
#>    <fct>                  <int>
#> 1 #trump2016               414
#> 2 hillary                  405
#> 3 people                   303
#> 4 #makeamericagreatagain   294
#> 5 america                  254
#> # ... with 5 more rows
```

Some exploration of the resulting words (not shown here) reveals a couple of unwanted characteristics in our tokens. First, some of our tokens are just numbers (years, for example). We want to remove these and we can find them using the regex ^\d+$. Second, some of our tokens come from a quote and they start with '. We want to remove the ' when it is at the start of a word so we will just `str_replace`. We add these two lines to the code above to generate our final table:

```
tweet_words <- campaign_tweets %>%
  mutate(text = str_replace_all(text, links, ""))  %>%
  unnest_tokens(word, text, token = "tweets") %>%
  filter(!word %in% stop_words$word &
           !str_detect(word, "^\\d+$")) %>%
  mutate(word = str_replace(word, "^'", ""))
```

Now that we have all our words in a table, along with information about what device was used to compose the tweet they came from, we can start exploring which words are more common when comparing Android to iPhone.

For each word, we want to know if it is more likely to come from an Android tweet or an iPhone tweet. In Section 15.10 we introduced the odds ratio as a summary statistic useful for quantifying these differences. For each device and a given word, let's call it y, we compute the odds or the ratio between the proportion of words that are y and not y and compute the ratio of those odds. Here we will have many proportions that are 0, so we use the 0.5 correction described in Section 15.10.

```
android_iphone_or <- tweet_words %>%
  count(word, source) %>%
  spread(source, n, fill = 0) %>%
  mutate(or = (Android + 0.5) / (sum(Android) - Android + 0.5) /
          ( (iPhone + 0.5) / (sum(iPhone) - iPhone + 0.5)))
```

Here are the highest odds ratios for Android

```
android_iphone_or %>% arrange(desc(or))
#> # A tibble: 5,914 x 4
#>    word        Android iPhone     or
#>    <chr>        <dbl>  <dbl>  <dbl>
#> 1 poor           13      0   23.1
#> 2 poorly         12      0   21.4
#> 3 turnberry      11      0   19.7
#> 4 @cbsnews       10      0   18.0
#> 5 angry          10      0   18.0
#> # ... with 5,909 more rows
```

and the top for iPhone:

```
android_iphone_or %>% arrange(or)
#> # A tibble: 5,914 x 4
#>    word               Android iPhone      or
#>    <chr>               <dbl>  <dbl>   <dbl>
#> 1 #makeamericagreatagain    0    294  0.00142
#> 2 #americafirst             0     71  0.00595
#> 3 #draintheswamp            0     63  0.00670
#> 4 #trump2016                3    411  0.00706
#> 5 #votetrump                0     56  0.00753
#> # ... with 5,909 more rows
```

Given that several of these words are overall low frequency words, we can impose a filter based on the total frequency like this:

```
android_iphone_or %>% filter(Android+iPhone > 100) %>%
  arrange(desc(or))
#> # A tibble: 30 x 4
#>    word     Android iPhone    or
#>    <chr>     <dbl>  <dbl> <dbl>
#> 1 @cnn         90     17  4.44
#> 2 bad         104     26  3.39
```

```
#> 3 crooked             156     49  2.72
#> 4 interviewed          76     25  2.57
#> 5 media                76     25  2.57
#> # ... with 25 more rows

android_iphone_or %>% filter(Android+iPhone > 100) %>%
  arrange(or)
#> # A tibble: 30 x 4
#>   word                  Android iPhone       or
#>   <chr>                   <dbl>  <dbl>    <dbl>
#> 1 #makeamericagreatagain      0    294  0.00142
#> 2 #trump2016                  3    411  0.00706
#> 3 join                        1    157  0.00805
#> 4 tomorrow                   24     99  0.209
#> 5 vote                       46     67  0.588
#> # ... with 25 more rows
```

We already see somewhat of a pattern in the types of words that are being tweeted more from one device versus the other. However, we are not interested in specific words but rather in the tone. Vaziri's assertion is that the Android tweets are more hyperbolic. So how can we check this with data? *Hyperbolic* is a hard sentiment to extract from words as it relies on interpreting phrases. However, words can be associated to more basic sentiment such as anger, fear, joy, and surprise. In the next section, we demonstrate basic sentiment analysis.

26.3   Sentiment analysis

In sentiment analysis, we assign a word to one or more "sentiments". Although this approach will miss context-dependent sentiments, such as sarcasm, when performed on large numbers of words, summaries can provide insights.

The first step in sentiment analysis is to assign a sentiment to each word. As we demonstrate, the **tidytext** package includes several maps or lexicons. We will also be using the **textdata** package.

```
library(tidytext)
library(textdata)
```

The bing lexicon divides words into `positive` and `negative` sentiments. We can see this using the *tidytext* function `get_sentiments`:

```
get_sentiments("bing")
```

The `AFINN` lexicon assigns a score between -5 and 5, with -5 the most negative and 5 the most positive. Note that this lexicon needs to be downloaded the first time you call the function `get_sentiment`:

```
get_sentiments("afinn")
```

The `loughran` and `nrc` lexicons provide several different sentiments. Note that these also have to be downloaded the first time you use them.

```
get_sentiments("loughran") %>% count(sentiment)
#> # A tibble: 6 x 2
#>   sentiment        n
#>   <chr>        <int>
#> 1 constraining   184
#> 2 litigious      904
#> 3 negative      2355
#> 4 positive       354
#> 5 superfluous     56
#> # ... with 1 more row
```

```
get_sentiments("nrc") %>% count(sentiment)
#> # A tibble: 10 x 2
#>   sentiment        n
#>   <chr>        <int>
#> 1 anger         1247
#> 2 anticipation   839
#> 3 disgust       1058
#> 4 fear          1476
#> 5 joy            689
#> # ... with 5 more rows
```

For our analysis, we are interested in exploring the different sentiments of each tweet so we will use the `nrc` lexicon:

```
nrc <- get_sentiments("nrc") %>%
  select(word, sentiment)
```

We can combine the words and sentiments using `inner_join`, which will only keep words associated with a sentiment. Here are 10 random words extracted from the tweets:

```
tweet_words %>% inner_join(nrc, by = "word") %>%
  select(source, word, sentiment) %>%
  sample_n(5)
#>     source     word     sentiment
#> 1   iPhone  failing          fear
#> 2  Android    proud         trust
#> 3  Android     time  anticipation
#> 4   iPhone horrible       disgust
#> 5  Android  failing         anger
```

Now we are ready to perform a quantitative analysis comparing Android and iPhone by comparing the sentiments of the tweets posted from each device. Here we could perform a tweet-by-tweet analysis, assigning a sentiment to each tweet. However, this will be challenging since each tweet will have several sentiments attached to it, one for each word appearing in the lexicon. For illustrative purposes, we will perform a much simpler analysis: we will count and compare the frequencies of each sentiment appearing in each device.

```
sentiment_counts <- tweet_words %>%
  left_join(nrc, by = "word") %>%
  count(source, sentiment) %>%
  spread(source, n) %>%
  mutate(sentiment = replace_na(sentiment, replace = "none"))
sentiment_counts
#> # A tibble: 11 x 3
#>    sentiment     Android iPhone
#>    <chr>           <int>  <int>
#> 1 anger             958    528
#> 2 anticipation      910    715
#> 3 disgust           638    322
#> 4 fear              795    486
#> 5 joy               688    535
#> # ... with 6 more rows
```

For each sentiment, we can compute the odds of being in the device: proportion of words with sentiment versus proportion of words without, and then compute the odds ratio comparing the two devices.

```
sentiment_counts %>%
  mutate(Android = Android / (sum(Android) - Android) ,
         iPhone = iPhone / (sum(iPhone) - iPhone),
         or = Android/iPhone) %>%
  arrange(desc(or))
#> # A tibble: 11 x 4
#>    sentiment Android iPhone     or
#>    <chr>       <dbl>  <dbl>  <dbl>
#> 1 disgust    0.0299 0.0186   1.61
#> 2 anger      0.0456 0.0309   1.47
#> 3 negative   0.0807 0.0556   1.45
#> 4 sadness    0.0424 0.0301   1.41
#> 5 fear       0.0375 0.0284   1.32
#> # ... with 6 more rows
```

So we do see some differences and the order is interesting: the largest three sentiments are disgust, anger, and negative! But are these differences just due to chance? How does this compare if we are just assigning sentiments at random? To answer this question we can compute, for each sentiment, an odds ratio and a confidence interval, as defined in Section 15.10. We will add the two values we need to form a two-by-two table and the odds ratio:

```
library(broom)
log_or <- sentiment_counts %>%
  mutate(log_or = log((Android / (sum(Android) - Android)) /
      (iPhone / (sum(iPhone) - iPhone))),
          se = sqrt(1/Android + 1/(sum(Android) - Android) +
                    1/iPhone + 1/(sum(iPhone) - iPhone)),
          conf.low = log_or - qnorm(0.975)*se,
          conf.high = log_or + qnorm(0.975)*se) %>%
  arrange(desc(log_or))
```

```
log_or
#> # A tibble: 11 x 7
#>   sentiment Android iPhone log_or      se conf.low conf.high
#>   <chr>       <int>  <int>  <dbl>   <dbl>    <dbl>     <dbl>
#> 1 disgust       638    322  0.474  0.0691    0.338     0.609
#> 2 anger         958    528  0.389  0.0552    0.281     0.497
#> 3 negative     1641    929  0.371  0.0424    0.288     0.454
#> 4 sadness       894    515  0.342  0.0563    0.232     0.452
#> 5 fear          795    486  0.280  0.0585    0.165     0.394
#> # ... with 6 more rows
```

A graphical visualization shows some sentiments that are clearly overrepresented:

```
log_or %>%
  mutate(sentiment = reorder(sentiment, log_or)) %>%
  ggplot(aes(x = sentiment, ymin = conf.low, ymax = conf.high)) +
  geom_errorbar() +
  geom_point(aes(sentiment, log_or)) +
  ylab("Log odds ratio for association between Android and sentiment") +
  coord_flip()
```

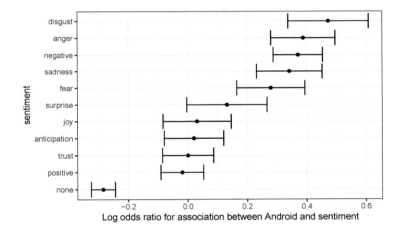

We see that the disgust, anger, negative, sadness, and fear sentiments are associated with the Android in a way that is hard to explain by chance alone. Words not associated to a sentiment were strongly associated with the iPhone source, which is in agreement with the original claim about hyperbolic tweets.

If we are interested in exploring which specific words are driving these differences, we can refer back to our android_iphone_or object:

```
android_iphone_or %>% inner_join(nrc) %>%
  filter(sentiment == "disgust" & Android + iPhone > 10) %>%
  arrange(desc(or))
#> Joining, by = "word"
#> # A tibble: 20 x 5
```

```
#>     word         Android iPhone    or sentiment
#>     <chr>         <dbl>   <dbl> <dbl> <chr>
#> 1 mess              13       2  4.62 disgust
#> 2 finally           12       2  4.28 disgust
#> 3 unfair            12       2  4.28 disgust
#> 4 bad              104      26  3.39 disgust
#> 5 terrible          31       8  3.17 disgust
#> # ... with 15 more rows
```

and we can make a graph:

```
android_iphone_or %>% inner_join(nrc, by = "word") %>%
  mutate(sentiment = factor(sentiment, levels = log_or$sentiment)) %>%
  mutate(log_or = log(or)) %>%
  filter(Android + iPhone > 10 & abs(log_or)>1) %>%
  mutate(word = reorder(word, log_or)) %>%
  ggplot(aes(word, log_or, fill = log_or < 0)) +
  facet_wrap(~sentiment, scales = "free_x", nrow = 2) +
  geom_bar(stat="identity", show.legend = FALSE) +
  theme(axis.text.x = element_text(angle = 90, hjust = 1))
```

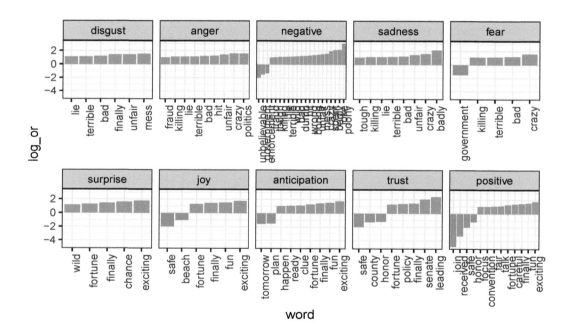

This is just a simple example of the many analyses one can perform with tidytext. To learn more, we again recommend the Tidy Text Mining book[4].

---

[4]https://www.tidytextmining.com/

26.4   Exercises

Project Gutenberg is a digital archive of public domain books. The R package **gutenbergr** facilitates the importation of these texts into R.

You can install and load by typing:

```
install.packages("gutenbergr")
library(gutenbergr)
```

You can see the books that are available like this:

```
gutenberg_metadata
```

1. Use `str_detect` to find the ID of the novel *Pride and Prejudice*.

2. We notice that there are several versions. The `gutenberg_works()` function filters this table to remove replicates and include only English language works. Read the help file and use this function to find the ID for *Pride and Prejudice*.

3. Use the `gutenberg_download` function to download the text for Pride and Prejudice. Save it to an object called `book`.

4. Use the **tidytext** package to create a tidy table with all the words in the text. Save the table in an object called `words`

5. We will later make a plot of sentiment versus location in the book. For this, it will be useful to add a column with the word number to the table.

6. Remove the stop words and numbers from the `words` object. Hint: use the `anti_join`.

7. Now use the `AFINN` lexicon to assign a sentiment value to each word.

8. Make a plot of sentiment score versus location in the book and add a smoother.

9. Assume there are 300 words per page. Convert the locations to pages and then compute the average sentiment in each page. Plot that average score by page. Add a smoother that appears to go through data.

Part V

Machine Learning

27

Introduction to machine learning

Perhaps the most popular data science methodologies come from the field of *machine learning*. Machine learning success stories include the handwritten zip code readers implemented by the postal service, speech recognition technology such as Apple's Siri, movie recommendation systems, spam and malware detectors, housing price predictors, and driverless cars. Although today Artificial Intelligence and machine learning are often used interchangeably, we make the following distinction: while the first artificial intelligence algorithms, such as those used by chess playing machines, implemented decision making based on programmable rules derived from theory or first principles, in machine learning decisions are based on algorithms **built with data**.

27.1 Notation

In machine learning, data comes in the form of:

1. the *outcome* we want to predict and
2. the *features* that we will use to predict the outcome

We want to build an algorithm that takes feature values as input and returns a prediction for the outcome when we don't know the outcome. The machine learning approach is to *train* an algorithm using a dataset for which we do know the outcome, and then apply this algorithm in the future to make a prediction when we don't know the outcome.

Here we will use Y to denote the outcome and $X_1, \ldots, X_p$ to denote features. Note that features are sometimes referred to as predictors or covariates. We consider all these to be synonyms.

Prediction problems can be divided into categorical and continuous outcomes. For categorical outcomes, Y can be any one of K classes. The number of classes can vary greatly across applications. For example, in the digit reader data, $K = 10$ with the classes being the digits 0, 1, 2, 3, 4, 5, 6, 7, 8, and 9. In speech recognition, the outcomes are all possible words or phrases we are trying to detect. Spam detection has two outcomes: spam or not spam. In this book, we denote the K categories with indexes $k = 1, \ldots, K$. However, for binary data we will use $k = 0, 1$ for mathematical conveniences that we demonstrate later.

The general setup is as follows. We have a series of features and an unknown outcome we want to predict:

outcome	feature 1	feature 2	feature 3	feature 4	feature 5
?	X_1	X_2	X_3	X_4	X_5

To *build a model* that provides a prediction for any set of observed values $X_1 = x_1, X_2 = x_2, \ldots X_5 = x_5$, we collect data for which we know the outcome:

outcome	feature 1	feature 2	feature 3	feature 4	feature 5
y_1	$x_{1,1}$	$x_{1,2}$	$x_{1,3}$	$x_{1,4}$	$x_{1,5}$
y_2	$x_{2,1}$	$x_{2,2}$	$x_{2,3}$	$x_{2,4}$	$x_{2,5}$
$\vdots$	$\vdots$	$\vdots$	$\vdots$	$\vdots$	$\vdots$
y_n	$x_{n,1}$	$x_{n,2}$	$x_{n,3}$	$x_{n,4}$	$x_{n,5}$

When the output is continuous we refer to the machine learning task as *prediction*, and the main output of the model is a function f that automatically produces a prediction, denoted with $\hat{y}$, for any set of predictors: $\hat{y} = f(x_1, x_2, \ldots, x_p)$. We use the term *actual outcome* to denote what we ended up observing. So we want the prediction $\hat{y}$ to match the actual outcome y as well as possible. Because our outcome is continuous, our predictions $\hat{y}$ will not be either exactly right or wrong, but instead we will determine an *error* defined as the difference between the prediction and the actual outcome $y - \hat{y}$.

When the outcome is categorical, we refer to the machine learning task as *classification*, and the main output of the model will be a *decision rule* which prescribes which of the K classes we should predict. In this scenario, most models provide functions of the predictors for each class k, $f_k(x_1, x_2, \ldots, x_p)$, that are used to make this decision. When the data is binary a typical decision rules looks like this: if $f_1(x_1, x_2, \ldots, x_p) > C$, predict category 1, if not the other category, with C a predetermined cutoff. Because the outcomes are categorical, our predictions will be either right or wrong.

Notice that these terms vary among courses, text books, and other publications. Often *prediction* is used for both categorical and continuous outcomes, and the term *regression* can be used for the continuous case. Here we avoid using *regression* to avoid confusion with our previous use of the term *linear regression*. In most cases it will be clear if our outcomes are categorical or continuous, so we will avoid using these terms when possible.

27.2   An example

Let's consider the zip code reader example. The first step in handling mail received in the post office is sorting letters by zip code:

Originally, humans had to sort these by hand. To do this, they had to read the zip codes on each letter. Today, thanks to machine learning algorithms, a computer can read zip codes and then a robot sorts the letters. In this part of the book, we will learn how to build algorithms that can read a digit.

The first step in building an algorithm is to understand what are the outcomes and features. Below are three images of written digits. These have already been read by a human and assigned an outcome Y. These are considered known and serve as the training set.

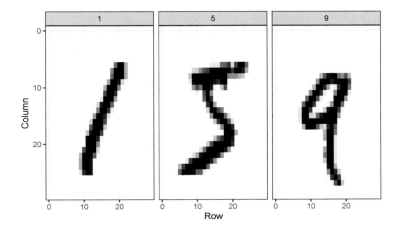

The images are converted into $28 \times 28 = 784$ pixels and, for each pixel, we obtain a grey scale intensity between 0 (white) and 255 (black), which we consider continuous for now. The following plot shows the individual features for each image:

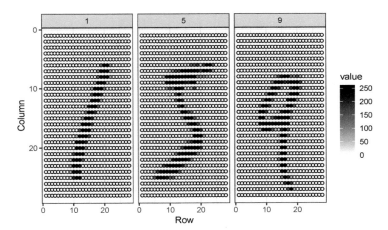

For each digitized image i, we have a categorical outcome Y_i which can be one of 10 values $(0, 1, 2, 3, 4, 5, 6, 7, 8, 9)$, and features $X_{i,1}, \ldots, X_{i,784}$. We use bold face $\mathbf{X}_i = (X_{i,1}, \ldots, X_{i,784})$ to distinguish the vector of predictors from the individual predictors. When referring to an arbitrary set of features rather than a specific image in our dataset, we drop the index i and use Y and $\mathbf{X} = (X_1, \ldots, X_{784})$. We use upper case variables because, in general, we think of the predictors as random variables. We use lower case, for example $\mathbf{X} = \mathbf{x}$, to denote observed values. When we code we stick to lower case.

The machine learning task is to build an algorithm that returns a prediction for any of the possible values of the features. Here, we will learn several approaches to building these algorithms. Although at this point it might seem impossible to achieve this, we will start with simple examples and build up our knowledge until we can attack more complex ones. In fact, we start with an artificially simple example with just one predictor and then move on to a slightly more realistic example with two predictors. Once we understand these, we will attack real-world machine learning challenges involving many predictors.

27.3   Exercises

1. For each of the following, determine if the outcome is continuous or categorical:

   a. Digit reader
   b. Movie recommendations
   c. Spam filter
   d. Hospitalizations
   e. Siri (speech recognition)

2. How many features are available to us for prediction in the digits dataset?

3. In the digit reader example, the outcomes are stored here:

```
library(dslabs)
y <- mnist$train$labels
```

Do the following operations have a practical meaning?

```
y[5] + y[6]
y[5] > y[6]
```

Pick the best answer:

   a. Yes, because $9 + 2 = 11$ and $9 > 2$.
   b. No, because y is not a numeric vector.
   c. No, because 11 is not a digit. It's two digits.
   d. No, because these are labels representing a category not a number. A 9 represents a class not the number 9.

27.4   Evaluation metrics

Before we start describing approaches to optimize the way we build algorithms, we first need to define what we mean when we say one approach is better than another. In this section, we focus on describing ways in which machine learning algorithms are evaluated. Specifically, we need to quantify what we mean by "better".

For our first introduction to machine learning concepts, we will start with a boring and simple example: how to predict sex using height. As we explain machine learning step by step, this example will let us set down the first building block. Soon enough, we will be attacking more interesting challenges. We use the **caret** package, which has several useful functions for building and assessing machine learning methods and we introduce in more detail in Section 30.

```
library(tidyverse)
library(caret)
```

For a first example, we use the height data in dslabs:

```
library(dslabs)
data(heights)
```

We start by defining the outcome and predictors.

```
y <- heights$sex
x <- heights$height
```

In this case, we have only one predictor, height, and `y` is clearly a categorical outcome since observed values are either `Male` or `Female`. We know that we will not be able to predict Y very accurately based on X because male and female average heights are not that different relative to within group variability. But can we do better than guessing? To answer this question, we need a quantitative definition of better.

27.4.1 Training and test sets

Ultimately, a machine learning algorithm is evaluated on how it performs in the real world with completely new datasets. However, when developing an algorithm, we usually have a dataset for which we know the outcomes, as we do with the heights: we know the sex of every student in our dataset. Therefore, to mimic the ultimate evaluation process, we typically split the data into two parts and act as if we don't know the outcome for one of these. We stop pretending we don't know the outcome to evaluate the algorithm, but only *after* we are done constructing it. We refer to the group for which we know the outcome, and use to develop the algorithm, as the *training* set. We refer to the group for which we pretend we don't know the outcome as the *test* set.

A standard way of generating the training and test sets is by randomly splitting the data. The **caret** package includes the function `createDataPartition` that helps us generates indexes for randomly splitting the data into training and test sets:

```
set.seed(2007)
test_index <- createDataPartition(y, times = 1, p = 0.5, list = FALSE)
```

The argument `times` is used to define how many random samples of indexes to return, the argument `p` is used to define what proportion of the data is represented by the index, and the argument `list` is used to decide if we want the indexes returned as a list or not. We can use the result of the `createDataPartition` function call to define the training and test sets like this:

```
test_set <- heights[test_index, ]
train_set <- heights[-test_index, ]
```

We will now develop an algorithm using **only** the training set. Once we are done developing the algorithm, we will *freeze* it and evaluate it using the test set. The simplest way to evaluate the algorithm when the outcomes are categorical is by simply reporting the proportion of cases that were correctly predicted **in the test set**. This metric is usually referred to as *overall accuracy*.

27.4.2   Overall accuracy

To demonstrate the use of overall accuracy, we will build two competing algorithms and compare them.

Let's start by developing the simplest possible machine algorithm: guessing the outcome.

```
y_hat <- sample(c("Male", "Female"), length(test_index), replace = TRUE)
```

Note that we are completely ignoring the predictor and simply guessing the sex.

In machine learning applications, it is useful to use factors to represent the categorical outcomes because R functions developed for machine learning, such as those in the **caret** package, require or recommend that categorical outcomes be coded as factors. So convert `y_hat` to factors using the `factor` function:

```
y_hat <- sample(c("Male", "Female"), length(test_index), replace = TRUE) %>%
  factor(levels = levels(test_set$sex))
```

The *overall accuracy* is simply defined as the overall proportion that is predicted correctly:

```
mean(y_hat == test_set$sex)
#> [1] 0.51
```

Not surprisingly, our accuracy is about 50%. We are guessing!

Can we do better? Exploratory data analysis suggests we can because, on average, males are slightly taller than females:

```
heights %>% group_by(sex) %>% summarize(mean(height), sd(height))
#> # A tibble: 2 x 3
#>   sex     `mean(height)` `sd(height)`
#>   <fct>            <dbl>        <dbl>
#> 1 Female           64.9         3.76
#> 2 Male             69.3         3.61
```

But how do we make use of this insight? Let's try another simple approach: predict `Male` if height is within two standard deviations from the average male:

```
y_hat <- ifelse(x > 62, "Male", "Female") %>%
  factor(levels = levels(test_set$sex))
```

The accuracy goes up from 0.50 to about 0.80:

```
mean(y == y_hat)
#> [1] 0.793
```

But can we do even better? In the example above, we used a cutoff of 62, but we can examine the accuracy obtained for other cutoffs and then pick the value that provides the best results. But remember, **it is important that we optimize the cutoff using only the training set**: the test set is only for evaluation. Although for this simplistic example it is not much of a problem, later we will learn that evaluating an algorithm on the training set can lead to *overfitting*, which often results in dangerously over-optimistic assessments.

Here we examine the accuracy of 10 different cutoffs and pick the one yielding the best result:

```
cutoff <- seq(61, 70)
accuracy <- map_dbl(cutoff, function(x){
  y_hat <- ifelse(train_set$height > x, "Male", "Female") %>%
    factor(levels = levels(test_set$sex))
  mean(y_hat == train_set$sex)
})
```

We can make a plot showing the accuracy obtained on the training set for males and females:

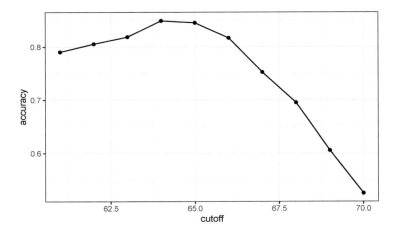

We see that the maximum value is:

```
max(accuracy)
#> [1] 0.85
```

which is much higher than 0.5. The cutoff resulting in this accuracy is:

```
best_cutoff <- cutoff[which.max(accuracy)]
best_cutoff
#> [1] 64
```

We can now test this cutoff on our test set to make sure our accuracy is not overly optimistic:

```
y_hat <- ifelse(test_set$height > best_cutoff, "Male", "Female") %>%
  factor(levels = levels(test_set$sex))
y_hat <- factor(y_hat)
mean(y_hat == test_set$sex)
#> [1] 0.804
```

We see that it is a bit lower than the accuracy observed for the training set, but it is still better than guessing. And by testing on a dataset that we did not train on, we know our result is not due to cherry-picking a good result.

27.4.3   The confusion matrix

The prediction rule we developed in the previous section predicts `Male` if the student is taller than 64 inches. Given that the average female is about 64 inches, this prediction rule seems wrong. What happened? If a student is the height of the average female, shouldn't we predict `Female`?

Generally speaking, overall accuracy can be a deceptive measure. To see this, we will start by constructing what is referred to as the *confusion matrix*, which basically tabulates each combination of prediction and actual value. We can do this in R using the function `table`:

```
table(predicted = y_hat, actual = test_set$sex)
#>            actual
#> predicted Female Male
#>    Female     48   32
#>    Male       71  374
```

If we study this table closely, it reveals a problem. If we compute the accuracy separately for each sex, we get:

```
test_set %>%
  mutate(y_hat = y_hat) %>%
  group_by(sex) %>%
  summarize(accuracy = mean(y_hat == sex))
#> # A tibble: 2 x 2
#>   sex    accuracy
#>   <fct>     <dbl>
#> 1 Female    0.403
#> 2 Male      0.921
```

There is an imbalance in the accuracy for males and females: too many females are predicted to be male. We are calling almost half of the females male! How can our overall accuracy be so high then? This is because the *prevalence* of males in this dataset is high. These heights were collected from three data sciences courses, two of which had more males enrolled:

```
prev <- mean(y == "Male")
prev
#> [1] 0.773
```

So when computing overall accuracy, the high percentage of mistakes made for females is outweighed by the gains in correct calls for men. **This can actually be a big problem in machine learning.** If your training data is biased in some way, you are likely to develop algorithms that are biased as well. The fact that we used a test set does not matter because it is also derived from the original biased dataset. This is one of the reasons we look at metrics other than overall accuracy when evaluating a machine learning algorithm.

There are several metrics that we can use to evaluate an algorithm in a way that prevalence does not cloud our assessment, and these can all be derived from the confusion matrix. A general improvement to using overall accuracy is to study *sensitivity* and *specificity* separately.

27.4.4 Sensitivity and specificity

To define sensitivity and specificity, we need a binary outcome. When the outcomes are categorical, we can define these terms for a specific category. In the digits example, we can ask for the specificity in the case of correctly predicting 2 as opposed to some other digit. Once we specify a category of interest, then we can talk about positive outcomes, $Y = 1$, and negative outcomes, $Y = 0$.

In general, *sensitivity* is defined as the ability of an algorithm to predict a positive outcome when the actual outcome is positive: $\hat{Y} = 1$ when $Y = 1$. Because an algorithm that calls everything positive ($\hat{Y} = 1$ no matter what) has perfect sensitivity, this metric on its own is not enough to judge an algorithm. For this reason, we also examine *specificity*, which is generally defined as the ability of an algorithm to not predict a positive $\hat{Y} = 0$ when the actual outcome is not a positive $Y = 0$. We can summarize in the following way:

- High sensitivity: $Y = 1 \implies \hat{Y} = 1$
- High specificity: $Y = 0 \implies \hat{Y} = 0$

Although the above is often considered the definition of specificity, another way to think of specificity is by the proportion of positive calls that are actually positive:

- High specificity: $\hat{Y} = 1 \implies Y = 1$.

To provide precise definitions, we name the four entries of the confusion matrix:

	Actually Positive	Actually Negative
Predicted positive	True positives (TP)	False positives (FP)
Predicted negative	False negatives (FN)	True negatives (TN)

Sensitivity is typically quantified by $TP/(TP + FN)$, the proportion of actual positives (the first column $= TP + FN$) that are called positives (TP). This quantity is referred to as the *true positive rate* (TPR) or *recall*.

Specificity is defined as $TN/(TN + FP)$ or the proportion of negatives (the second column $= FP + TN$) that are called negatives (TN). This quantity is also called the true negative rate (TNR). There is another way of quantifying specificity which is $TP/(TP + FP)$ or the proportion of outcomes called positives (the first row or $TP + FP$) that are actually positives (TP). This quantity is referred to as *positive predictive value (PPV)* and also as *precision*. Note that, unlike TPR and TNR, precision depends on prevalence since higher prevalence implies you can get higher precision even when guessing.

The multiple names can be confusing, so we include a table to help us remember the terms. The table includes a column that shows the definition if we think of the proportions as probabilities.

Measure of	Name 1	Name 2	Definition	Probability representation
sensitivity	TPR	Recall	$\frac{TP}{TP+FN}$	$\Pr(\hat{Y} = 1 \mid Y = 1)$
specificity	TNR	1-FPR	$\frac{TN}{TN+FP}$	$\Pr(\hat{Y} = 0 \mid Y = 0)$
specificity	PPV	Precision	$\frac{TP}{TP+FP}$	$\Pr(Y = 1 \mid \hat{Y} = 1)$

Here TPR is True Positive Rate, FPR is False Positive Rate, and PPV is Positive Predictive Value. The **caret** function `confusionMatrix` computes all these metrics for us once we define what category "positive" is. The function expects factors as input, and the first level is considered the positive outcome or $Y = 1$. In our example, `Female` is the first level because it comes before `Male` alphabetically. If you type this into R you will see several metrics including accuracy, sensitivity, specificity, and PPV.

```
cm <- confusionMatrix(data = y_hat, reference = test_set$sex)
```

You can acceess these directly, for example, like this:

```
cm$overall["Accuracy"]
#> Accuracy
#>    0.804
cm$byClass[c("Sensitivity","Specificity", "Prevalence")]
#> Sensitivity Specificity  Prevalence
#>       0.403       0.921       0.227
```

We can see that the high overall accuracy is possible despite relatively low sensitivity. As we hinted at above, the reason this happens is because of the low prevalence (0.23): the proportion of females is low. Because prevalence is low, failing to predict actual females as females (low sensitivity) does not lower the accuracy as much as failing to predict actual males as males (low specificity). This is an example of why it is important to examine sensitivity and specificity and not just accuracy. Before applying this algorithm to general datasets, we need to ask ourselves if prevalence will be the same.

27.4.5 Balanced accuracy and F_1 score

Although we usually recommend studying both specificity and sensitivity, very often it is useful to have a one-number summary, for example for optimization purposes. One metric that is preferred over overall accuracy is the average of specificity and sensitivity, referred to as *balanced accuracy*. Because specificity and sensitivity are rates, it is more appropriate to compute the *harmonic* average. In fact, the F_1-*score*, a widely used one-number summary, is the harmonic average of precision and recall:

$$\frac{1}{\frac{1}{2}\left(\frac{1}{\text{recall}} + \frac{1}{\text{precision}}\right)}$$

Because it is easier to write, you often see this harmonic average rewritten as:

$$2 \times \frac{\text{precision} \cdot \text{recall}}{\text{precision} + \text{recall}}$$

when defining F_1.

Remember that, depending on the context, some types of errors are more costly than others. For example, in the case of plane safety, it is much more important to maximize sensitivity over specificity: failing to predict a plane will malfunction before it crashes is a much more costly error than grounding a plane when, in fact, the plane is in perfect condition. In a capital murder criminal case, the opposite is true since a false positive can lead to executing an innocent person. The F_1-score can be adapted to weigh specificity and sensitivity differently. To do this, we define β to represent how much more important sensitivity is compared to specificity and consider a weighted harmonic average:

$$\frac{1}{\frac{\beta^2}{1+\beta^2}\frac{1}{\text{recall}} + \frac{1}{1+\beta^2}\frac{1}{\text{precision}}}$$

The `F_meas` function in the **caret** package computes this summary with `beta` defaulting to 1.

Let's rebuild our prediction algorithm, but this time maximizing the F-score instead of overall accuracy:

```
cutoff <- seq(61, 70)
F_1 <- map_dbl(cutoff, function(x){
  y_hat <- ifelse(train_set$height > x, "Male", "Female") %>%
    factor(levels = levels(test_set$sex))
  F_meas(data = y_hat, reference = factor(train_set$sex))
})
```

As before, we can plot these F_1 measures versus the cutoffs:

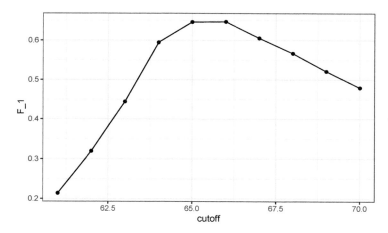

We see that it is maximized at F_1 value of:

```
max(F_1)
#> [1] 0.647
```

This maximum is achieved when we use the following cutoff:

```
best_cutoff <- cutoff[which.max(F_1)]
best_cutoff
#> [1] 66
```

A cutoff of 65 makes more sense than 64. Furthermore, it balances the specificity and sensitivity of our confusion matrix:

```
y_hat <- ifelse(test_set$height > best_cutoff, "Male", "Female") %>%
  factor(levels = levels(test_set$sex))
sensitivity(data = y_hat, reference = test_set$sex)
#> [1] 0.63
specificity(data = y_hat, reference = test_set$sex)
#> [1] 0.833
```

We now see that we do much better than guessing, that both sensitivity and specificity are relatively high, and that we have built our first machine learning algorithm. It takes height as a predictor and predicts female if you are 65 inches or shorter.

27.4.6 Prevalence matters in practice

A machine learning algorithm with very high sensitivity and specificity may not be useful in practice when prevalence is close to either 0 or 1. To see this, consider the case of a doctor that specializes in a rare disease and is interested in developing an algorithm for predicting who has the disease. The doctor shares data with you and you then develop an algorithm with very high sensitivity. You explain that this means that if a patient has the disease, the algorithm is very likely to predict correctly. You also tell the doctor that you are also concerned because, based on the dataset you analyzed, 1/2 the patients have the disease: $\Pr(\hat{Y} = 1)$. The doctor is neither concerned nor impressed and explains that what is important is the precision of the test: $\Pr(Y = 1|\hat{Y} = 1)$. Using Bayes theorem, we can connect the two measures:

$$\Pr(Y = 1 \mid \hat{Y} = 1) = \Pr(\hat{Y} = 1 \mid Y = 1)\frac{\Pr(Y = 1)}{\Pr(\hat{Y} = 1)}$$

The doctor knows that the prevalence of the disease is 5 in 1,000, which implies that $\Pr(Y = 1)/\Pr(\hat{Y} = 1) = 1/100$ and therefore the precision of your algorithm is less than 0.01. The doctor does not have much use for your algorithm.

27.4.7 ROC and precision-recall curves

When comparing the two methods (guessing versus using a height cutoff), we looked at accuracy and F_1. The second method clearly outperformed the first. However, while we considered several cutoffs for the second method, for the first we only considered one approach: guessing with equal probability. Note that guessing `Male` with higher probability would give us higher accuracy due to the bias in the sample:

```
p <- 0.9
n <- length(test_index)
y_hat <- sample(c("Male", "Female"), n, replace = TRUE, prob=c(p, 1-p)) %>%
  factor(levels = levels(test_set$sex))
mean(y_hat == test_set$sex)
#> [1] 0.739
```

But, as described above, this would come at the cost of lower sensitivity. The curves we describe in this section will help us see this.

Remember that for each of these parameters, we can get a different sensitivity and specificity. For this reason, a very common approach to evaluating methods is to compare them graphically by plotting both.

A widely used plot that does this is the *receiver operating characteristic* (ROC) curve. If you are wondering where this name comes from, you can consult the ROC Wikipedia page[1].

The ROC curve plots sensitivity (TPR) versus 1 - specificity or the false positive rate (FPR). Here we compute the TPR and FPR needed for different probabilities of guessing male:

```
probs <- seq(0, 1, length.out = 10)
guessing <- map_df(probs, function(p){
  y_hat <-
    sample(c("Male", "Female"), n, replace = TRUE, prob=c(p, 1-p)) %>%
    factor(levels = c("Female", "Male"))
  list(method = "Guessing",
       FPR = 1 - specificity(y_hat, test_set$sex),
       TPR = sensitivity(y_hat, test_set$sex))
})
```

---

[1] https://en.wikipedia.org/wiki/Receiver_operating_characteristic

We can use similar code to compute these values for our our second approach. By plotting both curves together, we are able to compare sensitivity for different values of specificity:

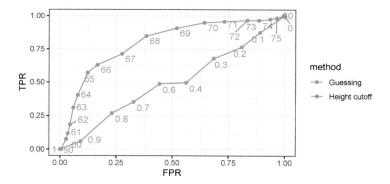

We can see that we obtain higher sensitivity with this approach for all values of specificity, which implies it is in fact a better method. Note that ROC curves for guessing always fall on the identiy line. Also note that when making ROC curves, it is often nice to add the cutoff associated with each point.

The packages **pROC** and **plotROC** are useful for generating these plots.

ROC curves have one weakness and it is that neither of the measures plotted depends on prevalence. In cases in which prevalence matters, we may instead make a precision-recall plot. The idea is similar, but we instead plot precision against recall:

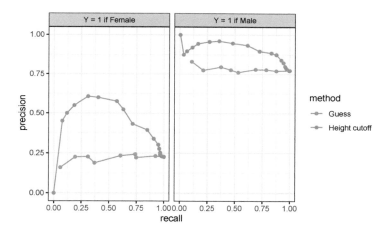

From this plot we immediately see that the precision of guessing is not high. This is because the prevalence is low. We also see that if we change positives to mean Male instead of Female, the ROC curve remains the same, but the precision recall plot changes.

27.4.8   The loss function

Up to now we have described evaluation metrics that apply exclusively to categorical data. Specifically, for binary outcomes, we have described how sensitivity, specificity, accuracy, and F_1 can be used as quantification. However, these metrics are not useful for continuous outcomes. In this section, we describe how the general approach to defining "best" in machine learning is to define a *loss function*, which can be applied to both categorical and continuous data.

The most commonly used loss function is the squared loss function. If $\hat{y}$ is our predictor and y is the observed outcome, the squared loss function is simply:

$$(\hat{y} - y)^2$$

Because we often have a test set with many observations, say N, we use the mean squared error (MSE):

$$\text{MSE} = \frac{1}{N}\text{RSS} = \frac{1}{N}\sum_{i=1}^{N}(\hat{y}_i - y_i)^2$$

In practice, we often report the root mean squared error (RMSE), which is $\sqrt{\text{MSE}}$, because it is in the same units as the outcomes. But doing the math is often easier with the MSE and it is therefore more commonly used in textbooks, since these usually describe theoretical properties of algorithms.

If the outcomes are binary, both RMSE and MSE are equivalent to accuracy, since $(\hat{y} - y)^2$ is 0 if the prediction was correct and 1 otherwise. In general, our goal is to build an algorithm that minimizes the loss so it is as close to 0 as possible.

Because our data is usually a random sample, we can think of the MSE as a random variable and the observed MSE can be thought of as an estimate of the expected MSE, which in mathematical notation we write like this:

$$\text{E}\left\{\frac{1}{N}\sum_{i=1}^{N}(\hat{Y}_i - Y_i)^2\right\}$$

This is a theoretical concept because in practice we only have one dataset to work with. But in theory, we think of having a very large number of random samples (call it B), apply our algorithm to each, obtain an MSE for each random sample, and think of the expected MSE as:

$$\frac{1}{B}\sum_{b=1}^{B}\frac{1}{N}\sum_{i=1}^{N}\left(\hat{y}_i^b - y_i^b\right)^2$$

with y_i^b denoting the ith observation in the bth random sample and $\hat{y}_i^b$ the resulting prediction obtained from applying the exact same algorithm to the bth random sample. Again, in practice we only observe one random sample, so the expected MSE is only theoretical. However, in Chapter 29 we describe an approach to estimating the MSE that tries to mimic this theoretical quantity.

Note that there are loss functions other than the squared loss. For example, the *Mean Absolute Error* uses absolute values, $|\hat{Y}_i - Y_i|$ instead of squaring the errors $(\hat{Y}_i - Y_i)^2$. However, in this book we focus on minimizing square loss since it is the most widely used.

27.5   Exercises

The `reported_height` and `height` datasets were collected from three classes taught in the Departments of Computer Science and Biostatistics, as well as remotely through the Extension School. The biostatistics class was taught in 2016 along with an online version offered by the Extension School. On 2016-01-25 at 8:15 AM, during one of the lectures, the instructors asked students to fill in the sex and height questionnaire that populated the `reported_height` dataset. The online students filled the survey during the next few days, after the lecture was posted online. We can use this insight to define a variable, call it `type`, to denote the type of student: `inclass` or `online`:

```
library(lubridate)
data("reported_heights")
dat <- mutate(reported_heights, date_time = ymd_hms(time_stamp)) %>%
  filter(date_time >= make_date(2016, 01, 25) &
           date_time < make_date(2016, 02, 1)) %>%
  mutate(type = ifelse(day(date_time) == 25 & hour(date_time) == 8 &
                         between(minute(date_time), 15, 30),
                       "inclass", "online")) %>% select(sex, type)
x <- dat$type
y <- factor(dat$sex, c("Female", "Male"))
```

1. Show summary statistics that indicate that the `type` is predictive of sex.

2. Instead of using height to predict sex, use the `type` variable.

3. Show the confusion matrix.

4. Use the `confusionMatrix` function in the **caret** package to report accuracy.

5. Now use the `sensitivity` and `specificity` functions to report specificity and sensitivity.

6. What is the prevalence (% of females) in the `dat` dataset defined above?

27.6   Conditional probabilities and expectations

In machine learning applications, we rarely can predict outcomes perfectly. For example, spam detectors often miss emails that are clearly spam, Siri often misunderstands the words we are saying, and your bank at times thinks your card was stolen when it was not. The most common reason for not being able to build perfect algorithms is that it is impossible. To see this, note that most datasets will include groups of observations with the same exact observed values for all predictors, but with different outcomes. Because our prediction rules are functions, equal inputs (the predictors) implies equal outputs (the predictions). Therefore, for a challenge in which the same predictors are associated with different outcomes across different individual observations, it is impossible to predict correctly for all these cases. We saw a simple example of this in the previous section: for any given height x, you will have both males and females that are x inches tall.

However, none of this means that we can't build useful algorithms that are much better than guessing, and in some cases better than expert opinions. To achieve this in an optimal way, we make use of probabilistic representations of the problem based on the ideas presented in Section 17.3. Observations with the same observed values for the predictors may not all be the same, but we can assume that they all have the same probability of this class or that class. We will write this idea out mathematically for the case of categorical data.

27.6.1 Conditional probabilities

We use the notation $(X_1 = x_1, \ldots, X_p = x_p)$ to represent the fact that we have observed values $x_1, \ldots, x_p$ for covariates $X_1, \ldots, X_p$. This does not imply that the outcome Y will take a specific value. Instead, it implies a specific probability. In particular, we denote the *conditional probabilities* for each class k:

$$\Pr(Y = k \mid X_1 = x_1, \ldots, X_p = x_p), \text{ for } k = 1, \ldots, K$$

To avoid writing out all the predictors, we will use the bold letters like this: $\mathbf{X} \equiv (X_1, \ldots, X_p)$ and $\mathbf{x} \equiv (x_1, \ldots, x_p)$. We will also use the following notation for the conditional probability of being class k:

$$p_k(\mathbf{x}) = \Pr(Y = k \mid \mathbf{X} = \mathbf{x}), \text{ for } k = 1, \ldots, K$$

Note: We will be using the $p(x)$ notation to represent conditional probabilities as functions of the predictors. Do not confuse it with the p that represents the number of predictors.

These probabilities guide the construction of an algorithm that makes the best prediction: for any given $\mathbf{x}$, we will predict the class k with the largest probability among $p_1(x), p_2(x), \ldots p_K(x)$. In mathematical notation, we write it like this: $\hat{Y} = \max_k p_k(\mathbf{x})$.

In machine learning, we refer to this as *Bayes' Rule*. But keep in mind that this is a theoretical rule since in practice we don't know $p_k(\mathbf{x}), k = 1, \ldots, K$. In fact, estimating these conditional probabilities can be thought of as the main challenge of machine learning. The better our probability estimates $\hat{p}_k(\mathbf{x})$, the better our predictor:

$$\hat{Y} = \max_k \hat{p}_k(\mathbf{x})$$

So what we will predict depends on two things: 1) how close are the $\max_k p_k(\mathbf{x})$ to 1 or 0 (perfect certainty) and 2) how close our estimates $\hat{p}_k(\mathbf{x})$ are to $p_k(\mathbf{x})$. We can't do anything about the first restriction as it is determined by the nature of the problem, so our energy goes into finding ways to best estimate conditional probabilities. The first restriction does imply that we have limits as to how well even the best possible algorithm can perform. You should get used to the idea that while in some challenges we will be able to achieve almost perfect accuracy, with digit readers for example, in others our success is restricted by the randomness of the process, with movie recommendations for example.

Before we continue, it is important to remember that defining our prediction by maximizing the probability is not always optimal in practice and depends on the context. As discussed above, sensitivity and specificity may differ in importance. But even in these cases, having a good estimate of the $p_k(x), k = 1, \ldots, K$ will suffice for us to build optimal prediction models, since we can control the balance between specificity and sensitivity however we wish. For instance, we can simply change the cutoffs used to predict one outcome or the other. In the plane example, we may ground the plane anytime the probability of malfunction is higher than 1 in a million as opposed to the default $1/2$ used when error types are equally undesired.

27.6.2   Conditional expectations

For binary data, you can think of the probability $\Pr(Y = 1 \mid \mathbf{X} = \mathbf{x})$ as the proportion of 1s in the stratum of the population for which $\mathbf{X} = \mathbf{x}$. Many of the algorithms we will learn can be applied to both categorical and continuous data due to the connection between *conditional probabilities* and *conditional expectations*.

Because the expectation is the average of values $y_1, \ldots, y_n$ in the population, in the case in which the ys are 0 or 1, the expectation is equivalent to the probability of randomly picking a one since the average is simply the proportion of ones:

$$ \mathrm{E}(Y \mid \mathbf{X} = \mathbf{x}) = \Pr(Y = 1 \mid \mathbf{X} = \mathbf{x}). $$

As a result, we often only use the expectation to denote both the conditional probability and conditional expectation.

Just like with categorical outcomes, in most applications the same observed predictors do not guarantee the same continuous outcomes. Instead, we assume that the outcome follows the same conditional distribution. We will now explain why we use the conditional expectation to define our predictors.

27.6.3   Conditional expectation minimizes squared loss function

Why do we care about the conditional expectation in machine learning? This is because the expected value has an attractive mathematical property: it minimizes the MSE. Specifically, of all possible predictions $\hat{Y}$,

$$ \hat{Y} = \mathrm{E}(Y \mid \mathbf{X} = \mathbf{x}) \ \text{ minimizes } \ \mathrm{E}\{(\hat{Y} - Y)^2 \mid \mathbf{X} = \mathbf{x}\} $$

Due to this property, a succinct description of the main task of machine learning is that we use data to estimate:

$$ f(\mathbf{x}) \equiv \mathrm{E}(Y \mid \mathbf{X} = \mathbf{x}) $$

for any set of features $\mathbf{x} = (x_1, \ldots, x_p)$. Of course this is easier said than done, since this function can take any shape and p can be very large. Consider a case in which we only have one predictor x. The expectation $\mathrm{E}\{Y \mid X = x\}$ can be any function of x: a line, a parabola, a sine wave, a step function, anything. It gets even more complicated when we consider instances with large p, in which case $f(\mathbf{x})$ is a function of a multidimensional vector $\mathbf{x}$. For example, in our digit reader example $p = 784$! **The main way in which competing machine learning algorithms differ is in their approach to estimating this expectation.**

27.7   Exercises

1. Compute conditional probabilities for being Male for the **heights** dataset. Round the heights to the closest inch. Plot the estimated conditional probability $P(x) = \Pr(\text{Male}|\text{height} = x)$ for each x.

2. In the plot we just made, we see high variability for low values of height. This is because we have few data points in these strata. This time use the **quantile** function for quantiles $0.1, 0.2, \ldots, 0.9$ and the **cut** function to assure each group has the same number of points. Hint: for any numeric vector x, you can create groups based on quantiles like this:

```
cut(x, quantile(x, seq(0, 1, 0.1)), include.lowest = TRUE)
```

3. Generate data from a bivariate normal distribution using the **MASS** package like this:

```
Sigma <- 9*matrix(c(1,0.5,0.5,1), 2, 2)
dat <- MASS::mvrnorm(n = 10000, c(69, 69), Sigma) %>%
  data.frame() %>% setNames(c("x", "y"))
```

You can make a quick plot of the data using **plot(dat)**. Use an approach similar to the previous exercise to estimate the conditional expectations and make a plot.

27.8   Case study: is it a 2 or a 7?

In the two simple examples above, we only had one predictor. We actually do not consider these machine learning challenges, which are characterized by cases with many predictors. Let's go back to the digits example in which we had 784 predictors. For illustrative purposes, we will start by simplifying this problem to one with two predictors and two classes. Specifically, we define the challenge as building an algorithm that can determine if a digit is a 2 or 7 from the predictors. We are not quite ready to build algorithms with 784 predictors, so we will extract two simple predictors from the 784: the proportion of dark pixels that are in the upper left quadrant (X_1) and the lower right quadrant (X_2).

We then select a random sample of 1,000 digits, 500 in the training set and 500 in the test set. We provide this dataset in the **dslabs** package:

```
library(tidyverse)
library(dslabs)
data("mnist_27")
```

We can explore the data by plotting the two predictors and using colors to denote the labels:

```
mnist_27$train %>% ggplot(aes(x_1, x_2, color = y)) + geom_point()
```

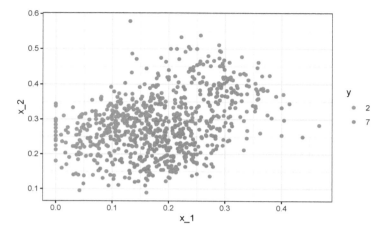

We can immediately see some patterns. For example, if X_1 (the upper left panel) is very large, then the digit is probably a 7. Also, for smaller values of X_1, the 2s appear to be in the mid range values of X_2.

These are the images of the digits with the largest and smallest values for X_1: And here are the original images corresponding to the largest and smallest value of X_2:

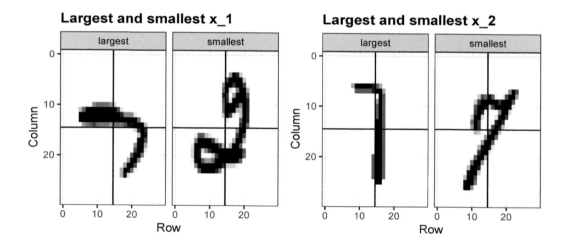

We can start getting a sense for why these predictors are useful, but also why the problem will be somewhat challenging.

We haven't really learned any algorithms yet, so let's try building an algorithm using regression. The model is simply:

$$p(x_1, x_2) = \Pr(Y = 1 \mid X_1 = x_1, X_2 = x_2) = \beta_0 + \beta_1 x_1 + \beta_2 x_2$$

We fit it like this:

```
fit <- mnist_27$train %>%
  mutate(y = ifelse(y==7, 1, 0)) %>%
  lm(y ~ x_1 + x_2, data = .)
```

We can now build a decision rule based on the estimate of $\hat{p}(x_1, x_2)$:

```
library(caret)
p_hat <- predict(fit, newdata = mnist_27$test)
y_hat <- factor(ifelse(p_hat > 0.5, 7, 2))
confusionMatrix(y_hat, mnist_27$test$y)$overall[["Accuracy"]]
#> [1] 0.75
```

We get an accuracy well above 50%. Not bad for our first try. But can we do better?

Because we constructed the `mnist_27` example and we had at our disposal 60,000 digits in just the MNIST dataset, we used this to build the *true* conditional distribution $p(x_1, x_2)$. Keep in mind that this is something we don't have access to in practice, but we include it in this example because it permits the comparison of $\hat{p}(x_1, x_2)$ to the true $p(x_1, x_2)$. This comparison teaches us the limitations of different algorithms. Let's do that here. We have stored the true $p(x_1, x_2)$ in the `mnist_27` object and can plot the image using the **ggplot2** function `geom_raster()`. We choose better colors and use the `stat_contour` function to draw a curve that separates pairs (x_1, x_2) for which $p(x_1, x_2) > 0.5$ and pairs for which $p(x_1, x_2) < 0.5$:

```
mnist_27$true_p %>% ggplot(aes(x_1, x_2, z = p, fill = p)) +
  geom_raster() +
  scale_fill_gradientn(colors=c("#F8766D", "white", "#00BFC4")) +
  stat_contour(breaks=c(0.5), color="black")
```

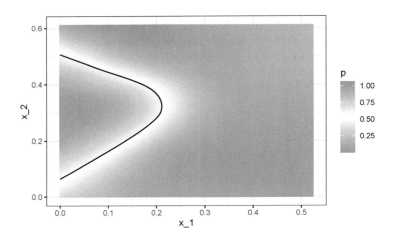

Above you see a plot of the true $p(x, y)$. To start understanding the limitations of logistic regression here, first note that with logistic regression $\hat{p}(x, y)$ has to be a plane, and as a result the boundary defined by the decision rule is given by: $\hat{p}(x, y) = 0.5$, which implies the boundary can't be anything other than a straight line:

$$\hat{\beta}_0 + \hat{\beta}_1 x_1 + \hat{\beta}_2 x_2 = 0.5 \implies \hat{\beta}_0 + \hat{\beta}_1 x_1 + \hat{\beta}_2 x_2 = 0.5 \implies x_2 = (0.5 - \hat{\beta}_0)/\hat{\beta}_2 - \hat{\beta}_1/\hat{\beta}_2 x_1$$

Note that for this boundary, x_2 is a linear function of x_1. This implies that our logistic regression approach has no chance of capturing the non-linear nature of the true $p(x_1, x_2)$. Below is a visual representation of $\hat{p}(x_1, x_2)$. We used the squish function from the **scales** package to constrain estimates to be between 0 and 1. We can see where the mistakes were made by also showing the data and the boundary. They mainly come from low values x_1 that have either high or low value of x_2. Regression can't catch this.

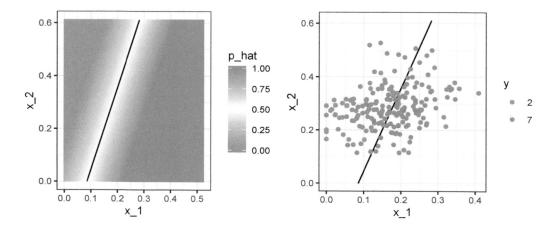

We need something more flexible: a method that permits estimates with shapes other than a plane.

We are going to learn a few new algorithms based on different ideas and concepts. But what they all have in common is that they permit more flexible approaches. We will start by describing nearest neighbor and kernel approaches. To introduce the concepts behinds these approaches, we will again start with a simple one-dimensional example and describe the concept of *smoothing*.

28

Smoothing

Before continuing learning about machine learning algorithms, we introduce the important concept of *smoothing*. Smoothing is a very powerful technique used all across data analysis. Other names given to this technique are *curve fitting* and *low pass filtering*. It is designed to detect trends in the presence of noisy data in cases in which the shape of the trend is unknown. The *smoothing* name comes from the fact that to accomplish this feat, we assume that the trend is *smooth*, as in a smooth surface. In contrast, the noise, or deviation from the trend, is unpredictably wobbly:

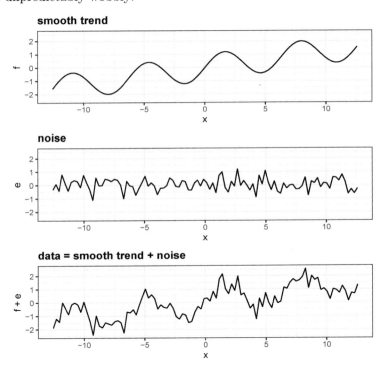

Part of what we explain in this section are the assumptions that permit us to extract the trend from the noise.

To understand why we cover this topic, remember that the concepts behind smoothing techniques are extremely useful in machine learning because conditional expectations/probabilities can be thought of as *trends* of unknown shapes that we need to estimate in the presence of uncertainty.

To explain these concepts, we will focus first on a problem with just one predictor. Specifically, we try to estimate the time trend in the 2008 US popular vote poll margin (difference between Obama and McCain).

```
library(tidyverse)
library(dslabs)
data("polls_2008")
qplot(day, margin, data = polls_2008)
```

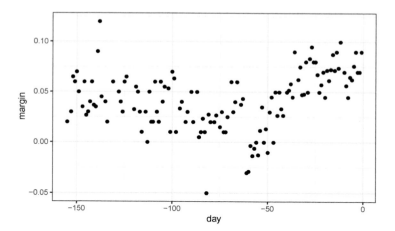

For the purposes of this example, do not think of it as a forecasting problem. Instead, we are simply interested in learning the shape of the trend *after* the election is over.

We assume that for any given day x, there is a true preference among the electorate $f(x)$, but due to the uncertainty introduced by the polling, each data point comes with an error ε. A mathematical model for the observed poll margin Y_i is:

$$Y_i = f(x_i) + \varepsilon_i$$

To think of this as a machine learning problem, consider that we want to predict Y given a day x. If we knew the conditional expectation $f(x) = \mathrm{E}(Y \mid X = x)$, we would use it. But since we don't know this conditional expectation, we have to estimate it. Let's use regression, since it is the only method we have learned up to now.

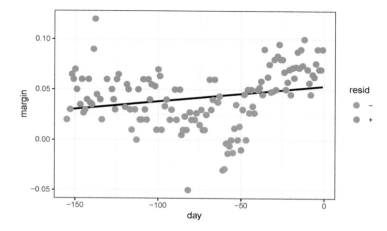

The line we see does not appear to describe the trend very well. For example, on September 4 (day -62), the Republican Convention was held and the data suggest that it gave John McCain a boost in the polls. However, the regression line does not capture this potential trend. To see the *lack of fit* more clearly, we note that points above the fitted line (blue) and those below (red) are not evenly distributed across days. We therefore need an alternative, more flexible approach.

28.1 Bin smoothing

The general idea of smoothing is to group data points into strata in which the value of $f(x)$ can be assumed to be constant. We can make this assumption because we think $f(x)$ changes slowly and, as a result, $f(x)$ is almost constant in small windows of time. An example of this idea for the `poll_2008` data is to assume that public opinion remained approximately the same within a week's time. With this assumption in place, we have several data points with the same expected value.

If we fix a day to be in the center of our week, call it x_0, then for any other day x such that $|x - x_0| \leq 3.5$, we assume $f(x)$ is a constant $f(x) = \mu$. This assumption implies that:

$$E[Y_i | X_i = x_i] \approx \mu \text{ if } |x_i - x_0| \leq 3.5$$

In smoothing, we call the size of the interval satisfying $|x_i - x_0| \leq 3.5$ the *window size*, *bandwidth* or *span*. Later we will see that we try to optimize this parameter.

This assumption implies that a good estimate for $f(x)$ is the average of the Y_i values in the window. If we define A_0 as the set of indexes i such that $|x_i - x_0| \leq 3.5$ and N_0 as the number of indexes in A_0, then our estimate is:

$$\hat{f}(x_0) = \frac{1}{N_0} \sum_{i \in A_0} Y_i$$

The idea behind *bin smoothing* is to make this calculation with each value of x as the center. In the poll example, for each day, we would compute the average of the values within a week with that day in the center. Here are two examples: $x_0 = -125$ and $x_0 = -55$. The blue segment represents the resulting average.

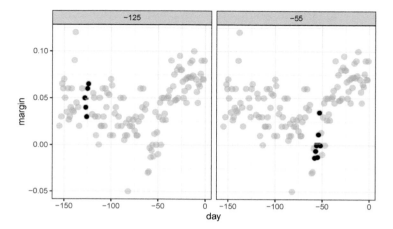

By computing this mean for every point, we form an estimate of the underlying curve $f(x)$. Below we show the procedure happening as we move from the -155 up to 0. At each value of x_0, we keep the estimate $\hat{f}(x_0)$ and move on to the next point:

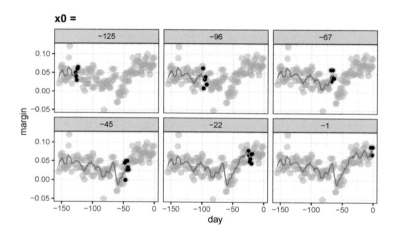

The final code and resulting estimate look like this:

```
span <- 7
fit <- with(polls_2008,
            ksmooth(day, margin, kernel = "box", bandwidth = span))

polls_2008 %>% mutate(smooth = fit$y) %>%
  ggplot(aes(day, margin)) +
    geom_point(size = 3, alpha = .5, color = "grey") +
  geom_line(aes(day, smooth), color="red")
```

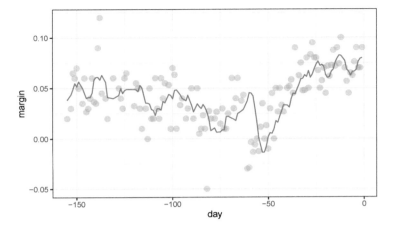

28.2   Kernels

The final result from the bin smoother is quite wiggly. One reason for this is that each time the window moves, two points change. We can attenuate this somewhat by taking weighted averages that give the center point more weight than far away points, with the two points at the edges receiving very little weight.

You can think of the bin smoother approach as a weighted average:

$$\hat{f}(x_0) = \sum_{i=1}^{N} w_0(x_i) Y_i$$

in which each point receives a weight of either 0 or $1/N_0$, with N_0 the number of points in the week. In the code above, we used the argument `kernel="box"` in our call to the function `ksmooth`. This is because the weight function looks like a box. The `ksmooth` function provides a "smoother" option which uses the normal density to assign weights.

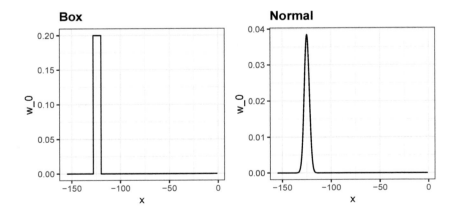

The final code and resulting plot for the normal kerenl look like this:

```
span <- 7
fit <- with(polls_2008,
            ksmooth(day, margin, kernel = "normal", bandwidth = span))

polls_2008 %>% mutate(smooth = fit$y) %>%
  ggplot(aes(day, margin)) +
  geom_point(size = 3, alpha = .5, color = "grey") +
  geom_line(aes(day, smooth), color="red")
```

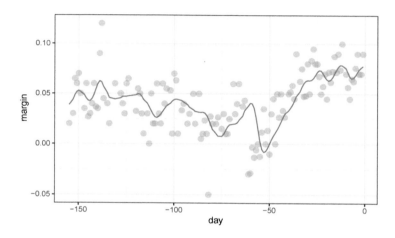

Notice that the final estimate now looks smoother.

There are several functions in R that implement bin smoothers. One example is ksmooth, shown above. In practice, however, we typically prefer methods that use slightly more complex models than fitting a constant. The final result above, for example, is still somewhat wiggly in parts we don't expect it to be (between -125 and -75, for example). Methods such as loess, which we explain next, improve on this.

28.3   Local weighted regression (loess)

A limitation of the bin smoother approach just described is that we need small windows for the approximately constant assumptions to hold. As a result, we end up with a small number of data points to average and obtain imprecise estimates $\hat{f}(x)$. Here we describe how *local weighted regression* (loess) permits us to consider larger window sizes. To do this, we will use a mathematical result, referred to as Taylor's theorem, which tells us that if you look closely enough at any smooth function $f(x)$, it will look like a line. To see why this makes sense, consider the curved edges gardeners make using straight-edged spades:

("Downing Street garden path edge"[1] by Flckr user Number 10[2]. CC-BY 2.0 license[3].)

Instead of assuming the function is approximately constant in a window, we assume the function is locally linear. We can consider larger window sizes with the linear assumption than with a constant. Instead of the one-week window, we consider a larger one in which the trend is approximately linear. We start with a three-week window and later consider and evaluate other options:

$$E[Y_i|X_i = x_i] = \beta_0 + \beta_1(x_i - x_0) \text{ if } |x_i - x_0| \le 21$$

For every point x_0, loess defines a window and fits a line within that window. Here is an example showing the fits for $x_0 = -125$ and $x_0 = -55$:

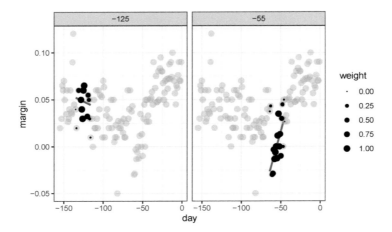

The fitted value at x_0 becomes our estimate $\hat{f}(x_0)$. Below we show the procedure happening as we move from the -155 up to 0.

[1] https://www.flickr.com/photos/49707497@N06/7361631644
[2] https://www.flickr.com/photos/number10gov/
[3] https://creativecommons.org/licenses/by/2.0/

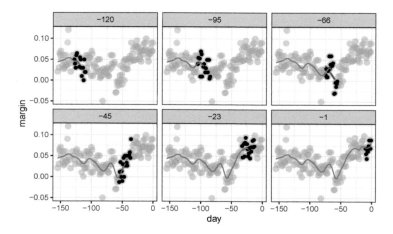

The final result is a smoother fit than the bin smoother since we use larger sample sizes to estimate our local parameters:

```
total_days <- diff(range(polls_2008$day))
span <- 21/total_days

fit <- loess(margin ~ day, degree=1, span = span, data=polls_2008)

polls_2008 %>% mutate(smooth = fit$fitted) %>%
  ggplot(aes(day, margin)) +
  geom_point(size = 3, alpha = .5, color = "grey") +
  geom_line(aes(day, smooth), color="red")
```

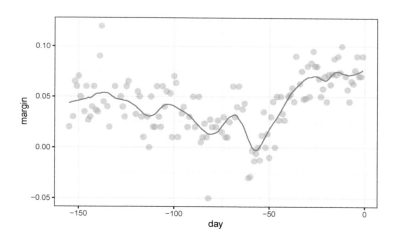

Different spans give us different estimates. We can see how different window sizes lead to different estimates:

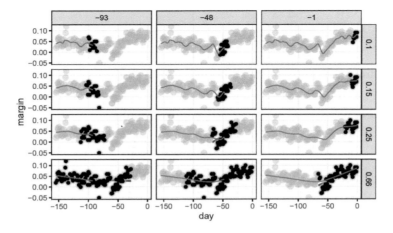

Here are the final estimates:

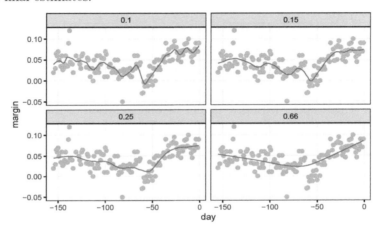

There are three other differences between `loess` and the typical bin smoother.

1. Rather than keeping the bin size the same, `loess` keeps the number of points used in the local fit the same. This number is controlled via the `span` argument, which expects a proportion. For example, if N is the number of data points and `span=0.5`, then for a given x, `loess` will use the `0.5 * N` closest points to x for the fit.

2. When fitting a line locally, `loess` uses a *weighted* approach. Basically, instead of using least squares, we minimize a weighted version:

$$\sum_{i=1}^{N} w_0(x_i) \left[Y_i - \{\beta_0 + \beta_1(x_i - x_0)\} \right]^2$$

However, instead of the Gaussian kernel, loess uses a function called the Tukey tri-weight:

$$W(u) = \left(1 - |u|^3\right)^3 \text{ if } |u| \le 1 \text{ and } W(u) = 0 \text{ if } |u| > 1$$

To define the weights, we denote $2h$ as the window size and define:

$$w_0(x_i) = W\left(\frac{x_i - x_0}{h}\right)$$

This kernel differs from the Gaussian kernel in that more points get values closer to the max:

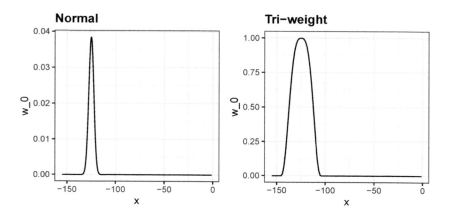

3. `loess` has the option of fitting the local model *robustly*. An iterative algorithm is implemented in which, after fitting a model in one iteration, outliers are detected and down-weighted for the next iteration. To use this option, we use the argument `family="symmetric"`.

28.3.1   Fitting parabolas

Taylor's theorem also tells us that if you look at any mathematical function closely enough, it looks like a parabola. The theorem also states that you don't have to look as closely when approximating with parabolas as you do when approximating with lines. This means we can make our windows even larger and fit parabolas instead of lines.

$$E[Y_i|X_i = x_i] = \beta_0 + \beta_1(x_i - x_0) + \beta_2(x_i - x_0)^2 \text{ if } |x_i - x_0| \le h$$

This is actually the default procedure of the function `loess`. You may have noticed that when we showed the code for using loess, we set `degree = 1`. This tells loess to fit polynomials of degree 1, a fancy name for lines. If you read the help page for loess, you will see that the argument `degree` defaults to 2. By default, loess fits parabolas not lines. Here is a comparison of the fitting lines (red dashed) and fitting parabolas (orange solid):

```
total_days <- diff(range(polls_2008$day))
span <- 28/total_days
fit_1 <- loess(margin ~ day, degree=1, span = span, data=polls_2008)

fit_2 <- loess(margin ~ day, span = span, data=polls_2008)


polls_2008 %>% mutate(smooth_1 = fit_1$fitted, smooth_2 = fit_2$fitted) %>%
```

```
ggplot(aes(day, margin)) +
geom_point(size = 3, alpha = .5, color = "grey") +
geom_line(aes(day, smooth_1), color="red", lty = 2) +
geom_line(aes(day, smooth_2), color="orange", lty = 1)
```

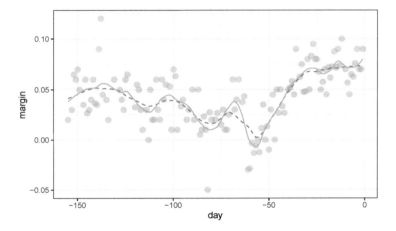

The `degree = 2` gives us more wiggly results. We actually prefer `degree = 1` as it is less prone to this kind of noise.

28.3.2 Beware of default smoothing parameters

ggplot uses loess in its `geom_smooth` function:

```
polls_2008 %>% ggplot(aes(day, margin)) +
  geom_point() +
  geom_smooth()
```

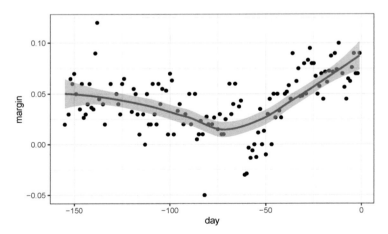

But be careful with default parameters as they are rarely optimal. However, you can conveniently change them:

```
polls_2008 %>% ggplot(aes(day, margin)) +
  geom_point() +
  geom_smooth(span = 0.15, method.args = list(degree=1))
```

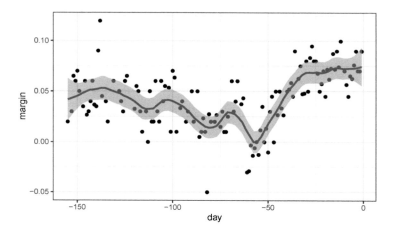

28.4   Connecting smoothing to machine learning

To see how smoothing relates to machine learning with a concrete example, consider our 27.8 example. If we define the outcome $Y = 1$ for digits that are seven and $Y = 0$ for digits that are 2, then we are interested in estimating the conditional probability:

$$p(x_1, x_2) = \Pr(Y = 1 \mid X_1 = x_1, X_2 = x_2).$$

with X_1 and X_2 the two predictors defined in Section 27.8. In this example, the 0s and 1s we observe are "noisy" because for some regions the probabilities $p(x_1, x_2)$ are not that close to 0 or 1. So we need to estimate $p(x_1, x_2)$. Smoothing is an alternative to accomplishing this. In Section 27.8 we saw that linear regression was not flexible enough to capture the non-linear nature of $p(x_1, x_2)$, thus smoothing approaches may provide an improvement. In the next chapter we describe a popular machine learning algorithm, k-nearest neighbors, which is based on bin smoothing.

28.5   Exercises

1. In the wrangling part of this book, we used the code below to obtain mortality counts for Puerto Rico for 2015-2018.

```
library(tidyverse)
library(purrr)
library(pdftools)
library(dslabs)

fn <- system.file("extdata", "RD-Mortality-Report_2015-18-180531.pdf",
                  package="dslabs")
dat <- map_df(str_split(pdf_text(fn), "\n"), function(s){
  s <- str_trim(s)
  header_index <- str_which(s, "2015")[1]
  tmp <- str_split(s[header_index], "\\s+", simplify = TRUE)
  month <- tmp[1]
  header <- tmp[-1]
  tail_index  <- str_which(s, "Total")
  n <- str_count(s, "\\d+")
  out <- c(1:header_index, which(n == 1),
           which(n >= 28), tail_index:length(s))
  s[-out] %>%  str_remove_all("[^\\d\\s]") %>% str_trim() %>%
    str_split_fixed("\\s+", n = 6) %>% .[,1:5] %>% as_tibble() %>%
    setNames(c("day", header)) %>%
    mutate(month = month, day = as.numeric(day)) %>%
    gather(year, deaths, -c(day, month)) %>%
    mutate(deaths = as.numeric(deaths))
}) %>%
  mutate(month = recode(month,
                        "JAN" = 1, "FEB" = 2, "MAR" = 3,
                        "APR" = 4, "MAY" = 5, "JUN" = 6,
                        "JUL" = 7, "AGO" = 8, "SEP" = 9,
                        "OCT" = 10, "NOV" = 11, "DEC" = 12)) %>%
  mutate(date = make_date(year, month, day)) %>%
  filter(date <= "2018-05-01")
```

Use the `loess` function to obtain a smooth estimate of the expected number of deaths as a function of date. Plot this resulting smooth function. Make the span about two months long.

2. Plot the smooth estimates against day of the year, all on the same plot but with different colors.

3. Suppose we want to predict 2s and 7s in our `mnist_27` dataset with just the second covariate. Can we do this? On first inspection it appears the data does not have much predictive power. In fact, if we fit a regular logistic regression, the coefficient for `x_2` is not significant!

```
library(broom)
library(dslabs)
data("mnist_27")
mnist_27$train %>%
  glm(y ~ x_2, family = "binomial", data = .) %>%
  tidy()
```

Plotting a scatterplot here is not useful since y is binary:

```
qplot(x_2, y, data = mnist_27$train)
```

Fit a loess line to the data above and plot the results. Notice that there is predictive power, except the conditional probability is not linear.

29

Cross validation

In this chapter we introduce cross validation, one of the most important ideas in machine learning. Here we focus on the conceptual and mathematical aspects. We will describe how to implement cross validation in practice with the **caret** package later, in Section 30.2 in the next chapter. To motivate the concept, we will use the two predictor digits data presented in Section 27.8 and introduce, for the first time, an actual machine learning algorithm: k-nearest neighbors (kNN).

29.1 Motivation with k-nearest neighbors

Let's start by loading the data and showing a plot of the predictors with outcome represented with color.

```
library(tidyverse)
library(dslabs)
data("mnist_27")
mnist_27$test%>% ggplot(aes(x_1, x_2, color = y)) + geom_point()
```

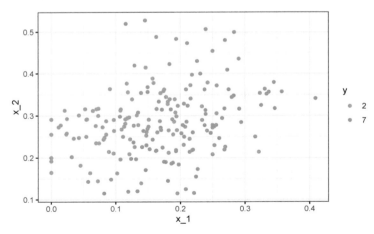

We will use these data to estimate the conditional probability function

$$p(x_1, x_2) = \Pr(Y = 1 \mid X_1 = x_1, X_2 = x_2).$$

as defined in Section 28.4. With k-nearest neighbors (kNN) we estimate $p(x_1, x_2)$ in a similar way to bin smoothing. However, as we will see, kNN is easier to adapt to multiple dimensions.

First we define the distance between all observations based on the features. Then, for any point (x_1, x_2) for which we want an estimate of $p(x_1, x_2)$, we look for the k nearest points to (x_1, x_2) and then take an average of the 0s and 1s associated with these points. We refer to the set of points used to compute the average as the *neighborhood*. Due to the connection we described earlier between conditional expectations and conditional probabilities, this gives us a $\hat{p}(x_1, x_2)$, just like the bin smoother gave us an estimate of a trend. As with bin smoothers, we can control the flexibility of our estimate, in this case through the k parameter: larger ks result in smoother estimates, while smaller ks result in more flexible and more wiggly estimates.

To implement the algorithm, we can use the `knn3` function from the **caret** package. Looking at the help file for this package, we see that we can call it in one of two ways. We will use the first in which we specify a *formula* and a data frame. The data frame contains all the data to be used. The formula has the form `outcome ~ predictor_1 + predictor_2 + predictor_3` and so on. Therefore, we would type `y ~ x_1 + x_2`. If we are going to use all the predictors, we can use the `.` like this `y ~ .`. The final call looks like this:

```
library(caret)
knn_fit <- knn3(y ~ ., data = mnist_27$train)
```

For this function, we also need to pick a parameter: the number of neighbors to include. Let's start with the default $k = 5$.

```
knn_fit <- knn3(y ~ ., data = mnist_27$train, k = 5)
```

In this case, since our dataset is balanced and we care just as much about sensitivity as we do about specificity, we will use accuracy to quantify performance.

The `predict` function for `knn` produces a probability for each class. We keep the probability of being a 7 as the estimate $\hat{p}(x_1, x_2)$

```
y_hat_knn <- predict(knn_fit, mnist_27$test, type = "class")
confusionMatrix(y_hat_knn, mnist_27$test$y)$overall["Accuracy"]
#> Accuracy
#>    0.815
```

In Section 27.8 we used linear regression to generate an estimate.

```
fit_lm <- mnist_27$train %>%
  mutate(y = ifelse(y == 7, 1, 0)) %>%
  lm(y ~ x_1 + x_2, data = .)
p_hat_lm <- predict(fit_lm, mnist_27$test)
y_hat_lm <- factor(ifelse(p_hat_lm > 0.5, 7, 2))
confusionMatrix(y_hat_lm, mnist_27$test$y)$overall["Accuracy"]
#> Accuracy
#>     0.75
```

And we see that kNN, with the default parameter, already beats regression. To see why this is the case, we will plot $\hat{p}(x_1, x_2)$ and compare it to the true conditional probability $p(x_1, x_2)$:

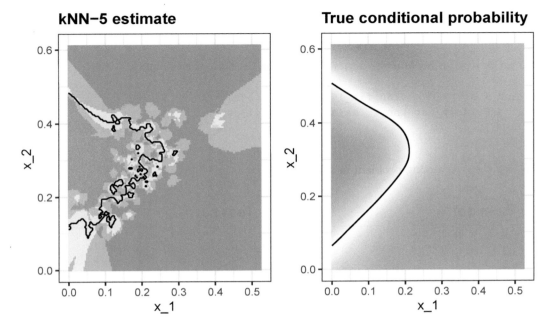

We see that kNN better adapts to the non-linear shape of $p(x_1, x_2)$. However, our estimate has some islands of blue in the red area, which intuitively does not make much sense. This is due to what we call *over-training*. We describe over-training in detail below. Over-training is the reason that we have higher accuracy in the train set compared to the test set:

```
y_hat_knn <- predict(knn_fit, mnist_27$train, type = "class")
confusionMatrix(y_hat_knn, mnist_27$train$y)$overall["Accuracy"]
#> Accuracy
#>    0.882
```

```
y_hat_knn <- predict(knn_fit, mnist_27$test, type = "class")
confusionMatrix(y_hat_knn, mnist_27$test$y)$overall["Accuracy"]
#> Accuracy
#>    0.815
```

29.1.1   Over-training

Over-training is at its worst when we set $k = 1$. With $k = 1$, the estimate for each (x_1, x_2) in the training set is obtained with just the y corresponding to that point. In this case, if the (x_1, x_2) are unique, we will obtain perfect accuracy in the training set because each point is used to predict itself. Remember that if the predictors are not unique and have different outcomes for at least one set of predictors, then it is impossible to predict perfectly.

Here we fit a kNN model with $k = 1$:

```
knn_fit_1 <- knn3(y ~ ., data = mnist_27$train, k = 1)
y_hat_knn_1 <- predict(knn_fit_1, mnist_27$train, type = "class")
```

```
confusionMatrix(y_hat_knn_1, mnist_27$train$y)$overall[["Accuracy"]]
#> [1] 0.996
```

However, the test set accuracy is actually worse than logistic regression:

```
y_hat_knn_1 <- predict(knn_fit_1, mnist_27$test, type = "class")
confusionMatrix(y_hat_knn_1, mnist_27$test$y)$overall["Accuracy"]
#> Accuracy
#>    0.735
```

We can see the over-fitting problem in this figure.

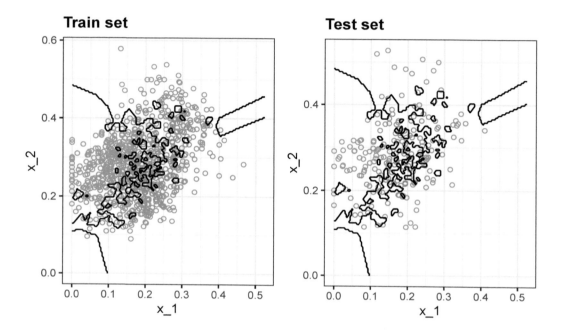

The black curves denote the decision rule boundaries.

The estimate $\hat{p}(x_1, x_2)$ follows the training data too closely (left). You can see that in the training set, boundaries have been drawn to perfectly surround a single red point in a sea of blue. Because most points (x_1, x_2) are unique, the prediction is either 1 or 0 and the prediction for that point is the associated label. However, once we introduce the training set (right), we see that many of these small islands now have the opposite color and we end up making several incorrect predictions.

29.1.2   Over-smoothing

Although not as badly as with the previous examples, we saw that with $k = 5$ we also over-trained. Hence, we should consider a larger k. Let's try, as an example, a much larger number: $k = 401$.

```
knn_fit_401 <- knn3(y ~ ., data = mnist_27$train, k = 401)
y_hat_knn_401 <- predict(knn_fit_401, mnist_27$test, type = "class")
confusionMatrix(y_hat_knn_401, mnist_27$test$y)$overall["Accuracy"]
#> Accuracy
#>    0.79
```

This turns out to be similar to regression:

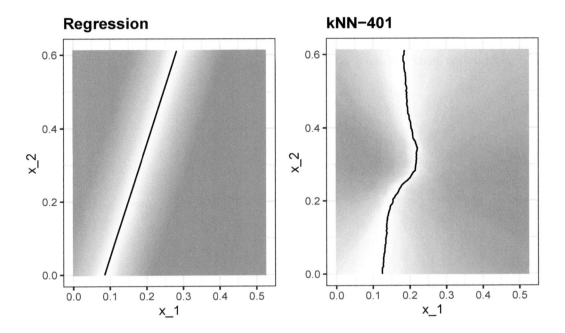

This size of k is so large that it does not permit enough flexibility. We call this *over-smoothing*.

29.1.3 Picking the k in kNN

So how do we pick k? In principle we want to pick the k that maximizes accuracy, or minimizes the expected MSE as defined in 27.4.8. The goal of cross validation is to estimate these quantities for any given algorithm and set of tuning parameters such as k. To understand why we need a special method to do this let's repeat what we did above but for different values of k:

```
ks <- seq(3, 251, 2)
```

We do this using `map_df` function to repeat the above for each one.

```
library(purrr)
accuracy <- map_df(ks, function(k){
  fit <- knn3(y ~ ., data = mnist_27$train, k = k)

  y_hat <- predict(fit, mnist_27$train, type = "class")
```

```
  cm_train <- confusionMatrix(y_hat, mnist_27$train$y)
  train_error <- cm_train$overall["Accuracy"]

  y_hat <- predict(fit, mnist_27$test, type = "class")
  cm_test <- confusionMatrix(y_hat, mnist_27$test$y)
  test_error <- cm_test$overall["Accuracy"]

  tibble(train = train_error, test = test_error)
})
```

Note that we estimate accuracy by using both the training set and the test set. We can now plot the accuracy estimates for each value of k:

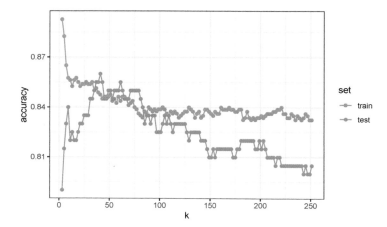

First, note that the estimate obtained on the training set is generally lower than the estimate obtained with the test set, with the difference larger for smaller values of k. This is due to over-training. Also note that the accuracy versus k plot is quite jagged. We do not expect this because small changes in k should not affect the algorithm's performance too much. The jaggedness is explained by the fact that the accuracy is computed on a sample and therefore is a random variable. This demonstrates why we prefer to minimize the expected loss rather than the loss we observe with one dataset.

If we were to use these estimates to pick the k that maximizes accuracy, we would use the estimates built on the test data:

```
ks[which.max(accuracy$test)]
#> [1] 41
max(accuracy$test)
#> [1] 0.86
```

Another reason we need a better estimate of accuracy is that if we use the test set to pick this k, we should not expect the accompanying accuracy estimate to extrapolate to the real world. This is because even here we broke a golden rule of machine learning: we selected the k using the test set. Cross validation also provides an estimate that takes this into account.

29.2 Mathematical description of cross validation

In Section 27.4.8, we described that a common goal of machine learning is to find an algorithm that produces predictors $\hat{Y}$ for an outcome Y that minimizes the MSE:

$$\text{MSE} = \text{E}\left\{\frac{1}{N}\sum_{i=1}^{N}(\hat{Y}_i - Y_i)^2\right\}$$

When all we have at our disposal is one dataset, we can estimate the MSE with the observed MSE like this:

$$\hat{\text{MSE}} = \frac{1}{N}\sum_{i=1}^{N}(\hat{y}_i - y_i)^2$$

These two are often referred to as the *true error* and *apparent error*, respectively.

There are two important characteristics of the apparent error we should always keep in mind:

1. Because our data is random, the apparent error is a random variable. For example, the dataset we have may be a random sample from a larger population. An algorithm may have a lower apparent error than another algorithm due to luck.

2. If we train an algorithm on the same dataset that we use to compute the apparent error, we might be overtraining. In general, when we do this, the apparent error will be an underestimate of the true error. We will see an extreme example of this with k-nearest neighbors.

Cross validation is a technique that permits us to alleviate both these problems. To understand cross validation, it helps to think of the true error, a theoretical quantity, as the average of many apparent errors obtained by applying the algorithm to B new random samples of the data, none of them used to train the algorithm. As shown in a previous chapter, we think of the true error as:

$$\frac{1}{B}\sum_{b=1}^{B}\frac{1}{N}\sum_{i=1}^{N}\left(\hat{y}_i^b - y_i^b\right)^2$$

with B a large number that can be thought of as practically infinite. As already mentioned, this is a theoretical quantity because we only have available one set of outcomes: $y_1, \ldots, y_n$. Cross validation is based on the idea of imitating the theoretical setup above as best we can with the data we have. To do this, we have to generate a series of different random samples. There are several approaches we can use, but the general idea for all of them is to randomly generate smaller datasets that are not used for training, and instead used to estimate the true error.

29.3   K-fold cross validation

The first one we describe is *K-fold cross validation*. Generally speaking, a machine learning challenge starts with a dataset (blue in the image below). We need to build an algorithm using this dataset that will eventually be used in completely independent datasets (yellow).

But we don't get to see these independent datasets.

So to imitate this situation, we carve out a piece of our dataset and pretend it is an independent dataset: we divide the dataset into a *training set* (blue) and a *test set* (red). We will train our algorithm exclusively on the training set and use the test set only for evaluation purposes.

We usually try to select a small piece of the dataset so that we have as much data as possible to train. However, we also want the test set to be large so that we obtain a stable estimate of the loss without fitting an impractical number of models. Typical choices are to use 10%-20% of the data for testing.

Let's reiterate that it is indispensable that we not use the test set at all: not for filtering out rows, not for selecting features, nothing!

Now this presents a new problem because for most machine learning algorithms we need to select parameters, for example the number of neighbors k in k-nearest neighbors. Here, we will refer to the set of parameters as λ. We need to optimize algorithm parameters without using our test set and we know that if we optimize and evaluate on the same dataset, we will overtrain. This is where cross validation is most useful.

For each set of algorithm parameters being considered, we want an estimate of the MSE and then we will choose the parameters with the smallest MSE. Cross validation provides this estimate.

First, before we start the cross validation procedure, it is important to fix all the algorithm parameters. Although we will train the algorithm on the set of training sets, the parameters λ will be the same across all training sets. We will use $\hat{y}_i(\lambda)$ to denote the predictors obtained when we use parameters λ.

So, if we are going to imitate this definition:

$$\text{MSE}(\lambda) = \frac{1}{B} \sum_{b=1}^{B} \frac{1}{N} \sum_{i=1}^{N} \left(\hat{y}_i^b(\lambda) - y_i^b \right)^2$$

we want to consider datasets that can be thought of as an independent random sample and we want to do this several times. With K-fold cross validation, we do it K times. In the cartoons, we are showing an example that uses $K = 5$.

We will eventually end up with K samples, but let's start by describing how to construct the first: we simply pick $M = N/K$ observations at random (we round if M is not a round number) and think of these as a random sample $y_1^b, \ldots, y_M^b$, with $b = 1$. We call this the validation set:

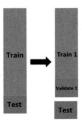

Now we can fit the model in the training set, then compute the apparent error on the independent set:

$$\hat{\text{MSE}}_b(\lambda) = \frac{1}{M} \sum_{i=1}^{M} \left(\hat{y}_i^b(\lambda) - y_i^b \right)^2$$

Note that this is just one sample and will therefore return a noisy estimate of the true error. This is why we take K samples, not just one. In K-cross validation, we randomly split the observations into K non-overlapping sets:

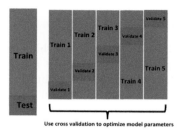

Now we repeat the calculation above for each of these sets $b = 1, \ldots, K$ and obtain $\hat{\text{MSE}}_1(\lambda), \ldots, \hat{\text{MSE}}_K(\lambda)$. Then, for our final estimate, we compute the average:

$$\hat{\text{MSE}}(\lambda) = \frac{1}{B} \sum_{b=1}^{K} \hat{\text{MSE}}_b(\lambda)$$

and obtain an estimate of our loss. A final step would be to select the λ that minimizes the MSE.

We have described how to use cross validation to optimize parameters. However, we now have to take into account the fact that the optimization occurred on the training data and therefore we need an estimate of our final algorithm based on data that was not used to optimize the choice. Here is where we use the test set we separated early on:

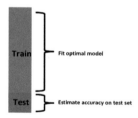

We can do cross validation again:

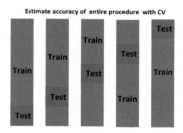

and obtain a final estimate of our expected loss. However, note that this means that our entire compute time gets multiplied by K. You will soon learn that performing this task takes time because we are performing many complex computations. As a result, we are always looking for ways to reduce this time. For the final evaluation, we often just use the one test set.

Once we are satisfied with this model and want to make it available to others, we could refit the model on the entire dataset, without changing the optimized parameters.

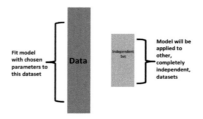

Now how do we pick the cross validation K? Large values of K are preferable because the training data better imitates the original dataset. However, larger values of K will have much slower computation time: for example, 100-fold cross validation will be 10 times slower than 10-fold cross validation. For this reason, the choices of $K = 5$ and $K = 10$ are popular.

One way we can improve the variance of our final estimate is to take more samples. To do this, we would no longer require the training set to be partitioned into non-overlapping sets. Instead, we would just pick K sets of some size at random.

One popular version of this technique, at each fold, picks observations at random with replacement (which means the same observation can appear twice). This approach has some advantages (not discussed here) and is generally referred to as the *bootstrap*. In fact, this is the default approach in the **caret** package. We describe how to implement cross validation with the **caret** package in the next chapter. In the next section, we include an explanation of how the bootstrap works in general.

29.4 Exercises

Generate a set of random predictors and outcomes like this:

```
set.seed(1996)
n <- 1000
p <- 10000
x <- matrix(rnorm(n * p), n, p)
colnames(x) <- paste("x", 1:ncol(x), sep = "_")
y <- rbinom(n, 1, 0.5) %>% factor()

x_subset <- x[ ,sample(p, 100)]
```

1. Because x and y are completely independent, you should not be able to predict y using x with accuracy larger than 0.5. Confirm this by running cross validation using logistic regression to fit the model. Because we have so many predictors, we selected a random

sample x_subset. Use the subset when training the model. Hint: use the caret train function. The results component of the output of train shows you the accuracy. Ignore the warnings.

2. Now, instead of a random selection of predictors, we are going to search for those that are most predictive of the outcome. We can do this by comparing the values for the $y = 1$ group to those in the $y = 0$ group, for each predictor, using a t-test. You can perform this step like this:

```
devtools::install_bioc("genefilter")
install.packages("genefilter")
library(genefilter)
tt <- colttests(x, y)
```

Create a vector of the p-values and call it pvals.

3. Create an index ind with the column numbers of the predictors that were "statistically significantly" associated with y. Use a p-value cutoff of 0.01 to define "statistically significant". How many predictors survive this cutoff?

4. Re-run the cross validation but after redefining x_subset to be the subset of x defined by the columns showing "statistically significant" association with y. What is the accuracy now?

5. Re-run the cross validation again, but this time using kNN. Try out the following grid of tuning parameters: k = seq(101, 301, 25). Make a plot of the resulting accuracy.

6. In exercises 3 and 4, we see that despite the fact that x and y are completely independent, we were able to predict y with accuracy higher than 70%. We must be doing something wrong then. What is it?

   a. The function train estimates accuracy on the same data it uses to train the algorithm.
   b. We are over-fitting the model by including 100 predictors.
   c. We used the entire dataset to select the columns used in the model. This step needs to be included as part of the algorithm. The cross validation was done **after** this selection.
   d. The high accuracy is just due to random variability.

7. Advanced. Re-do the cross validation but this time include the selection step in the cross validation. The accuracy should now be close to 50%.

8. Load the tissue_gene_expression dataset. Use the train function to predict tissue from gene expression. Use kNN. What k works best?

29.5   Bootstrap

Suppose the income distribution of your population is as follows:

```
set.seed(1995)
n <- 10^6
income <- 10^(rnorm(n, log10(45000), log10(3)))
qplot(log10(income), bins = 30, color = I("black"))
```

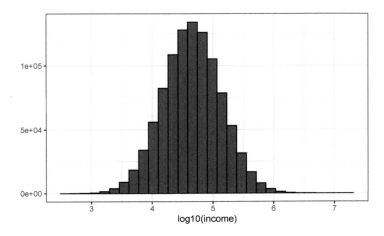

The population median is:

```
m <- median(income)
m
#> [1] 44939
```

Suppose we don't have access to the entire population, but want to estimate the median m. We take a sample of 100 and estimate the population median m with the sample median M:

```
N <- 100
X <- sample(income, N)
median(X)
#> [1] 38461
```

Can we construct a confidence interval? What is the distribution of M ?

Because we are simulating the data, we can use a Monte Carlo simulation to learn the distribution of M.

```
library(gridExtra)
B <- 10^4
M <- replicate(B, {
  X <- sample(income, N)
  median(X)
})
p1 <- qplot(M, bins = 30, color = I("black"))
p2 <- qplot(sample = scale(M), xlab = "theoretical", ylab = "sample") +
  geom_abline()
grid.arrange(p1, p2, ncol = 2)
```

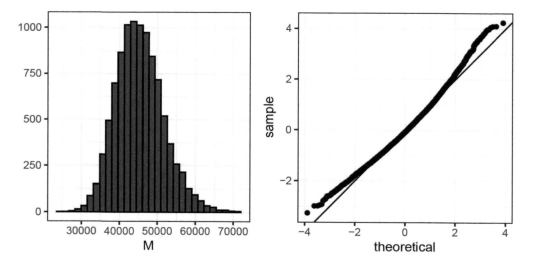

If we know this distribution, we can construct a confidence interval. The problem here is that, as we have already described, in practice we do not have access to the distribution. In the past, we have used the Central Limit Theorem, but the CLT we studied applies to averages and here we are interested in the median. We can see that the 95% confidence interval based on CLT

```
median(X) + 1.96 * sd(X) / sqrt(N) * c(-1, 1)
#> [1] 21018 55905
```

is quite different from the confidence interval we would generate if we know the actual distribution of M:

```
quantile(M, c(0.025, 0.975))
#>  2.5% 97.5%
#> 34438 59050
```

The bootstrap permits us to approximate a Monte Carlo simulation without access to the entire distribution. The general idea is relatively simple. We act as if the observed sample is the population. We then sample (with replacement) datasets, of the same sample size as the original dataset. Then we compute the summary statistic, in this case the median, on these *bootstrap samples*.

Theory tells us that, in many situations, the distribution of the statistics obtained with bootstrap samples approximate the distribution of our actual statistic. This is how we construct bootstrap samples and an approximate distribution:

```
B <- 10^4
M_star <- replicate(B, {
  X_star <- sample(X, N, replace = TRUE)
  median(X_star)
})
```

Note a confidence interval constructed with the bootstrap is much closer to one constructed with the theoretical distribution:

```
quantile(M_star, c(0.025, 0.975))
#>   2.5% 97.5%
#> 30253 56909
```

For more on the Bootstrap, including corrections one can apply to improve these confidence intervals, please consult the book *An introduction to the bootstrap* by Efron, B., & Tibshirani, R. J.

Note that we can use ideas similar to those used in the bootstrap in cross validation: instead of dividing the data into equal partitions, we simply bootstrap many times.

29.6 Exercises

1. The `createResample` function can be used to create bootstrap samples. For example, we can create 10 bootstrap samples for the `mnist_27` dataset like this:

```
set.seed(1995)
indexes <- createResample(mnist_27$train$y, 10)
```

How many times do 3, 4, and 7 appear in the first re-sampled index?

2. We see that some numbers appear more than once and others appear no times. This has to be this way for each dataset to be independent. Repeat the exercise for all the re-sampled indexes.

3. Generate a random dataset like this:

```
y <- rnorm(100, 0, 1)
```

Estimate the 75th quantile, which we know is:

```
qnorm(0.75)
```

with the sample quantile:

```
quantile(y, 0.75)
```

Run a Monte Carlo simulation to learn the expected value and standard error of this random variable.

4. In practice, we can't run a Monte Carlo simulation because we don't know if `rnorm` is being used to simulate the data. Use the bootstrap to estimate the standard error using just the initial sample y. Use 10 bootstrap samples.

5. Redo exercise 4, but with 10,000 bootstrap samples.

30

The caret package

We have already learned about regression and kNN as machine learning algorithms. In later sections, we learn several others, and this is just a small subset of all the algorithms out there. Many of these algorithms are implemented in R. However, they are distributed via different packages, developed by different authors, and often use different syntax. The **caret** package tries to consolidate these differences and provide consistency. It currently includes 237 different methods which are summarized in the **caret** package manual[1]. Keep in mind that **caret** does not include the needed packages and, to implement a package through **caret**, you still need to install the library. The required packages for each method are described in the package manual.

The **caret** package also provides a function that performs cross validation for us. Here we provide some examples showing how we use this incredibly helpful package. We will use the 2 or 7 example to illustrate:

```
library(tidyverse)
library(dslabs)
data("mnist_27")
```

30.1   The caret train functon

The **caret train** function lets us train different algorithms using similar syntax. So, for example, we can type:

```
library(caret)
train_glm <- train(y ~ ., method = "glm", data = mnist_27$train)
train_knn <- train(y ~ ., method = "knn", data = mnist_27$train)
```

To make predictions, we can use the output of this function directly without needing to look at the specifics of `predict.glm` and `predict.knn`. Instead, we can learn how to obtain predictions from `predict.train`.

The code looks the same for both methods:

```
y_hat_glm <- predict(train_glm, mnist_27$test, type = "raw")
y_hat_knn <- predict(train_knn, mnist_27$test, type = "raw")
```

---

[1] https://topepo.github.io/caret/available-models.html

This permits us to quickly compare the algorithms. For example, we can compare the accuracy like this:

```
confusionMatrix(y_hat_glm, mnist_27$test$y)$overall[["Accuracy"]]
#> [1] 0.75
confusionMatrix(y_hat_knn, mnist_27$test$y)$overall[["Accuracy"]]
#> [1] 0.84
```

30.2 Cross validation

When an algorithm includes a tuning parameter, `train` automatically uses cross validation to decide among a few default values. To find out what parameter or parameters are optimized, you can read the manual [2] or study the output of:

```
getModelInfo("knn")
```

We can also use a quick lookup like this:

```
modelLookup("knn")
```

If we run it with default values:

```
train_knn <- train(y ~ ., method = "knn", data = mnist_27$train)
```

you can quickly see the results of the cross validation using the `ggplot` function. The argument `highlight` highlights the max:

```
ggplot(train_knn, highlight = TRUE)
```

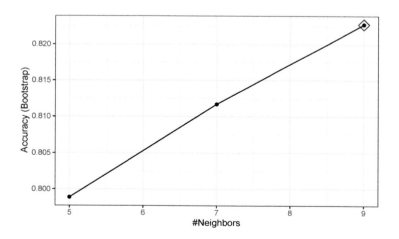

By default, the cross validation is performed by taking 25 bootstrap samples comprised of 25% of the observations. For the kNN method, the default is to try $k = 5, 7, 9$. We change this using the tuneGrid parameter. The grid of values must be supplied by a data frame with the parameter names as specified in the modelLookup output.

Here, we present an example where we try out 30 values between 9 and 67. To do this with **caret**, we need to define a column named k, so we use this: data.frame(k = seq(9, 67, 2)).

Note that when running this code, we are fitting 30 versions of kNN to 25 bootstrapped samples. Since we are fitting $30 \times 25 = 750$ kNN models, running this code will take several seconds. We set the seed because cross validation is a random procedure and we want to make sure the result here is reproducible.

```
set.seed(2008)
train_knn <- train(y ~ ., method = "knn",
                   data = mnist_27$train,
                   tuneGrid = data.frame(k = seq(9, 71, 2)))
ggplot(train_knn, highlight = TRUE)
```

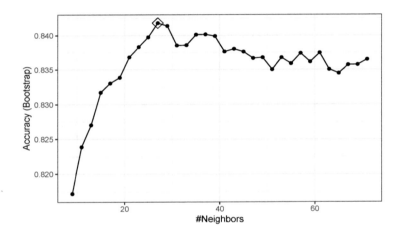

To access the parameter that maximized the accuracy, you can use this:

```
train_knn$bestTune
#>     k
#> 10 27
```

and the best performing model like this:

```
train_knn$finalModel
#> 27-nearest neighbor model
#> Training set outcome distribution:
#>
#>   2   7
#> 379 421
```

The function **predict** will use this best performing model. Here is the accuracy of the best model when applied to the test set, which we have not used at all yet because the cross validation was done on the training set:

```
confusionMatrix(predict(train_knn, mnist_27$test, type = "raw"),
                mnist_27$test$y)$overall["Accuracy"]
#> Accuracy
#>    0.835
```

If we want to change how we perform cross validation, we can use the **trainControl** function. We can make the code above go a bit faster by using, for example, 10-fold cross validation. This means we have 10 samples using 10% of the observations each. We accomplish this using the following code:

```
control <- trainControl(method = "cv", number = 10, p = .9)
train_knn_cv <- train(y ~ ., method = "knn",
              data = mnist_27$train,
              tuneGrid = data.frame(k = seq(9, 71, 2)),
              trControl = control)
ggplot(train_knn_cv, highlight = TRUE)
```

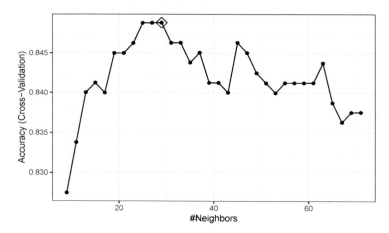

We notice that the accuracy estimates are more variable, which is expected since we changed the number of samples used to estimate accuracy.

Note that **results** component of the **train** output includes several summary statistics related to the variability of the cross validation estimates:

```
names(train_knn$results)
#> [1] "k"         "Accuracy"   "Kappa"      "AccuracySD" "KappaSD"
```

30.3 Example: fitting with loess

The best fitting kNN model approximates the true conditional probability:

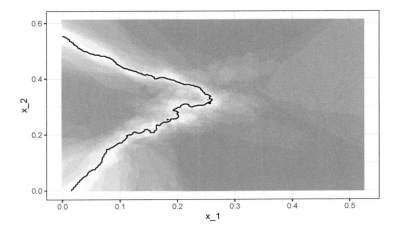

However, we do see that the boundary is somewhat wiggly. This is because kNN, like the basic bin smoother, does not use a kernel. To improve this we could try loess. By reading through the available models part of the manual[3] we see that we can use the `gamLoess` method. In the manual[4] we also see that we need to install the **gam** package if we have not done so already:

```
install.packages("gam")
```

Then we see that we have two parameters to optimize:

```
modelLookup("gamLoess")
#>       model parameter   label forReg forClass probModel
#> 1 gamLoess      span    Span   TRUE     TRUE      TRUE
#> 2 gamLoess    degree  Degree   TRUE     TRUE      TRUE
```

We will stick to a degree of 1. But to try out different values for the span, we still have to include a column in the table with the name **degree** so we can do this:

```
grid <- expand.grid(span = seq(0.15, 0.65, len = 10), degree = 1)
```

We will use the default cross validation control parameters.

```
train_loess <- train(y ~ .,
                method = "gamLoess",
                tuneGrid=grid,
                data = mnist_27$train)
ggplot(train_loess, highlight = TRUE)
```

---

[3] https://topepo.github.io/caret/available-models.html
[4] https://topepo.github.io/caret/train-models-by-tag.html

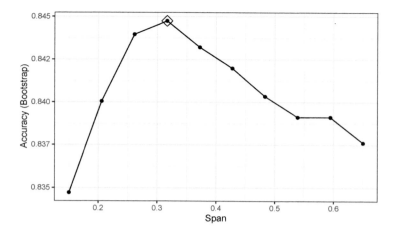

We can see that the method performs similar to kNN:

```
confusionMatrix(data = predict(train_loess, mnist_27$test),
                reference = mnist_27$test$y)$overall["Accuracy"]
#> Accuracy
#>     0.85
```

and produces a smoother estimate of the conditional probability:

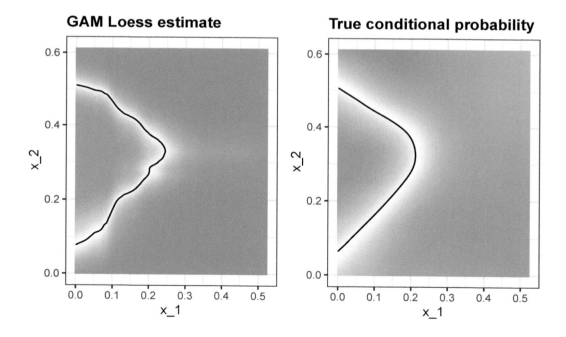

31

Examples of algorithms

There are dozens of machine learning algorithms. Here we provide a few examples spanning rather different approaches. Throughout the chapter we will be using the two predictor digits data introduced in Section 27.8 to demonstrate how the algorithms work.

```
library(tidyverse)
library(dslabs)
library(caret)
data("mnist_27")
```

31.1 Linear regression

Linear regression can be considered a machine learning algorithm. In Section 27.8 we demonstrated how linear regression can be too rigid to be useful. This is generally true, but for some challenges it works rather well. It also serves as a baseline approach: if you can't beat it with a more complex approach, you probably want to stick to linear regression. To quickly make the connection between regression and machine learning, we will reformulate Galton's study with heights, a continuous outcome.

```
library(HistData)

set.seed(1983)
galton_heights <- GaltonFamilies %>%
  filter(gender == "male") %>%
  group_by(family) %>%
  sample_n(1) %>%
  ungroup() %>%
  select(father, childHeight) %>%
  rename(son = childHeight)
```

Suppose you are tasked with building a machine learning algorithm that predicts the son's height Y using the father's height X. Let's generate testing and training sets:

```
y <- galton_heights$son
test_index <- createDataPartition(y, times = 1, p = 0.5, list = FALSE)

train_set <- galton_heights %>% slice(-test_index)
test_set <- galton_heights %>% slice(test_index)
```

In this case, if we were just ignoring the father's height and guessing the son's height, we would guess the average height of sons.

```
m <- mean(train_set$son)
m
#> [1] 69.2
```

Our squared loss is:

```
mean((m - test_set$son)^2)
#> [1] 7.65
```

Can we do better? In the regression chapter, we learned that if the pair (X, Y) follow a bivariate normal distribution, the conditional expectation (what we want to estimate) is equivalent to the regression line:

$$f(x) = \mathrm{E}(Y \mid X = x) = \beta_0 + \beta_1 x$$

In Section 18.3 we introduced least squares as a method for estimating the slope β_0 and intercept β_1:

```
fit <- lm(son ~ father, data = train_set)
fit$coef
#> (Intercept)        father
#>      35.976         0.482
```

This gives us an estimate of the conditional expectation:

$$\hat{f}(x) = 52 + 0.25x$$

We can see that this does indeed provide an improvement over our guessing approach.

```
y_hat <- fit$coef[1] + fit$coef[2]*test_set$father
mean((y_hat - test_set$son)^2)
#> [1] 6.47
```

31.1.1   The `predict` function

The `predict` function is very useful for machine learning applications. This function takes a fitted object from functions such as `lm` or `glm` (we learn about `glm` soon) and a data frame with the new predictors for which to predict. So in our current example, we would use `predict` like this:

```
y_hat <- predict(fit, test_set)
```

Using `predict`, we can get the same results as we did previously:

```
y_hat <- predict(fit, test_set)
mean((y_hat - test_set$son)^2)
#> [1] 6.47
```

predict does not always return objects of the same types; it depends on what type of object is sent to it. To learn about the specifics, you need to look at the help file specific for the type of fit object that is being used. The `predict` is actually a special type of function in R (called a *generic function*) that calls other functions depending on what kind of object it receives. So if `predict` receives an object coming out of the `lm` function, it will call `predict.lm`. If it receives an object coming out of `glm`, it calls `predict.glm`. These two functions are similar but different. You can learn more about the differences by reading the help files:

```
?predict.lm
?predict.glm
```

There are many other versions of `predict` and many machine learning algorithms have a predict function.

31.2 Exercises

1. Create a dataset using the following code.

```
n <- 100
Sigma <- 9*matrix(c(1.0, 0.5, 0.5, 1.0), 2, 2)
dat <- MASS::mvrnorm(n = 100, c(69, 69), Sigma) %>%
  data.frame() %>% setNames(c("x", "y"))
```

Use the **caret** package to partition into a test and training set of equal size. Train a linear model and report the RMSE. Repeat this exercise 100 times and make a histogram of the RMSEs and report the average and standard deviation. Hint: adapt the code shown earlier like this:

```
y <- dat$y
test_index <- createDataPartition(y, times = 1, p = 0.5, list = FALSE)
train_set <- dat %>% slice(-test_index)
test_set <- dat %>% slice(test_index)
fit <- lm(y ~ x, data = train_set)
y_hat <- fit$coef[1] + fit$coef[2]*test_set$x
mean((y_hat - test_set$y)^2)
```

and put it inside a call to `replicate`.

2. Now we will repeat the above but using larger datasets. Repeat exercise 1 but for datasets with `n <- c(100, 500, 1000, 5000, 10000)`. Save the average and standard deviation of RMSE from the 100 repetitions. Hint: use the `sapply` or `map` functions.

3. Describe what you observe with the RMSE as the size of the dataset becomes larger.

   a. On average, the RMSE does not change much as n gets larger, while the variability
       of RMSE does decrease.
   b. Because of the law of large numbers, the RMSE decreases: more data, more precise
       estimates.
   c. n = 10000 is not sufficiently large. To see a decrease in RMSE, we need to make
       it larger.
   d. The RMSE is not a random variable.

4. Now repeat exercise 1, but this time make the correlation between x and y larger by
changing `Sigma` like this:

```
n <- 100
Sigma <- 9*matrix(c(1, 0.95, 0.95, 1), 2, 2)
dat <- MASS::mvrnorm(n = 100, c(69, 69), Sigma) %>%
  data.frame() %>% setNames(c("x", "y"))
```

Repeat the exercise and note what happens to the RMSE now.

5. Which of the following best explains why the RMSE in exercise 4 is so much lower than
exercise 1.

   a. It is just luck. If we do it again, it will be larger.
   b. The Central Limit Theorem tells us the RMSE is normal.
   c. When we increase the correlation between x and y, x has more predictive power
       and thus provides a better estimate of y. This correlation has a much bigger effect
       on RMSE than n. Large n simply provide us more precise estimates of the linear
       model coefficients.
   d. These are both examples of regression, so the RMSE has to be the same.

6. Create a dataset using the following code:

```
n <- 1000
Sigma <- matrix(c(1, 3/4, 3/4, 3/4, 1, 0, 3/4, 0, 1), 3, 3)
dat <- MASS::mvrnorm(n = 100, c(0, 0, 0), Sigma) %>%
  data.frame() %>% setNames(c("y", "x_1", "x_2"))
```

Note that y is correlated with both x_1 and x_2, but the two predictors are independent of
each other.

```
cor(dat)
```

Use the **caret** package to partition into a test and training set of equal size. Compare the
RMSE when using just x_1, just x_2, and both x_1 and x_2. Train a linear model and
report the RMSE.

7. Repeat exercise 6 but now create an example in which x_1 and x_2 are highly correlated:

```
n <- 1000
Sigma <- matrix(c(1.0, 0.75, 0.75, 0.75, 1.0, 0.95, 0.75, 0.95, 1.0), 3, 3)
dat <- MASS::mvrnorm(n = 100, c(0, 0, 0), Sigma) %>%
  data.frame() %>% setNames(c("y", "x_1", "x_2"))
```

Use the **caret** package to partition into a test and training set of equal size. Compare the RMSE when using just x_1, just x_2, and both x_1 and x_2 Train a linear model and report the RMSE.

8. Compare the results in 6 and 7 and choose the statement you agree with:

    a. Adding extra predictors can improve RMSE substantially, but not when they are highly correlated with another predictor.

    b. Adding extra predictors improves predictions equally in both exercises.

    c. Adding extra predictors results in over fitting.

    d. Unless we include all predictors, we have no predicting power.

31.3 Logistic regression

The regression approach can be extended to categorical data. In this section we first illustrate how, for binary data, one can simply assign numeric values of 0 and 1 to the outcomes y, and apply regression as if the data were continuous. We will then point out a limitation with this approach and introduce *logistic regression* as a solution. Logistic regression is a specific case of a set of *generalized linear models*. To illustrate logistic regression, we will apply it to our previous predicting sex example:

If we define the outcome Y as 1 for females and 0 for males, and X as the height, we are interested in the conditional probability:

$$\Pr(Y = 1 \mid X = x)$$

As an example, let's provide a prediction for a student that is 66 inches tall. What is the conditional probability of being female if you are 66 inches tall? In our dataset, we can estimate this by rounding to the nearest inch and computing:

```
train_set %>%
  filter(round(height)==66) %>%
  summarize(y_hat = mean(sex=="Female"))
#>    y_hat
#> 1 0.327
```

To construct a prediction algorithm, we want to estimate the proportion of the population that is female for any given height $X = x$, which we write as the conditional probability described above: $\Pr(Y = 1|X = x)$. Let's see what this looks like for several values of x (we will remove strata of x with few data points):

```
heights %>%
  mutate(x = round(height)) %>%
  group_by(x) %>%
  filter(n() >= 10) %>%
  summarize(prop = mean(sex == "Female")) %>%
  ggplot(aes(x, prop)) +
  geom_point()
```

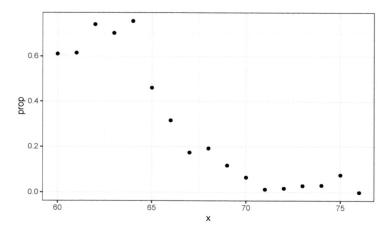

Since the results from the plot above look close to linear, and it is the only approach we currently know, we will try regression. We assume that:

$$p(x) = \Pr(Y = 1 | X = x) = \beta_0 + \beta_1 x$$

Note: because $p_0(x) = 1 - p_1(x)$, we will only estimate $p_1(x)$ and drop the $_1$ index.

If we convert the factors to 0s and 1s, we can estimate β_0 and β_1 with least squares.

```
lm_fit <- mutate(train_set, y = as.numeric(sex == "Female")) %>%
  lm(y ~ height, data = .)
```

Once we have estimates $\hat{\beta}_0$ and $\hat{\beta}_1$, we can obtain an actual prediction. Our estimate of the conditional probability $p(x)$ is:

$$\hat{p}(x) = \hat{\beta}_0 + \hat{\beta}_1 x$$

To form a prediction, we define a *decision rule*: predict female if $\hat{p}(x) > 0.5$. We can compare our predictions to the outcomes using:

```
p_hat <- predict(lm_fit, test_set)
y_hat <- ifelse(p_hat > 0.5, "Female", "Male") %>% factor()
confusionMatrix(y_hat, test_set$sex)[["Accuracy"]]
#> NULL
```

We see this method does substantially better than guessing.

31.3.1   Generalized linear models

The function $\beta_0 + \beta_1 x$ can take any value including negatives and values larger than 1. In fact, the estimate $\hat{p}(x)$ computed in the linear regression section does indeed become negative at around 76 inches.

```
heights %>%
  mutate(x = round(height)) %>%
  group_by(x) %>%
  filter(n() >= 10) %>%
  summarize(prop = mean(sex == "Female")) %>%
  ggplot(aes(x, prop)) +
  geom_point() +
  geom_abline(intercept = lm_fit$coef[1], slope = lm_fit$coef[2])
```

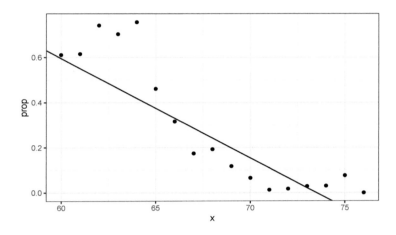

The range is:

```
range(p_hat)
#> [1] -0.331  1.036
```

But we are estimating a probability: $\Pr(Y = 1 \mid X = x)$ which is constrained between 0 and 1.

The idea of generalized linear models (GLM) is to 1) define a distribution of Y that is consistent with it's possible outcomes and 2) find a function g so that $g(\Pr(Y = 1 \mid X = x))$ can be modeled as a linear combination of predictors. Logistic regression is the most commonly used GLM. It is an extension of linear regression that assures that the estimate of $\Pr(Y = 1 \mid X = x)$ is between 0 and 1. This approach makes use of the *logistic* transformation introduced in Section 9.8.1:

$$g(p) = \log \frac{p}{1 - p}$$

This logistic transformation converts probability to log odds. As discussed in the data visualization lecture, the odds tell us how much more likely it is something will happen compared to not happening. $p = 0.5$ means the odds are 1 to 1, thus the odds are 1. If $p = 0.75$, the odds are 3 to 1. A nice characteristic of this transformation is that it converts probabilities to be symmetric around 0. Here is a plot of $g(p)$ versus p:

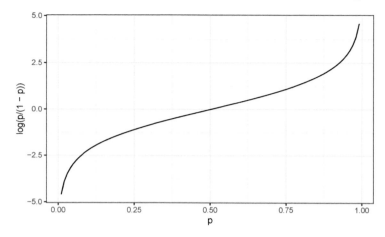

With *logistic regression*, we model the conditional probability directly with:

$$g\left\{\Pr(Y = 1 \mid X = x)\right\} = \beta_0 + \beta_1 x$$

With this model, we can no longer use least squares. Instead we compute the *maximum likelihood estimate* (MLE). You can learn more about this concept in a statistical theory textbook[1].

In R, we can fit the logistic regression model with the function `glm`: generalized linear models. This function is more general than logistic regression so we need to specify the model we want through the `family` parameter:

```
glm_fit <- train_set %>%
  mutate(y = as.numeric(sex == "Female")) %>%
  glm(y ~ height, data=., family = "binomial")
```

We can obtain prediction using the predict function:

```
p_hat_logit <- predict(glm_fit, newdata = test_set, type = "response")
```

When using `predict` with a `glm` object, we have to specify that we want `type="response"` if we want the conditional probabilities, since the default is to return the logistic transformed values.

This model fits the data slightly better than the line:

---

[1]http://www.amazon.com/Mathematical-Statistics-Analysis-Available-Enhanced/dp/0534399428

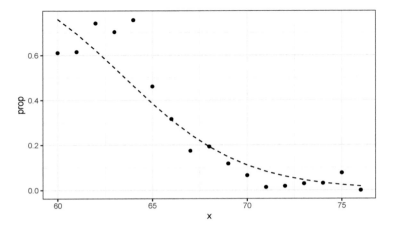

Because we have an estimate $\hat{p}(x)$, we can obtain predictions:

```
y_hat_logit <- ifelse(p_hat_logit > 0.5, "Female", "Male") %>% factor
confusionMatrix(y_hat_logit, test_set$sex)[["Accuracy"]]
#> NULL
```

The resulting predictions are similar. This is because the two estimates of $p(x)$ are larger than $1/2$ in about the same region of x:

```
data.frame(x = seq(min(tmp$x), max(tmp$x))) %>%
  mutate(logistic = plogis(glm_fit$coef[1] + glm_fit$coef[2]*x),
         regression = lm_fit$coef[1] + lm_fit$coef[2]*x) %>%
  gather(method, p_x, -x) %>%
  ggplot(aes(x, p_x, color = method)) +
  geom_line() +
  geom_hline(yintercept = 0.5, lty = 5)
```

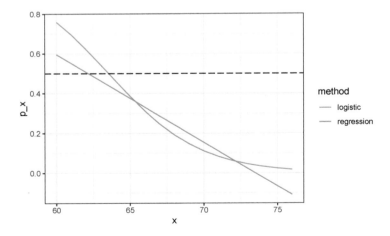

Both linear and logistic regressions provide an estimate for the conditional expectation:

$$\mathrm{E}(Y \mid X = x)$$

which in the case of binary data is equivalent to the conditional probability:

$$\Pr(Y = 1 \mid X = x)$$

31.3.2   Logistic regression with more than one predictor

In this section we apply logistic regression to the two or seven data introduced in Section 27.8. In this case, we are interested in estimating a conditional probability that depends on two variables. The standard logistic regression model in this case will assume that

$$g\{p(x_1, x_2)\} = g\{\Pr(Y = 1 \mid X_1 = x_1, X_2 = x_2)\} = \beta_0 + \beta_1 x_1 + \beta_2 x_2$$

with $g(p) = \log \frac{p}{1-p}$ the logistic function described in the previous section. To fit the model we use the following code:

```
fit_glm <- glm(y ~ x_1 + x_2, data=mnist_27$train, family = "binomial")
p_hat_glm <- predict(fit_glm, mnist_27$test)
y_hat_glm <- factor(ifelse(p_hat_glm > 0.5, 7, 2))
confusionMatrix(y_hat_glm, mnist_27$test$y)$overall["Accuracy"]
#> Accuracy
#>      0.76
```

Comparing to the results we obtained in Section 27.8, we see that logistic regression performs similarly to regression. This is not surprising, given that the estimate of $\hat{p}(x_1, x_2)$ looks similar as well:

```
p_hat <- predict(fit_glm, newdata = mnist_27$true_p, type = "response")
mnist_27$true_p %>% mutate(p_hat = p_hat) %>%
  ggplot(aes(x_1, x_2,  z=p_hat, fill=p_hat)) +
  geom_raster() +
  scale_fill_gradientn(colors=c("#F8766D","white","#00BFC4")) +
  stat_contour(breaks=c(0.5), color="black")
```

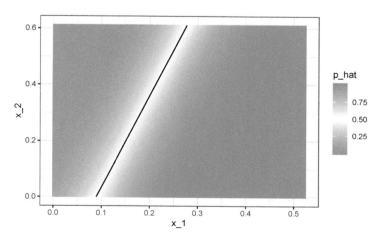

Just like regression, the decision rule is a line, a fact that can be corroborated mathematically since

$$g^{-1}(\hat{\beta}_0 + \hat{\beta}_1 x_1 + \hat{\beta}_2 x_2) = 0.5 \implies \hat{\beta}_0 + \hat{\beta}_1 x_1 + \hat{\beta}_2 x_2 = g(0.5) = 0 \implies x_2 = -\hat{\beta}_0/\hat{\beta}_2 - \hat{\beta}_1/\hat{\beta}_2 x_1$$

Thus x_2 is a linear function of x_1. This implies that, just like regression, our logistic regression approach has no chance of capturing the non-linear nature of the true $p(x_1, x_2)$. Once we move on to more complex examples, we will see that linear regression and generalized linear regression are limited and not flexible enough to be useful for most machine learning challenges. The new techniques we learn are essentially approaches to estimating the conditional probability in a way that is more flexible.

31.4 Exercises

1. Define the following dataset:

```
make_data <- function(n = 1000, p = 0.5,
                      mu_0 = 0, mu_1 = 2,
                      sigma_0 = 1,  sigma_1 = 1){
  y <- rbinom(n, 1, p)
  f_0 <- rnorm(n, mu_0, sigma_0)
  f_1 <- rnorm(n, mu_1, sigma_1)
  x <- ifelse(y == 1, f_1, f_0)
  test_index <- createDataPartition(y, times = 1, p = 0.5, list = FALSE)
  list(train = data.frame(x = x, y = as.factor(y)) %>%
           slice(-test_index),
       test = data.frame(x = x, y = as.factor(y)) %>%
           slice(test_index))
}
dat <- make_data()
```

Note that we have defined a variable x that is predictive of a binary outcome y.

```
dat$train %>% ggplot(aes(x, color = y)) + geom_density()
```

Compare the accuracy of linear regression and logistic regression.

2. Repeat the simulation from exercise 1 100 times and compare the average accuracy for each method and notice they give practically the same answer.

3. Generate 25 different datasets changing the difference between the two class: delta <- seq(0, 3, len = 25). Plot accuracy versus delta.

31.5   k-nearest neighbors

We introduced the kNN algorithm in Section 29.1 and demonstrated how we use cross validation to pick k in Section 30.2. Here we quickly review how we fit a kNN model using the **caret** package. In Section 30.2 we introduced the following code to fit a kNN model:

```
train_knn <- train(y ~ ., method = "knn",
                   data = mnist_27$train,
                   tuneGrid = data.frame(k = seq(9, 71, 2)))
```

We saw that the parameter that maximized the estimated accuracy was:

```
train_knn$bestTune
#>      k
#> 10 27
```

This model improves the accuracy over regression and logistic regression:

```
confusionMatrix(predict(train_knn, mnist_27$test, type = "raw"),
                mnist_27$test$y)$overall["Accuracy"]
#> Accuracy
#>    0.835
```

A plot of the estimated conditional probability shows that the kNN estimate is flexible enough and does indeed capture the shape of the true conditional probability.

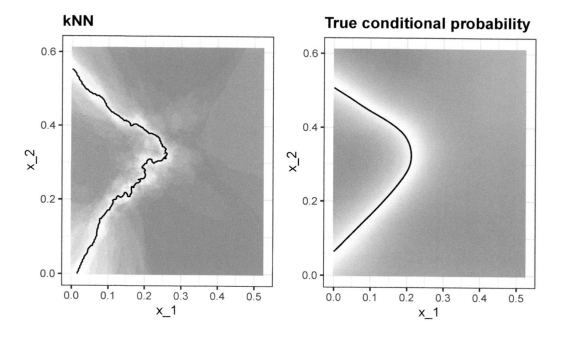

31.6   Exercises

1. Earlier we used logistic regression to predict sex from height. Use kNN to do the same. Use the code described in this chapter to select the F_1 measure and plot it against k. Compare to the F_1 of about 0.6 we obtained with regression.

2. Load the following dataset:

```
data("tissue_gene_expression")
```

This dataset includes a matrix **x**:

```
dim(tissue_gene_expression$x)
```

with the gene expression measured on 500 genes for 189 biological samples representing seven different tissues. The tissue type is stored in **y**:

```
table(tissue_gene_expression$y)
```

Split the data in training and test sets, then use kNN to predict tissue type and see what accuracy you obtain. Try it for $k = 1, 3, \ldots, 11$.

31.7   Generative models

We have described how, when using squared loss, the conditional expectation/probabilities provide the best approach to developing a decision rule. In a binary case, the smallest true error we can achieve is determined by Bayes' rule, which is a decision rule based on the true conditional probability:

$$p(\mathbf{x}) = \Pr(Y = 1 \mid \mathbf{X} = \mathbf{x})$$

We have described several approaches to estimating $p(\mathbf{x})$. In all these approaches, we estimate the conditional probability directly and do not consider the distribution of the predictors. In machine learning, these are referred to as *discriminative* approaches.

However, Bayes' theorem tells us that knowing the distribution of the predictors **X** may be useful. Methods that model the joint distribution of Y and **X** are referred to as *generative models* (we model how the entire data, **X** and Y, are generated). We start by describing the most general generative model, Naive Bayes, and then proceed to describe two specific cases, quadratic discriminant analysis (QDA) and linear discriminant analysis (LDA).

31.7.1   Naive Bayes

Recall that Bayes rule tells us that we can rewrite $p(\mathbf{x})$ like this:

$$p(\mathbf{x}) = \Pr(Y = 1 | \mathbf{X} = \mathbf{x}) = \frac{f_{\mathbf{X}|Y=1}(\mathbf{x})\Pr(Y = 1)}{f_{\mathbf{X}|Y=0}(\mathbf{x})\Pr(Y = 0) + f_{\mathbf{X}|Y=1}(\mathbf{x})\Pr(Y = 1)}$$

with $f_{\mathbf{X}|Y=1}$ and $f_{\mathbf{X}|Y=0}$ representing the distribution functions of the predictor $\mathbf{X}$ for the two classes $Y = 1$ and $Y = 0$. The formula implies that if we can estimate these conditional distributions of the predictors, we can develop a powerful decision rule. However, this is a big *if*. As we go forward, we will encounter examples in which $\mathbf{X}$ has many dimensions and we do not have much information about the distribution. In these cases, Naive Bayes will be practically impossible to implement. However, there are instances in which we have a small number of predictors (not much more than 2) and many categories in which generative models can be quite powerful. We describe two specific examples and use our previously described case studies to illustrate them.

Let's start with a very simple and uninteresting, yet illustrative, case: the example related to predicting sex from height.

```
library(tidyverse)
library(caret)

library(dslabs)
data("heights")

y <- heights$height
set.seed(1995)
test_index <- createDataPartition(y, times = 1, p = 0.5, list = FALSE)
train_set <- heights %>% slice(-test_index)
test_set <- heights %>% slice(test_index)
```

In this case, the Naive Bayes approach is particularly appropriate because we know that the normal distribution is a good approximation for the conditional distributions of height given sex for both classes $Y = 1$ (female) and $Y = 0$ (male). This implies that we can approximate the conditional distributions $f_{X|Y=1}$ and $f_{X|Y=0}$ by simply estimating averages and standard deviations from the data:

```
params <- train_set %>%
  group_by(sex) %>%
  summarize(avg = mean(height), sd = sd(height))
params
#> # A tibble: 2 x 3
#>   sex      avg     sd
#>   <fct>  <dbl>  <dbl>
#> 1 Female  64.8   4.14
#> 2 Male    69.2   3.57
```

The prevalence, which we will denote with $\pi = \Pr(Y = 1)$, can be estimated from the data with:

```
pi <- train_set %>% summarize(pi=mean(sex=="Female")) %>% pull(pi)
pi
#> [1] 0.212
```

Now we can use our estimates of average and standard deviation to get an actual rule:

```
x <- test_set$height
```

```
f0 <- dnorm(x, params$avg[2], params$sd[2])
f1 <- dnorm(x, params$avg[1], params$sd[1])
```

```
p_hat_bayes <- f1*pi / (f1*pi + f0*(1 - pi))
```

Our Naive Bayes estimate $\hat{p}(x)$ looks a lot like our logistic regression estimate:

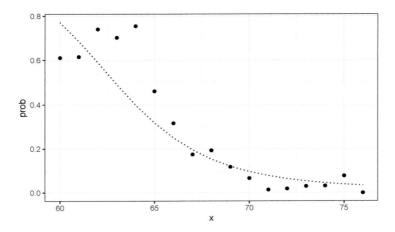

In fact, we can show that the Naive Bayes approach is similar to the logistic regression prediction mathematically. However, we leave the demonstration to a more advanced text, such as the Elements of Statistical Learning[2]. We can see that they are similar empirically by comparing the two resulting curves.

31.7.2   Controlling prevalence

One useful feature of the Naive Bayes approach is that it includes a parameter to account for differences in prevalence. Using our sample, we estimated $f_{X|Y=1}$, $f_{X|Y=0}$ and π. If we use hats to denote the estimates, we can write $\hat{p}(x)$ as:

$$\hat{p}(x) = \frac{\hat{f}_{X|Y=1}(x)\hat{\pi}}{\hat{f}_{X|Y=0}(x)(1-\hat{\pi}) + \hat{f}_{X|Y=1}(x)\hat{\pi}}$$

As we discussed earlier, our sample has a much lower prevalence, 0.21, than the general population. So if we use the rule $\hat{p}(x) > 0.5$ to predict females, our accuracy will be affected due to the low sensitivity:

---

[2]https://web.stanford.edu/~hastie/Papers/ESLII.pdf

```
y_hat_bayes <- ifelse(p_hat_bayes > 0.5, "Female", "Male")
sensitivity(data = factor(y_hat_bayes), reference = factor(test_set$sex))
#> [1] 0.213
```

Again, this is because the algorithm gives more weight to specificity to account for the low prevalence:

```
specificity(data = factor(y_hat_bayes), reference = factor(test_set$sex))
#> [1] 0.967
```

This is due mainly to the fact that $\hat{\pi}$ is substantially less than 0.5, so we tend to predict `Male` more often. It makes sense for a machine learning algorithm to do this in our sample because we do have a higher percentage of males. But if we were to extrapolate this to a general population, our overall accuracy would be affected by the low sensitivity.

The Naive Bayes approach gives us a direct way to correct this since we can simply force $\hat{\pi}$ to be whatever value we want it to be. So to balance specificity and sensitivity, instead of changing the cutoff in the decision rule, we could simply change $\hat{\pi}$ to 0.5 like this:

```
p_hat_bayes_unbiased <- f1 * 0.5 / (f1 * 0.5 + f0 * (1 - 0.5))
y_hat_bayes_unbiased <- ifelse(p_hat_bayes_unbiased> 0.5, "Female", "Male")
```

Note the difference in sensitivity with a better balance:

```
sensitivity(factor(y_hat_bayes_unbiased), factor(test_set$sex))
#> [1] 0.693
specificity(factor(y_hat_bayes_unbiased), factor(test_set$sex))
#> [1] 0.832
```

The new rule also gives us a very intuitive cutoff between 66-67, which is about the middle of the female and male average heights:

```
qplot(x, p_hat_bayes_unbiased, geom = "line") +
  geom_hline(yintercept = 0.5, lty = 2) +
  geom_vline(xintercept = 67, lty = 2)
```

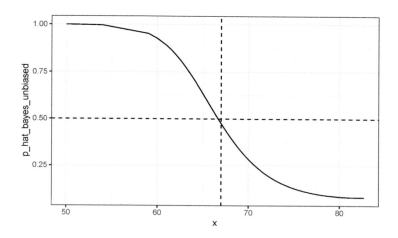

31.7.3 Quadratic discriminant analysis

Quadratic Discriminant Analysis (QDA) is a version of Naive Bayes in which we assume that the distributions $p_{\mathbf{X}|Y=1}(x)$ and $p_{\mathbf{X}|Y=0}(\mathbf{x})$ are multivariate normal. The simple example we described in the previous section is actually QDA. Let's now look at a slightly more complicated case: the 2 or 7 example.

```
data("mnist_27")
```

In this case, we have two predictors so we assume each one is bivariate normal. This implies that we need to estimate two averages, two standard deviations, and a correlation for each case $Y = 1$ and $Y = 0$. Once we have these, we can approximate the distributions $f_{X_1,X_2|Y=1}$ and $f_{X_1,X_2|Y=0}$. We can easily estimate parameters from the data:

```
params <- mnist_27$train %>%
  group_by(y) %>%
  summarize(avg_1 = mean(x_1), avg_2 = mean(x_2),
            sd_1= sd(x_1), sd_2 = sd(x_2),
            r = cor(x_1, x_2))
params
#> # A tibble: 2 x 6
#>   y     avg_1 avg_2   sd_1   sd_2     r
#>   <fct> <dbl> <dbl>  <dbl>  <dbl> <dbl>
#> 1 2     0.129 0.283 0.0702 0.0578 0.401
#> 2 7     0.234 0.288 0.0719 0.105  0.455
```

Here we provide a visual way of showing the approach. We plot the data and use contour plots to give an idea of what the two estimated normal densities look like (we show the curve representing a region that includes 95% of the points):

```
mnist_27$train %>% mutate(y = factor(y)) %>%
  ggplot(aes(x_1, x_2, fill = y, color=y)) +
  geom_point(show.legend = FALSE) +
  stat_ellipse(type="norm", lwd = 1.5)
```

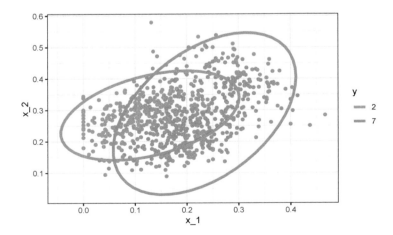

This defines the following estimate of $f(x_1, x_2)$.

We can use the `train` function from the **caret** package to fit the model and obtain predictors:

```
library(caret)
train_qda <- train(y ~ ., method = "qda", data = mnist_27$train)
```

We see that we obtain relatively good accuracy:

```
y_hat <- predict(train_qda, mnist_27$test)
confusionMatrix(y_hat, mnist_27$test$y)$overall["Accuracy"]
#> Accuracy
#>     0.82
```

The estimated conditional probability looks relatively good, although it does not fit as well as the kernel smoothers:

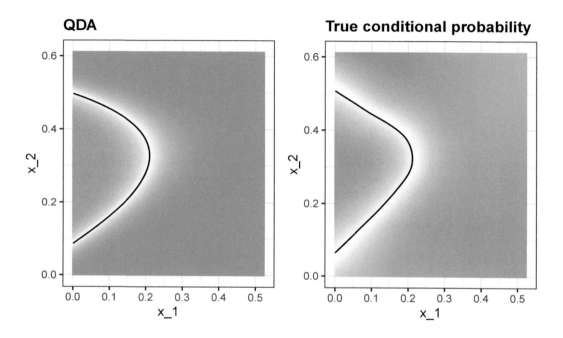

One reason QDA does not work as well as the kernel methods is perhaps because the assumption of normality does not quite hold. Although for the 2s it seems reasonable, for the 7s it does seem to be off. Notice the slight curvature in the points for the 7s:

```
mnist_27$train %>% mutate(y = factor(y)) %>%
  ggplot(aes(x_1, x_2, fill = y, color=y)) +
  geom_point(show.legend = FALSE) +
  stat_ellipse(type="norm") +
  facet_wrap(~y)
```

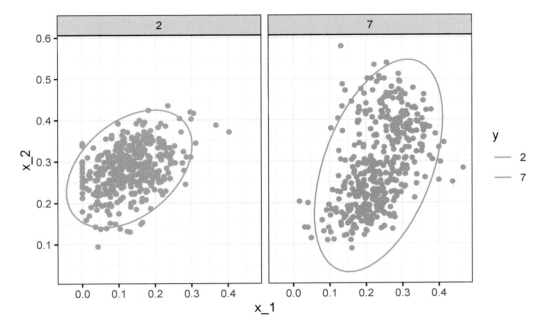

QDA can work well here, but it becomes harder to use as the number of predictors increases. Here we have 2 predictors and had to compute 4 means, 4 SDs, and 2 correlations. How many parameters would we have if instead of 2 predictors, we had 10? The main problem comes from estimating correlations for 10 predictors. With 10, we have 45 correlations for each class. In general, the formula is $K \times p(p-1)/2$, which gets big fast. Once the number of parameters approaches the size of our data, the method becomes impractical due to overfitting.

31.7.4 Linear discriminant analysis

A relatively simple solution to the problem of having too many parameters is to assume that the correlation structure is the same for all classes, which reduces the number of parameters we need to estimate.

In this case, we would compute just one pair of standard deviations and one correlation, and the distributions looks like this:

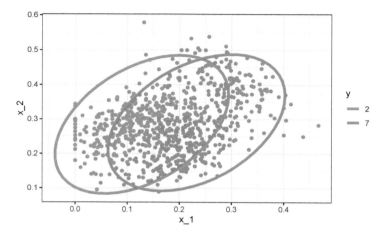

Now the size of the ellipses as well as the angle are the same. This is because they have the same standard deviations and correlations.

We can fit the LDA model using **caret**:

```
train_lda <- train(y ~ ., method = "lda", data = mnist_27$train)
y_hat <- predict(train_lda, mnist_27$test)
confusionMatrix(y_hat, mnist_27$test$y)$overall["Accuracy"]
#> Accuracy
#>     0.75
```

When we force this assumption, we can show mathematically that the boundary is a line, just as with logistic regression. For this reason, we call the method *linear* discriminant analysis (LDA). Similarly, for QDA, we can show that the boundary must be a quadratic function.

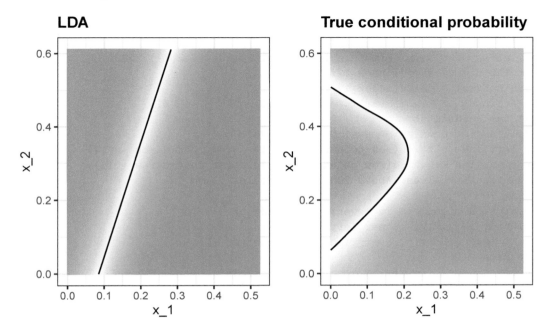

In the case of LDA, the lack of flexibility does not permit us to capture the non-linearity in the true conditional probability function.

31.7.5   Connection to distance

The normal density is:

$$p(x) = \frac{1}{\sqrt{2\pi}\sigma} \exp\left\{-\frac{(x-\mu)^2}{\sigma^2}\right\}$$

If we remove the constant $1/(\sqrt{2\pi}\sigma)$ and then take the log, we get:

$$-\frac{(x-\mu)^2}{\sigma^2}$$

which is the negative of a distance squared scaled by the standard deviation. For higher dimensions, the same is true except the scaling is more complex and involves correlations.

31.8   Case study: more than three classes

We can generate an example with three categories like this:

```
if(!exists("mnist")) mnist <- read_mnist()
set.seed(3456)
index_127 <- sample(which(mnist$train$labels %in% c(1,2,7)), 2000)
y <- mnist$train$labels[index_127]
x <- mnist$train$images[index_127,]
index_train <- createDataPartition(y, p=0.8, list = FALSE)
## get the quadrants
row_column <- expand.grid(row=1:28, col=1:28)
upper_left_ind <- which(row_column$col <= 14 & row_column$row <= 14)
lower_right_ind <- which(row_column$col > 14 & row_column$row > 14)
## binarize the values. Above 200 is ink, below is no ink
x <- x > 200
## proportion of pixels in lower right quadrant
x <- cbind(rowSums(x[ ,upper_left_ind])/rowSums(x),
           rowSums(x[ ,lower_right_ind])/rowSums(x))
##save data
train_set <- data.frame(y = factor(y[index_train]),
                        x_1 = x[index_train,1], x_2 = x[index_train,2])
test_set <- data.frame(y = factor(y[-index_train]),
                       x_1 = x[-index_train,1], x_2 = x[-index_train,2])
```

Here is the training data:

```
train_set %>% ggplot(aes(x_1, x_2, color=y)) + geom_point()
```

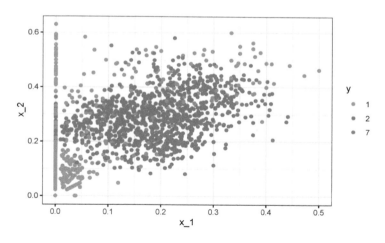

We can use the **caret** package to train the QDA model:

```
train_qda <- train(y ~ ., method = "qda", data = train_set)
```

Now we estimate three conditional probabilities (although they have to add to 1):

```
predict(train_qda, test_set, type = "prob") %>% head()
#>         1       2       7
#> 1 0.7655 0.23043 0.00405
```

```
#> 2 0.2031 0.72514 0.07175
#> 3 0.5396 0.45909 0.00132
#> 4 0.0393 0.09419 0.86655
#> 5 0.9600 0.00936 0.03063
#> 6 0.9865 0.00724 0.00623
```

Our predictions are one of the three classes:

```
predict(train_qda, test_set) %>% head()
#> [1] 1 2 1 7 1 1
#> Levels: 1 2 7
```

The confusion matrix is therefore a 3 by 3 table:

```
confusionMatrix(predict(train_qda, test_set), test_set$y)$table
#>           Reference
#> Prediction   1   2   7
#>          1 111   9  11
#>          2  10  86  21
#>          7  21  28 102
```

The accuracy is 0.749

Note that for sensitivity and specificity, we have a pair of values for **each** class. To define these terms, we need a binary outcome. We therefore have three columns: one for each class as the positives and the other two as the negatives.

To visualize what parts of the region are called 1, 2, and 7 we now need three colors:

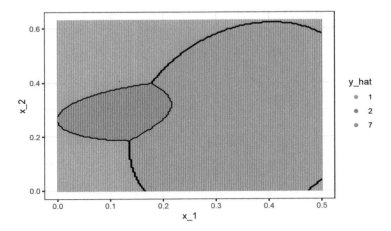

The accuracy for LDA, 0.629, is much worse because the model is more rigid. This is what the decision rule looks like:

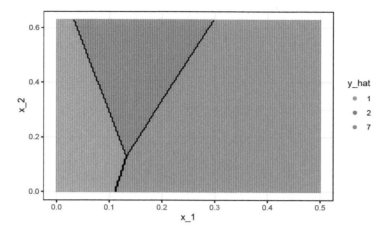

The results for kNN

```
train_knn <- train(y ~ ., method = "knn", data = train_set,
                   tuneGrid = data.frame(k = seq(15, 51, 2)))
```

are much better with an accuracy of 0.749. The decision rule looks like this:

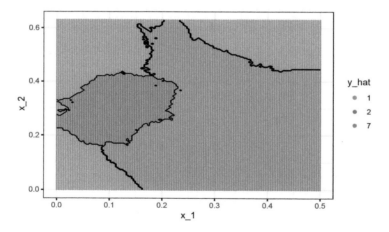

Note that one of the limitations of generative models here is due to the lack of fit of the normal assumption, in particular for class 1.

```
train_set %>% mutate(y = factor(y)) %>%
  ggplot(aes(x_1, x_2, fill = y, color=y)) +
  geom_point(show.legend = FALSE) +
  stat_ellipse(type="norm")
```

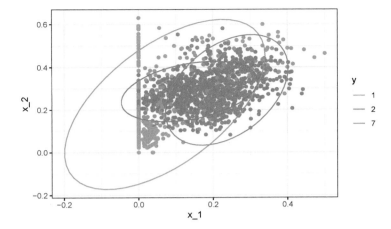

Generative models can be very powerful, but only when we are able to successfully approximate the joint distribution of predictors conditioned on each class.

31.9 Exercises

We are going to apply LDA and QDA to the `tissue_gene_expression` dataset. We will start with simple examples based on this dataset and then develop a realistic example.

1. Create a dataset with just the classes "cerebellum" and "hippocampus" (two parts of the brain) and a predictor matrix with 10 randomly selected columns.

```
set.seed(1993)
data("tissue_gene_expression")
tissues <- c("cerebellum", "hippocampus")
ind <- which(tissue_gene_expression$y %in% tissues)
y <- droplevels(tissue_gene_expression$y[ind])
x <- tissue_gene_expression$x[ind, ]
x <- x[, sample(ncol(x), 10)]
```

Use the `train` function to estimate the accuracy of LDA.

2. In this case, LDA fits two 10-dimensional normal distributions. Look at the fitted model by looking at the `finalModel` component of the result of train. Notice there is a component called `means` that includes the estimate `means` of both distributions. Plot the mean vectors against each other and determine which predictors (genes) appear to be driving the algorithm.

3. Repeat exercises 1 with QDA. Does it have a higher accuracy than LDA?

4. Are the same predictors (genes) driving the algorithm? Make a plot as in exercise 2.

5. One thing we see in the previous plot is that the value of predictors correlate in both groups: some predictors are low in both groups while others are high in both groups. The mean value of each predictor, `colMeans(x)`, is not informative or useful for prediction, and

often for interpretation purposes it is useful to center or scale each column. This can be achieved with the prePprocessing argument in train. Re-run LDA with prePprocessing = "scale". Note that accuracy does not change but see how it is easier to identify the predictors that differ more between groups in the plot made in exercise 4.

6. In the previous exercises we saw that both approaches worked well. Plot the predictor values for the two genes with the largest differences between the two groups in a scatterplot to see how they appear to follow a bivariate distribution as assumed by the LDA and QDA approaches. Color the points by the outcome.

7. Now we are going to increase the complexity of the challenge slightly: we will consider all the tissue types.

```
set.seed(1993)
data("tissue_gene_expression")
y <- tissue_gene_expression$y
x <- tissue_gene_expression$x
x <- x[, sample(ncol(x), 10)]
```

What accuracy do you get with LDA?

8. We see that the results are slightly worse. Use the confusionMatrix function to learn what type of errors we are making.

9. Plot an image of the centers of the seven 10-dimensional normal distributions.

31.10   Classification and regression trees (CART)

31.10.1   The curse of dimensionality

We described how methods such as LDA and QDA are not meant to be used with many predictors p because the number of parameters that we need to estimate becomes too large. For example, with the digits example $p = 784$, we would have over 600,000 parameters with LDA, and we would multiply that by the number of classes for QDA. Kernel methods such as kNN or local regression do not have model parameters to estimate. However, they also face a challenge when multiple predictors are used due to what is referred to as the *curse of dimensionality*. The *dimension* here refers to the fact that when we have p predictors, the distance between two observations is computed in p-dimensional space.

A useful way of understanding the curse of dimensionality is by considering how large we have to make a span/neighborhood/window to include a given percentage of the data. Remember that with larger neighborhoods, our methods lose flexibility.

For example, suppose we have one continuous predictor with equally spaced points in the [0,1] interval and we want to create windows that include 1/10th of data. Then it's easy to see that our windows have to be of size 0.1:

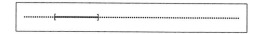

Now, for two predictors, if we decide to keep the neighborhood just as small, 10% for each dimension, we include only 1 point. If we want to include 10% of the data, then we need to increase the size of each side of the square to $\sqrt{.10} \approx .316$:

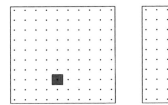

Using the same logic, if we want to include 10% of the data in a three-dimensional space, then the side of each cube is $\sqrt[3]{.10} \approx 0.464$. In general, to include 10% of the data in a case with p dimensions, we need an interval with each side of size $\sqrt[p]{.10}$ of the total. This proportion gets close to 1 quickly, and if the proportion is 1 it means we include all the data and are no longer smoothing.

```
library(tidyverse)
p <- 1:100
qplot(p, .1^(1/p), ylim = c(0,1))
```

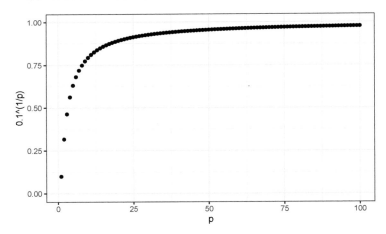

By the time we reach 100 predictors, the neighborhood is no longer very local, as each side covers almost the entire dataset.

Here we look at a set of elegant and versatile methods that adapt to higher dimensions and also allow these regions to take more complex shapes while still producing models that are interpretable. These are very popular, well-known and studied methods. We will concentrate on regression and decision trees and their extension to random forests.

31.10.2 CART motivation

To motivate this section, we will use a new dataset that includes the breakdown of the composition of olive oil into 8 fatty acids:

```r
library(tidyverse)
library(dslabs)
data("olive")
names(olive)
#> [1] "region"    "area"       "palmitic"   "palmitoleic"
#> [5] "stearic"   "oleic"      "linoleic"   "linolenic"
#> [9] "arachidic" "eicosenoic"
```

For illustrative purposes, we will try to predict the region using the fatty acid composition values as predictors.

```r
table(olive$region)
#>
#> Northern Italy      Sardinia Southern Italy
#>            151            98            323
```

We remove the `area` column because we won't use it as a predictor.

```r
olive <- select(olive, -area)
```

Let's very quickly try to predict the region using kNN:

```r
library(caret)
fit <- train(region ~ .,  method = "knn",
             tuneGrid = data.frame(k = seq(1, 15, 2)),
             data = olive)
ggplot(fit)
```

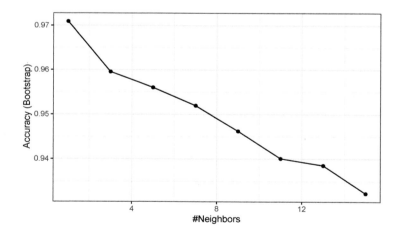

We see that using just one neighbor, we can predict relatively well. However, a bit of data exploration reveals that we should be able to do even better. For example, if we look at the distribution of each predictor stratified by region we see that eicosenoic is only present in Southern Italy and that linoleic separates Northern Italy from Sardinia.

```
olive %>% gather(fatty_acid, percentage, -region) %>%
  ggplot(aes(region, percentage, fill = region)) +
  geom_boxplot() +
  facet_wrap(~fatty_acid, scales = "free", ncol = 4) +
  theme(axis.text.x = element_blank(), legend.position="bottom")
```

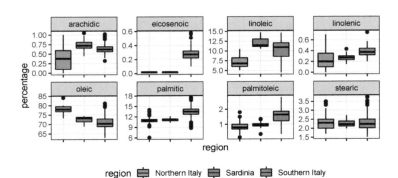

This implies that we should be able to build an algorithm that predicts perfectly! We can see this clearly by plotting the values for eicosenoic and linoleic.

```
olive %>%
  ggplot(aes(eicosenoic, linoleic, color = region)) +
  geom_point() +
  geom_vline(xintercept = 0.065, lty = 2) +
  geom_segment(x = -0.2, y = 10.54, xend = 0.065, yend = 10.54,
               color = "black", lty = 2)
```

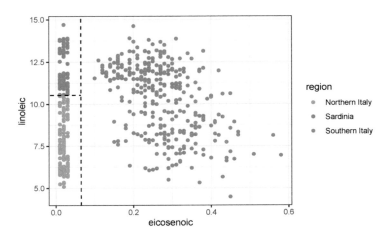

In Section 33.3.4 we define predictor spaces. The predictor space here consists of eight-dimensional points with values between 0 and 100. In the plot above, we show the space defined by the two predictors eicosenoic and linoleic, and, by eye, we can construct a prediction rule that partitions the predictor space so that each partition contains only outcomes of a one category. This in turn can be used to define an algorithm with perfect accuracy. Specifically, we define the following decision rule. If eicosenoic is larger than 0.065, predict Southern Italy. If not, then if linoleic is larger than 10.535, predict Sardinia, and if lower, predict Northern Italy. We can draw this decision tree like this:

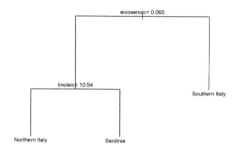

Decision trees like this are often used in practice. For example, to decide on a person's risk of poor outcome after having a heart attack, doctors use the following:

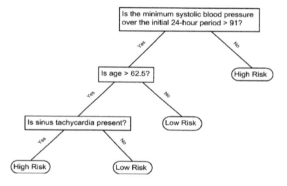

(Source: Walton 2010 Informal Logic, Vol. 30, No. 2, pp. 159-184[3].)

A tree is basically a flow chart of yes or no questions. The general idea of the methods we are describing is to define an algorithm that uses data to create these trees with predictions at the ends, referred to as *nodes*. Regression and decision trees operate by predicting an outcome variable Y by partitioning the predictors.

31.10.3 Regression trees

When the outcome is continuous, we call the method a *regression* tree. To introduce regression trees, we will use the 2008 poll data used in previous sections to describe the basic idea of how we build these algorithms. As with other machine learning algorithms, we will try to estimate the conditional expectation $f(x) = E(Y|X = x)$ with Y the poll margin and x the day.

---

[3]https://papers.ssrn.com/sol3/Delivery.cfm/SSRN__ID1759289__code1486039.pdf?abstractid=1759289&mirid=1&type=2

```
data("polls_2008")
qplot(day, margin, data = polls_2008)
```

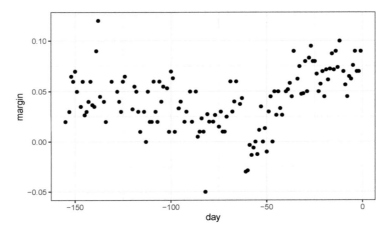

The general idea here is to build a decision tree and, at the end of each *node*, obtain a predictor $\hat{y}$. A mathematical way to describe this is to say that we are partitioning the predictor space into J non-overlapping regions, $R_1, R_2, \ldots, R_J$, and then for any predictor x that falls within region R_j, estimate $f(x)$ with the average of the training observations y_i for which the associated predictor x_i is also in R_j.

But how do we decide on the partition $R_1, R_2, \ldots, R_J$ and how do we choose J? Here is where the algorithm gets a bit complicated.

Regression trees create partitions recursively. We start the algorithm with one partition, the entire predictor space. In our simple first example, this space is the interval [-155, 1]. But after the first step we will have two partitions. After the second step we will split one of these partitions into two and will have three partitions, then four, then five, and so on. We describe how we pick the partition to further partition, and when to stop, later.

Once we select a partition $\mathbf{x}$ to split in order to create the new partitions, we find a predictor j and value s that define two new partitions, which we will call $R_1(j, s)$ and $R_2(j, s)$, that split our observations in the current partition by asking if x_j is bigger than s:

$$R_1(j, s) = \{\mathbf{x} \mid x_j < s\} \text{ and } R_2(j, s) = \{\mathbf{x} \mid x_j \geq s\}$$

In our current example we only have one predictor, so we will always choose $j = 1$, but in general this will not be the case. Now, after we define the new partitions R_1 and R_2, and we decide to stop the partitioning, we compute predictors by taking the average of all the observations y for which the associated $\mathbf{x}$ is in R_1 and R_2. We refer to these two as $\hat{y}_{R_1}$ and $\hat{y}_{R_2}$ respectively.

But how do we pick j and s? Basically we find the pair that minimizes the residual sum of square (RSS):

$$\sum_{i:\, x_i \in R_1(j,s)} (y_i - \hat{y}_{R_1})^2 + \sum_{i:\, x_i \in R_2(j,s)} (y_i - \hat{y}_{R_2})^2$$

This is then applied recursively to the new regions R_1 and R_2. We describe how we stop later, but once we are done partitioning the predictor space into regions, in each region a prediction is made using the observations in that region.

Let's take a look at what this algorithm does on the 2008 presidential election poll data. We will use the **rpart** function in the **rpart** package.

```
library(rpart)
fit <- rpart(margin ~ ., data = polls_2008)
```

Here, there is only one predictor. Thus we do not have to decide which predictor j to split by, we simply have to decide what value s we use to split. We can visually see where the splits were made:

```
plot(fit, margin = 0.1)
text(fit, cex = 0.75)
```

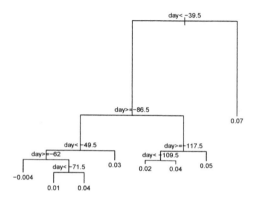

The first split is made on day 39.5. One of those regions is then split at day 86.5. The two resulting new partitions are split on days 49.5 and 117.5, respectively, and so on. We end up with 8 partitions. The final estimate $\hat{f}(x)$ looks like this:

```
polls_2008 %>%
  mutate(y_hat = predict(fit)) %>%
  ggplot() +
  geom_point(aes(day, margin)) +
  geom_step(aes(day, y_hat), col="red")
```

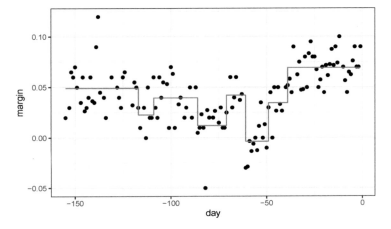

Note that the algorithm stopped partitioning at 8. Now we explain how this decision is made.

First we need to define the term *complexity parameter* (cp). Every time we split and define two new partitions, our training set RSS decreases. This is because with more partitions, our model has more flexibility to adapt to the training data. In fact, if you split until every point is its own partition, then RSS goes all the way down to 0 since the average of one value is that same value. To avoid this, the algorithm sets a minimum for how much the RSS must improve for another partition to be added. This parameter is referred to as the *complexity parameter* (cp). The RSS must improve by a factor of cp for the new partition to be added. Large values of cp will therefore force the algorithm to stop earlier which results in fewer nodes.

However, cp is not the only parameter used to decide if we should partition a current partition or not. Another common parameter is the minimum number of observations required in a partition before partitioning it further. The argument used in the `rpart` function is `minsplit` and the default is 20. The `rpart` implementation of regression trees also permits users to determine a minimum number of observations in each node. The argument is `minbucket` and defaults to `round(minsplit/3)`.

As expected, if we set `cp = 0` and `minsplit = 2`, then our prediction is as flexible as possible and our predictor is our original data:

```
fit <- rpart(margin ~ ., data = polls_2008,
             control = rpart.control(cp = 0, minsplit = 2))
polls_2008 %>%
  mutate(y_hat = predict(fit)) %>%
  ggplot() +
  geom_point(aes(day, margin)) +
  geom_step(aes(day, y_hat), col="red")
```

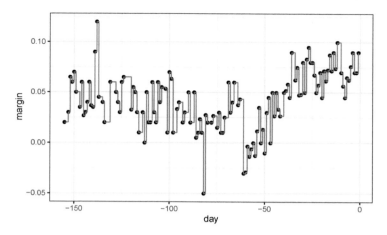

Intuitively we know that this is not a good approach as it will generally result in over-training. These `cp`, `minsplit`, and `minbucket`, three parameters can be used to control the variability of the final predictors. The larger these values are the more data is averaged to compute a predictor and thus reduce variability. The drawback is that it restricts flexibility.

So how do we pick these parameters? We can use cross validation, described in Chapter 29, just like with any tuning parameter. Here is an example of using cross validation to chose cp.

```
library(caret)
train_rpart <- train(margin ~ .,
                     method = "rpart",
                     tuneGrid = data.frame(cp = seq(0, 0.05, len = 25)),
                     data = polls_2008)
ggplot(train_rpart)
```

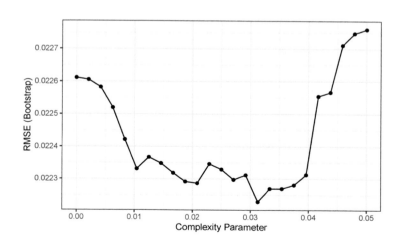

To see the resulting tree, we access the `finalModel` and plot it:

```
plot(train_rpart$finalModel, margin = 0.1)
text(train_rpart$finalModel, cex = 0.75)
```

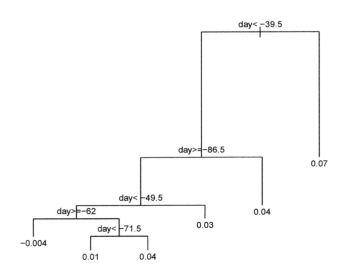

And because we only have one predictor, we can actually plot $\hat{f}(x)$:

```
polls_2008 %>%
  mutate(y_hat = predict(train_rpart)) %>%
  ggplot() +
  geom_point(aes(day, margin)) +
  geom_step(aes(day, y_hat), col="red")
```

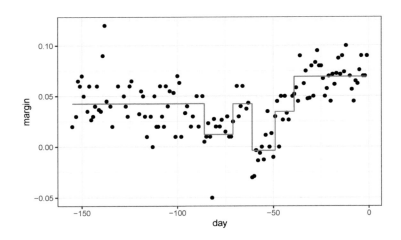

Note that if we already have a tree and want to apply a higher cp value, we can use the prune function. We call this *pruning* a tree because we are snipping off partitions that do not meet a cp criterion. We previously created a tree that used a cp = 0 and saved it to fit. We can prune it like this:

```
pruned_fit <- prune(fit, cp = 0.01)
```

31.10.4   Classification (decision) trees

Classification trees, or decision trees, are used in prediction problems where the outcome is categorical. We use the same partitioning principle with some differences to account for the fact that we are now working with a categorical outcome.

The first difference is that we form predictions by calculating which class is the most common among the training set observations within the partition, rather than taking the average in each partition (as we can't take the average of categories).

The second is that we can no longer use RSS to choose the partition. While we could use the naive approach of looking for partitions that minimize training error, better performing approaches use more sophisticated metrics. Two of the more popular ones are the *Gini Index* and *Entropy*.

In a perfect scenario, the outcomes in each of our partitions are all of the same category since this will permit perfect accuracy. The *Gini Index* is going to be 0 in this scenario, and become larger the more we deviate from this scenario. To define the Gini Index, we define $\hat{p}_{j,k}$ as the proportion of observations in partition j that are of class k. The Gini Index is defined as

$$\text{Gini}(j) = \sum_{k=1}^{K} \hat{p}_{j,k}(1 - \hat{p}_{j,k})$$

If you study the formula carefully you will see that it is in fact 0 in the perfect scenario described above.

Entropy is a very similar quantity, defined as

$$\text{entropy}(j) = -\sum_{k=1}^{K} \hat{p}_{j,k} \log(\hat{p}_{j,k}), \text{ with } 0 \times \log(0) \text{ defined as } 0$$

Let us look at how a classification tree performs on the digits example we examined before:

We can use this code to run the algorithm and plot the resulting tree:

```
train_rpart <- train(y ~ .,
                     method = "rpart",
                     tuneGrid = data.frame(cp = seq(0.0, 0.1, len = 25)),
                     data = mnist_27$train)
plot(train_rpart)
```

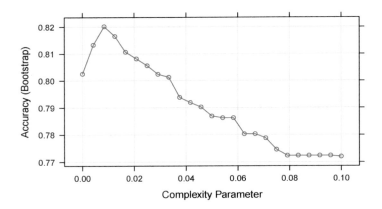

The accuracy achieved by this approach is better than what we got with regression, but is not as good as what we achieved with kernel methods:

```
y_hat <- predict(train_rpart, mnist_27$test)
confusionMatrix(y_hat, mnist_27$test$y)$overall["Accuracy"]
#> Accuracy
#>     0.82
```

The plot of the estimated conditional probability shows us the limitations of classification trees:

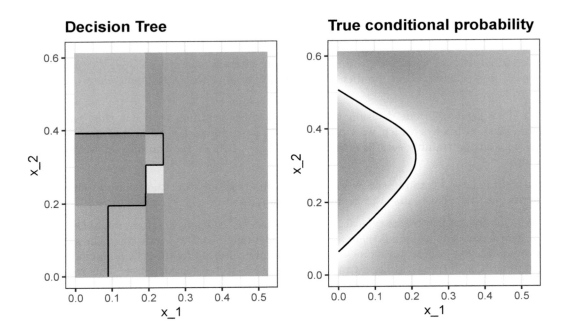

Note that with decision trees, it is difficult to make the boundaries smooth since each partition creates a discontinuity.

Classification trees have certain advantages that make them very useful. They are highly interpretable, even more so than linear models. They are easy to visualize (if small enough). Finally, they can model human decision processes and don't require use of dummy predictors for categorical variables. On the other hand, the approach via recursive partitioning can easily over-train and is therefore a bit harder to train than, for example, linear regression or kNN. Furthermore, in terms of accuracy, it is rarely the best performing method since it is not very flexible and is highly unstable to changes in training data. Random forests, explained next, improve on several of these shortcomings.

31.11   Random forests

Random forests are a **very popular** machine learning approach that addresses the shortcomings of decision trees using a clever idea. The goal is to improve prediction performance and reduce instability by *averaging* multiple decision trees (a forest of trees constructed with randomness). It has two features that help accomplish this.

The first step is *bootstrap aggregation* or *bagging*. The general idea is to generate many predictors, each using regression or classification trees, and then forming a final prediction based on the average prediction of all these trees. To assure that the individual trees are not the same, we use the bootstrap to induce randomness. These two features combined explain the name: the bootstrap makes the individual trees **randomly** different, and the combination of trees is the **forest**. The specific steps are as follows.

1. Build B decision trees using the training set. We refer to the fitted models as $T_1, T_2, \ldots, T_B$. We later explain how we ensure they are different.

2. For every observation in the test set, form a prediction $\hat{y}_j$ using tree T_j.

3. For continuous outcomes, form a final prediction with the average $\hat{y} = \frac{1}{B} \sum_{j=1}^{B} \hat{y}_j$. For categorical data classification, predict $\hat{y}$ with majority vote (most frequent class among $\hat{y}_1, \ldots, \hat{y}_T$).

So how do we get different decision trees from a single training set? For this, we use randomness in two ways which we explain in the steps below. Let N be the number of observations in the training set. To create T_j, $j = 1, \ldots, B$ from the training set we do the following:

1. Create a bootstrap training set by sampling N observations from the training set **with replacement**. This is the first way to induce randomness.

2. A large number of features is typical in machine learning challenges. Often, many features can be informative but including them all in the model may result in overfitting. The second way random forests induce randomness is by randomly selecting features to be included in the building of each tree. A different random subset is selected for each tree. This reduces correlation between trees in the forest, thereby improving prediction accuracy.

To illustrate how the first steps can result in smoother estimates we will demonstrate by fitting a random forest to the 2008 polls data. We will use the `randomForest` function in the **randomForest** package:

```
library(randomForest)
fit <- randomForest(margin~., data = polls_2008)
```

Note that if we apply the function `plot` to the resulting object, stored in `fit`, we see how the error rate of our algorithm changes as we add trees.

```
rafalib::mypar()
plot(fit)
```

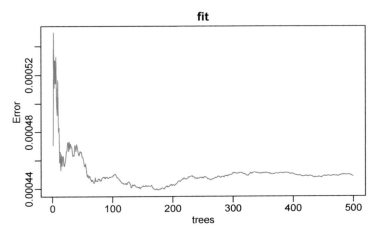

We can see that in this case, the accuracy improves as we add more trees until about 30 trees where accuracy stabilizes.

The resulting estimate for this random forest can be seen like this:

```
polls_2008 %>%
  mutate(y_hat = predict(fit, newdata = polls_2008)) %>%
  ggplot() +
  geom_point(aes(day, margin)) +
  geom_line(aes(day, y_hat), col="red")
```

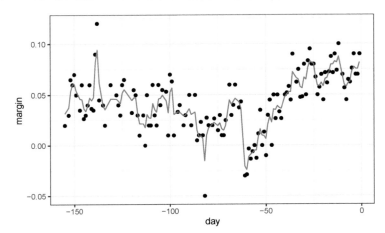

Notice that the random forest estimate is much smoother than what we achieved with the regression tree in the previous section. This is possible because the average of many step functions can be smooth. We can see this by visually examining how the estimate changes as we add more trees. In the following figure you see each of the bootstrap samples for several values of b and for each one we see the tree that is fitted in grey, the previous trees that were fitted in lighter grey, and the result of averaging all the trees estimated up to that point.

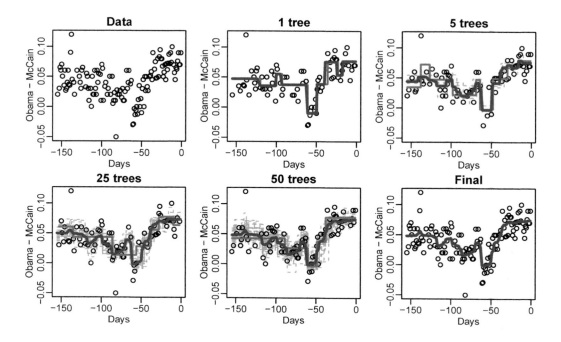

Here is the random forest fit for our digits example based on two predictors:

```
library(randomForest)
train_rf <- randomForest(y ~ ., data=mnist_27$train)

confusionMatrix(predict(train_rf, mnist_27$test),
                mnist_27$test$y)$overall["Accuracy"]
#> Accuracy
#>     0.79
```

Here is what the conditional probabilities look like:

Random Forest

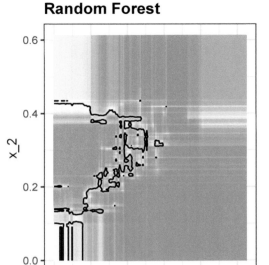

True conditional probability

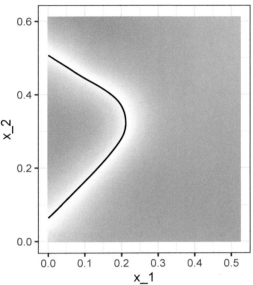

Visualizing the estimate shows that, although we obtain high accuracy, it appears that there is room for improvement by making the estimate smoother. This could be achieved by changing the parameter that controls the minimum number of data points in the nodes of the tree. The larger this minimum, the smoother the final estimate will be. We can train the parameters of the random forest. Below, we use the **caret** package to optimize over the minimum node size. Because, this is not one of the parameters that the **caret** package optimizes by default we will write our own code:

```
nodesize <- seq(1, 51, 10)
acc <- sapply(nodesize, function(ns){
  train(y ~ ., method = "rf", data = mnist_27$train,
                tuneGrid = data.frame(mtry = 2),
                nodesize = ns)$results$Accuracy
})
qplot(nodesize, acc)
```

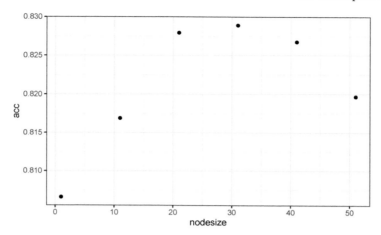

We can now fit the random forest with the optimized minimun node size to the entire training data and evaluate performance on the test data.

```
train_rf_2 <- randomForest(y ~ ., data=mnist_27$train,
                           nodesize = nodesize[which.max(acc)])

confusionMatrix(predict(train_rf_2, mnist_27$test),
               mnist_27$test$y)$overall["Accuracy"]
#> Accuracy
#>     0.82
```

The selected model improves accuracy and provides a smoother estimate.

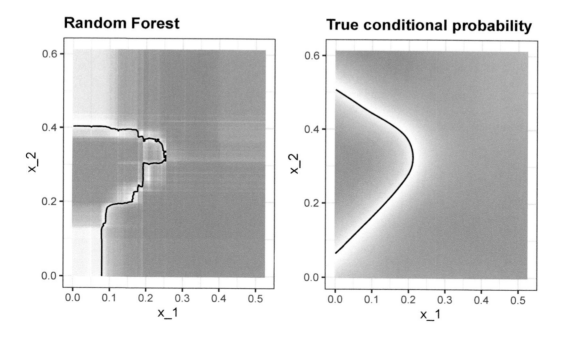

Note that we can avoid writing our own code by using other random forest implementations as described in the **caret** manual[4].

Random forest performs better in all the examples we have considered. However, a disadvantage of random forests is that we lose interpretability. An approach that helps with interpretability is to examine *variable importance*. To define *variable importance* we count how often a predictor is used in the individual trees. You can learn more about *variable importance* in an advanced machine learning book[5]. The **caret** package includes the function `varImp` that extracts variable importance from any model in which the calculation is implemented. We give an example on how we use variable importance in the next section.

31.12 Exercises

1. Create a simple dataset where the outcome grows 0.75 units on average for every increase in a predictor:

```
n <- 1000
sigma <- 0.25
x <- rnorm(n, 0, 1)
y <- 0.75 * x + rnorm(n, 0, sigma)
dat <- data.frame(x = x, y = y)
```

Use `rpart` to fit a regression tree and save the result to `fit`.

2. Plot the final tree so that you can see where the partitions occurred.

3. Make a scatterplot of y versus x along with the predicted values based on the fit.

4. Now model with a random forest instead of a regression tree using `randomForest` from the **randomForest** package, and remake the scatterplot with the prediction line.

5. Use the function `plot` to see if the random forest has converged or if we need more trees.

6. It seems that the default values for the random forest result in an estimate that is too flexible (not smooth). Re-run the random forest but this time with `nodesize` set at 50 and `maxnodes` set at 25. Remake the plot.

7. We see that this yields smoother results. Let's use the `train` function to help us pick these values. From the **caret** manual[6] we see that we can't tune the `maxnodes` parameter or the `nodesize` argument with `randomForest`, so we will use the **Rborist** package and tune the `minNode` argument. Use the `train` function to try values `minNode <- seq(5, 250, 25)`. See which value minimizes the estimated RMSE.

8. Make a scatterplot along with the prediction from the best fitted model.

9. Use the `rpart` function to fit a classification tree to the `tissue_gene_expression` dataset. Use the `train` function to estimate the accuracy. Try out `cp` values of `seq(0, 0.05, 0.01)`. Plot the accuracy to report the results of the best model.

---

[4]http://topepo.github.io/caret/available-models.html
[5]https://web.stanford.edu/~hastie/Papers/ESLII.pdf
[6]https://topepo.github.io/caret/available-models.html

10. Study the confusion matrix for the best fitting classification tree. What do you observe happening for placenta?

11. Notice that placentas are called endometrium more often than placenta. Note also that the number of placentas is just six, and that, by default, `rpart` requires 20 observations before splitting a node. Thus it is not possible with these parameters to have a node in which placentas are the majority. Rerun the above analysis but this time permit `rpart` to split any node by using the argument `control = rpart.control(minsplit = 0)`. Does the accuracy increase? Look at the confusion matrix again.

12. Plot the tree from the best fitting model obtained in exercise 11.

13. We can see that with just six genes, we are able to predict the tissue type. Now let's see if we can do even better with a random forest. Use the `train` function and the `rf` method to train a random forest. Try out values of `mtry` ranging from, at least, `seq(50, 200, 25)`. What `mtry` value maximizes accuracy? To permit small `nodesize` to grow as we did with the classification trees, use the following argument: `nodesize = 1`. This will take several seconds to run. If you want to test it out, try using smaller values with `ntree`. Set the seed to 1990.

14. Use the function `varImp` on the output of `train` and save it to an object called `imp`.

15. The `rpart` model we ran above produced a tree that used just six predictors. Extracting the predictor names is not straightforward, but can be done. If the output of the call to train was `fit_rpart`, we can extract the names like this:

```
ind <- !(fit_rpart$finalModel$frame$var == "<leaf>")
tree_terms <-
  fit_rpart$finalModel$frame$var[ind] %>%
  unique() %>%
  as.character()
tree_terms
```

What is the variable importance in the random forest call for these predictors? Where do they rank?

16. Advanced: Extract the top 50 predictors based on importance, take a subset of x with just these predictors and apply the function `heatmap` to see how these genes behave across the tissues. We will introduce the `heatmap` function in Chapter 34.

32

Machine learning in practice

Now that we have learned several methods and explored them with illustrative examples, we are going to try them out on a real example: the MNIST digits.

We can load this data using the following **dslabs** package:

```
library(tidyverse)
library(dslabs)
mnist <- read_mnist()
```

The dataset includes two components, a training set and test set:

```
names(mnist)
#> [1] "train" "test"
```

Each of these components includes a matrix with features in the columns:

```
dim(mnist$train$images)
#> [1] 60000   784
```

and vector with the classes as integers:

```
class(mnist$train$labels)
#> [1] "integer"
table(mnist$train$labels)
#>
#>    0    1    2    3    4    5    6    7    8    9
#> 5923 6742 5958 6131 5842 5421 5918 6265 5851 5949
```

Because we want this example to run on a small laptop and in less than one hour, we will consider a subset of the dataset. We will sample 10,000 random rows from the training set and 1,000 random rows from the test set:

```
set.seed(1990)
index <- sample(nrow(mnist$train$images), 10000)
x <- mnist$train$images[index,]
y <- factor(mnist$train$labels[index])

index <- sample(nrow(mnist$test$images), 1000)
x_test <- mnist$test$images[index,]
y_test <- factor(mnist$test$labels[index])
```

32.1   Preprocessing

In machine learning, we often transform predictors before running the machine algorithm. We also remove predictors that are clearly not useful. We call these steps *preprocessing*.

Examples of preprocessing include standardizing the predictors, taking the log transform of some predictors, removing predictors that are highly correlated with others, and removing predictors with very few non-unique values or close to zero variation. We show an example below.

We can run the `nearZero` function from the **caret** package to see that several features do not vary much from observation to observation. We can see that there is a large number of features with 0 variability:

```
library(matrixStats)
sds <- colSds(x)
qplot(sds, bins = 256)
```

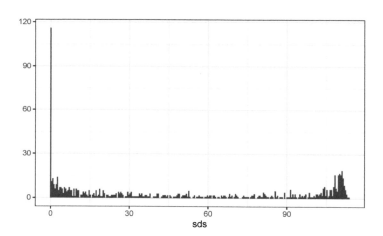

This is expected because there are parts of the image that rarely contain writing (dark pixels).

The **caret** packages includes a function that recommends features to be removed due to *near zero variance*:

```
library(caret)
nzv <- nearZeroVar(x)
```

We can see the columns recommended for removal:

```
image(matrix(1:784 %in% nzv, 28, 28))
```

```
rafalib::mypar()
image(matrix(1:784 %in% nzv, 28, 28))
```

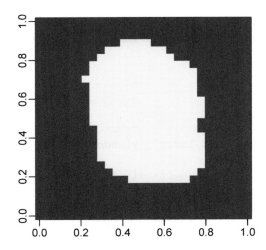

So we end up keeping this number of columns:

```
col_index <- setdiff(1:ncol(x), nzv)
length(col_index)
#> [1] 252
```

Now we are ready to fit some models. Before we start, we need to add column names to the feature matrices as these are required by **caret**:

```
colnames(x) <- 1:ncol(mnist$train$images)
colnames(x_test) <- colnames(x)
```

32.2   k-nearest neighbor and random forest

Let's start with kNN. The first step is to optimize for k. Keep in mind that when we run the algorithm, we will have to compute a distance between each observation in the test set and each observation in the training set. There are a lot of computations. We will therefore use k-fold cross validation to improve speed.

If we run the following code, the computing time on a standard laptop will be several minutes.

```
control <- trainControl(method = "cv", number = 10, p = .9)
train_knn <- train(x[ ,col_index], y,
```

```
                  method = "knn",
                  tuneGrid = data.frame(k = c(3,5,7)),
                  trControl = control)
train_knn
```

In general, it is a good idea to try a test run with a subset of the data to get an idea of timing before we start running code that might take hours to complete. We can do this as follows:

```
n <- 1000
b <- 2
index <- sample(nrow(x), n)
control <- trainControl(method = "cv", number = b, p = .9)
train_knn <- train(x[index, col_index], y[index],
                  method = "knn",
                  tuneGrid = data.frame(k = c(3,5,7)),
                  trControl = control)
```

We can then increase n and b and try to establish a pattern of how they affect computing time to get an idea of how long the fitting process will take for larger values of n and b. You want to know if a function is going to take hours, or even days, before you run it.

Once we optimize our algorithm, we can fit it to the entire dataset:

```
fit_knn <- knn3(x[, col_index], y,   k = 3)
```

The accuracy is almost 0.95!

```
y_hat_knn <- predict(fit_knn, x_test[, col_index], type="class")
cm <- confusionMatrix(y_hat_knn, factor(y_test))
cm$overall["Accuracy"]
#> Accuracy
#>    0.953
```

We now achieve an accuracy of about 0.95. From the specificity and sensitivity, we also see that 8s are the hardest to detect and the most commonly incorrectly predicted digit is 7.

```
cm$byClass[,1:2]
#>            Sensitivity Specificity
#> Class: 0      0.990       0.996
#> Class: 1      1.000       0.993
#> Class: 2      0.965       0.997
#> Class: 3      0.950       0.999
#> Class: 4      0.930       0.997
#> Class: 5      0.921       0.993
#> Class: 6      0.977       0.996
#> Class: 7      0.956       0.989
#> Class: 8      0.887       0.999
#> Class: 9      0.951       0.990
```

Now let's see if we can do even better with the random forest algorithm.

With random forest, computation time is a challenge. For each forest, we need to build hundreds of trees. We also have several parameters we can tune.

Because with random forest the fitting is the slowest part of the procedure rather than the predicting (as with kNN), we will use only five-fold cross validation. We will also reduce the number of trees that are fit since we are not yet building our final model.

Finally, to compute on a smaller dataset, we will take a random sample of the observations when constructing each tree. We can change this number with the **nSamp** argument.

```
library(randomForest)
control <- trainControl(method="cv", number = 5)
grid <- data.frame(mtry = c(1, 5, 10, 25, 50, 100))

train_rf <-  train(x[, col_index], y,
                   method = "rf",
                   ntree = 150,
                   trControl = control,
                   tuneGrid = grid,
                   nSamp = 5000)
```

```
ggplot(train_rf)
train_rf$bestTune
#>    mtry
#> 3    10
```

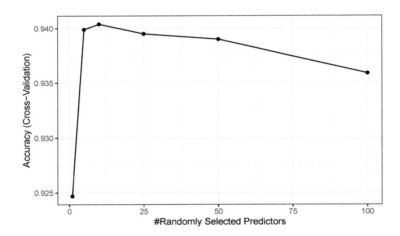

Now that we have optimized our algorithm, we are ready to fit our final model:

```
fit_rf <- randomForest(x[, col_index], y,
                       minNode = train_rf$bestTune$mtry)
```

To check that we ran enough trees we can use the plot function:

```
plot(fit_rf)
```

We see that we achieve high accuracy:

```
y_hat_rf <- predict(fit_rf, x_test[ ,col_index])
cm <- confusionMatrix(y_hat_rf, y_test)
cm$overall["Accuracy"]
#> Accuracy
#>    0.952
```

With some further tuning, we can get even higher accuracy.

32.3   Variable importance

The following function computes the importance of each feature:

```
imp <- importance(fit_rf)
```

We can see which features are being used most by plotting an image:

```
mat <- rep(0, ncol(x))
mat[col_index] <- imp
image(matrix(mat, 28, 28))

rafalib::mypar()
mat <- rep(0, ncol(x))
mat[col_index] <- imp
image(matrix(mat, 28, 28))
```

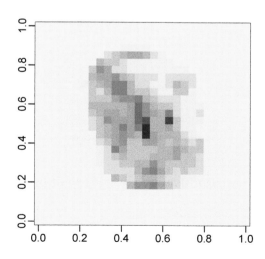

32.4 Visual assessments

An important part of data analysis is visualizing results to determine why we are failing. How we do this depends on the application. Below we show the images of digits for which we made an incorrect prediction. We can compare what we get with kNN to random forest.

Here are some errors for the random forest:

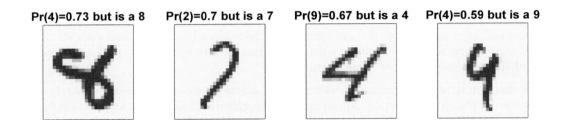

By examining errors like this we often find specific weaknesses to algorithms or parameter choices and can try to correct them.

32.5 Ensembles

The idea of an ensemble is similar to the idea of combining data from different pollsters to obtain a better estimate of the true support for each candidate.

In machine learning, one can usually greatly improve the final results by combining the results of different algorithms.

Here is a simple example where we compute new class probabilities by taking the average of random forest and kNN. We can see that the accuracy improves to 0.96:

```
p_rf <- predict(fit_rf, x_test[,col_index], type = "prob")
p_rf<- p_rf / rowSums(p_rf)
p_knn  <- predict(fit_knn, x_test[,col_index])
p <- (p_rf + p_knn)/2
y_pred <- factor(apply(p, 1, which.max)-1)
confusionMatrix(y_pred, y_test)$overall["Accuracy"]
#> Accuracy
#>    0.962
```

In the exercises we are going to build several machine learning models for the mnist_27 dataset and then build an ensemble.

32.6   Exercises

1. Use the `mnist_27` training set to build a model with several of the models available from the **caret** package. For example, you can try these:

```
models <- c("glm", "lda",  "naive_bayes",  "svmLinear", "gamboost",
            "gamLoess", "qda", "knn", "kknn", "loclda", "gam", "rf",
            "ranger","wsrf", "Rborist", "avNNet", "mlp", "monmlp", "gbm",
            "adaboost", "svmRadial", "svmRadialCost", "svmRadialSigma")
```

We have not explained many of these, but apply them anyway using `train` with all the default parameters. Keep the results in a list. You might need to install some packages. Keep in mind that you will likely get some warnings.

2. Now that you have all the trained models in a list, use `sapply` or `map` to create a matrix of predictions for the test set. You should end up with a matrix with `length(mnist_27$test$y)` rows and `length(models)` columns.

3. Now compute accuracy for each model on the test set.

4. Now build an ensemble prediction by majority vote and compute its accuracy.

5. Earlier we computed the accuracy of each method on the training set and noticed they varied. Which individual methods do better than the ensemble?

6. It is tempting to remove the methods that do not perform well and re-do the ensemble. The problem with this approach is that we are using the test data to make a decision. However, we could use the accuracy estimates obtained from cross validation with the training data. Obtain these estimates and save them in an object.

7. Now let's only consider the methods with an estimated accuracy of 0.8 when constructing the ensemble. What is the accuracy now?

8. **Advanced**: If two methods give results that are the same, ensembling them will not change the results at all. For each pair of metrics compare the percent of time they call the same thing. Then use the `heatmap` function to visualize the results. Hint: use the `method = "binary"` argument in the `dist` function.

9. **Advanced**: Note that each method can also produce an estimated conditional probability. Instead of majority vote we can take the average of these estimated conditional probabilities. For most methods, we can the use the `type = "prob"` in the train function. However, some of the methods require you to use the argument `trControl=trainControl(classProbs=TRUE)` when calling train. Also these methods do not work if classes have numbers as names. Hint: change the levels like this:

```
dat$train$y <- recode_factor(dat$train$y, "2"="two", "7"="seven")
dat$test$y <- recode_factor(dat$test$y, "2"="two", "7"="seven")
```

10. In this chapter, we illustrated a couple of machine learning algorithms on a subset of the MNIST dataset. Try fitting a model to the entire dataset.

33

Large datasets

Machine learning problems often involve datasets that are as large or larger than the MNIST dataset. There is a variety of computational techniques and statistical concepts that are useful for the analysis of large datasets. In this chapter we scratch the surface of these techniques and concepts by describing matrix algebra, dimension reduction, regularization and matrix factorization. We use recommendation systems related to movie ratings as a motivating example.

33.1 Matrix algebra

In machine learning, situations in which all predictors are numeric, or can be converted to numeric in a meaningful way, are common. The digits data set is an example: every pixel records a number between 0 and 255. Let's load the data:

```
library(tidyverse)
library(dslabs)
if(!exists("mnist")) mnist <- read_mnist()
```

In these cases, it is often convenient to save the predictors in a matrix and the outcome in a vector rather than using a data frame. You can see that the predictors are saved as a matrix:

```
class(mnist$train$images)
#> [1] "matrix"
```

This matrix represents 60,000 digits, so for the examples in this chapter, we will take a more manageable subset. We will take the first 1,000 predictors x and labels y:

```
x <- mnist$train$images[1:1000,]
y <- mnist$train$labels[1:1000]
```

The main reason for using matrices is that certain mathematical operations needed to develop efficient code can be performed using techniques from a branch of mathematics called *linear algebra*. In fact, linear algebra and matrix notation are key elements of the language used in academic papers describing machine learning techniques. We will not cover linear algebra in detail here, but will demonstrate how to use matrices in R so that you can apply the linear algebra techniques already implemented in base R or other packages.

To motivate the use of matrices, we will pose five questions/challenges:

1. Do some digits require more ink than others? Study the distribution of the total pixel darkness and how it varies by digits.

2. Are some pixels uninformative? Study the variation of each pixel and remove predictors (columns) associated with pixels that don't change much and thus can't provide much information for classification.

3. Can we remove smudges? First, look at the distribution of all pixel values. Use this to pick a cutoff to define unwritten space. Then, set anything below that cutoff to 0.

4. Binarize the data. First, look at the distribution of all pixel values. Use this to pick a cutoff to distinguish between writing and no writing. Then, convert all entries into either 1 or 0, respectively.

5. Scale each of the predictors in each entry to have the same average and standard deviation.

To complete these, we will have to perform mathematical operations involving several variables. The **tidyverse** is not developed to perform these types of mathematical operations. For this task, it is convenient to use matrices.

Before we do this, we will introduce matrix notation and basic R code to define and operate on matrices.

33.1.1   Notation

In matrix algebra, we have three main types of objects: scalars, vectors, and matrices. A scalar is just one number, for example $a = 1$. To denote scalars in matrix notation, we usually use a lower case letter and do not bold.

Vectors are like the numeric vectors we define in R: they include several scalar entries. For example, the column containing the first pixel:

```
length(x[,1])
#> [1] 1000
```

has 1,000 entries. In matrix algebra, we use the following notation for a vector representing a feature/predictor:

$$\begin{pmatrix} x_1 \\ x_2 \\ \vdots \\ x_N \end{pmatrix}$$

Similarly, we can use math notation to represent different features mathematically by adding an index:

$$\mathbf{X}_1 = \begin{pmatrix} x_{1,1} \\ \vdots \\ x_{N,1} \end{pmatrix} \text{ and } \mathbf{X}_2 = \begin{pmatrix} x_{1,2} \\ \vdots \\ x_{N,2} \end{pmatrix}$$

If we are writing out a column, such as $\mathbf{X}_1$, in a sentence we often use the notation:

$\mathbf{X}_1 = (x_{1,1}, \ldots x_{N,1})^\top$ with $^\top$ the transpose operation that converts columns into rows and rows into columns.

A matrix can be defined as a series of vectors of the same size joined together as columns:

```
x_1 <- 1:5
x_2 <- 6:10
cbind(x_1, x_2)
#>      x_1 x_2
#> [1,]   1   6
#> [2,]   2   7
#> [3,]   3   8
#> [4,]   4   9
#> [5,]   5  10
```

Mathematically, we represent them with bold upper case letters:

$$\mathbf{X} = [\mathbf{X}_1 \mathbf{X}_2] = \begin{pmatrix} x_{1,1} & x_{1,2} \\ \vdots & \\ x_{N,1} & x_{N,2} \end{pmatrix}$$

The *dimension* of a matrix is often an important characteristic needed to assure that certain operations can be performed. The dimension is a two-number summary defined as the number of rows × the number of columns. In R, we can extract the dimension of a matrix with the function `dim`:

```
dim(x)
#> [1] 1000  784
```

Vectors can be thought of as $N \times 1$ matrices. However, in R, a vector does not have dimensions:

```
dim(x_1)
#> NULL
```

Yet we explicitly convert a vector into a matrix using the function `as.matrix`:

```
dim(as.matrix(x_1))
#> [1] 5 1
```

We can use this notation to denote an arbitrary number of predictors with the following $N \times p$ matrix, for example, with $p = 784$:

$$\mathbf{X} = \begin{pmatrix} x_{1,1} & \cdots & x_{1,p} \\ x_{2,1} & \cdots & x_{2,p} \\ & \vdots & \\ x_{N,1} & \cdots & x_{N,p} \end{pmatrix}$$

We stored this matrix in x:

```
dim(x)
#> [1] 1000  784
```

We will now learn several useful operations related to matrix algebra. We use three of the motivating questions listed above.

33.1.2   Converting a vector to a matrix

It is often useful to convert a vector to a matrix. For example, because the variables are pixels on a grid, we can convert the rows of pixel intensities into a matrix representing this grid.

We can convert a vector into a matrix with the `matrix` function and specifying the number of rows and columns that the resulting matrix should have. The matrix is filled in **by column**: the first column is filled first, then the second and so on. This example helps illustrate:

```
my_vector <- 1:15
mat <- matrix(my_vector, 5, 3)
mat
#>      [,1] [,2] [,3]
#> [1,]    1    6   11
#> [2,]    2    7   12
#> [3,]    3    8   13
#> [4,]    4    9   14
#> [5,]    5   10   15
```

We can fill by row by using the `byrow` argument. So, for example, to *transpose* the matrix `mat`, we can use:

```
mat_t <- matrix(my_vector, 3, 5, byrow = TRUE)
mat_t
#>      [,1] [,2] [,3] [,4] [,5]
#> [1,]    1    2    3    4    5
#> [2,]    6    7    8    9   10
#> [3,]   11   12   13   14   15
```

When we turn the columns into rows, we refer to the operations as *transposing* the matrix. The function `t` can be used to directly transpose a matrix:

```
identical(t(mat), mat_t)
#> [1] TRUE
```

Warning: The `matrix` function recycles values in the vector **without warning** if the product of columns and rows does not match the length of the vector:

```
matrix(my_vector, 4, 5)
#> Warning in matrix(my_vector, 4, 5): data length [15] is not a sub-
#> multiple or multiple of the number of rows [4]
#>      [,1] [,2] [,3] [,4] [,5]
#> [1,]    1    5    9   13    2
```

```
#> [2,]    2    6   10   14    3
#> [3,]    3    7   11   15    4
#> [4,]    4    8   12    1    5
```

To put the pixel intensities of our, say, 3rd entry, which is a 4 into grid, we can use:

```
grid <- matrix(x[3,], 28, 28)
```

To confirm that in fact we have done this correctly, we can use the function `image`, which shows an image of its third argument. The top of this plot is pixel 1, which is shown at the bottom so the image is flipped. To code below includes code showing how to flip it back:

```
image(1:28, 1:28, grid)
image(1:28, 1:28, grid[, 28:1])
```

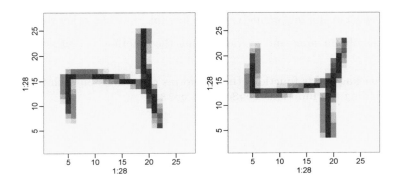

33.1.3   Row and column summaries

For the first task, related to total pixel darkness, we want to sum the values of each row and then visualize how these values vary by digit.

The function `rowSums` takes a matrix as input and computes the desired values:

```
sums <- rowSums(x)
```

We can also compute the averages with `rowMeans` if we want the values to remain between 0 and 255:

```
avg <- rowMeans(x)
```

Once we have this, we can simply generate a boxplot:

```
tibble(labels = as.factor(y), row_averages = avg) %>%
  qplot(labels, row_averages, data = ., geom = "boxplot")
```

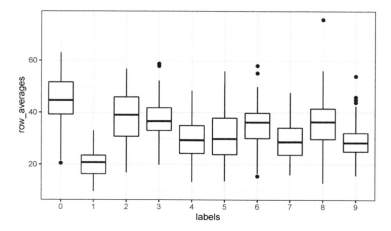

From this plot we see that, not surprisingly, 1s use less ink than the other digits.

We can compute the column sums and averages using the function `colSums` and `colMeans`, respectively.

The **matrixStats** package adds functions that performs operations on each row or column very efficiently, including the functions `rowSds` and `colSds`.

33.1.4 apply

The functions just described are performing an operation similar to what `sapply` and the **purrr** function `map` do: apply the same function to a part of your object. In this case, the function is applied to either each row or each column. The `apply` function lets you apply any function, not just `sum` or `mean`, to a matrix. The first argument is the matrix, the second is the dimension, 1 for rows, 2 for columns, and the third is the function. So, for example, `rowMeans` can be written as:

```
avgs <- apply(x, 1, mean)
```

But notice that just like with `sapply` and `map`, we can perform any function. So if we wanted the standard deviation for each column, we could write:

```
sds <- apply(x, 2, sd)
```

The tradeoff for this flexibility is that these operations are not as fast as dedicated functions such as `rowMeans`.

33.1.5 Filtering columns based on summaries

We now turn to task 2: studying the variation of each pixel and removing columns associated with pixels that don't change much and thus do not inform the classification. Although a simplistic approach, we will quantify the variation of each pixel with its standard deviation across all entries. Since each column represents a pixel, we use the `colSds` function from the **matrixStats** package:

```
library(matrixStats)
sds <- colSds(x)
```

A quick look at the distribution of these values shows that some pixels have very low entry to entry variability:

```
qplot(sds, bins = "30", color = I("black"))
```

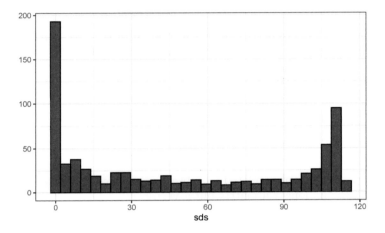

This makes sense since we don't write in some parts of the box. Here is the variance plotted by location:

```
image(1:28, 1:28, matrix(sds, 28, 28)[, 28:1])
```

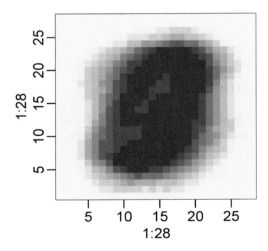

We see that there is little variation in the corners.

We could remove features that have no variation since these can't help us predict. In Section 2.4.7, we described the operations used to extract columns:

```
x[ ,c(351,352)]
```

and rows:

```
x[c(2,3),]
```

We can also use logical indexes to determine which columns or rows to keep. So if we wanted to remove uninformative predictors from our matrix, we could write this one line of code:

```
new_x <- x[ ,colSds(x) > 60]
dim(new_x)
#> [1] 1000   314
```

Only the columns for which the standard deviation is above 60 are kept, which removes over half the predictors.

Here we add an important warning related to subsetting matrices: if you select one column or one row, the result is no longer a matrix but a vector.

```
class(x[,1])
#> [1] "integer"
dim(x[1,])
#> NULL
```

However, we can preserve the matrix class by using the argument `drop=FALSE`:

```
class(x[ , 1, drop=FALSE])
#> [1] "matrix"
dim(x[, 1, drop=FALSE])
#> [1] 1000     1
```

33.1.6   Indexing with matrices

We can quickly make a histogram of all the values in our dataset. We saw how we can turn vectors into matrices. We can also undo this and turn matrices into vectors. The operation will happen by row:

```
mat <- matrix(1:15, 5, 3)
as.vector(mat)
#>  [1]  1  2  3  4  5  6  7  8  9 10 11 12 13 14 15
```

To see a histogram of all our predictor data, we can use:

```
qplot(as.vector(x), bins = 30, color = I("black"))
```

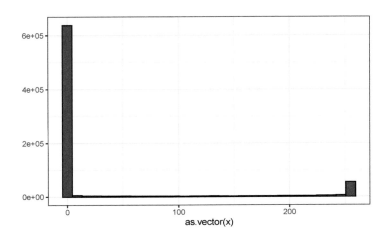

We notice a clear dichotomy which is explained as parts of the image with ink and parts without. If we think that values below, say, 50 are smudges, we can quickly make them zero using:

```
new_x <- x
new_x[new_x < 50] <- 0
```

To see what this does, we look at a smaller matrix:

```
mat <- matrix(1:15, 5, 3)
mat[mat < 3] <- 0
mat
#>      [,1] [,2] [,3]
#> [1,]    0    6   11
#> [2,]    0    7   12
#> [3,]    3    8   13
#> [4,]    4    9   14
#> [5,]    5   10   15
```

We can also use logical operations with matrix logical:

```
mat <- matrix(1:15, 5, 3)
mat[mat > 6 & mat < 12] <- 0
mat
#>      [,1] [,2] [,3]
#> [1,]    1    6    0
#> [2,]    2    0   12
#> [3,]    3    0   13
#> [4,]    4    0   14
#> [5,]    5    0   15
```

33.1.7 Binarizing the data

The histogram above seems to suggest that this data is mostly binary. A pixel either has ink or does not. Using what we have learned, we can binarize the data using just matrix operations:

```
bin_x <- x
bin_x[bin_x < 255/2] <- 0
bin_x[bin_x > 255/2] <- 1
```

We can also convert to a matrix of logicals and then coerce to numbers like this:

```
bin_X <- (x > 255/2)*1
```

33.1.8 Vectorization for matrices

In R, if we subtract a vector from a matrix, the first element of the vector is subtracted from the first row, the second element from the second row, and so on. Using mathematical notation, we would write it as follows:

$$\begin{pmatrix} X_{1,1} & \dots & X_{1,p} \\ X_{2,1} & \dots & X_{2,p} \\ & \vdots & \\ X_{N,1} & \dots & X_{N,p} \end{pmatrix} - \begin{pmatrix} a_1 \\ a_2 \\ \vdots \\ a_N \end{pmatrix} = \begin{pmatrix} X_{1,1} - a_1 & \dots & X_{1,p} - a_1 \\ X_{2,1} - a_2 & \dots & X_{2,p} - a_2 \\ & \vdots & \\ X_{N,1} - a_n & \dots & X_{N,p} - a_n \end{pmatrix}$$

The same holds true for other arithmetic operations. This implies that we can scale each row of a matrix like this:

```
(x - rowMeans(x)) / rowSds(x)
```

If you want to scale each column, be careful since this approach does not work for columns. To perform a similar operation, we convert the columns to rows using the transpose `t`, proceed as above, and then transpose back:

```
t(t(X) - colMeans(X))
```

We can also use a function called `sweep` that works similarly to `apply`. It takes each entry of a vector and subtracts it from the corresponding row or column.

```
X_mean_0 <- sweep(x, 2, colMeans(x))
```

The function `sweep` actually has another argument that lets you define the arithmetic operation. So to divide by the standard deviation, we do the following:

```
x_mean_0 <- sweep(x, 2, colMeans(x))
x_standardized <- sweep(x_mean_0, 2, colSds(x), FUN = "/")
```

33.1.9 Matrix algebra operations

Finally, although we do not cover matrix algebra operations such as matrix multiplication, we share here the relevant commands for those that know the mathematics and want to learn the code:

1. Matrix multiplication is done with %*%. For example, the cross product is:

```
t(x) %*% x
```

2. We can compute the cross product directly with the function:

```
crossprod(x)
```

3. To compute the inverse of a function, we use `solve`. Here it is applied to the cross product:

```
solve(crossprod(x))
```

4. The QR decomposition is readily available by using the `qr` function:

```
qr(x)
```

33.2 Exercises

1. Create a 100 by 10 matrix of randomly generated normal numbers. Put the result in x.

2. Apply the three R functions that give you the dimension of x, the number of rows of x, and the number of columns of x, respectively.

3. Add the scalar 1 to row 1, the scalar 2 to row 2, and so on, to the matrix x.

4. Add the scalar 1 to column 1, the scalar 2 to column 2, and so on, to the matrix x. Hint: use `sweep` with `FUN = "+"`.

5. Compute the average of each row of x.

6. Compute the average of each column of x.

7. For each digit in the MNIST training data, compute the proportion of pixels that are in a *grey area*, defined as values between 50 and 205. Make boxplot by digit class. Hint: use logical operators and `rowMeans`.

33.3 Distance

Many of the analyses we perform with high-dimensional data relate directly or indirectly to distance. Most clustering and machine learning techniques rely on being able to define distance between observations, using features or predictors.

33.3.1   Euclidean distance

As a review, let's define the distance between two points, A and B, on a Cartesian plane.

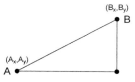

The Euclidean distance between A and B is simply:

$$\text{dist}(A, B) = \sqrt{(A_x - B_x)^2 + (A_y - B_y)^2}$$

This definition applies to the case of one dimension, in which the distance between two numbers is simply the absolute value of their difference. So if our two one-dimensional numbers are A and B, the distance is:

$$\text{dist}(A, B) = \sqrt{(A - B)^2} = |A - B|$$

33.3.2   Distance in higher dimensions

Earlier we introduced a training dataset with feature matrix measurements for 784 features. For illustrative purposes, we will look at a random sample of 2s and 7s.

```
library(tidyverse)
library(dslabs)
```

```
if(!exists("mnist")) mnist <- read_mnist()
```

```
set.seed(1995)
ind <- which(mnist$train$labels %in% c(2,7)) %>% sample(500)
x <- mnist$train$images[ind,]
y <- mnist$train$labels[ind]
```

The predictors are in x and the labels in y.

For the purposes of, for example, smoothing, we are interested in describing distance between observation; in this case, digits. Later, for the purposes of selecting features, we might also be interested in finding pixels that *behave similarly* across samples.

To define distance, we need to know what *points* are since mathematical distance is computed between points. With high dimensional data, points are no longer on the Cartesian plane. Instead, points are in higher dimensions. We can no longer visualize them and need to think abstractly. For example, predictors $\mathbf{X}_i$ are defined as a point in 784 dimensional space: $\mathbf{X}_i = (x_{i,1}, \ldots, x_{i,784})^{\top}$.

Once we define points this way, the Euclidean distance is defined very similarly as it was for two dimensions. For example, the distance between the predictors for two observations, say observations $i = 1$ and $i = 2$, is:

$$\text{dist}(1, 2) = \sqrt{\sum_{j=1}^{784} (x_{1,j} - x_{2,j})^2}$$

This is just one non-negative number, just as it is for two dimensions.

33.3.3 Euclidean distance example

The labels for the first three observations are:

```
y[1:3]
#> [1] 7 2 7
```

The vectors of predictors for each of these observations are:

```
x_1 <- x[1,]
x_2 <- x[2,]
x_3 <- x[3,]
```

The first two numbers are seven and the third one is a 2. We expect the distances between the same number:

```
sqrt(sum((x_1 - x_2)^2))
#> [1] 3273
```

to be smaller than between different numbers:

```
sqrt(sum((x_1 - x_3)^2))
#> [1] 2311
sqrt(sum((x_2 - x_3)^2))
#> [1] 2636
```

As expected, the 7s are closer to each other.

A faster way to compute this is using matrix algebra:

```
sqrt(crossprod(x_1 - x_2))
#>         [,1]
#> [1,] 3273
sqrt(crossprod(x_1 - x_3))
#>         [,1]
#> [1,] 2311
sqrt(crossprod(x_2 - x_3))
#>         [,1]
#> [1,] 2636
```

We can also compute **all** the distances at once relatively quickly using the function `dist`, which computes the distance between each row and produces an object of class `dist`:

```
d <- dist(x)
class(d)
#> [1] "dist"
```

There are several machine learning related functions in R that take objects of class `dist` as input. To access the entries using row and column indices, we need to coerce it into a matrix. We can see the distance we calculated above like this:

```
as.matrix(d)[1:3,1:3]
#>      1    2    3
#> 1    0 3273 2311
#> 2 3273    0 2636
#> 3 2311 2636    0
```

We can quickly see an image of these distances using this code:

```
image(as.matrix(d))
```

If we order this distance by the labels, we can see that, in general, the twos are closer to each other and the sevens are closer to each other:

```
image(as.matrix(d)[order(y), order(y)])
```

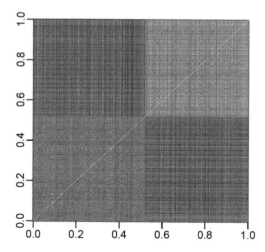

One thing we notice here is that there appears to be more uniformity in how the sevens are drawn, since they appear to be closer (more red) to other sevens than twos are to other twos.

33.3.4 Predictor space

Predictor space is a concept that is often used to describe machine learning algorithms. The term *space* refers to a mathematical definition that we don't describe in detail here. Instead, we provide a simplified explanation to help understand the term predictor space when used in the context of machine learning algorithms.

The predictor space can be thought of as the collection of all possible vectors of predictors that should be considered for the machine learning challenge in question. Each member of the space is referred to as a *point*. For example, in the 2 or 7 dataset, the predictor space consists of all pairs (x_1, x_2) such that both x_1 and x_2 are within 0 and 1. This particular *space* can be represented graphically as a square. In the MNIST dataset the predictor space consists of all 784-th dimensional vectors with each vector element an integer between 0 and 256. An essential element of a predictor space is that we need to define a function that provides the distance between any two points. In most cases we use Euclidean distance, but there are other possibilities. A particular case in which we can't simply use Euclidean distance is when we have categorical predictors.

Defining a predictor space is useful in machine learning because we do things like define neighborhoods of points, as required by many smoothing techniques. For example, we can define a neighborhood as all the points that are within 2 units away from a predefined center. If the points are two-dimensional and we use Euclidean distance, this neighborhood is graphically represented as a circle with radius 2. In three dimensions the neighborhood is a sphere. We will soon learn about algorithms that partition the space into non-overlapping regions and then make different predictions for each region using the data in the region.

33.3.5 Distance between predictors

We can also compute distances between predictors. If N is the number of observations, the distance between two predictors, say 1 and 2, is:

$$\text{dist}(1, 2) = \sqrt{\sum_{i=1}^{N} (x_{i,1} - x_{i,2})^2}$$

To compute the distance between all pairs of the 784 predictors, we can transpose the matrix first and then use `dist`:

```
d <- dist(t(x))
dim(as.matrix(d))
#> [1] 784 784
```

33.4 Exercises

1. Load the following dataset:

```
data("tissue_gene_expression")
```

This dataset includes a matrix x

```
dim(tissue_gene_expression$x)
```

with the gene expression measured on 500 genes for 189 biological samples representing seven different tissues. The tissue type is stored in y

```
table(tissue_gene_expression$y)
```

Compute the distance between each observation and store it in an object d.

2. Compare the distance between the first two observations (both cerebellums), the 39th and 40th (both colons), and the 73rd and 74th (both endometriums). See if the observations of the same tissue type are closer to each other.

3. We see that indeed observations of the same tissue type are closer to each other in the six tissue examples we just examined. Make a plot of all the distances using the `image` function to see if this pattern is general. Hint: convert d to a matrix first.

33.5   Dimension reduction

A typical machine learning challenge will include a large number of predictors, which makes visualization somewhat challenging. We have shown methods for visualizing univariate and paired data, but plots that reveal relationships between many variables are more complicated in higher dimensions. For example, to compare each of the 784 features in our predicting digits example, we would have to create, for example, 306,936 scatterplots. Creating one single scatter-plot of the data is impossible due to the high dimensionality.

Here we describe powerful techniques useful for exploratory data analysis, among other things, generally referred to as *dimension reduction*. The general idea is to reduce the dimension of the dataset while preserving important characteristics, such as the distance between features or observations. With fewer dimensions, visualization then becomes more feasible. The technique behind it all, the singular value decomposition, is also useful in other contexts. Principal component analysis (PCA) is the approach we will be showing. Before applying PCA to high-dimensional datasets, we will motivate the ideas behind with a simple example.

33.5.1   Preserving distance

We consider an example with twin heights. Some pairs are adults, the others are children. Here we simulate 100 two-dimensional points that represent the number of standard deviations each individual is from the mean height. Each point is a pair of twins. We use the `mvrnorm` function from the **MASS** package to simulate bivariate normal data.

```
set.seed(1988)
library(MASS)
n <- 100
Sigma <- matrix(c(9, 9 * 0.9, 9 * 0.92, 9 * 1), 2, 2)
x <- rbind(mvrnorm(n / 2, c(69, 69), Sigma),
           mvrnorm(n / 2, c(55, 55), Sigma))
```

A scatterplot quickly reveals that the correlation is high and that there are two groups of twins, the adults (upper right points) and the children (lower left points):

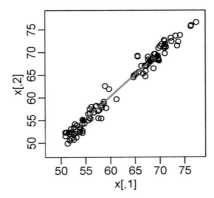

Our features are N two-dimensional points, the two heights, and, for illustrative purposes, we will act as if visualizing two dimensions is too challenging. We therefore want to reduce the dimensions from two to one, but still be able to understand important characteristics of the data, for example that the observations cluster into two groups: adults and children.

Let's consider a specific challenge: we want a one-dimensional summary of our predictors from which we can approximate the distance between any two observations. In the figure above we show the distance between observation 1 and 2 (blue), and observation 1 and 51 (red). Note that the blue line is shorter, which implies 1 and 2 are closer.

We can compute these distances using `dist`:

```
d <- dist(x)
as.matrix(d)[1, 2]
#> [1] 1.98
as.matrix(d)[2, 51]
#> [1] 18.7
```

This distance is based on two dimensions and we need a distance approximation based on just one.

Let's start with the naive approach of simply removing one of the two dimensions. Let's compare the actual distances to the distance computed with just the first dimension:

```
z <- x[,1]
```

Here are the approximate distances versus the original distances:

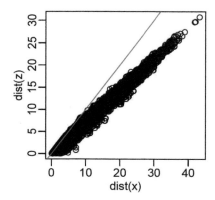

The plot looks about the same if we use the second dimension. We obtain a general underestimation. This is to be expected because we are adding more positive quantities in the distance calculation as we increase the number of dimensions. If instead we use an average, like this

$$\sqrt{\frac{1}{2}\sum_{j=1}^{2}(X_{i,j}-X_{i,j})^2},$$

then the underestimation goes away. We divide the distance by $\sqrt{2}$ to achieve the correction.

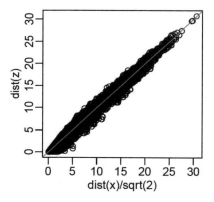

This actually works pretty well and we get a typical difference of:

```
sd(dist(x) - dist(z)*sqrt(2))
#> [1] 1.21
```

Now, can we pick a one-dimensional summary that makes this approximation even better?

If we look back at the previous scatterplot and visualize a line between any pair of points, the length of this line is the distance between the two points. These lines tend to go along the direction of the diagonal. Notice that if we instead plot the difference versus the average:

```
z   <- cbind((x[,2] + x[,1])/2,   x[,2] - x[,1])
```

we can see how the distance between points is mostly explained by the first dimension: the average.

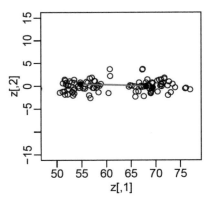

This means that we can ignore the second dimension and not lose too much information. If the line is completely flat, we lose no information at all. Using the first dimension of this transformed matrix we obtain an even better approximation:

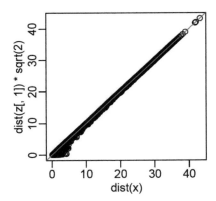

with the typical difference improved by about 35%:

```
sd(dist(x) - dist(z[,1])*sqrt(2))
#> [1] 0.315
```

Later we learn that z[,1] is the first principal component of the matrix x.

33.5.2   Linear transformations (advanced)

Note that each row of X was transformed using a linear transformation. For any row i, the first entry was:

$$Z_{i,1} = a_{1,1}X_{i,1} + a_{2,1}X_{i,2}$$

with $a_{1,1} = 0.5$ and $a_{2,1} = 0.5$.

The second entry was also a linear transformation:

$$Z_{i,2} = a_{1,2}X_{i,1} + a_{2,2}X_{i,2}$$

with $a_{1,2} = 1$ and $a_{2,2} = -1$.

We can also use linear transformation to get X back from Z:

$$X_{i,1} = b_{1,1}Z_{i,1} + b_{2,1}Z_{i,2}$$

with $b_{1,2} = 1$ and $b_{2,1} = 0.5$ and

$$X_{i,2} = b_{2,1}Z_{i,1} + b_{2,2}Z_{i,2}$$

with $b_{2,1} = 1$ and $a_{1,2} = -0.5$.

If you are familiar with linear algebra, we can write the operation we just performed like this:

$$Z = XA \text{ with } A = \begin{pmatrix} 1/2 & 1 \\ 1/2 & -1 \end{pmatrix}.$$

And that we can transform back by simply multiplying by A^{-1} as follows:

$$X = ZA^{-1} \text{ with } A^{-1} = \begin{pmatrix} 1 & 1 \\ 1/2 & -1/2 \end{pmatrix}.$$

Dimension reduction can often be described as applying a transformation A to a matrix X with many columns that *moves* the information contained in X to the first few columns of $Z = AX$, then keeping just these few informative columns, thus reducing the dimension of the vectors contained in the rows.

33.5.3   Orthogonal transformations (advanced)

Note that we divided the above by $\sqrt{2}$ to account for the differences in dimensions when comparing a 2 dimension distance to a 1 dimension distance. We can actually guarantee that the distance scales remain the same if we re-scale the columns of A to assure that the sum of squares is 1

$$a_{1,1}^2 + a_{2,1}^2 = 1 \text{ and } a_{1,2}^2 + a_{2,2}^2 = 1,$$

and that the correlation of the columns is 0:

$$a_{1,1}a_{1,2} + a_{2,1}a_{2,2} = 0.$$

Remember that if the columns are centered to have average 0, then the sum of squares is equivalent to the variance or standard deviation squared.

In our example, to achieve orthogonality, we multiply the first set of coefficients (first column of A) by $\sqrt{2}$ and the second by $1/\sqrt{2}$, then we get the same exact distance if we use both dimensions:

```
z[,1] <- (x[,1] + x[,2]) / sqrt(2)
z[,2] <- (x[,2] - x[,1]) / sqrt(2)
```

This gives us a transformation that preserves the distance between any two points:

```
max(dist(z) - dist(x))
#> [1] 3.24e-14
```

and an improved approximation if we use just the first dimension:

```
sd(dist(x) - dist(z[,1]))
#> [1] 0.315
```

In this case Z is called an orthogonal rotation of X: it preserves the distances between rows.

Note that by using the transformation above we can summarize the distance between any two pairs of twins with just one dimension. For example, one-dimensional data exploration of the first dimension of Z clearly shows that there are two groups, adults and children:

```
library(tidyverse)
qplot(z[,1], bins = 20, color = I("black"))
```

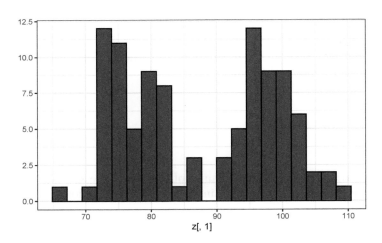

We successfully reduced the number of dimensions from two to one with very little loss of information.

The reason we were able to do this is because the columns of X were very correlated:

```
cor(x[,1], x[,2])
#> [1] 0.988
```

and the transformation produced uncorrelated columns with "independent" information in each column:

```
cor(z[,1], z[,2])
#> [1] 0.0876
```

One way this insight may be useful in a machine learning application is that we can reduce the complexity of a model by using just Z_1 rather than both X_1 and X_2.

It is actually common to obtain data with several highly correlated predictors. In these cases PCA, which we describe next, can be quite useful for reducing the complexity of the model being fit.

33.5.4   Principal component analysis

In the computation above, the total variability in our data can be defined as the sum of the sum of squares of the columns. We assume the columns are centered, so this sum is equivalent to the sum of the variances of each column:

$$v_1 + v_2, \text{ with } v_1 = \frac{1}{N} \sum_{i=1}^{N} X_{i,1}^2 \text{ and } v_2 = \frac{1}{N} \sum_{i=1}^{N} X_{i,2}^2$$

We can compute v_1 and v_2 using:

```
colMeans(x^2)
#> [1] 3904 3902
```

and we can show mathematically that if we apply an orthogonal transformation as above, then the total variation remains the same:

```
sum(colMeans(x^2))
#> [1] 7806
sum(colMeans(z^2))
#> [1] 7806
```

However, while the variability in the two columns of X is about the same, in the transformed version Z 99% of the variability is included in just the first dimension:

```
v <- colMeans(z^2)
v/sum(v)
#> [1] 1.00e+00 9.93e-05
```

The *first principal component (PC)* of a matrix X is the linear orthogonal transformation of X that maximizes this variability. The function `prcomp` provides this info:

```
pca <- prcomp(x)
pca$rotation
#>           PC1    PC2
#> [1,] -0.702  0.712
#> [2,] -0.712 -0.702
```

Note that the first PC is almost the same as that provided by the $(X_1 + X_2)/\sqrt{2}$ we used earlier (except perhaps for a sign change that is arbitrary).

The function PCA returns both the rotation needed to transform X so that the variability of the columns is decreasing from most variable to least (accessed with $rotation) as well as the resulting new matrix (accessed with $x). By default the columns of X are first centered.

So, using the matrix multiplication shown above, we have that the following are the same (demonstrated by a difference between elements of essentially zero):

```
a <- sweep(x, 2, colMeans(x))
b <- pca$x %*% t(pca$rotation)
max(abs(a - b))
#> [1] 3.55e-15
```

The rotation is orthogonal which means that the inverse is its transpose. So we also have that these two are identical:

```
a <- sweep(x, 2, colMeans(x)) %*% pca$rotation
b <- pca$x
max(abs(a - b))
#> [1] 0
```

We can visualize these to see how the first component summarizes the data. In the plot below red represents high values and blue negative values (later we learn why we call these weights and patterns):

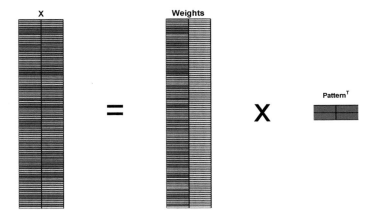

It turns out that we can find this linear transformation not just for two dimensions but for matrices of any dimension p.

For a multidimensional matrix with X with p columns, we can find a transformation that creates Z that preserves distance between rows, but with the variance of the columns in decreasing order. The second column is the second principal component, the third column is the third principal component, and so on. As in our example, if after a certain number of columns, say k, the variances of the columns of Z_j, $j > k$ are very small, it means these dimensions have little to contribute to the distance and we can approximate distance between any two points with just k dimensions. If k is much smaller than p, then we can achieve a very efficient summary of our data.

33.5.5   Iris example

The iris data is a widely used example in data analysis courses. It includes four botanical measurements related to three flower species:

```
names(iris)
#> [1] "Sepal.Length" "Sepal.Width"  "Petal.Length" "Petal.Width"
#> [5] "Species"
```

If you print `iris$Species` you will see that the data is ordered by the species.

Let's compute the distance between each observation. You can clearly see the three species with one species very different from the other two:

```
x <- iris[,1:4] %>% as.matrix()
d <- dist(x)
image(as.matrix(d), col = rev(RColorBrewer::brewer.pal(9, "RdBu")))
```

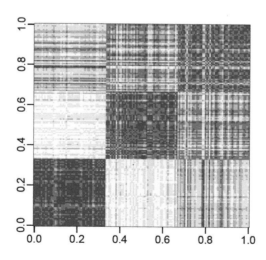

Our predictors here have four dimensions, but three are very correlated:

```
cor(x)
#>              Sepal.Length Sepal.Width Petal.Length Petal.Width
```

```
#> Sepal.Length        1.000      -0.118      0.872      0.818
#> Sepal.Width        -0.118       1.000     -0.428     -0.366
#> Petal.Length        0.872      -0.428      1.000      0.963
#> Petal.Width         0.818      -0.366      0.963      1.000
```

If we apply PCA, we should be able to approximate this distance with just two dimensions, compressing the highly correlated dimensions. Using the `summary` function we can see the variability explained by each PC:

```
pca <- prcomp(x)
summary(pca)
#> Importance of components:
#>                           PC1    PC2    PC3     PC4
#> Standard deviation       2.056 0.4926 0.2797 0.15439
#> Proportion of Variance   0.925 0.0531 0.0171 0.00521
#> Cumulative Proportion    0.925 0.9777 0.9948 1.00000
```

The first two dimensions account for 97% of the variability. Thus we should be able to approximate the distance very well with two dimensions. We can visualize the results of PCA:

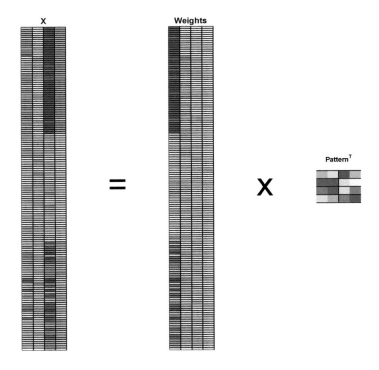

And see that the first pattern is sepal length, petal length, and petal width (red) in one direction and sepal width (blue) in the other. The second pattern is the sepal length and petal width in one direction (blue) and petal length and petal width in the other (red). You can see from the weights that the first PC1 drives most of the variability and it clearly separates the first third of samples (setosa) from the second two thirds (versicolor and virginica). If you look at the second column of the weights, you notice that it somewhat separates versicolor (red) from virginica (blue).

We can see this better by plotting the first two PCs with color representing the species:

```
data.frame(pca$x[,1:2], Species=iris$Species) %>%
  ggplot(aes(PC1,PC2, fill = Species))+
  geom_point(cex=3, pch=21) +
  coord_fixed(ratio = 1)
```

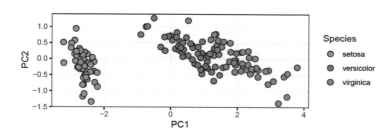

We see that the first two dimensions preserve the distance:

```
d_approx <- dist(pca$x[, 1:2])
qplot(d, d_approx) + geom_abline(color="red")
```

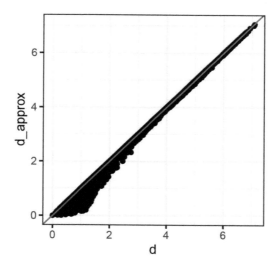

This example is more realistic than the first artificial example we used, since we showed how we can visualize the data using two dimensions when the data was four-dimensional.

33.5.6   MNIST example

The written digits example has 784 features. Is there any room for data reduction? Can we create simple machine learning algorithms using fewer features?

Let's load the data:

```
library(dslabs)
if(!exists("mnist")) mnist <- read_mnist()
```

Because the pixels are so small, we expect pixels close to each other on the grid to be correlated, meaning that dimension reduction should be possible.

Let's try PCA and explore the variance of the PCs. This will take a few seconds as it is a rather large matrix.

```
col_means <- colMeans(mnist$test$images)
pca <- prcomp(mnist$train$images)

pc <- 1:ncol(mnist$test$images)
qplot(pc, pca$sdev)
```

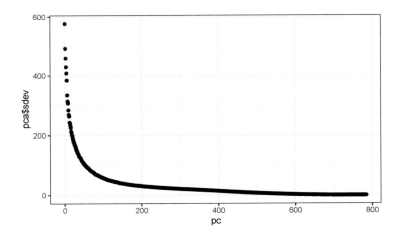

We can see that the first few PCs already explain a large percent of the variability:

```
summary(pca)$importance[,1:5]
#>                              PC1       PC2       PC3       PC4       PC5
#> Standard deviation       576.823   493.238   459.8993  429.8562  408.5668
#> Proportion of Variance     0.097     0.071     0.0617    0.0539    0.0487
#> Cumulative Proportion      0.097     0.168     0.2297    0.2836    0.3323
```

And just by looking at the first two PCs we see information about the class. Here is a random sample of 2,000 digits:

```
data.frame(PC1 = pca$x[,1], PC2 = pca$x[,2],
           label=factor(mnist$train$label)) %>%
  sample_n(2000) %>%
  ggplot(aes(PC1, PC2, fill=label))+
  geom_point(cex=3, pch=21)
```

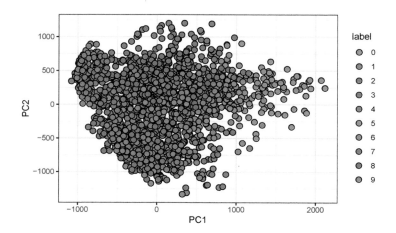

We can also *see* the linear combinations on the grid to get an idea of what is getting weighted:

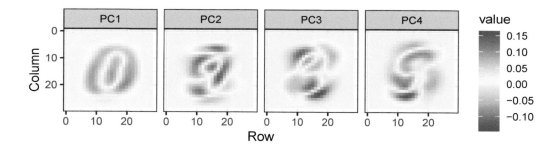

The lower variance PCs appear related to unimportant variability in the corners:

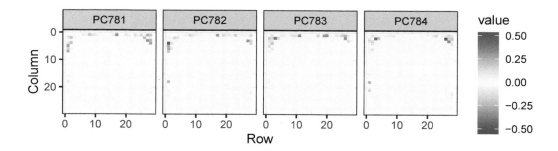

Now let's apply the transformation we learned with the training data to the test data, reduce the dimension and run knn on just a small number of dimensions.

We try 36 dimensions since this explains about 80% of the data. First fit the model:

```
library(caret)
k <- 36
x_train <- pca$x[,1:k]
y <- factor(mnist$train$labels)
fit <- knn3(x_train, y)
```

Now transform the test set:

```
x_test <- sweep(mnist$test$images, 2, col_means) %*% pca$rotation
x_test <- x_test[,1:k]
```

And we are ready to predict and see how we do:

```
y_hat <- predict(fit, x_test, type = "class")
confusionMatrix(y_hat, factor(mnist$test$labels))$overall["Accuracy"]
#> Accuracy
#>    0.975
```

With just 36 dimensions we get an accuracy well above 0.95.

33.6   Exercises

1. We want to explore the `tissue_gene_expression` predictors by plotting them.

```
data("tissue_gene_expression")
dim(tissue_gene_expression$x)
```

We want to get an idea of which observations are close to each other, but the predictors are 500-dimensional so plotting is difficult. Plot the first two principal components with color representing tissue type.

2. The predictors for each observation are measured on the same device and experimental procedure. This introduces biases that can affect all the predictors from one observation. For each observation, compute the average across all predictors and then plot this against the first PC with color representing tissue. Report the correlation.

3. We see an association with the first PC and the observation averages. Redo the PCA but only after removing the center.

4. For the first 10 PCs, make a boxplot showing the values for each tissue.

5. Plot the percent variance explained by PC number. Hint: use the **summary** function.

33.7   Recommendation systems

Recommendation systems use ratings that *users* have given *items* to make specific recommendations. Companies that sell many products to many customers and permit these customers to rate their products, like Amazon, are able to collect massive datasets that can be used to predict what rating a particular user will give a specific item. Items for which a high rating is predicted for a given user are then recommended to that user.

Netflix uses a recommendation system to predict how many *stars* a user will give a specific movie. One star suggests it is not a good movie, whereas five stars suggests it is an excellent movie. Here, we provide the basics of how these recommendations are made, motivated by some of the approaches taken by the winners of the *Netflix challenges*.

In October 2006, Netflix offered a challenge to the data science community: improve our recommendation algorithm by 10% and win a million dollars. In September 2009, the winners were announced[1]. You can read a good summary of how the winning algorithm was put together here: http://blog.echen.me/2011/10/24/winning-the-netflix-prize-a-summary/ and a more detailed explanation here: http://www.netflixprize.com/assets/GrandPrize2009_ BPC_BellKor.pdf. We will now show you some of the data analysis strategies used by the winning team.

33.7.1   Movielens data

The Netflix data is not publicly available, but the GroupLens research lab[2] generated their own database with over 20 million ratings for over 27,000 movies by more than 138,000 users. We make a small subset of this data available via the **dslabs** package:

```
library(tidyverse)
library(dslabs)
data("movielens")
```

We can see this table is in tidy format with thousands of rows:

```
movielens %>% as_tibble()
#> # A tibble: 100,004 x 7
#>    movieId title                year genres          userId rating timestamp
#>      <int> <chr>               <int> <fct>            <int>  <dbl>     <int>
#> 1       31 Dangerous Minds      1995 Drama                1    2.5    1.26e9
#> 2     1029 Dumbo                1941 Animation|Chi~       1    3      1.26e9
#> 3     1061 Sleepers             1996 Thriller             1    3      1.26e9
#> 4     1129 Escape from New ~    1981 Action|Advent~       1    2      1.26e9
#> 5     1172 Cinema Paradiso ~    1989 Drama                1    4      1.26e9
#> # ... with 1e+05 more rows
```

Each row represents a rating given by one user to one movie.

---

[1]http://bits.blogs.nytimes.com/2009/09/21/netflix-awards-1-million-prize-and-starts-a-new-contest/
[2]https://grouplens.org/

We can see the number of unique users that provided ratings and how many unique movies were rated:

```
movielens %>%
    summarize(n_users = n_distinct(userId),
                n_movies = n_distinct(movieId))
#>    n_users n_movies
#> 1     671     9066
```

If we multiply those two numbers, we get a number larger than 5 million, yet our data table has about 100,000 rows. This implies that not every user rated every movie. So we can think of these data as a very large matrix, with users on the rows and movies on the columns, with many empty cells. The `gather` function permits us to convert it to this format, but if we try it for the entire matrix, it will crash R. Let's show the matrix for seven users and four movies.

userId	Forrest Gump	Pulp Fiction	Shawshank Redemption	Silence of the Lambs
13	5.0	3.5	4.5	NA
15	1.0	5.0	2.0	5.0
16	NA	NA	4.0	NA
17	2.5	5.0	5.0	4.5
19	5.0	5.0	4.0	3.0
20	2.0	0.5	4.5	0.5

You can think of the task of a recommendation system as filling in the NAs in the table above. To see how *sparse* the matrix is, here is the matrix for a random sample of 100 movies and 100 users with yellow indicating a user/movie combination for which we have a rating.

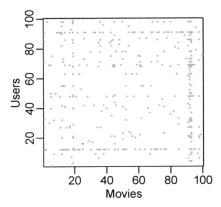

This machine learning challenge is more complicated than what we have studied up to now because each outcome Y has a different set of predictors. To see this, note that if we are predicting the rating for movie i by user u, in principle, all other ratings related to movie i and by user u may be used as predictors, but different users rate different movies and a different number of movies. Furthermore, we may be able to use information from other movies that we have determined are similar to movie i or from users determined to be similar to user u. In essence, the entire matrix can be used as predictors for each cell.

Let's look at some of the general properties of the data to better understand the challenges.

The first thing we notice is that some movies get rated more than others. Below is the distribution. This should not surprise us given that there are blockbuster movies watched by millions and artsy, independent movies watched by just a few. Our second observation is that some users are more active than others at rating movies:

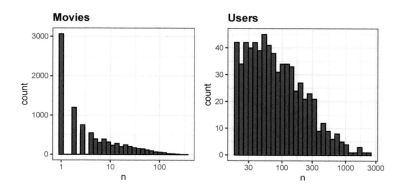

33.7.2   Recommendation systems as a machine learning challenge

To see how this is a type of machine learning, notice that we need to build an algorithm with data we have collected that will then be applied outside our control, as users look for movie recommendations. So let's create a test set to assess the accuracy of the models we implement.

```
library(caret)
set.seed(755)
test_index <- createDataPartition(y = movielens$rating, times = 1, p = 0.2,
                                  list = FALSE)
train_set <- movielens[-test_index,]
test_set <- movielens[test_index,]
```

To make sure we don't include users and movies in the test set that do not appear in the training set, we remove these entries using the `semi_join` function:

```
test_set <- test_set %>%
  semi_join(train_set, by = "movieId") %>%
  semi_join(train_set, by = "userId")
```

33.7.3   Loss function

The Netflix challenge used the typical error loss: they decided on a winner based on the residual mean squared error (RMSE) on a test set. We define $y_{u,i}$ as the rating for movie i by user u and denote our prediction with $\hat{y}_{u,i}$. The RMSE is then defined as:

$$\text{RMSE} = \sqrt{\frac{1}{N} \sum_{u,i} (\hat{y}_{u,i} - y_{u,i})^2}$$

with N being the number of user/movie combinations and the sum occurring over all these combinations.

Remember that we can interpret the RMSE similarly to a standard deviation: it is the typical error we make when predicting a movie rating. If this number is larger than 1, it means our typical error is larger than one star, which is not good.

Let's write a function that computes the RMSE for vectors of ratings and their corresponding predictors:

```
RMSE <- function(true_ratings, predicted_ratings){
    sqrt(mean((true_ratings - predicted_ratings)^2))
}
```

33.7.4 A first model

Let's start by building the simplest possible recommendation system: we predict the same rating for all movies regardless of user. What number should this prediction be? We can use a model based approach to answer this. A model that assumes the same rating for all movies and users with all the differences explained by random variation would look like this:

$$Y_{u,i} = \mu + \varepsilon_{u,i}$$

with $\varepsilon_{i,u}$ independent errors sampled from the same distribution centered at 0 and μ the "true" rating for all movies. We know that the estimate that minimizes the RMSE is the least squares estimate of μ and, in this case, is the average of all ratings:

```
mu_hat <- mean(train_set$rating)
mu_hat
#> [1] 3.54
```

If we predict all unknown ratings with $\hat{\mu}$ we obtain the following RMSE:

```
naive_rmse <- RMSE(test_set$rating, mu_hat)
naive_rmse
#> [1] 1.05
```

Keep in mind that if you plug in any other number, you get a higher RMSE. For example:

```
predictions <- rep(3, nrow(test_set))
RMSE(test_set$rating, predictions)
#> [1] 1.19
```

From looking at the distribution of ratings, we can visualize that this is the standard deviation of that distribution. We get a RMSE of about 1. To win the grand prize of $1,000,000, a participating team had to get an RMSE of about 0.857. So we can definitely do better!

As we go along, we will be comparing different approaches. Let's start by creating a results table with this naive approach:

```
rmse_results <- tibble(method = "Just the average", RMSE = naive_rmse)
```

33.7.5   Modeling movie effects

We know from experience that some movies are just generally rated higher than others. This intuition, that different movies are rated differently, is confirmed by data. We can augment our previous model by adding the term b_i to represent average ranking for movie i:

$$Y_{u,i} = \mu + b_i + \varepsilon_{u,i}$$

Statistics textbooks refer to the bs as effects. However, in the Netflix challenge papers, they refer to them as "bias", thus the b notation.

We can again use least squares to estimate the b_i in the following way:

```
fit <- lm(rating ~ as.factor(movieId), data = movielens)
```

Because there are thousands of b_i as each movie gets one, the `lm()` function will be very slow here. We therefore don't recommend running the code above. But in this particular situation, we know that the least squares estimate $\hat{b}_i$ is just the average of $Y_{u,i} - \hat{\mu}$ for each movie i. So we can compute them this way (we will drop the `hat` notation in the code to represent estimates going forward):

```
mu <- mean(train_set$rating)
movie_avgs <- train_set %>%
  group_by(movieId) %>%
  summarize(b_i = mean(rating - mu))
```

We can see that these estimates vary substantially:

```
qplot(b_i, data = movie_avgs, bins = 10, color = I("black"))
```

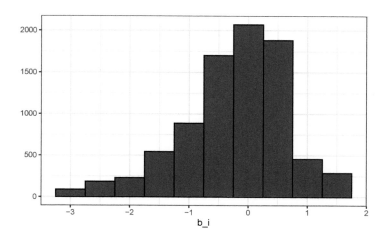

Remember $\hat{\mu} = 3.5$ so a $b_i = 1.5$ implies a perfect five star rating.

Let's see how much our prediction improves once we use $\hat{y}_{u,i} = \hat{\mu} + \hat{b}_i$:

```
predicted_ratings <- mu + test_set %>%
  left_join(movie_avgs, by='movieId') %>%
  pull(b_i)
RMSE(predicted_ratings, test_set$rating)
#> [1] 0.989
```

We already see an improvement. But can we make it better?

33.7.6   User effects

Let's compute the average rating for user u for those that have rated over 100 movies:

```
train_set %>%
  group_by(userId) %>%
  summarize(b_u = mean(rating)) %>%
  filter(n()>=100) %>%
  ggplot(aes(b_u)) +
  geom_histogram(bins = 30, color = "black")
```

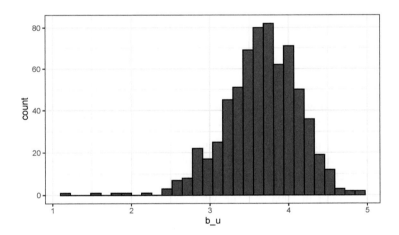

Notice that there is substantial variability across users as well: some users are very cranky and others love every movie. This implies that a further improvement to our model may be:

$$Y_{u,i} = \mu + b_i + b_u + \varepsilon_{u,i}$$

where b_u is a user-specific effect. Now if a cranky user (negative b_u) rates a great movie (positive b_i), the effects counter each other and we may be able to correctly predict that this user gave this great movie a 3 rather than a 5.

To fit this model, we could again use lm like this:

```
lm(rating ~ as.factor(movieId) + as.factor(userId))
```

but, for the reasons described earlier, we won't. Instead, we will compute an approximation by computing $\hat{\mu}$ and $\hat{b}_i$ and estimating $\hat{b}_u$ as the average of $y_{u,i} - \hat{\mu} - \hat{b}_i$:

```
user_avgs <- train_set %>%
  left_join(movie_avgs, by='movieId') %>%
  group_by(userId) %>%
  summarize(b_u = mean(rating - mu - b_i))
```

We can now construct predictors and see how much the RMSE improves:

```
predicted_ratings <- test_set %>%
  left_join(movie_avgs, by='movieId') %>%
  left_join(user_avgs, by='userId') %>%
  mutate(pred = mu + b_i + b_u) %>%
  pull(pred)
RMSE(predicted_ratings, test_set$rating)
#> [1] 0.905
```

33.8   Exercises

1. Load the `movielens` data.

```
data("movielens")
```

Compute the number of ratings for each movie and then plot it against the year the movie came out. Use the square root transformation on the counts.

2. We see that, on average, movies that came out after 1993 get more ratings. We also see that with newer movies, starting in 1993, the number of ratings decreases with year: the more recent a movie is, the less time users have had to rate it.

Among movies that came out in 1993 or later, what are the 25 movies with the most ratings per year? Also report their average rating.

3. From the table constructed in the previous example, we see that the most rated movies tend to have above average ratings. This is not surprising: more people watch popular movies. To confirm this, stratify the post 1993 movies by ratings per year and compute their average ratings. Make a plot of average rating versus ratings per year and show an estimate of the trend.

4. In the previous exercise, we see that the more a movie is rated, the higher the rating. Suppose you are doing a predictive analysis in which you need to fill in the missing ratings with some value. Which of the following strategies would you use?

    a. Fill in the missing values with average rating of all movies.
    b. Fill in the missing values with 0.

   c. Fill in the value with a lower value than the average since lack of rating is associated with lower ratings. Try out different values and evaluate prediction in a test set.

   d. None of the above.

5. The `movielens` dataset also includes a time stamp. This variable represents the time and data in which the rating was provided. The units are seconds since January 1, 1970. Create a new column `date` with the date. Hint: use the `as_datetime` function in the **lubridate** package.

6. Compute the average rating for each week and plot this average against day. Hint: use the `round_date` function before you `group_by`.

7. The plot shows some evidence of a time effect. If we define $d_{u,i}$ as the day for user's u rating of movie i, which of the following models is most appropriate:

   a. $Y_{u,i} = \mu + b_i + b_u + d_{u,i} + \varepsilon_{u,i}$.
   b. $Y_{u,i} = \mu + b_i + b_u + d_{u,i}\beta + \varepsilon_{u,i}$.
   c. $Y_{u,i} = \mu + b_i + b_u + d_{u,i}\beta_i + \varepsilon_{u,i}$.
   d. $Y_{u,i} = \mu + b_i + b_u + f(d_{u,i}) + \varepsilon_{u,i}$, with f a smooth function of $d_{u,i}$.

8. The `movielens` data also has a **genres** column. This column includes every genre that applies to the movie. Some movies fall under several genres. Define a category as whatever combination appears in this column. Keep only categories with more than 1,000 ratings. Then compute the average and standard error for each category. Plot these as error bar plots.

9. The plot shows strong evidence of a genre effect. If we define $g_{u,i}$ as the genre for user's u rating of movie i, which of the following models is most appropriate:

   a. $Y_{u,i} = \mu + b_i + b_u + d_{u,i} + \varepsilon_{u,i}$.
   b. $Y_{u,i} = \mu + b_i + b_u + d_{u,i}\beta + \varepsilon_{u,i}$.
   c. $Y_{u,i} = \mu + b_i + b_u + \sum_{k=1}^{K} x_{u,i}\beta_k + \varepsilon_{u,i}$, with $x_{u,i}^k = 1$ if $g_{u,i}$ is genre k.
   d. $Y_{u,i} = \mu + b_i + b_u + f(d_{u,i}) + \varepsilon_{u,i}$, with f a smooth function of $d_{u,i}$.

33.9 Regularization

33.9.1 Motivation

Despite the large movie to movie variation, our improvement in RMSE was only about 5%. Let's explore where we made mistakes in our first model, using only movie effects b_i. Here are the 10 largest mistakes:

```
test_set %>%
  left_join(movie_avgs, by='movieId') %>%
  mutate(residual = rating - (mu + b_i)) %>%
  arrange(desc(abs(residual))) %>%
```

```
   slice(1:10) %>%
   pull(title)
#> [1] "Kingdom, The (Riget)"        "Heaven Knows, Mr. Allison"
#> [3] "American Pimp"               "Chinatown"
#> [5] "American Beauty"             "Apocalypse Now"
#> [7] "Taxi Driver"                 "Wallace & Gromit: A Close Shave"
#> [9] "Down in the Delta"           "Stalag 17"
```

These all seem like obscure movies. Many of them have large predictions. Let's look at the top 10 worst and best movies based on $\hat{b}_i$. First, let's create a database that connects movieId to movie title:

```
movie_titles <- movielens %>%
  select(movieId, title) %>%
  distinct()
```

Here are the 10 best movies according to our estimate:

```
movie_avgs %>% left_join(movie_titles, by="movieId") %>%
  arrange(desc(b_i)) %>%
  slice(1:10) %>%
  pull(title)
#>  [1] "When Night Is Falling"
#>  [2] "Lamerica"
#>  [3] "Mute Witness"
#>  [4] "Picture Bride (Bijo photo)"
#>  [5] "Red Firecracker, Green Firecracker (Pao Da Shuang Deng)"
#>  [6] "Paris, France"
#>  [7] "Faces"
#>  [8] "Maya Lin: A Strong Clear Vision"
#>  [9] "Heavy"
#> [10] "Gate of Heavenly Peace, The"
```

And here are the 10 worst:

```
movie_avgs %>% left_join(movie_titles, by="movieId") %>%
  arrange(b_i) %>%
  slice(1:10) %>%
  pull(title)
#>  [1] "Children of the Corn IV: The Gathering"
#>  [2] "Barney's Great Adventure"
#>  [3] "Merry War, A"
#>  [4] "Whiteboyz"
#>  [5] "Catfish in Black Bean Sauce"
#>  [6] "Killer Shrews, The"
#>  [7] "Horrors of Spider Island (Ein Toter Hing im Netz)"
#>  [8] "Monkeybone"
#>  [9] "Arthur 2: On the Rocks"
#> [10] "Red Heat"
```

They all seem to be quite obscure. Let's look at how often they are rated.

```
train_set %>% count(movieId) %>%
  left_join(movie_avgs, by="movieId") %>%
  left_join(movie_titles, by="movieId") %>%
  arrange(desc(b_i)) %>%
  slice(1:10) %>%
  pull(n)
#>  [1] 1 1 1 1 3 1 1 2 1 1

train_set %>% count(movieId) %>%
  left_join(movie_avgs) %>%
  left_join(movie_titles, by="movieId") %>%
  arrange(b_i) %>%
  slice(1:10) %>%
  pull(n)
#> Joining, by = "movieId"
#>  [1] 1 1 1 1 1 1 1 1 1 1
```

The supposed "best" and "worst" movies were rated by very few users, in most cases just 1. These movies were mostly obscure ones. This is because with just a few users, we have more uncertainty. Therefore, larger estimates of b_i, negative or positive, are more likely.

These are noisy estimates that we should not trust, especially when it comes to prediction. Large errors can increase our RMSE, so we would rather be conservative when unsure.

In previous sections, we computed standard error and constructed confidence intervals to account for different levels of uncertainty. However, when making predictions, we need one number, one prediction, not an interval. For this, we introduce the concept of regularization.

Regularization permits us to penalize large estimates that are formed using small sample sizes. It has commonalities with the Bayesian approach that shrunk predictions described in Section 16.4.

33.9.2   Penalized least squares

The general idea behind regularization is to constrain the total variability of the effect sizes. Why does this help? Consider a case in which we have movie $i = 1$ with 100 user ratings and 4 movies $i = 2, 3, 4, 5$ with just one user rating. We intend to fit the model

$$Y_{u,i} = \mu + b_i + \varepsilon_{u,i}$$

Suppose we know the average rating is, say, $\mu = 3$. If we use least squares, the estimate for the first movie effect b_1 is the average of the 100 user ratings, $1/100 \sum_{i=1}^{100}(Y_{i,1} - \mu)$, which we expect to be a quite precise. However, the estimate for movies 2, 3, 4, and 5 will simply be the observed deviation from the average rating $\hat{b}_i = Y_{u,i} - \hat{\mu}$ which is an estimate based on just one number so it won't be precise at all. Note these estimates make the error $Y_{u,i} - \mu + \hat{b}_i$ equal to 0 for $i = 2, 3, 4, 5$, but this is a case of over-training. In fact, ignoring the one user and guessing that movies 2,3,4, and 5 are just average movies ($b_i = 0$) might provide a better prediction. The general idea of penalized regression is to control the total variability of the movie effects: $\sum_{i=1}^{5} b_i^2$. Specifically, instead of minimizing the least squares equation, we minimize an equation that adds a penalty:

$$\frac{1}{N} \sum_{u,i} (y_{u,i} - \mu - b_i)^2 + \lambda \sum_i b_i^2$$

The first term is just least squares and the second is a penalty that gets larger when many b_i are large. Using calculus we can actually show that the values of b_i that minimize this equation are:

$$\hat{b}_i(\lambda) = \frac{1}{\lambda + n_i} \sum_{u=1}^{n_i} (Y_{u,i} - \hat{\mu})$$

where n_i is the number of ratings made for movie i. This approach will have our desired effect: when our sample size n_i is very large, a case which will give us a stable estimate, then the penalty λ is effectively ignored since $n_i + \lambda \approx n_i$. However, when the n_i is small, then the estimate $\hat{b}_i(\lambda)$ is shrunken towards 0. The larger λ, the more we shrink.

Let's compute these regularized estimates of b_i using $\lambda = 3$. Later, we will see why we picked 3.

```
lambda <- 3
mu <- mean(train_set$rating)
movie_reg_avgs <- train_set %>%
  group_by(movieId) %>%
  summarize(b_i = sum(rating - mu)/(n()+lambda), n_i = n())
```

To see how the estimates shrink, let's make a plot of the regularized estimates versus the least squares estimates.

```
tibble(original = movie_avgs$b_i,
       regularlized = movie_reg_avgs$b_i,
       n = movie_reg_avgs$n_i) %>%
  ggplot(aes(original, regularlized, size=sqrt(n))) +
  geom_point(shape=1, alpha=0.5)
```

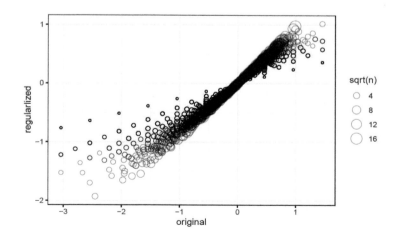

Now, let's look at the top 10 best movies based on the penalized estimates $\hat{b}_i(\lambda)$:

```
train_set %>%
  count(movieId) %>%
  left_join(movie_reg_avgs, by = "movieId") %>%
  left_join(movie_titles, by = "movieId") %>%
  arrange(desc(b_i)) %>%
  slice(1:10) %>%
  pull(title)
#>  [1] "Paris Is Burning"          "Shawshank Redemption, The"
#>  [3] "Godfather, The"            "African Queen, The"
#>  [5] "Band of Brothers"          "Paperman"
#>  [7] "On the Waterfront"         "All About Eve"
#>  [9] "Usual Suspects, The"       "Ikiru"
```

These make much more sense! These movies are watched more and have more ratings. Here are the top 10 worst movies:

```
train_set %>%
  count(movieId) %>%
  left_join(movie_reg_avgs, by = "movieId") %>%
  left_join(movie_titles, by="movieId") %>%
  arrange(b_i) %>%
  select(title, b_i, n) %>%
  slice(1:10) %>%
  pull(title)
#>  [1] "Battlefield Earth"
#>  [2] "Joe's Apartment"
#>  [3] "Super Mario Bros."
#>  [4] "Speed 2: Cruise Control"
#>  [5] "Dungeons & Dragons"
#>  [6] "Batman & Robin"
#>  [7] "Police Academy 6: City Under Siege"
#>  [8] "Cats & Dogs"
#>  [9] "Disaster Movie"
#> [10] "Mighty Morphin Power Rangers: The Movie"
```

Do we improve our results?

```
predicted_ratings <- test_set %>%
  left_join(movie_reg_avgs, by = "movieId") %>%
  mutate(pred = mu + b_i) %>%
  pull(pred)
RMSE(predicted_ratings, test_set$rating)
#> [1] 0.97
```

```
#> # A tibble: 4 x 2
#>   method                RMSE
#>   <chr>                <dbl>
#> 1 Just the average      1.05
#> 2 Movie Effect Model    0.989
```

```
#> 3 Movie + User Effects Model        0.905
#> 4 Regularized Movie Effect Model 0.970
```

The penalized estimates provide a large improvement over the least squares estimates.

33.9.3   Choosing the penalty terms

Note that λ is a tuning parameter. We can use cross-validation to choose it.

```
lambdas <- seq(0, 10, 0.25)

mu <- mean(train_set$rating)
just_the_sum <- train_set %>%
  group_by(movieId) %>%
  summarize(s = sum(rating - mu), n_i = n())

rmses <- sapply(lambdas, function(l){
  predicted_ratings <- test_set %>%
    left_join(just_the_sum, by='movieId') %>%
    mutate(b_i = s/(n_i+l)) %>%
    mutate(pred = mu + b_i) %>%
    pull(pred)
  return(RMSE(predicted_ratings, test_set$rating))
})
qplot(lambdas, rmses)
lambdas[which.min(rmses)]
#> [1] 3
```

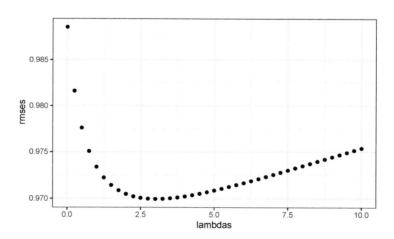

However, while we show this as an illustration, in practice we should be using full cross-validation just on the train set, without using the test set until the final assessment. The test set should never be used for tuning.

We can use regularization for the estimate user effects as well. We are minimizing:

$$\frac{1}{N}\sum_{u,i}\left(y_{u,i} - \mu - b_i - b_u\right)^2 + \lambda\left(\sum_i b_i^2 + \sum_u b_u^2\right)$$

The estimates that minimize this can be found similarly to what we did above. Here we use cross-validation to pick a λ:

```
lambdas <- seq(0, 10, 0.25)

rmses <- sapply(lambdas, function(l){

  mu <- mean(train_set$rating)

  b_i <- train_set %>%
    group_by(movieId) %>%
    summarize(b_i = sum(rating - mu)/(n()+l))

  b_u <- train_set %>%
    left_join(b_i, by="movieId") %>%
    group_by(userId) %>%
    summarize(b_u = sum(rating - b_i - mu)/(n()+l))

  predicted_ratings <-
    test_set %>%
    left_join(b_i, by = "movieId") %>%
    left_join(b_u, by = "userId") %>%
    mutate(pred = mu + b_i + b_u) %>%
    pull(pred)

    return(RMSE(predicted_ratings, test_set$rating))
})

qplot(lambdas, rmses)
```

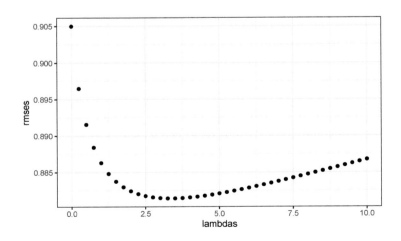

For the full model, the optimal λ is:

```
lambda <- lambdas[which.min(rmses)]
lambda
#> [1] 3.25
```

method	RMSE
Just the average	1.053
Movie Effect Model	0.989
Movie + User Effects Model	0.905
Regularized Movie Effect Model	0.970
Regularized Movie + User Effect Model	0.881

33.10   Exercises

An education expert is advocating for smaller schools. The expert bases this recommendation on the fact that among the best performing schools, many are small schools. Let's simulate a dataset for 100 schools. First, let's simulate the number of students in each school.

```
set.seed(1986)
n <- round(2^rnorm(1000, 8, 1))
```

Now let's assign a *true* quality for each school completely independent from size. This is the parameter we want to estimate.

```
mu <- round(80 + 2 * rt(1000, 5))
range(mu)
schools <- data.frame(id = paste("PS",1:100),
                      size = n,
                      quality = mu,
                      rank = rank(-mu))
```

We can see that the top 10 schools are:

```
schools %>% top_n(10, quality) %>% arrange(desc(quality))
```

Now let's have the students in the school take a test. There is random variability in test taking so we will simulate the test scores as normally distributed with the average determined by the school quality and standard deviations of 30 percentage points:

```
scores <- sapply(1:nrow(schools), function(i){
  scores <- rnorm(schools$size[i], schools$quality[i], 30)
  scores
})
schools <- schools %>% mutate(score = sapply(scores, mean))
```

1. What are the top schools based on the average score? Show just the ID, size, and the average score.

2. Compare the median school size to the median school size of the top 10 schools based on the score.

3. According to this test, it appears small schools are better than large schools. Five out of the top 10 schools have 100 or fewer students. But how can this be? We constructed the simulation so that quality and size are independent. Repeat the exercise for the worst 10 schools.

4. The same is true for the worst schools! They are small as well. Plot the average score versus school size to see what's going on. Highlight the top 10 schools based on the *true* quality. Use the log scale transform for the size.

5. We can see that the standard error of the score has larger variability when the school is smaller. This is a basic statistical reality we learned in the probability and inference sections. In fact, note that 4 of the top 10 schools are in the top 10 schools based on the exam score.

Let's use regularization to pick the best schools. Remember regularization *shrinks* deviations from the average towards 0. So to apply regularization here, we first need to define the overall average for all schools:

```
overall <- mean(sapply(scores, mean))
```

and then define, for each school, how it deviates from that average. Write code that estimates the score above average for each school but dividing by $n + \lambda$ instead of n, with n the school size and λ a regularization parameter. Try $\lambda = 3$.

6. Notice that this improves things a bit. The number of small schools that are not highly ranked is now 4. Is there a better λ? Find the λ that minimizes the RMSE $= 1/100 \sum_{i=1}^{100} (\text{quality} - \text{estimate})^2$.

7. Rank the schools based on the average obtained with the best α. Note that no small school is incorrectly included.

8. A common mistake to make when using regularization is shrinking values towards 0 that are not centered around 0. For example, if we don't subtract the overall average before shrinking, we actually obtain a very similar result. Confirm this by re-running the code from exercise 6 but without removing the overall mean.

33.11 Matrix factorization

Matrix factorization is a widely used concept in machine learning. It is very much related to factor analysis, singular value decomposition (SVD), and principal component analysis (PCA). Here we describe the concept in the context of movie recommendation systems.

We have described how the model:

$$Y_{u,i} = \mu + b_i + b_u + \varepsilon_{u,i}$$

accounts for movie to movie differences through the b_i and user to user differences through the b_u. But this model leaves out an important source of variation related to the fact that groups of movies have similar rating patterns and groups of users have similar rating patterns as well. We will discover these patterns by studying the residuals:

$$r_{u,i} = y_{u,i} - \hat{b}_i - \hat{b}_u$$

To see this, we will convert the data into a matrix so that each user gets a row, each movie gets a column, and $y_{u,i}$ is the entry in row u and column i. For illustrative purposes, we will only consider a small subset of movies with many ratings and users that have rated many movies. We also keep Scent of a Woman (`movieId == 3252`) because we use it for a specific example:

```
train_small <- movielens %>%
  group_by(movieId) %>%
  filter(n() >= 50 | movieId == 3252) %>% ungroup() %>%
  group_by(userId) %>%
  filter(n() >= 50) %>% ungroup()
```

```
y <- train_small %>%
  select(userId, movieId, rating) %>%
  spread(movieId, rating) %>%
  as.matrix()
```

We add row names and column names:

```
rownames(y)<- y[,1]
y <- y[,-1]
```

```
movie_titles <- movielens %>%
  select(movieId, title) %>%
  distinct()
```

```
colnames(y) <- with(movie_titles, title[match(colnames(y), movieId)])
```

and convert them to residuals by removing the column and row effects:

```
y <- sweep(y, 2, colMeans(y, na.rm=TRUE))
y <- sweep(y, 1, rowMeans(y, na.rm=TRUE))
```

If the model above explains all the signals, and the ε are just noise, then the residuals for different movies should be independent from each other. But they are not. Here are some examples:

```
m_1 <- "Godfather, The"
m_2 <- "Godfather: Part II, The"
p1 <- qplot(y[ ,m_1], y[,m_2], xlab = m_1, ylab = m_2)
```

```
m_1 <- "Godfather, The"
m_3 <- "Goodfellas"
p2 <- qplot(y[ ,m_1], y[,m_3], xlab = m_1, ylab = m_3)
```

```
m_4 <- "You've Got Mail"
m_5 <- "Sleepless in Seattle"
p3 <- qplot(y[ ,m_4], y[,m_5], xlab = m_4, ylab = m_5)

gridExtra::grid.arrange(p1, p2 ,p3, ncol = 3)
```

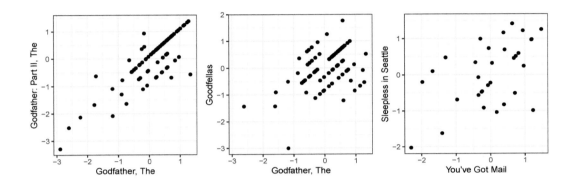

This plot says that users that liked The Godfather more than what the model expects them to, based on the movie and user effects, also liked The Godfather II more than expected. A similar relationship is seen when comparing The Godfather and Goodfellas. Although not as strong, there is still correlation. We see correlations between You've Got Mail and Sleepless in Seattle as well

By looking at the correlation between movies, we can see a pattern (we rename the columns to save print space):

```
x <- y[, c(m_1, m_2, m_3, m_4, m_5)]
short_names <- c("Godfather", "Godfather2", "Goodfellas",
                 "You've Got", "Sleepless")
colnames(x) <- short_names
cor(x, use="pairwise.complete")
#>           Godfather Godfather2 Goodfellas You've Got Sleepless
#> Godfather     1.000      0.829      0.444     -0.440    -0.378
#> Godfather2    0.829      1.000      0.521     -0.331    -0.358
#> Goodfellas    0.444      0.521      1.000     -0.481    -0.402
#> You've Got   -0.440     -0.331     -0.481      1.000     0.533
#> Sleepless    -0.378     -0.358     -0.402      0.533     1.000
```

There seems to be people that like romantic comedies more than expected, while others that like gangster movies more than expected.

These results tell us that there is structure in the data. But how can we model this?

33.11.1   Factors analysis

Here is an illustration, using a simulation, of how we can use some structure to predict the $r_{u,i}$. Suppose our residuals **r** look like this:

```
round(r, 1)
#>      Godfather Godfather2 Goodfellas You've Got Sleepless
#> 1        2.0        2.3        2.2       -1.8      -1.9
#> 2        2.0        1.7        2.0       -1.9      -1.7
#> 3        1.9        2.4        2.1       -2.3      -2.0
#> 4       -0.3        0.3        0.3       -0.4      -0.3
#> 5       -0.3       -0.4        0.3        0.2       0.3
#> 6       -0.1        0.1        0.2       -0.3       0.2
#> 7       -0.1        0.0       -0.2       -0.2       0.3
#> 8        0.2        0.2        0.1        0.0       0.4
#> 9       -1.7       -2.1       -1.8        2.0       2.4
#> 10      -2.3       -1.8       -1.7        1.8       1.7
#> 11      -1.7       -2.0       -2.1        1.9       2.3
#> 12      -1.8       -1.7       -2.1        2.3       2.0
```

There seems to be a pattern here. In fact, we can see very strong correlation patterns:

```
cor(r)
#>              Godfather Godfather2 Goodfellas You've Got Sleepless
#> Godfather        1.000      0.980      0.978     -0.974    -0.966
#> Godfather2       0.980      1.000      0.983     -0.987    -0.992
#> Goodfellas       0.978      0.983      1.000     -0.986    -0.989
#> You've Got      -0.974     -0.987     -0.986      1.000     0.986
#> Sleepless       -0.966     -0.992     -0.989      0.986     1.000
```

We can create vectors **q** and **p**, that can explain much of the structure we see. The **q** would look like this:

```
t(q)
#>      Godfather Godfather2 Goodfellas You've Got Sleepless
#> [1,]         1          1          1         -1        -1
```

and it narrows down movies to two groups: gangster (coded with 1) and romance (coded with -1). We can also reduce the users to three groups:

```
t(p)
#>      1 2 3 4 5 6 7 8  9 10 11 12
#> [1,] 2 2 2 0 0 0 0 0 -2 -2 -2 -2
```

those that like gangster movies and dislike romance movies (coded as 2), those that like romance movies and dislike gangster movies (coded as -2), and those that don't care (coded as 0). The main point here is that we can almost reconstruct r, which has 60 values, with a couple of vectors totaling 17 values. If r contains the residuals for users $u = 1, \ldots, 12$ for movies $i = 1, \ldots, 5$ we can write the following mathematical formula for our residuals $r_{u,i}$.

$$r_{u,i} \approx p_u q_i$$

This implies that we can explain more variability by modifying our previous model for movie recommendations to:

$$Y_{u,i} = \mu + b_i + b_u + p_u q_i + \varepsilon_{u,i}$$

However, we motivated the need for the $p_u q_i$ term with a simple simulation. The structure found in data is usually more complex. For example, in this first simulation we assumed there were was just one factor p_u that determined which of the two genres movie u belongs to. But the structure in our movie data seems to be much more complicated than gangster movie versus romance. We may have many other factors. Here we present a slightly more complex simulation. We now add a sixth movie.

```
round(r, 1)
#>      Godfather Godfather2 Goodfellas You've Got Sleepless Scent
#> 1        0.5       0.6        1.6       -0.5     -0.5  -1.6
#> 2        1.5       1.4        0.5       -1.5     -1.4  -0.4
#> 3        1.5       1.6        0.5       -1.6     -1.5  -0.5
#> 4       -0.1       0.1        0.1       -0.1     -0.1   0.1
#> 5       -0.1      -0.1        0.1        0.0      0.1  -0.1
#> 6        0.5       0.5       -0.4       -0.6     -0.5   0.5
#> 7        0.5       0.5       -0.5       -0.6     -0.4   0.4
#> 8        0.5       0.6       -0.5       -0.5     -0.4   0.4
#> 9       -0.9      -1.0       -0.9        1.0      1.1   0.9
#> 10      -1.6      -1.4       -0.4        1.5      1.4   0.5
#> 11      -1.4      -1.5       -0.5        1.5      1.6   0.6
#> 12      -1.4      -1.4       -0.5        1.6      1.5   0.6
```

By exploring the correlation structure of this new dataset

```
colnames(r)[4:6] <- c("YGM", "SS", "SW")
cor(r)
#>             Godfather Godfather2 Goodfellas    YGM     SS     SW
#> Godfather      1.000      0.997      0.562  -0.997 -0.996 -0.571
#> Godfather2     0.997      1.000      0.577  -0.998 -0.999 -0.583
#> Goodfellas     0.562      0.577      1.000  -0.552 -0.583 -0.994
#> YGM           -0.997     -0.998     -0.552   1.000  0.998  0.558
#> SS            -0.996     -0.999     -0.583   0.998  1.000  0.588
#> SW            -0.571     -0.583     -0.994   0.558  0.588  1.000
```

We note that we perhaps need a second factor to account for the fact that some users like Al Pacino, while others dislike him or don't care. Notice that the overall structure of the correlation obtained from the simulated data is not that far off the real correlation:

```
six_movies <- c(m_1, m_2, m_3, m_4, m_5, m_6)
x <- y[, six_movies]
colnames(x) <- colnames(r)
cor(x, use="pairwise.complete")
#>            Godfather Godfather2 Goodfellas    YGM     SS     SW
#> Godfather     1.0000      0.829      0.444  -0.440 -0.378  0.0589
#> Godfather2    0.8285      1.000      0.521  -0.331 -0.358  0.1186
```

```
#> Goodfellas    0.4441      0.521      1.000 -0.481 -0.402 -0.1230
#> YGM          -0.4397     -0.331     -0.481  1.000  0.533 -0.1699
#> SS           -0.3781     -0.358     -0.402  0.533  1.000 -0.1822
#> SW            0.0589      0.119     -0.123 -0.170 -0.182  1.0000
```

To explain this more complicated structure, we need two factors. For example something like this:

```
t(q)
#>        Godfather Godfather2 Goodfellas You've Got Sleepless Scent
#> [1,]        1          1          1         -1        -1    -1
#> [2,]        1          1         -1         -1        -1     1
```

With the first factor (the first row) used to code the gangster versus romance groups and a second factor (the second row) to explain the Al Pacino versus no Al Pacino groups. We will also need two sets of coefficients to explain the variability introduced by the 3×3 types of groups:

```
t(p)
#>          1   2   3 4 5   6   7   8 9   10   11   12
#> [1,]   1.0 1.0 1.0 0 0 0.0 0.0 0.0 -1 -1.0 -1.0 -1.0
#> [2,]  -0.5 0.5 0.5 0 0 0.5 0.5 0.5  0 -0.5 -0.5 -0.5
```

The model with two factors has 36 parameters that can be used to explain much of the variability in the 72 ratings:

$$Y_{u,i} = \mu + b_i + b_u + p_{u,1}q_{1,i} + p_{u,2}q_{2,i} + \varepsilon_{u,i}$$

Note that in an actual data application, we need to fit this model to data. To explain the complex correlation we observe in real data, we usually permit the entries of p and q to be continuous values, rather than discrete ones as we used in the simulation. For example, rather than dividing movies into gangster or romance, we define a continuum. Also note that this is not a linear model and to fit it we need to use an algorithm other than the one used by `lm` to find the parameters that minimize the least squares. The winning algorithms for the Netflix challenge fit a model similar to the above and used regularization to penalize for large values of p and q, rather than using least squares. Implementing this approach is beyond the scope of this book.

33.11.2   Connection to SVD and PCA

The decomposition:

$$r_{u,i} \approx p_{u,1}q_{1,i} + p_{u,2}q_{2,i}$$

is very much related to SVD and PCA. SVD and PCA are complicated concepts, but one way to understand them is that SVD is an algorithm that finds the vectors p and q that permit us to rewrite the matrix r with m rows and n columns as:

$$r_{u,i} = p_{u,1}q_{1,i} + p_{u,2}q_{2,i} + \cdots + p_{u,m}q_{m,i}$$

with the variability of each term decreasing and with the ps uncorrelated. The algorithm also computes this variability so that we can know how much of the matrices, total variability is explained as we add new terms. This may permit us to see that, with just a few terms, we can explain most of the variability.

Let's see an example with the movie data. To compute the decomposition, we will make the residuals with NAs equal to 0:

```
y[is.na(y)] <- 0
pca <- prcomp(y)
```

The q vectors are called the principal components and they are stored in this matrix:

```
dim(pca$rotation)
#> [1] 454 292
```

While the p, or the user effects, are here:

```
dim(pca$x)
#> [1] 292 292
```

We can see the variability of each of the vectors:

```
qplot(1:nrow(x), pca$sdev, xlab = "PC")
```

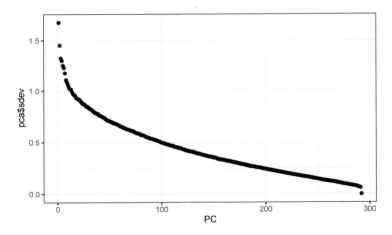

We also notice that the first two principal components are related to the structure in opinions about movies:

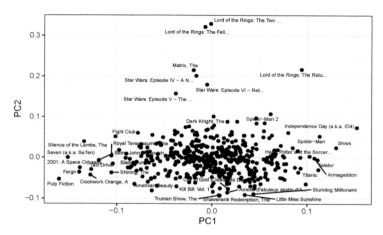

Just by looking at the top 10 in each direction, we see a meaningful pattern. The first PC shows the difference between critically acclaimed movies on one side:

```
#>  [1] "Pulp Fiction"              "Seven (a.k.a. Se7en)"
#>  [3] "Fargo"                     "2001: A Space Odyssey"
#>  [5] "Silence of the Lambs, The" "Clockwork Orange, A"
#>  [7] "Taxi Driver"               "Being John Malkovich"
#>  [9] "Royal Tenenbaums, The"     "Shining, The"
```

and Hollywood blockbusters on the other:

```
#>  [1] "Independence Day (a.k.a. ID4)"  "Shrek"
#>  [3] "Spider-Man"                     "Titanic"
#>  [5] "Twister"                        "Armageddon"
#>  [7] "Harry Potter and the Sorcer..." "Forrest Gump"
#>  [9] "Lord of the Rings: The Retu..." "Enemy of the State"
```

While the second PC seems to go from artsy, independent films:

```
#>  [1] "Shawshank Redemption, The"      "Truman Show, The"
#>  [3] "Little Miss Sunshine"           "Slumdog Millionaire"
#>  [5] "Amelie (Fabuleux destin d'A..." "Kill Bill: Vol. 1"
#>  [7] "American Beauty"                "City of God (Cidade de Deus)"
#>  [9] "Mars Attacks!"                  "Beautiful Mind, A"
```

to nerd favorites:

```
#>  [1] "Lord of the Rings: The Two ..." "Lord of the Rings: The Fell..."
#>  [3] "Lord of the Rings: The Retu..." "Matrix, The"
#>  [5] "Star Wars: Episode IV - A N..." "Star Wars: Episode VI - Ret..."
#>  [7] "Star Wars: Episode V - The ..." "Spider-Man 2"
#>  [9] "Dark Knight, The"               "Speed"
```

Fitting a model that incorporates these estimates is complicated. For those interested in implementing an approach that incorporates these ideas, we recommend trying the **recommenderlab** package. The details are beyond the scope of this book.

33.12 Exercises

In this exercise set, we will be covering a topic useful for understanding matrix factorization: the singular value decomposition (SVD). SVD is a mathematical result that is widely used in machine learning, both in practice and to understand the mathematical properties of some algorithms. This is a rather advanced topic and to complete this exercise set you will have to be familiar with linear algebra concepts such as matrix multiplication, orthogonal matrices, and diagonal matrices.

The SVD tells us that we can *decompose* an $N \times p$ matrix Y with $p < N$ as

$$Y = UDV^\top$$

With U and V *orthogonal* of dimensions $N \times p$ and $p \times p$, respectively, and D a $p \times p$ *diagonal* matrix with the values of the diagonal decreasing:

$$d_{1,1} \geq d_{2,2} \geq \ldots d_{p,p}.$$

In this exercise, we will see one of the ways that this decomposition can be useful. To do this, we will construct a dataset that represents grade scores for 100 students in 24 different subjects. The overall average has been removed so this data represents the percentage point each student received above or below the average test score. So a 0 represents an average grade (C), a 25 is a high grade (A+), and a -25 represents a low grade (F). You can simulate the data like this:

```
set.seed(1987)
n <- 100
k <- 8
Sigma <- 64  * matrix(c(1, .75, .5, .75, 1, .5, .5, .5, 1), 3, 3)
m <- MASS::mvrnorm(n, rep(0, 3), Sigma)
m <- m[order(rowMeans(m), decreasing = TRUE),]
y <- m %x% matrix(rep(1, k), nrow = 1) +
  matrix(rnorm(matrix(n * k * 3)), n, k * 3)
colnames(y) <- c(paste(rep("Math",k), 1:k, sep="_"),
                 paste(rep("Science",k), 1:k, sep="_"),
                 paste(rep("Arts",k), 1:k, sep="_"))
```

Our goal is to describe the student performances as succinctly as possible. For example, we want to know if these test results are all just random independent numbers. Are all students just about as good? Does being good in one subject imply you will be good in another? How does the SVD help with all this? We will go step by step to show that with just three relatively small pairs of vectors we can explain much of the variability in this 100×24 dataset.

You can visualize the 24 test scores for the 100 students by plotting an image:

```
my_image <- function(x, zlim = range(x), ...){
  colors = rev(RColorBrewer::brewer.pal(9, "RdBu"))
```

```
  cols <- 1:ncol(x)
  rows <- 1:nrow(x)
  image(cols, rows, t(x[rev(rows),,drop=FALSE]), xaxt = "n", yaxt = "n",
        xlab="", ylab="",  col = colors, zlim = zlim, ...)
  abline(h=rows + 0.5, v = cols + 0.5)
  axis(side = 1, cols, colnames(x), las = 2)
}
```

```
my_image(y)
```

1. How would you describe the data based on this figure?

    a.   The test scores are all independent of each other.
    b.   The students that test well are at the top of the image and there seem to be three groupings by subject.
    c.   The students that are good at math are not good at science.
    d.   The students that are good at math are not good at humanities.

2. You can examine the correlation between the test scores directly like this:

```
my_image(cor(y), zlim = c(-1,1))
range(cor(y))
axis(side = 2, 1:ncol(y), rev(colnames(y)), las = 2)
```

Which of the following best describes what you see?

    a.   The test scores are independent.
    b.   Math and science are highly correlated but the humanities are not.
    c.   There is high correlation between tests in the same subject but no correlation across subjects.
    d.   There is a correlation among all tests, but higher if the tests are in science and math and even higher within each subject.

3. Remember that orthogonality means that $U^\top U$ and $V^\top V$ are equal to the identity matrix. This implies that we can also rewrite the decomposition as

$$YV = UD \text{ or } U^\top Y = DV^\top$$

We can think of YV and $U^\top V$ as two transformations of Y that preserve the total variability of Y since U and V are orthogonal.

Use the function svd to compute the SVD of y. This function will return U, V and the diagonal entries of D.

```
s <- svd(y)
names(s)
```

You can check that the SVD works by typing:

```
y_svd <- s$u %*% diag(s$d) %*% t(s$v)
max(abs(y - y_svd))
```

Compute the sum of squares of the columns of Y and store them in `ss_y`. Then compute the sum of squares of columns of the transformed YV and store them in `ss_yv`. Confirm that `sum(ss_y)` is equal to `sum(ss_yv)`.

4. We see that the total sum of squares is preserved. This is because V is orthogonal. Now to start understanding how YV is useful, plot `ss_y` against the column number and then do the same for `ss_yv`. What do you observe?

5. We see that the variability of the columns of YV is decreasing. Furthermore, we see that, relative to the first three, the variability of the columns beyond the third is almost 0. Now notice that we didn't have to compute `ss_yv` because we already have the answer. How? Remember that $YV = UD$ and because U is orthogonal, we know that the sum of squares of the columns of UD are the diagonal entries of D squared. Confirm this by plotting the square root of `ss_yv` versus the diagonal entries of D.

6. From the above we know that the sum of squares of the columns of Y (the total sum of squares) add up to the sum of `s$d^2` and that the transformation YV gives us columns with sums of squares equal to `s$d^2`. Now compute what percent of the total variability is explained by just the first three columns of YV.

7. We see that almost 99% of the variability is explained by the first three columns of $YV = UD$. So we get the sense that we should be able to explain much of the variability and structure we found while exploring the data with a few columns. Before we continue, let's show a useful computational trick to avoid creating the matrix `diag(s$d)`. To motivate this, we note that if we write U out in its columns $[U_1, U_2, \ldots, U_p]$ then UD is equal to

$$UD = [U_1 d_{1,1}, U_2 d_{2,2}, \ldots, U_p d_{p,p}]$$

Use the `sweep` function to compute UD without constructing `diag(s$d)` nor matrix multiplication.

8. We know that $U_1 d_{1,1}$, the first column of UD, has the most variability of all the columns of UD. Earlier we saw an image of Y:

`my_image(y)`

in which we can see that the student to student variability is quite large and that it appears that students that are good in one subject are good in all. This implies that the average (across all subjects) for each student should explain a lot of the variability. Compute the average score for each student and plot it against $U_1 d_{1,1}$ and describe what you find.

9. We note that the signs in SVD are arbitrary because:

$$UDV^\top = (-U)D(-V)^\top$$

With this in mind we see that the first column of UD is almost identical to the average score for each student except for the sign.

This implies that multiplying Y by the first column of V must be performing a similar operation to taking the average. Make an image plot of V and describe the first column relative to others and how this relates to taking an average.

10. We already saw that we can rewrite UD as

$$U_1 d_{1,1} + U_2 d_{2,2} + \cdots + U_p d_{p,p}$$

with U_j the j-th column of U. This implies that we can rewrite the entire SVD as:

$$Y = U_1 d_{1,1} V_1^\top + U_2 d_{2,2} V_2^\top + \cdots + U_p d_{p,p} V_p^\top$$

with V_j the jth column of V. Plot U_1, then plot $V_1^\top$ using the same range for the y-axis limits, then make an image of $U_1 d_{1,1} V_1^\top$ and compare it to the image of Y. Hint: use the `my_image` function defined above and use the `drop=FALSE` argument to assure the subsets of matrices are matrices.

11. We see that with just a vector of length 100, a scalar, and a vector of length 24, we actually come close to reconstructing the original 100×24 matrix. This is our first matrix factorization:

$$Y \approx d_{1,1} U_1 V_1^\top$$

We know it explains `s$d[1]^2/sum(s$d^2) * 100` percent of the total variability. Our approximation only explains the observation that good students tend to be good in all subjects. But another aspect of the original data that our approximation does not explain was the higher similarity we observed within subjects. We can see this by computing the difference between our approximation and original data and then computing the correlations. You can see this by running this code:

```
resid <- y - with(s,(u[,1, drop=FALSE]*d[1]) %*% t(v[,1, drop=FALSE]))
my_image(cor(resid), zlim = c(-1,1))
axis(side = 2, 1:ncol(y), rev(colnames(y)), las = 2)
```

Now that we have removed the overall student effect, the correlation plot reveals that we have not yet explained the within subject correlation nor the fact that math and science are closer to each other than to the arts. So let's explore the second column of the SVD. Repeat the previous exercise but for the second column: Plot U_2, then plot $V_2^\top$ using the same range for the y-axis limits, then make an image of $U_2 d_{2,2} V_2^\top$ and compare it to the image of `resid`.

12. The second column clearly relates to a student's difference in ability in math/science versus the arts. We can see this most clearly from the plot of `s$v[,2]`. Adding the matrix we obtain with these two columns will help with our approximation:

$$Y \approx d_{1,1} U_1 V_1^\top + d_{2,2} U_2 V_2^\top$$

We know it will explain

```
sum(s$d[1:2]^2)/sum(s$d^2) * 100
```

percent of the total variability. We can compute new residuals like this:

```
resid <- y - with(s,sweep(u[,1:2], 2, d[1:2], FUN="*") %*% t(v[,1:2]))
my_image(cor(resid), zlim = c(-1,1))
axis(side = 2, 1:ncol(y), rev(colnames(y)), las = 2)
```

and see that the structure that is left is driven by the differences between math and science. Confirm this by plotting U_3, then plot $V_3^\top$ using the same range for the y-axis limits, then make an image of $U_3 d_{3,3} V_3^\top$ and compare it to the image of `resid`.

13. The third column clearly relates to a student's difference in ability in math and science. We can see this most clearly from the plot of `s$v[,3]`. Adding the matrix we obtain with these two columns will help with our approximation:

$$Y \approx d_{1,1} U_1 V_1^\top + d_{2,2} U_2 V_2^\top + d_{3,3} U_3 V_3^\top$$

We know it will explain:

```
sum(s$d[1:3]^2)/sum(s$d^2) * 100
```

percent of the total variability. We can compute new residuals like this:

```
resid <- y - with(s,sweep(u[,1:3], 2, d[1:3], FUN="*") %*% t(v[,1:3]))
my_image(cor(resid), zlim = c(-1,1))
axis(side = 2, 1:ncol(y), rev(colnames(y)), las = 2)
```

We no longer see structure in the residuals: they seem to be independent of each other. This implies that we can describe the data with the following model:

$$Y = d_{1,1} U_1 V_1^\top + d_{2,2} U_2 V_2^\top + d_{3,3} U_3 V_3^\top + \varepsilon$$

with ε a matrix of independent identically distributed errors. This model is useful because we summarize of 100×24 observations with $3 \times (100 + 24 + 1) = 375$ numbers. Furthermore, the three components of the model have useful interpretations: 1) the overall ability of a student, 2) the difference in ability between the math/sciences and arts, and 3) the remaining differences between the three subjects. The sizes $d_{1,1}, d_{2,2}$ and $d_{3,3}$ tell us the variability explained by each component. Finally, note that the components $d_{j,j} U_j V_j^\top$ are equivalent to the jth principal component.

Finish the exercise by plotting an image of Y, an image of $d_{1,1} U_1 V_1^\top + d_{2,2} U_2 V_2^\top + d_{3,3} U_3 V_3^\top$ and an image of the residuals, all with the same `zlim`.

14. **Advanced.** The `movielens` dataset included in the **dslabs** package is a small subset of a larger dataset with millions of ratings. You can find the entire latest dataset here https://grouplens.org/datasets/movielens/20m/. Create your own recommendation system using all the tools we have shown you.

34

Clustering

The algorithms we have described up to now are examples of a general approach referred to as *supervised* machine learning. The name comes from the fact that we use the outcomes in a training set to *supervise* the creation of our prediction algorithm. There is another subset of machine learning referred to as *unsupervised*. In this subset we do not necessarily know the outcomes and instead are interested in discovering groups. These algorithms are also referred to as *clustering* algorithms since predictors are used to define *clusters*.

In the two examples we have shown here, clustering would not be very useful. In the first example, if we are simply given the heights we will not be able to discover two groups, males and females, because the intersection is large. In the second example, we can see from plotting the predictors that discovering the two digits, two and seven, will be challenging:

```
library(tidyverse)
library(dslabs)
data("mnist_27")
mnist_27$train %>% qplot(x_1, x_2, data = .)
```

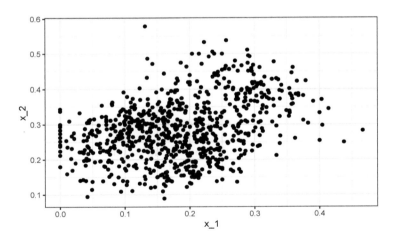

However, there are applications in which unsupervised learning can be a powerful technique, in particular as an exploratory tool.

A first step in any clustering algorithm is defining a distance between observations or groups of observations. Then we need to decide how to join observations into clusters. There are many algorithms for doing this. Here we introduce two as examples: hierarchical and k-means.

We will construct a simple example based on movie ratings. Here we quickly construct a matrix x that has ratings for the 50 movies with the most ratings.

```
data("movielens")
top <- movielens %>%
  group_by(movieId) %>%
  summarize(n=n(), title = first(title)) %>%
  top_n(50, n) %>%
  pull(movieId)

x <- movielens %>%
  filter(movieId %in% top) %>%
  group_by(userId) %>%
  filter(n() >= 25) %>%
  ungroup() %>%
  select(title, userId, rating) %>%
  spread(userId, rating)

row_names <- str_remove(x$title, ": Episode") %>% str_trunc(20)
x <- x[,-1] %>% as.matrix()
x <- sweep(x, 2, colMeans(x, na.rm = TRUE))
x <- sweep(x, 1, rowMeans(x, na.rm = TRUE))
rownames(x) <- row_names
```

We want to use these data to find out if there are clusters of movies based on the ratings from 139 movie raters. A first step is to find the distance between each pair of movies using the `dist` function:

```
d <- dist(x)
```

34.1   Hierarchical clustering

With the distance between each pair of movies computed, we need an algorithm to define groups from these. Hierarchical clustering starts by defining each observation as a separate group, then the two closest groups are joined into a group iteratively until there is just one group including all the observations. The `hclust` function implements this algorithm and it takes a distance as input.

```
h <- hclust(d)
```

We can see the resulting groups using a *dendrogram*.

```
plot(h, cex = 0.65, main = "", xlab = "")
```

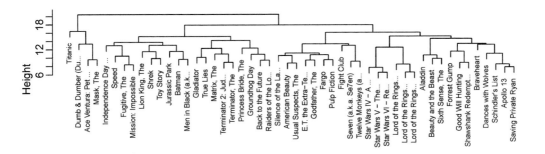

To interpret this graph we do the following. To find the distance between any two movies, find the first location from top to bottom that the movies split into two different groups. The height of this location is the distance between the groups. So the distance between the *Star Wars* movies is 8 or less, while the distance between *Raiders of the Lost of Ark* and *Silence of the Lambs* is about 17.

To generate actual groups we can do one of two things: 1) decide on a minimum distance needed for observations to be in the same group or 2) decide on the number of groups you want and then find the minimum distance that achieves this. The function `cutree` can be applied to the output of `hclust` to perform either of these two operations and generate groups.

```
groups <- cutree(h, k = 10)
```

Note that the clustering provides some insights into types of movies. Group 4 appears to be blockbusters:

```
names(groups)[groups==4]
#> [1] "Apollo 13"            "Braveheart"          "Dances with Wolves"
#> [4] "Forrest Gump"         "Good Will Hunting"   "Saving Private Ryan"
#> [7] "Schindler's List"     "Shawshank Redempt..."
```

And group 9 appears to be nerd movies:

```
names(groups)[groups==9]
#> [1] "Lord of the Rings..." "Lord of the Rings..." "Lord of the Rings..."
#> [4] "Star Wars IV - A ..." "Star Wars V - The..." "Star Wars VI - Re..."
```

We can change the size of the group by either making `k` larger or `h` smaller. We can also explore the data to see if there are clusters of movie raters.

```
h_2 <- dist(t(x)) %>% hclust()
```

34.2   k-means

To use the k-means clustering algorithm we have to pre-define k, the number of clusters we want to define. The k-means algorithm is iterative. The first step is to define k centers. Then each observation is assigned to the cluster with the closest center to that observation. In a second step the centers are redefined using the observation in each cluster: the column means are used to define a *centroid*. We repeat these two steps until the centers converge.

The `kmeans` function included in R-base does not handle NAs. For illustrative purposes we will fill out the NAs with 0s. In general, the choice of how to fill in missing data, or if one should do it at all, should be made with care.

```
x_0 <- x
x_0[is.na(x_0)] <- 0
k <- kmeans(x_0, centers = 10)
```

The cluster assignments are in the `cluster` component:

```
groups <- k$cluster
```

Note that because the first center is chosen at random, the final clusters are random. We impose some stability by repeating the entire function several times and averaging the results. The number of random starting values to use can be assigned through the `nstart` argument.

```
k <- kmeans(x_0, centers = 10, nstart = 25)
```

34.3   Heatmaps

A powerful visualization tool for discovering clusters or patterns in your data is the heatmap. The idea is simple: plot an image of your data matrix with colors used as the visual cue and both the columns and rows ordered according to the results of a clustering algorithm. We will demonstrate this with the `tissue_gene_expression` dataset. We will scale the rows of the gene expression matrix.

The first step is compute:

```
data("tissue_gene_expression")
x <- sweep(tissue_gene_expression$x, 2, colMeans(tissue_gene_expression$x))
h_1 <- hclust(dist(x))
h_2 <- hclust(dist(t(x)))
```

Now we can use the results of this clustering to order the rows and columns.

```
image(x[h_1$order, h_2$order])
```

But there is `heatmap` function that does it for us:

```
heatmap(x, col = RColorBrewer::brewer.pal(11, "Spectral"))
```

We do not show the results of the heatmap function because there are too many features for the plot to be useful. We will therefore filter some columns and remake the plots.

34.4 Filtering features

If the information about clusters in included in just a few features, including all the features can add enough noise that detecting clusters becomes challenging. One simple approach to try to remove features with no information is to only include those with high variance. In the movie example, a user with low variance in their ratings is not really informative: all the movies seem about the same to them. Here is an example of how we can include only the features with high variance.

```
library(matrixStats)
sds <- colSds(x, na.rm = TRUE)
o <- order(sds, decreasing = TRUE)[1:25]
heatmap(x[,o], col = RColorBrewer::brewer.pal(11, "Spectral"))
```

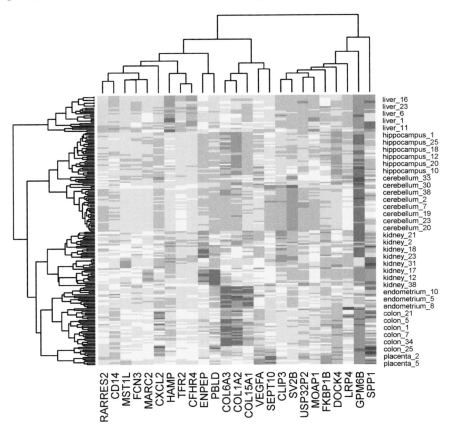

34.5 Exercises

1. Load the `tissue_gene_expression` dataset. Remove the row means and compute the distance between each observation. Store the result in `d`.

2. Make a hierarchical clustering plot and add the tissue types as labels.

3. Run a k-means clustering on the data with $K = 7$. Make a table comparing the identified clusters to the actual tissue types. Run the algorithm several times to see how the answer changes.

4. Select the 50 most variable genes. Make sure the observations show up in the columns, that the predictors are centered, and add a color bar to show the different tissue types. Hint: use the `ColSideColors` argument to assign colors. Also, use `col = RColorBrewer::brewer.pal(11, "RdBu")` for a better use of colors.

Part VI

Productivity Tools

35

Introduction to productivity tools

Generally speaking, we do not recommend using point-and-click approaches for data analysis. Instead, we recommend scripting languages, such as R, since they are more flexible and greatly facilitate reproducibility. Similarly, we recommend against the use of point-and-click approaches to organizing files and document preparation. In this chapter, we demonstrate alternative approaches. Specifically, we will learn to use freely available tools that, although at first may seem cumbersome and non-intuitive, will eventually make you a much more efficient and productive data scientist.

Three general guiding principles that motivate what we learn here are 1) be systematic when organizing your filesystem, 2) automate when possible, and 3) minimize the use of the mouse. As you become more proficient at coding, you will find that 1) you want to minimize the time you spend remembering what you called a file or where you put it, 2) if you find yourself repeating the same task over and over, there is probably a way to automate, and 3) anytime your fingers leave the keyboard, it results in loss of productivity.

A data analysis project is not always a dataset and a script. A typical data analysis challenge may involve several parts, each involving several data files, including files containing the scripts we use to analyze data. Keeping all this organized can be challenging. We will learn to use the *Unix shell* as a tool for managing files and directories on your computer system. Using Unix will permit you to use the keyboard, rather than the mouse, when creating folders, moving from directory to directory, and renaming, deleting, or moving files. We also provide specific suggestions on how to keep the filesystem organized.

The data analysis process is also iterative and adaptive. As a result, we are constantly editing our scripts and reports. In this chapter, we introduce you to the version control system *Git*, which is a powerful tool for keeping track of these changes. We also introduce you to GitHub[1], a service that permits you to host and share your code. We will demonstrate how you can use this service to facilitate collaborations. **Keep in mind that another positive benefit of using GitHub is that you can easily showcase your work to potential employers**.

Finally, we learn to write reports in R markdown, which permits you to incorporate text and code into a single document. We will demonstrate how, using the `knitr` package, we can write reproducible and aesthetically pleasing reports by running the analysis and generating the report simultaneously.

We will put all this together using the powerful integrated desktop environment RStudio[2]. Throughout the chapter we will be building up an example on US gun murders. The final project, which includes several files and folders, can be seen here: https://github. com/rairizarry/murders. Note that one of the files in that project is the final report: https://github.com/rairizarry/murders/blob/master/report.md.

---

[1]http://github.com
[2]https://www.rstudio.com/

36

Organizing with Unix

Unix is the operating system of choice in data science. We will introduce you to the Unix way of thinking using an example: how to keep a data analysis project organized. We will learn some of the most commonly used commands along the way. However, we won't go into the details here. We highly encourage you to learn more, especially when you find yourself using the mouse too much or performing a repetitive task often. In those cases, there is probably a more efficient way to do it in Unix. Here are some basic courses to get you started:

- https://www.codecademy.com/learn/learn-the-command-line
- https://www.edx.org/course/introduction-linux-linuxfoundationx-lfs101x-1
- https://www.coursera.org/learn/unix

There are many reference books[1] as well. Bite Size Linux[2] and Bite Size Command Line[3] are two particularly clear, succinct, and complete examples.

When searching for Unix resources, keep in mind that other terms used to describe what we will learn here are *Linux, the shell* and *the command line*. Basically, what we are learning is a series of commands and a way of thinking that facilitates the organization of files without using the mouse.

To serve as motivation, we are going to start constructing a directory using Unix tools and RStudio.

36.1 Naming convention

Before you start organizing projects with Unix you want to pick a name convention that you will use to systematically name your files and directories. This will help you find files and know what is in them.

In general you want to name your files in a way that is related to their contents and specifies how they relate to other files. The Smithsonian Data Management Best Practices[4] has "five precepts of file naming and organization" and they are:

- Have a distinctive, human-readable name that gives an indication of the content.
- Follow a consistent pattern that is machine-friendly.

---

[1] https://www.quora.com/Which-are-the-best-Unix-Linux-reference-books
[2] https://gumroad.com/l/bite-size-linux
[3] https://jvns.ca/blog/2018/08/05/new-zine--bite-size-command-line/
[4] https://library.si.edu/sites/default/files/tutorial/pdf/filenamingorganizing20180227.pdf

- Organize files into directories (when necessary) that follow a consistent pattern.
- Avoid repetition of semantic elements among file and directory names.
- Have a file extension that matches the file format (no changing extensions!)

For specific recommendations we highly recommend you follow The Tidyverse Style Guide[5].

36.2   The terminal

Instead of clicking, dragging, and dropping to organize our files and folders, we will be typing Unix commands into the terminal. The way we do this is similar to how we type commands into the R console, but instead of generating plots and statistical summaries, we will be organizing files on our system.

You will need access to a terminal[6] Once you have a terminal open, you can start typing commands. You should see a blinking cursor at the spot where what you type will show up. This position is called the *command line*. Once you type something and hit enter on Windows or return on the Mac, Unix will try to execute this command. If you want to try out an example, type this command into your command line:

```
echo "hello world"
```

The command `echo` is similar to `cat` in R. Executing this line should print out `hello world`, then return back to the command line.

Notice that you can't use the mouse to move around in the terminal. You have to use the keyboard. To go back to a command you previously typed, you can use the up arrow.

Note that above we included a chunk of code showing Unix commands in the same way we have previously shown R commands. We will make sure to distinguish when the command is meant for R and when it is meant for Unix.

36.3   The filesystem

We refer to all the files, folders, and programs on your computer as *the filesystem*. Keep in mind that folders and programs are also files, but this is a technicality we rarely think about and ignore in this book. We will focus on files and folders for now and discuss programs, or *executables*, in a later section.

---

[5]https://style.tidyverse.org/
[6]https://rafalab.github.io/dsbook/accessing-the-terminal-and-installing-git.html

36.3.1 Directories and subdirectories

The first concept you need to grasp to become a Unix user is how your filesystem is organized. You should think of it as a series of nested folders, each containing files, folders, and executables.

Here is a visual representation of the structure we are describing:

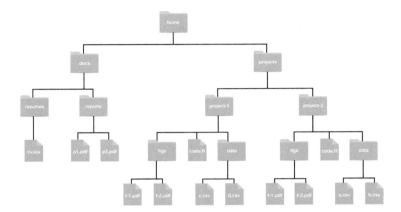

In Unix, we refer to folders as *directories*. Directories that are inside other directories are often referred to as *subdirectories*. So, for example, in the figure above, the directory *docs* has two subdirectories: *reports* and *resumes*, and *docs* is a subdirectory of *home*.

36.3.2 The home directory

The *home* directory is where all your stuff is kept, as opposed to the system files that come with your computer, which are kept elsewhere. In the figure above, the directory called *home* represents your home directory, but that is rarely the name used. On your system, the name of your home directory is likely the same as your username on that system. Below are an example on Windows and Mac showing a home directory, in this case, named *rafa*:

Now, look back at the figure showing a filesystem. Suppose you are using a point-and-click system and you want to remove the file *cv.tex*. Imagine that on your screen you can see the *home* directory. To erase this file, you would double click on the *home* directory, then *docs*, then *resumes*, and then drag *cv.tex* to the trash. Here you are experiencing the hierarchical nature of the system: *cv.tex* is a file inside the *resumes* directory, which is a subdirectory inside the *docs* directory, which is a subdirectory of the *home* directory.

Now suppose you can't see your home directory on your screen. You would somehow need to make it appear on your screen. One way to do this is to navigate from what is called the *root* directory all the way to your home directory. Any filesystem will have what is called a *root* directory, which is the directory that contains all directories. The *home* directory shown in the figure above will usually be two or more levels from the root. On Windows, you will have a structure like this:

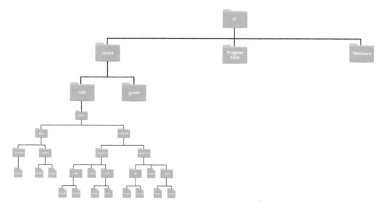

while on the Mac, it will be like this:

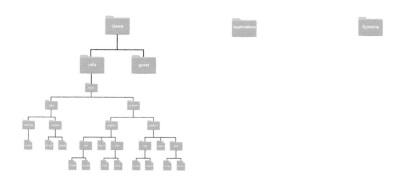

Note for Windows Users: The typical R installation will make your *Documents* directory your home directory in R. This will likely be different from your home directory in Git Bash. Generally, when we discuss home directories, we refer to the Unix home directory which for Windows, in this book, is the Git Bash Unix directory.

36.3.3   Working directory

The concept of a *current location* is part of the point-and-click experience: at any given moment we are *in a folder* and see the content of that folder. As you search for a file, as we did above, you are experiencing the concept of a current location: once you double click on a directory, you change locations and are now *in that folder*, as opposed to the folder you were in before.

In Unix, we don't have the same visual cues, but the concept of a *current location* is indispensable. We refer to this as the *working directory*. Each terminal window you have open has a working directory associated with it.

How do we know what is our working directory? To answer this, we learn our first Unix command: `pwd`, which stands for *print working directory*. This command returns the working directory.

Open a terminal and type:

```
pwd
```

We do not show the result of running this command because it will be quite different on your system compared to others. If you open a terminal and type `pwd` as your first command, you should see something like `/Users/yourusername` on a Mac or something like `/c/Users/yourusername` on Windows. The character string returned by calling `pwd` represents your working directory. When we first open a terminal, it will start in our home directory so in this case the working directory is the home directory.

Notice that the forward slashes / in the strings above separate directories. So, for example, the location `/c/Users/rafa` implies that our working directory is called `rafa` and it is a subdirectory of `Users`, which is a subdirectory of `c`, which is a subdirectory of the root directory. The root directory is therefore represented by just a forward slash: `/`.

36.3.4 Paths

We refer to the string returned by `pwd` as the *full path* of the working directory. The name comes from the fact that this string spells out the *path* you need to follow to get to the directory in question from the root directory. Every directory has a full path. Later, we will learn about *relative paths*, which tell us how to get to a directory from the working directory.

In Unix, we use the shorthand ~ as a nickname for your home directory. So, for example, if `docs` is a directory in your home directory, the full path for *docs* can be written like this `~/docs`.

Most terminals will show the path to your working directory right on the command line. If you are using default settings and open a terminal on the Mac, you will see that right at the command line you have something like `computername:~ username` with ~ representing your working directory, which in this example is the home directory ~. The same is true for the Git Bash terminal where you will see something like `username@computername MINGW64 ~`, with the working directory at the end. When we change directories, we will see this change on both Macs and Windows.

36.4 Unix commands

We will now learn a series of Unix commands that will permit us to prepare a directory for a data science project. We also provide examples of commands that, if you type into your terminal, will return an error. This is because we are assuming the filesystem in the earlier diagram. Your filesystem is different. In the next section, we will provide examples that you can type in.

36.4.1   `ls`: Listing directory content

In a point-and-click system, we know what is in a directory because we see it. In the terminal, we do not see the icons. Instead, we use the command `ls` to list the directory content.

To see the content of your home directory, open a terminal and type:

```
ls
```

We will see more examples soon.

36.4.2   `mkdir` and `rmdir`: make and remove a directory

When we are preparing for a data science project, we will need to create directories. In Unix, we can do this with the command `mkdir`, which stands for *make directory*.

Because you will soon be working on several projects, we highly recommend creating a directory called *projects* in your home directory.

You can try this particular example on your system. Open a terminal and type:

```
mkdir projects
```

If you do this correctly, nothing will happen: no news is good news. If the directory already exists, you will get an error message and the existing directory will remain untouched.

To confirm that you created these directories, you can list the directories:

```
ls
```

You should see the directories we just created listed. Perhaps you can also see many other directories that come pre-installed on your computer.

For illustrative purposes, let's make a few more directories. You can list more than one directory name like this:

```
mkdir docs teaching
```

You can check to see if the three directories were created:

```
ls
```

If you made a mistake and need to remove the directory, you can use the command `rmdir` to remove it.

```
mkdir junk
rmdir junk
```

This will remove the directory as long as it is empty. If it is not empty, you will get an error message and the directory will remain untouched. To remove directories that are not empty, we will learn about the command `rm` later.

36.4.3   cd: navigating the filesystem by changing directories

Next we want to create directories inside directories that we have already created. We also want to avoid pointing and clicking our way through the filesystem. We explain how to do this in Unix, using the command line.

Suppose we open a terminal and our working directory is our home directory. We want to change our working directory to `projects`. We do this using the `cd` command, which stands for *change directory*:

```
cd projects
```

To check that the working directory changed, we can use a command we previously learned to see our location:

```
pwd
```

Our working directory should now be `~/projects`. Note that on your computer the home directory `~` will be spelled out to something like `/c/Users/yourusername`).

Important Pro Tip: In Unix you can auto-complete by hitting tab. This means that we can type `cd d` then hit tab. Unix will either auto-complete if `docs` is the only directory/file starting with `d` or show you the options. Try it out! Using Unix without auto-complete will make it unbearable.

When using `cd`, we can either type a full path, which will start with `/` or `~`, or a *relative path*. In the example above, in which we typed `cd projects`, we used a relative path. **If the path you type does not start with `/` or `~`, Unix will assume you are typing a relative path, meaning that it will look for the directory in your current working directory**. So something like this will give you an error:

```
cd Users
```

because there is no `Users` directory in your working directory.

Now suppose we want to move back to the directory in which `projects` is a subdirectory, referred to as the *parent directory*. We could use the full path of the parent directory, but Unix provides a shortcut for this: the parent directory of the working directory is represented with two dots: `..`, so to move back we simply type:

```
cd ..
```

You should now be back in your home directory which you can confirm using `pwd`.

Because we can use full paths with `cd`, the following command:

```
cd ~
```

will always take us back to the home directory, no matter where we are in the filesystem.

The working directory also has a nickname, which is a single ., so if you type

```
cd .
```

you will not move. Although this particular use of . is not useful, this nickname does come in handy sometimes. The reasons are not relevant for this section, but you should still be aware of this fact.

In summary, we have learned that when using `cd` we either stay put, move to a new directory using the desired directory name, or move back to the parent directory using ...

When typing directory names, we can concatenate directories with the forward-slashes. So if we want a command that takes us to the `projects` directory no matter where we are in the filesystem, we can type:

```
cd ~/projects
```

which is equivalent to writing the entire path out. For example, in Windows we would write something like

```
cd /c/Users/yourusername/projects
```

The last two commands are equivalent and in both cases we are typing the full path.

When typing out the path of the directory we want, either full or relative, we can concatenate directories with the forward-slashes. We already saw that we can move to the `projects` directory regardless of where we are by typing the full path like this:

```
cd ~/projects
```

We can also concatenate directory names for relative paths. For instance, if we want to move back to the parent directory of the parent directory of the working directory, we can type:

```
cd ../..
```

Here are a couple of final tips related to the `cd` command. First, you can go back to whatever directory you just left by typing:

```
cd -
```

This can be useful if you type a very long path and then realize you want to go back to where you were, and that too has a very long path.

Second, if you just type:

```
cd
```

you will be returned to your home directory.

36.5   Some examples

Let's explore some examples of using `cd`. To help visualize, we will show the graphical representation of our filesystem vertically:

Suppose our working directory is `~/projects` and we want to move to `figs` in `project-1`. Here it is convenient to use relative paths:

```
cd project-1/figs
```

Now suppose our working directory is `~/projects` and we want to move to `reports` in `docs`, how can we do this?

One way is to use relative paths:

```
cd ../docs/reports
```

Another is to use the full path:

```
cd ~/docs/reports
```

If you are trying this out on your system, remember to use auto-complete.

Let's examine one more example. Suppose we are in `~/projects/project-1/figs` and want to change to `~/projects/project-2`. Again, there are two ways.

With relative paths:

```
cd ../../proejct-2
```

and with full paths:

```
cd ~/projects/project-2
```

36.6 More Unix commands

36.6.1 mv: moving files

In a point-and-click system, we move files from one directory to another by dragging and dropping. In Unix, we use the `mv` command.

Warning: `mv` will not ask "are you sure?" if your move results in overwriting a file.

Now that you know how to use full and relative paths, using `mv` is relatively straightforward. The general form is:

```
mv path-to-file path-to-destination-directory
```

For example, if we want to move the file `cv.tex` from `resumes` to `reports`, you could use the full paths like this:

```
mv ~/docs/resumes/cv.tex ~/docs/reports/
```

You can also use relative paths. So you could do this:

```
cd ~/docs/resumes
mv cv.tex ../reports/
```

or this:

```
cd ~/docs/reports/
mv ../cv.tex ./
```

Notice that in the last one we used the working directory shortcut . to give a relative path as the destination directory.

We can also use `mv` to change the name of a file. To do this, instead of the second argument being the destination directory, it also includes a filename. So, for example, to change the name from `cv.tex` to `resume.tex`, we simply type:

```
cd ~/docs/resumes
mv cv.tex resume.tex
```

We can also combine the move and a rename. For example:

```
cd ~/docs/resumes
mv cv.tex ../reports/resume.tex
```

And we can move entire directories. To move the `resumes` directory into `reports`, we do as follows:

```
mv ~/docs/resumes ~/docs/reports/
```

It is important to add the last / to make it clear you do not want to rename the `resumes` directory to `reports`, but rather move it into the `reports` directory.

36.6.2   cp: copying files

The command `cp` behaves similar to `mv` except instead of moving, we copy the file, meaning that the original file stays untouched.

So in all the `mv` examples above, you can switch `mv` to `cp` and they will copy instead of move with one exception: we can't copy entire directories without learning about arguments, which we do later.

36.6.3   rm: removing files

In point-and-click systems, we remove files by dragging and dropping them into the trash or using a special click on the mouse. In Unix, we use the `rm` command.

Warning: Unlike throwing files into the trash, `rm` is permanent. Be careful!

The general way it works is as follows:

```
rm filename
```

You can actually list files as well like this:

```
rm filename-1 filename-2 filename-3
```

You can use full or relative paths. To remove directories, you will have to learn about arguments, which we do later.

36.6.4   less: looking at a file

Often you want to quickly look at the content of a file. If this file is a text file, the quickest way to do is by using the command `less`. To look a the file `cv.tex`, you do this:

```
cd ~/docs/resumes
less cv.tex
```

To exit the viewer, you type `q`. If the files are long, you can use the arrow keys to move up and down. There are many other keyboard commands you can use within `less` to, for example, search or jump pages. You will learn more about this in a later section. If you are wondering why the command is called `less`, it is because the original was called `more`, as in "show me more of this file". The second version was called `less` because of the saying "less is more".

36.7   Preparing for a data science project

We are now ready to prepare a directory for a project. We will use the US murders project[7] as an example.

You should start by creating a directory where you will keep all your projects. We recommend a directory called *projects* in your home directory. To do this you would type:

```
cd ~
mkdir projects
```

Our project relates to gun violence murders so we will call the directory for our project `murders`. It will be a subdirectory in our projects directories. In the `murders` directory, we will create two subdirectories to hold the raw data and intermediate data. We will call these `data` and `rda`, respectively.

Open a terminal and make sure you are in the home directory:

```
cd ~
```

Now run the following commands to create the directory structure we want. At the end, we use `ls` and `pwd` to confirm we have generated the correct directories in the correct working directory:

```
cd projects
mkdir murders
cd murders
mkdir data rdas
ls
pwd
```

Note that the full path of our `murders` dataset is `~/projects/murders`.

So if we open a new terminal and want to navigate into that directory we type:

```
cd projects/murders
```

In Section 38.3 we will describe how we can use RStudio to organize a data analysis project, once these directories have been created.

---

[7]https://github.com/rairizarry/murders

36.8 Advanced Unix

Most Unix implementations include a large number of powerful tools and utilities. We have just learned the very basics here. We recommend that you use Unix as your main file management tool. It will take time to become comfortable with it, but as you struggle, you will find yourself learning just by looking up solutions on the internet. In this section, we superficially cover slightly more advanced topics. The main purpose of the section is to make you aware of what is available rather than explain everything in detail.

36.8.1 Arguments

Most Unix commands can be run with arguments. Arguments are typically defined by using a dash - or two dashes -- (depending on the command) followed by a letter or a word. An example of an argument is the -r behind `rm`. The r stands for recursive and the result is that files and directories are removed recursively, which means that if you type:

```
rm -r directory-name
```

all files, subdirectories, files in subdirectories, subdirectories in subdirectories, and so on, will be removed. This is equivalent to throwing a folder in the trash, except you can't recover it. Once you remove it, it is deleted for good. Often, when you are removing directories, you will encounter files that are protected. In such cases, you can use the argument -f which stands for `force`.

You can also combine arguments. For instance, to remove a directory regardless of protected files, you type:

```
rm -rf directory-name
```

Remember that once you remove there is no going back, so use this command very carefully.

A command that is often called with argument is `ls`. Here are some examples:

```
ls -a
```

The a stands for all. This argument makes `ls` show you all files in the directory, including hidden files. In Unix, all files starting with a . are hidden. Many applications create hidden directories to store important information without getting in the way of your work. An example is `git` (which we cover in depth later). Once you initialize a directory as a git directory with `git init`, a hidden directory called `.git` is created. Another hidden file is the `.gitignore` file.

Another example of using an argument is:

```
ls -l
```

The l stands for long and the result is that more information about the files is shown.

It is often useful to see files in chronological order. For that we use:

```
ls -t
```

and to reverse the order of how files are shown you can use:

```
ls -r
```

We can combine all these arguments to show more information for all files in reverse chronological order:

```
ls -lart
```

Each command has a different set of arguments. In the next section, we learn how to find out what they each do.

36.8.2   Getting help

As you may have noticed, Unix uses an extreme version of abbreviations. This makes it very efficient, but hard to guess how to call commands. To make up for this weakness, Unix includes complete help files or *man pages* (man is short for manual). In most systems, you can type `man` followed by the command name to get help. So for `ls`, we would type:

```
man ls
```

This command is not available in some of the compact implementations of Unix, such as Git Bash. An alternative way to get help that works on Git Bash is to type the command followed by `--help`. So for `ls`, it would be as follows:

```
ls --help
```

36.8.3   Pipes

The help pages are typically long and if you type the commands above to see the help, it scrolls all the way to the end. It would be useful if we could save the help to a file and then use `less` to see it. The `pipe`, written like this `|`, does something similar. It *pipes* the results of a command to the command after the `pipe`. This is similar to the pipe `%>%` that we use in R. To get more help we thus can type:

```
man ls | less
```

or in Git Bash:

```
ls --help | less
```

This is also useful when listing files with many files. We can type:

```
ls -lart | less
```

36.8.4 Wild cards

Some of the most powerful aspects of Unix are the *wild cards*. Suppose we want to remove all the temporary html files produced while trouble shooting for a project. Imagine there are dozens of files. It would be quite painful to remove them one by one. In Unix, we can actually write an expression that means all the files that end in `.html`. To do this we type *wild card*: `*`. As discussed in the data wrangling part of this book, this character means any number of any combination of characters. Specifically, to list all html files, we would type:

```
ls *.html
```

To remove all html files in a directory, we would type:

```
rm *.html
```

The other useful wild card is the `?` symbol. This means any single character. So if all the files we want to erase have the form `file-001.html` with the numbers going from 1 to 999, we can type:

```
rm file-???.html
```

This will only remove files with that format.

We can combine wild cards. For example, to remove all files with the name `file-001` regardless of suffix, we can type:

```
rm file-001.*
```

Warning: Combining rm with the `*` wild card can be dangerous. There are combinations of these commands that will erase your entire filesystem without asking "are you sure?". Make sure you understand how it works before using this wild card with the rm command.

36.8.5 Environment variables

Unix has settings that affect your command line *environment*. These are called environment variables. The home directory is one of them. We can actually change some of these. In Unix, variables are distinguished from other entities by adding a `$` in front. The home directory is stored in `$HOME`.

Earlier we saw that `echo` is the Unix command for print. So we can see our home directory by typing:

```
echo $HOME
```

You can see them all by typing:

```
env
```

You can change some of these environment variables. But their names vary across different *shells*. We describe shells in the next section.

36.8.6 Shells

Much of what we use in this chapter is part of what is called the *Unix shell*. There are actually different shells, but the differences are almost unnoticeable. They are also important, although we do not cover those here. You can see what shell you are using by typing:

```
echo $SHELL
```

The most common one is bash.

Once you know the shell, you can change environmental variables. In Bash Shell, we do it using export variable value. To change the path, described in more detail soon, type: (**Don't actually run this command though!**)

```
export PATH = /usr/bin/
```

There is a program that is run before each terminal starts where you can edit variables so they change whenever you call the terminal. This changes in different implementations, but if using bash, you can create a file called .bashrc, .bash_profile,.bash_login, or .profile. You might already have one.

36.8.7 Executables

In Unix, all programs are files. They are called executables. So ls, mv and git are all files. But where are these program files? You can find out using the command which:

```
which git
#> /usr/bin/git
```

That directory is probably full of program files. The directory /usr/bin usually holds many program files. If you type:

```
ls /usr/bin
```

in your terminal, you will see several executable files.

There are other directories that usually hold program files. The Application directory in the Mac or Program Files directory in Windows are examples.

When you type ls, Unix knows to run a program which is an executable that is stored in some other directory. So how does Unix know where to find it? This information is included in the environmental variable $PATH. If you type:

```
echo $PATH
```

you will see a list of directories separated by :. The directory /usr/bin is probably one of the first ones on the list.

Unix looks for program files in those directories in that order. Although we don't teach it here, you can actually create executables yourself. However, if you put it in your working directory and this directory is not on the path, you can't run it just by typing the command. You get around this by typing the full path. So if your command is called my-ls, you can type:

```
./my-ls
```

Once you have mastered the basics of Unix, you should consider learning to write your own executables as they can help alleviate repetitive work.

36.8.8 Permissions and file types

If you type:

```
ls -l
```

At the beginning, you will see a series of symbols like this `-rw-r--r--`. This string indicates the type of file: regular file `-`, directory `d`, or executable `x`. This string also indicates the permission of the file: is it readable? writable? executable? Can other users on the system read the file? Can other users on the system edit the file? Can other users execute if the file is executable? This is more advanced than what we cover here, but you can learn much more in a Unix reference book.

36.8.9 Commands you should learn

There are many commands that we do not teach in this book, but we want to make you aware of them and what they do. They are:

- open/start - On the Mac `open filename` tries to figure out the right application of the filename and open it with that application. This is a very useful command. On Git Bash, you can try `start filename`. Try opening an `R` or `Rmd` file with `open` or `start`: it should open them with RStudio.

- nano - A bare-bones text editor.

- ln - create a symbolic link. We do not recommend its use, but you should be familiar with it.

- tar - archive files and subdirectories of a directory into one file.

- ssh - connect to another computer.

- grep - search for patterns in a file.

- awk/sed - These are two very powerful commands that permit you to find specific strings in files and change them.

36.8.10 File manipulation in R

We can also perform file management from within R. The key functions to learn about can be seen by looking at the help file for `?files`. Another useful function is `unlink`.

Although not generally recommended, note that you can run Unix commands in R using `system`.

37

Git and GitHub

Here we provide some details on Git and GitHub. However, we are only scratching the surface. To learn more about this topic, we highly recommend the following resources:

- Codeacademy: https://www.codecademy.com/learn/learn-git
- GitHub Guides: https://guides.github.com/activities/hello-world/
- Try Git tutorial: https://try.github.io/levels/1/challenges/1
- Happy Git and GitHub for the useR: http://happygitwithr.com/

37.1 Why use Git and GitHub?

There are three main reasons to use Git and GitHub.

1. Sharing: Even if we do not take advantage of the advanced and powerful version control functionality, we can still use Git and GitHub to share our code. We have already shown how we can do this with RStudio.

2. Collaborating: Once you set up a central repo, you can have multiple people make changes to code and keep versions synched. GitHub provides a free service for centralized repos. GitHub also has a special utility, called a *pull request*, that can be used by anybody to suggest changes to your code. You can easily either accept or deny the request.

3. Version control: The version control capabilities of Git permit us to keep track of changes we make to our code. We can also revert back to previous versions of files. Git also permits us to create *branches* in which we can test out ideas, then decide if we *merge* the new branch with the original.

Here we focus on the sharing aspects of Git and GitHub and refer the reader to the links above to learn more about this powerful tool.

37.2 GitHub accounts

After installing git[1], the first step is to get a GitHub account. Basic GitHub accounts are free. To do this, go to GitHub where you will see a box in which you can sign up.

---

[1] https://rafalab.github.io/dsbook/accessing-the-terminal-and-installing-git.html

You want to pick a name carefully. It should be short, easy to remember and to spell, somehow related to your name, and professional. This last one is important since you might be sending potential employers a link to your GitHub account. In the example below, I am sacrificing on the ease of spelling to incorporate my name. Your initials and last name are usually a good choice. If you have a very common name, then this may have to be taken into account. A simple solution would be to add numbers or spell out part of your name.

The account I use for my research, *rafalab*, is the same one I use for my webpage[2] and Twitter[3], which makes it easy to remember for those that follow my work.

Once you have a GitHub account, you are ready to connect Git and RStudio to this account.

A first step is to let Git know who we are. This will make it easier to connect with GitHub. We start by opening a terminal window in RStudio (remember you can get one through *Tools* in the menu bar). Now we use the `git config` command to tell Git who we are. We will type the following two commands in our terminal window:

```
git config --global user.name "Your Name"
git config --global user.mail "your@email.com"
```

You need to use the email account that you used to open your GitHub account. The RStudio session should look something like this:

You start by going to the *Global Options*, selecting *Git/SVN*, and then you enter a path for the Git executable we just installed.

---

[2]http://rafalab.org
[3]http://twitter.com/rafalab

On the Windows default installation, this will be *C:/Program File/Git/bin/git.exe*, but you should find it by browsing your system as this can change from system to system. Now to avoid entering our GitHub password every time we try to access our repository, we will create what is called an *SSH RSA Key*. RStudio can do this for us automatically if we click on the *Create RSA Key* button:

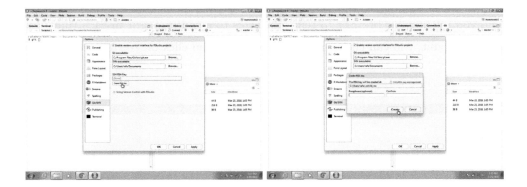

You can follow the default instructions as shown below:

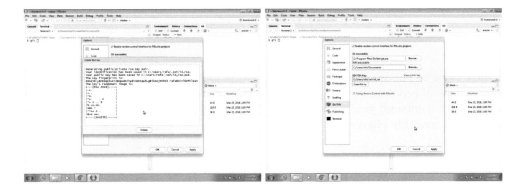

Git, RStudio and GitHub should now be able to connect and we are ready to create a first GitHub code repository.

37.3   GitHub repositories

You are now ready to create a GitHub repository (repo). The general idea is that you will have at least two copies of your code: one on your computer and one on GitHub. If you add collaborators to this project, then each will have a copy on their computer. The GitHub copy is usually considered the *master* copy that each collaborator syncs to. Git will help you keep all the different copies synced.

As mentioned, one of the advantages of keeping code on a GitHub repository is that you can easily share it with potential employers interested in seeing examples of your work. Because many data science companies use version control systems, like Git, to collaborate on projects, they might also be impressed that you already know at least the basics.

The first step in creating a repo for your code is to initialize on GitHub. Because you already created an account, you will have a page on GitHub with the URL `http://github.com/username`.

To create a repo, first log in to your account by clicking the *Sign In* button on https://github.com. You might already be signed in, in which case the *Sign In* button will not show up. If signing in, you will have to enter your username and password. We recommend you set up your browser to remember this to avoid typing it in each time.

Once on your account, you can click on *Repositories* and then click on *New* to create a new repo:

You will then want to choose a good descriptive name for the project. In the future, you might have dozens of repos so keep that in mind when choosing a name. Here we will use `homework-0`. We recommend you make the repo public. If you want to keep it private, you will have to pay a monthly charge.

You now have your first repo on GitHub. The next step will be to *clone it* on your computer and start editing and syncing using Git.

To do this, it is convenient to copy the link provided by GitHub specifically to connect to this repo, using Git as shown below. We will later need to copy and paste this so make sure to remember this step.

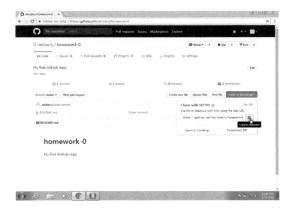

37.4 Overview of Git

The main actions in Git are to:

1. **pull** changes from the remote repo, in this case the GitHub repo
2. **add** files, or as we say in the Git lingo *stage* files
3. **commit** changes to the local repo
4. **push** changes to the *remote* repo, in our case the GitHub repo

To effectively permit version control and collaboration in Git, files move across four different areas:

But how does it all get started? There are two ways: we can clone an existing repo or initialize one. We will explore cloning first.

37.4.1 Clone

We are going to *clone* an existing *Upstream Repository*. You can see it on GitHub here: https://github.com/rairizarry/murders. By visiting this page, you can see multiple files and directories. This is the Upstream Repository. By clicking the green clone button, we can copy the repo's URL `https://github.com/rairizarry/murders.git`.

But what does *clone* mean? Rather than download all these files to your computer, we are going to actually copy the entire Git structure, which means we will add the files and directories to each of the three local stages: Working Directory, Staging Area, and Local Repository. When you clone, all three are exactly the same to start.

You can quickly see an example of this by doing the following. Open a terminal and type:

```
pwd
mkdir git-example
cd git-example
git clone https://github.com/rairizarry/murders.git
cd murders
#> /Users/rafa/myDocuments/teaching/data-science/dsbook
#> Cloning into 'murders'...
```

You now have cloned a GitHub repo and have a working Git directory, with all the files, on your system.

```
ls
#> README.txt
#> analysis.R
#> data
#> download-data.R
#> murders.Rproj
#> rdas
#> report.Rmd
#> report.md
#> report_files
#> wrangle-data.R
```

The *Working Directory* is the same as your Unix working directory. When you edit files using an editor such as RStudio, you change the files in this area and only in this area. Git can tell you how these files relate to the versions of the files in other areas with the command `git status`:

If you check the status now, you will see that nothing has changed and you get the following message:

```
git status
#> On branch master
#> Your branch is up to date with 'origin/master'.
#>
#> nothing to commit, working tree clean
```

Now we are going to make changes to these files. Eventually, we want these new versions of the files to be tracked and synched with the upstream repo. But we don't want to keep track of every little change: we don't want to sync until we are sure these versions are final enough to share. For this reason, edits in the staging area are not kept by the version control system.

To demonstrate, we add a file to the staging area with the `git add` command. Below we create a file using the Unix `echo` command just as an example (in reality you would use RStudio):

```
echo "test" >> new-file.txt
```

We are also adding a temporary file that we do not want to track at all:

```
echo "temporary" >> tmp.txt
```

Now we can stage the file we eventually want to add to our repository:

```
git add new-file.txt
```

Notice what the status says now:

```
git status
#> On branch master
#> Your branch is up to date with 'origin/master'.
#>
#> Changes to be committed:
#>   (use "git reset HEAD <file>..." to unstage)
#>
#>   new file:   new-file.txt
#>
#> Untracked files:
#>   (use "git add <file>..." to include in what will be committed)
```

```
#>
#>  tmp.txt
```

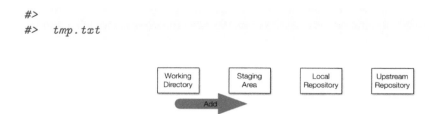

Because `new-file.txt` is staged, the current version of the file will get added to the local repository next time we commit, which we do as follows:

```
git commit -m "adding a new file"
#> [master b4eb2bb] adding a new file
#>  1 file changed, 1 insertion(+)
#>  create mode 100644 new-file.txt
```

We have now changed the local repo, which you can confirm using

```
git status
```

However, if we edit that file again, it changes only in the working directory. To add to the local repo, we need to stage it and commit the changes that are added to the local repo:

```
echo "adding a line" >> new-file.txt
git add new-file.txt
git commit -m "adding a new line to new-file"
#> [master 6593042] adding a new line to new-file
#>  1 file changed, 1 insertion(+)
```

Note that this step is often unnecessary in our uses of Git. We can skip the staging part if we add the file name to the commit command like this:

```
echo "adding a second line" >> new-file.txt
git commit -m "minor change to new-file" new-file.txt
#> [master d7f481a] minor change to new-file
#>  1 file changed, 1 insertion(+)
```

We can keep track of all the changes we have made with:

```
git log new-file.txt
#> commit d7f481a987d146b1db849ece15030b7e99dac0bd
#> Author: Rafael A. Irizarry <rairizarry@gmail.com>
```

```
#> Date:    Mon Sep 16 17:15:53 2019 -0400
#>
#>       minor change to new-file
#>
#> commit 6593042bf59d285ceaf7fd4092247d6489d7e9cd
#> Author: Rafael A. Irizarry <rairizarry@gmail.com>
#> Date:    Mon Sep 16 17:15:53 2019 -0400
#>
#>       adding a new line to new-file
#>
#> commit b4eb2bb3391271ce6a7b10cec558b6bd86a98859
#> Author: Rafael A. Irizarry <rairizarry@gmail.com>
#> Date:    Mon Sep 16 17:15:53 2019 -0400
#>
#>       adding a new file
```

To keep everything synced, the final step is to push the changes to the upstream repo. This is done with the `git push` command like this:

```
git push
```

However, in this particular example, you will not be able to do this because you do not have permission to edit the upstream repo. If this was your repo, you could.

If this is a collaborative project, the upstream repo may change and become different than our version. To update our local repository to be like the upstream repo, we use the command `fetch`:

```
git fetch
```

And then to make these copies to the staging and working directory areas, we use the command:

```
git merge
```

However, we often just want to change both with one command. For this, we use:

`git pull`

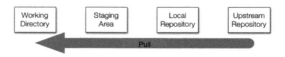

We will learn in Section 37.6 how RStudio has buttons to do all this. The details provided here should help you understand what happens in the background.

37.5   Initializing a Git directory

Now let's learn the second way we can get started: by initializing a directory on our own computer rather than cloning.

Suppose we already have a populated local directory and we want to turn this directory into a collaborative GitHub repository. The most efficient way of achieving this is by *initializing* the local directory.

To demonstrate how to do this we will initialize the gun murders directory we created in Section 36.7. Note that we already created a directory with several subdirectories on our computer but we do not yet have a Git local repo or GitHub upstream repo.

We start by creating a new repo on our GitHub page. We click on the *New* button:

We call it `murders` here to match the name of the directory on our local system. But if you are doing this for another project, please choose an appropriate name.

We then get a series of instructions on how to get started, but we can instead use what we have learned. The main thing we need from this page is to copy the repo's URL, in this case: https://github.com/rairizarry/murders.git.

At this moment, we can start a terminal and `cd` into our local projects directory. In our example, it would be:

```
cd ~/projects/murders
```

We then *intialize* the directory. This turns the directory into a Git directory and Git starts tracking:

```
git init
```

All the files are now **only** in our working directory; no files are in our local repo or on GitHub.

The next step is to connect the local repo with the GitHub repo. In a previous example, we had RStudio do this for us. Now we need to do it ourselves. We can by adding any of the files and committing it:

```
git add README.txt
git commit -m "First commit. Adding README.txt file just to get started"
```

We now have a file in our local repo and can connect it to the upstream repo, which has url: https://github.com/rairizarry/murders.git.

To do this, we use the command `git remote add`.

```
git remote add origin `https://github.com/rairizarry/murders.git`
```

We can now use `git push` since there is a connection to an upstream repo:

```
git push
```

In Section 38.3 we continue to work with this example, as we demonstrate how we can use RStudio to work with Git and keep a project synced on GitHub.

37.6   Using Git and GitHub in RStudio

While command line Git is a powerful and flexible tool, it can be somewhat daunting when we are getting started. RStudio provides a graphical interface that facilitates the use of Git in the context of a data analysis project. We describe how to use this RStudio feature to do this here.

Now we are ready to start an RStudio project that uses version control and stores the code on a GitHub repo. To do this, we start a project but, instead of *New Directory*, we will select *Version Control* and then we will select *Git* as our version control system:

The repository URL is the link you used to clone. In Section 37.3, we used `https://github.com/username/homework-0.git` as an example. In the project directory name, you need to put the name of the folder that was generated, which in our example will be the name of the repo `homework-0`. This will create a folder called `homework-0` on your local system. Once you do this, the project is created and it is aware of the connection to a GitHub repo. You will see on the top right corner the name and type of project as well as a new tab on the upper right pane titled *Git*.

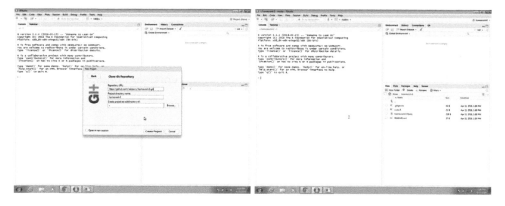

If you select this tab, it will show you the files on your project with some icons that give you information about these files and their relationship to the repo. In the example below, we already added a file to the folder, called *code.R* which you can see in the editing pane.

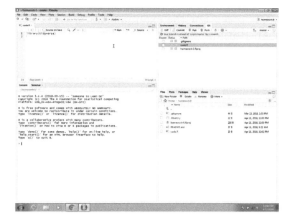

We now need to pay attention to the Git pane. It is important to know that **your local files and the GitHub repo will not be synced automatically**. As described in Section 37.4, you have to sync using git push when you are ready. We show you can do this through RStudio rather than the terminal below.

Before we start working on a collaborative project, usually the first thing we do is *pull* in the changes from the remote repo, in our case the one on GitHub. However, for the example shown here, since we are starting with an empty repo and we are the only ones making changes, we don't need to start by pulling.

In RStudio, the status of the file as it relates to the remote and local repos are represented in the status symbols with colors. A yellow square means that Git knows nothing about this file. To sync with the GitHub repo, we need to *add* the file, then *commit* the change to our local Git repo, then *push* the change to the GitHub repo. Right now, the file is just on our computer. To add the file using RStudio, we click the *Stage* box. You will see that the status icon now changes to a green A.

Note: we are only adding the *code.R* file. We don't necessarily need to add all the files in our local repo to the GitHub repo, only the ones we want to keep track of or the ones we want to share. If our work is producing files of a certain type that we do not want to keep track of, we can add the suffix that defines these files to the .gitignore file. More details on using .gitignore are included here: https://git-scm.com/docs/gitignore. These files will stop appearing in your RStudio Git pane. For the example shown here, we will only be adding *code.R*. But, in general, for an RStudio project, we recommend adding both the .gitignore and .Rproj files.

Now we are ready to commit the file to our local repo. In RStudio, we can use the *Commit* button. This will open a new dialog window. With Git, whenever we commit a change, we are required to enter a comment describing the changes being *committed*.

In this case, we will simply describe that we are adding a new script. In this dialog box, RStudio also gives you a summary of what you are changing to the GitHub repo. In this case, because it is a new file, the entire file is highlighted as green, which highlights the changes.

Once we hit the commit button, we should see a message from Git with a summary of the changes that were committed. Now we are ready to *push* these changes to the GitHub repo. We can do this by clicking on the *Push* button on the top right corner:

We now see a message from Git letting us know that the push has succeeded. In the pop-up window we no longer see the `code.R` file. This is because no new changes have been performed since we last pushed. We can exit this pop-up window now and continue working on our code.

If we now visit our repo on the web, we will see that it matches our local copy.

Congratulations, you have successfully shared code on a GitHub repository!

38

Reproducible projects with RStudio and R markdown

The final product of a data analysis project is often a report. Many scientific publications can be thought of as a final report of a data analysis. The same is true for news articles based on data, an analysis report for your company, or lecture notes for a class on how to analyze data. The reports are often on paper or in a PDF that includes a textual description of the findings along with some figures and tables resulting from the analysis.

Imagine that after you finish the analysis and the report, you are told that you were given the wrong dataset, you are sent a new one and you are asked to run the same analysis with this new dataset. Or what if you realize that a mistake was made and need to re-examine the code, fix the error, and re-run the analysis? Or imagine that someone you are training wants to see the code and be able to reproduce the results to learn about your approach?

Situations like the ones just described are actually quite common for a data scientist. Here, we describe how you can keep your data science projects organized with RStudio so that re-running an analysis is straight-forward. We then demonstrate how to generate reproducible reports with R markdown and the **knitR** package in a way that will greatly help with recreating reports with minimal work. This is possible due to the fact that R markdown documents permit code and textual descriptions to be combined into the same document, and the figures and tables produced by the code are automatically added to the document.

38.1 RStudio projects

RStudio provides a way to keep all the components of a data analysis project organized into one folder and to keep track of information about this project, such as the Git status of files, in one file. In Section 37.6 we demonstrate how RStudio facilitates the use of Git and GitHub through RStudio projects. In this section we quickly demonstrate how to start a new a project and some recommendations on how to keep these organized. RStudio projects also permit you to have several RStudio sessions open and keep track of which is which.

To start a project, click on *File* and then *New Project*. Often we have already created a folder to save the work, as we did in Section 36.7 and we select *Existing Directory*. Here we show an example in which we have not yet created a folder and select the *New Directory* option.

Then, for a data analysis project, you usually select the *New Project* option:

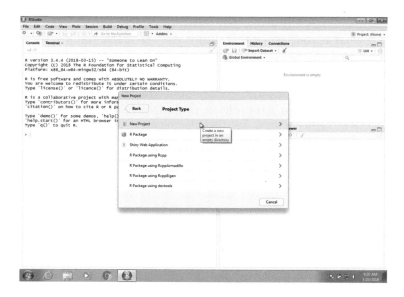

Now you will have to decide on the location of the folder that will be associated with your project, as well as the name of the folder. When choosing a folder name, just like with file names, make sure it is a meaningful name that will help you remember what the project is about. As with files, we recommend using lower case letters, no spaces, and hyphens to separate words. We will call the folder for this project *my-first-project*. This will then generate a *Rproj* file called *my-first-project.Rproj* in the folder associated with the project. We will see how this is useful a few lines below.

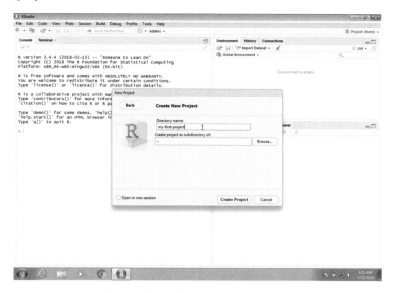

You will be given options on where this folder should be on your filesystem. In this example, we will place it in our home folder, but this is generally not good practice. As we described in Section 36.7 in the Unix chapter, you want to organize your filesystem following a hierarchical approach and with a folder called *projects* where you keep a folder for each project.

When you start using RStudio with a project, you will see the project name in the upper left corner. This will remind you what project this particular RStudio session belongs to. When you open an RStudio session with no project, it will say *Project: (None)*.

When working on a project, all files will be saved and searched for in the folder associated with the project. Below, we show an example of a script that we wrote and saved with the name *code.R*. Because we used a meaningful name for the project, we can be a bit less informative when we name the files. Although we do not do it here, you can have several scripts open at once. You simply need to click *File*, then *New File* and pick the type of file you want to edit.

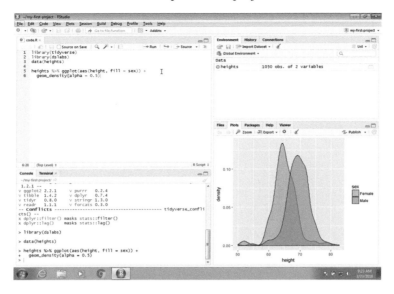

One of the main advantages of using Projects is that after closing RStudio, if we wish to continue where we left off on the project, we simply double click or open the file saved when we first created the RStudio project. In this case, the file is called *my-first-project.Rproj*. If we open this file, RStudio will start up and open the scripts we were editing.

Another advantage is that if you click on two or more different Rproj files, you start new RStudio and R sessions for each.

38.2   R markdown

R markdown is a format for *literate programming* documents. It is based on *markdown*, a markup language that is widely used to generate html pages. You can learn more about markdown here: https://www.markdowntutorial.com/. Literate programming weaves instructions, documentation, and detailed comments in between machine executable code, producing a document that describes the program that is best for human understanding

(Knuth 1984). Unlike a word processor, such as Microsoft Word, where what you see is what you get, with R markdown, you need to *compile* the document into the final report. The R markdown document looks different than the final product. This seems like a disadvantage at first, but it is not because, for example, instead of producing plots and inserting them one by one into the word processing document, the plots are automatically added.

In RStudio, you can start an R markdown document by clicking on *File*, *New File*, the *R Markdown*. You will then be asked to enter a title and author for your document. We are going to prepare a report on gun murders so we will give it an appropriate name. You can also decide what format you would like the final report to be in: HTML, PDF, or Microsoft Word. Later, we can easily change this, but here we select html as it is the preferred format for debugging purposes:

This will generate a template file:

As a convention, we use the `Rmd` suffix for these files.

Once you gain experience with R Markdown, you will be able to do this without the template and can simply start from a blank template.

In the template, you will see several things to note.

38.2.1 The header

At the top you see:

```
---
title: "Report on Gun Murders"
author: "Rafael Irizarry"
date: "April 16, 2018"
output: html_document
---
```

The things between the `---` is the header. We actually don't need a header, but it is often useful. You can define many other things in the header than what is included in the template. We don't discuss those here, but much information is available online. The one parameter that we will highlight is `output`. By changing this to, say, `pdf_document`, we can control the type of output that is produced when we compile.

38.2.2 R code chunks

In various places in the document, we see something like this:

```
```{r}
summary(pressure)
```
```

These are the code chunks. When you compile the document, the R code inside the chunk, in this case `summary(pressure)`, will be evaluated and the result included in that position in the final document.

To add your own R chunks, you can type the characters above quickly with the key binding command-option-I on the Mac and Ctrl-Alt-I on Windows.

This applies to plots as well; the plot will be placed in that position. We can write something like this:

```
```{r}
plot(pressure)
```
```

By default, the code will show up as well. To avoid having the code show up, you can use an argument. To avoid this, you can use the argument `echo=FALSE`. For example:

```
```{r echo=FALSE}
summary(pressure)
```
```

We recommend getting into the habit of adding a label to the R code chunks. This will be very useful when debugging, among other situations. You do this by adding a descriptive word like this:

```
```{r pressure-summary}
summary(pressure)
```
```

38.2.3 Global options

One of the R chunks contains a complex looking call:

```
```{r setup, include=FALSE}
knitr::opts_chunk$set(echo = TRUE)
```
```

We will not cover this here, but as you become more experienced with R Markdown, you will learn the advantages of setting global options for the compilation process.

38.2.4 knitR

We use the **knitR** package to compile R markdown documents. The specific function used to compile is the `knit` function, which takes a filename as input. RStudio provides a button that makes it easier to compile the document. For the screenshot below, we have edited the document so that a report on gun murders is produced. You can see the file here: https://raw.githubusercontent.com/rairizarry/murders/master/report.Rmd. You can now click on the Knit button:

The first time you click on the *Knit* button, a dialog box may appear asking you to install packages you need.

Once you have installed the packages, clicking the *Knit* will compile your R markdown file and the resulting document will pop up:

This produces an html document which you can see in your working directory. To view it, open a terminal and list the files. You can open the file in a browser and use this to present your analysis. You can also produce a PDF or Microsoft document by changing:

`output: html_document` to `output: pdf_document` or `output: word_document`.

We can also produce documents that render on GitHub using `output: github_document`. This will produce a markdown file, with suffix `md`, that renders in GitHub. Because we have uploaded these files to GitHub, you can click on the `md` file and you will see the report as a webpage:

This is a convenient way to share your reports.

38.2.5   More on R markdown

There is a lot more you can do with R markdown. We highly recommend you continue learning as you gain more experience writing reports in R. There are many free resources on the internet including:

- RStudio's tutorial: https://rmarkdown.rstudio.com
- The cheat sheet: https://www.rstudio.com/wp-content/uploads/2015/02/rmarkdown-cheatsheet.pdf
- The knitR book: https://yihui.name/knitr/

38.3   Organizing a data science project

In this section we put it all together to create the US murders project and share it on GitHub.

38.3.1   Create directories in Unix

In Section 36.7 we demonstrated how to use Unix to prepare for a data science project using an example. Here we continue this example and show how to use RStudio. In Section 36.7 we created the following directories using Unix:

```
cd ~
cd projects
mkdir murders
cd murders
mkdir data rdas
```

38.3.2 Create an RStudio project

In the next section we will use create an RStudio project. In RStudio we go to *File* and then *New Project...* and when given the options we pick *Existing Directory*. We then write the full path of the `murders` directory created above.

Once you do this, you will see the `rdas` and `data` directories you created in the RStudio *Files* tab.

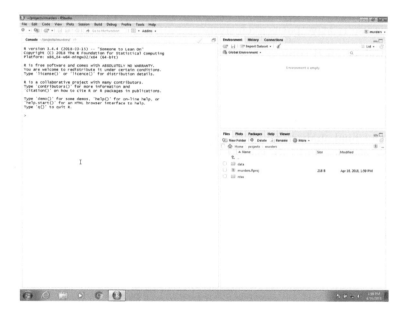

Keep in mind that when we are in this project, our default working directory will be `~/projects/murders`. You can confirm this by typing `getwd()` into your R session. This is important because it will help us organize the code when we need to write file paths. **Pro tip: always use relative paths in code for data science projects. These should be relative to the default working directory.** The problem with using full paths is that your code is unlikely to work on filesystems other than yours since the directory structures will be different. This includes using the home directory `~` as part of your path.

38.3.3   Edit some R scripts

Let's now write a script that downloads a file into the data directory. We will call this file download-data.R.

The content of this file will be:

```
url <- "https://raw.githubusercontent.com/rafalab/dslabs/master/inst/
extdata/murders.csv"
dest_file <- "data/murders.csv"
download.file(url, destfile = dest_file)
```

Notice that we are using the relative path data/murders.csv.

Run this code in R and you will see that a file is added to the data directory.

Now we are ready to write a script to read this data and prepare a table that we can use for analysis. Call the file wrangle-data.R. The content of this file will be:

```
library(tidyverse)
murders <- read_csv("data/murders.csv")
murders <-murders %>% mutate(region = factor(region),
                             rate = total / population * 10^5)
save(murders, file = "rdas/murders.rda")
```

Again note that we use relative paths exclusively.

In this file, we introduce an R command we have not seen: save. The save command in R saves objects into what is called an *rda file*: *rda* is short for R data. We recommend using the .rda suffix on files saving R objects. You will see that .RData is also used.

If you run this code above, the processed data object will be saved in a file in the rda directory. Although not the case here, this approach is often practical because generating the data object we use for final analyses and plots can be a complex and time-consuming process. So we run this process once and save the file. But we still want to be able to generate the entire analysis from the raw data.

Now we are ready to write the analysis file. Let's call it analysis.R. The content should be the following:

```
library(tidyverse)
load("rdas/murders.rda")

murders %>% mutate(abb = reorder(abb, rate)) %>%
  ggplot(aes(abb, rate)) +
  geom_bar(width = 0.5, stat = "identity", color = "black") +
  coord_flip()
```

If you run this analysis, you will see that it generates a plot.

38.3.4 Create some more directories using Unix

Now suppose we want to save the generated plot for use in a report or presentation. We can do this with the **ggplot** command **ggsave**. But where do we put the graph? We should be systematically organized so we will save plots to a directory called `figs`. Start by creating a directory by typing the following in the terminal:

```
mkdir figs
```

and then you can add the line:

```
ggsave("figs/barplot.png")
```

to your R script. If you run the script now, a png file will be saved into the `figs` directory. If we wanted to copy that file to some other directory where we are developing a presentation, we can avoid using the mouse by using the `cp` command in our terminal.

38.3.5 Add a README file

You now have a self-contained analysis in one directory. One final recommendation is to create a `README.txt` file describing what each of these files does for the benefit of others reading your code, including your future self. This would not be a script but just some notes. One of the options provided when opening a new file in RStudio is a text file. You can save something like this into the text file:

```
We analyze US gun murder data collected by the FBI.

download-data.R - Downloads csv file to data directory

wrangle-data.R - Creates a derived dataset and saves as R object in rdas
directory

analysis.R - A plot is generated and saved in the figs directory.
```

38.3.6 Initializing a Git directory

In Section 37.5 we demonstrated how to initialize a Git directory and connect it to the upstream repository on GitHub, which we already created in that section.

We can do this in the Unix terminal:

```
cd ~/projects/murders
git init
git add README.txt
git commit -m "First commit. Adding README.txt file just to get started"
git remote add origin `https://github.com/rairizarry/murders.git`
git push
```

38.3.7   Add, commit, and push files using RStudio

We can continue adding and committing each file, but it might be easier to use RStudio. To do this, start the project by opening the Rproj file. The git icons should appear and you can add, commit and push using these.

We can now go to GitHub and confirm that our files are there. You can see a version of this project, organized with Unix directories, on GitHub[1]. You can download a copy to your computer by using the `git clone` command on your terminal. This command will create a directory called `murders` in your working directory, so be careful where you call it from.

```
git clone https://github.com/rairizarry/murders.git
```

[1]https://github.com/rairizarry/murders

Index